CONVERSION FACTORS FROM ENGLISH TO SI UNITS

Length:	1 ft	= 0.3048 m
	1 ft	= 30.48 cm
	1 ft	= 304.8 mm
	1 in.	= 0.0254 m
	1 in.	= 2.54 cm
	1 in.	= 25.4 mm
Area:	1 ft^2	= 929.03 $\times$ 10^{-4} m^2
	1 ft^2	= 929.03 cm^2
	1 ft^2	= 929.03 $\times$ 10^2 mm^2
	1 in^2	= 6.452 $\times$ 10^{-4} m^2
	1 in^2	= 6.452 cm^2
	1 in^2	= 645.16 mm^2
Volume:	1 ft^3	= 28.317 $\times$ 10^{-3} m^3
	1 ft^3	= 28.317 $\times$ 10^3 cm^3
	1 in^3	= 16.387 $\times$ 10^{-6} m^3
	1 in^3	= 16.387 cm^3
Force:	1 lb	= 4.448 N
	1 lb	= 4.448 $\times$ 10^{-3} kN
	1 lb	= 0.4536 kgf
	1 kip	= 4.448 kN
	1 U.S. ton	= 8.896 kN
	1 lb	= 0.4536 $\times$ 10^{-3} metric ton
	1 lb/ft	= 14.593 N/m
Stress:	1 lb/ft^2	= 47.88 N/m^2
	1 lb/ft^2	= 0.04788 kN/m^2
	1 U.S. ton/ft^2	= 95.76 kN/m^2
	1 kip/ft^2	= 47.88 kN/m^2
	1 lb/in^2	= 6.895 kN/m^2
Unit weight:	1 lb/ft^3	= 0.1572 kN/m^3
	1 lb/in^3	= 271.43 kN/m^3
Moment:	1 lb-ft	= 1.3558 N·m
	1 lb-in.	= 0.11298 N·m
Energy:	1 ft-lb	= 1.3558 J
Moment of inertia:	1 in^4	= 0.4162 $\times$ 10^6 mm^4
	1 in^4	= 0.4162 $\times$ 10^{-6} m^4
Section modulus:	1 in^3	= 0.16387 $\times$ 10^5 mm^3
	1 in^3	= 0.16387 $\times$ 10^{-4} m^3
Hydraulic conductivity:	1 ft/min	= 0.3048 m/min
	1 ft/min	= 30.48 cm/min
	1 ft/min	= 304.8 mm/min
	1 ft/sec	= 0.3048 m/sec
	1 ft/sec	= 304.8 mm/sec
	1 in./min	= 0.0254 m/min
	1 in./sec	= 2.54 cm/sec
	1 in./sec	= 25.4 mm/sec
Coefficient of consolidation:	1 in^2/sec	= 6.452 cm^2/sec
	1 in^2/sec	= 20.346 $\times$ 10^3 m^2/yr
	1 ft^2/sec	= 929.03 cm^2/sec

Principles of Foundation Engineering

Principles of Foundation Engineering

Sixth Edition

Braja M. Das

Australia • Brazil • Japan • Korea • Mexico • Singapore • Spain • United Kingdom • United States

CENGAGE
Learning™

Principles of Foundation Engineering, Sixth Edition
Braja M. Das

Director, Global Engineering Program:
   Chris Carson

Senior Developmental Editor: Hilda Gowans

Developmental Editor: Kamilah Reid Burrell

Marketing Services Coordinator:
   Lauren Bestos

Production Service: RPK Editorial Services

Production Manager: Renate McCloy

Copyeditor: Patricia Daly

Proofreader: Erin Wagner

Indexer: RPK Editorial Services

Compositor: ICC MacMillan Inc.

Senior Art Director: Michelle Kunkler

Internal Designer: Carmela Pereira

Cover Designer: Andrew Adams

Text Permissions Researcher: Vicki Gould

Senior First Print Buyer: Doug Wilke

For product information and technology assistance, contact us at
Cengage Learning Customer & Sales Support, 1-800-354-9706.
For permission to use material from this text or product,
submit all requests online at **www.cengage.com/permissions**.
Further permissions questions can be emailed to
permissionrequest@cengage.com.

Library of Congress Control Number: 2006904569

U.S. Student Edition:
ISBN-13: 978-0-495-08246-0
ISBN-10: 0-495-08246-5

Cengage Learning
200 First Stamford Place, Suite 400
Stamford, CT 06902
USA

Cengage Learning is a leading provider of customized learning solutions with office locations around the globe, including Singapore, the United Kingdom, Australia, Mexico, Brazil, and Japan. Locate your local office at: **international.cengage.com/region**.

Cengage Learning products are represented in Canada by Nelson Education Ltd.

For your course and learning solutions,
visit **academic.cengage.com/engineering**.

Purchase any of our products at your local college store or at our preferred online store **www.ichapters.com**.

Printed in the United States of America
5 6 7 8 10 09 08

To our granddaughter, Elizabeth Madison

Contents

1 Geotechnical Properties of Soil   **1**

1.1 Introduction   1
1.2 Grain-Size Distribution   2
1.3 Size Limits for Soils   5
1.4 Weight–Volume Relationships   5
1.5 Relative Density   9
1.6 Atterberg Limits   12
1.7 Soil Classification Systems   13
1.8 Hydraulic Conductivity of Soil   21
1.9 Steady-State Seepage   23
1.10 Effective Stress   25
1.11 Consolidation   27
1.12 Calculation of Primary Consolidation Settlement   32
1.13 Time Rate of Consolidation   33
1.14 Degree of Consolidation Under Ramp Loading   40
1.15 Shear Strength   43
1.16 Unconfined Compression Test   48
1.17 Comments on Friction Angle, φ' 49
1.18 Correlations for Undrained Shear Strength, c_u 52
1.19 Sensitivity   53
Problems   54
References   58

2 Natural Soil Deposits and Subsoil Exploration   **60**

2.1 Introduction   60

Natural Soil Deposits   60

2.2 Soil Origin   60

2.3   Residual Soil   61
2.4   Gravity Transported Soil   62
2.5   Alluvial Deposits   62
2.6   Lacustrine Deposits   65
2.7   Glacial Deposits   65
2.8   Aeolian Soil Deposits   67
2.9   Organic Soil   68

Subsurface Exploration   68

2.10   Purpose of Subsurface Exploration   68
2.11   Subsurface Exploration Program   68
2.12   Exploratory Borings in the Field   71
2.13   Procedures for Sampling Soil   74
2.14   Observation of Water Tables   85
2.15   Vane Shear Test   86
2.16   Cone Penetration Test   90
2.17   Pressuremeter Test (PMT)   97
2.18   Dilatometer Test   99
2.19   Coring of Rocks   101
2.20   Preparation of Boring Logs   105
2.21   Geophysical Exploration   105
2.22   Subsoil Exploration Report   113
Problems   114
References   119

3   Shallow Foundations: Ultimate Bearing Capacity   121

3.1   Introduction   121
3.2   General Concept   121
3.3   Terzaghi's Bearing Capacity Theory   124
3.4   Factor of Safety   128
3.5   Modification of Bearing Capacity Equations for Water Table   130
3.6   The General Bearing Capacity Equation   131
3.7   Meyerhof's Bearing Capacity, Shape, Depth, and Inclination Factors   136
3.8   Some Comments on Bearing Capacity Factor, N_γ, and Shape Factors   138
3.9   A Case History for Bearing Capacity Failure   140
3.10   Effect of Soil Compressibility   142
3.11   Eccentrically Loaded Foundations   146
3.12   Ultimate Bearing Capacity under Eccentric Loading—Meyerhof's Theory   148
3.13   Eccentrically Loaded Foundation—Prakash and Saran's Theory   154
3.14   Bearing Capacity of a Continuous Foundation Subjected to Eccentric Inclined Loading   161
Problems   165
References   168

4   Ultimate Bearing Capacity of Shallow Foundations: Special Cases   **170**

4.1   Introduction   170
4.2   Foundation Supported by a Soil with a Rigid Base at Shallow Depth   170
4.3   Bearing Capacity of Layered Soils: Stronger Soil Underlain by Weaker Soil   177
4.4   Closely Spaced Foundations—Effect on Ultimate Bearing Capacity   185
4.5   Bearing Capacity of Foundations on Top of a Slope   188
4.6   Bearing Capacity of Foundations on a Slope   191
4.7   Uplift Capacity of Foundations   193
Problems   199
References   202

5   Shallow Foundations: Allowable Bearing Capacity and Settlement   **203**

5.1   Introduction   203

Vertical Stress Increase in a Soil Mass Caused by Foundation Load   204

5.2   Stress Due to a Concentrated Load   204
5.3   Stress Due to a Circularly Loaded Area   205
5.4   Stress below a Rectangular Area   206
5.5   Average Vertical Stress Increase Due to a Rectangularly Loaded Area   213
5.6   Stress Increase under an Embankment   216

Elastic Settlement   220

5.7   Elastic Settlement Based on the Theory of Elasticity   220
5.8   Elastic Settlement of Foundations on Saturated Clay   230
5.9   Improved Equation for Elastic Settlement   230
5.10   Settlement of Sandy Soil: Use of Strain Influence Factor   236
5.11   Range of Material Parameters for Computing Elastic Settlement   240
5.12   Settlement of Foundation on Sand Based on Standard Penetration Resistance   241
5.13   General Comments on Elastic Settlement Prediction   246
5.14   Seismic Bearing Capacity and Settlement in Granular Soil   247

Consolidation Settlement   252

5.15   Primary Consolidation Settlement Relationships   252
5.16   Three-Dimensional Effect on Primary Consolidation Settlement   254
5.17   Settlement Due to Secondary Consolidation   258
5.18   Field Load Test   260

5.19   Presumptive Bearing Capacity   263
5.20   Tolerable Settlement of Buildings   264
Problems   266
References   269

6   Mat Foundations   272

6.1   Introduction   272
6.2   Combined Footings   272
6.3   Common Types of Mat Foundations   275
6.4   Bearing Capacity of Mat Foundations   277
6.5   Differential Settlement of Mats   280
6.6   Field Settlement Observations for Mat Foundations   281
6.7   Compensated Foundation   281
6.8   Structural Design of Mat Foundations   285
Problems   304
References   307

7   Lateral Earth Pressure   308

7.1   Introduction   308
7.2   Lateral Earth Pressure at Rest   309

Active Pressure   312

7.3   Rankine Active Earth Pressure   312
7.4   A Generalized Case for Rankine Active Pressure   315
7.5   Coulomb's Active Earth Pressure   323
7.6   Active Earth Pressure for Earthquake Conditions   328
7.7   Active Pressure for Wall Rotation about the Top: Braced Cut   333
7.8   Active Earth Pressure for Translation of Retaining Wall—Granular Backfill   334
7.9   General Comments on Active Earth Pressure   338

Passive Pressure   338

7.10   Rankine Passive Earth Pressure   338
7.11   Rankine Passive Earth Pressure: Inclined Backfill   344
7.12   Coulomb's Passive Earth Pressure   345
7.13   Comments on the Failure Surface Assumption for Coulomb's Pressure Calculations   347
7.14   Passive Pressure under Earthquake Conditions   348
Problems   349
References   352

8   Retaining Walls   **353**

8.1   Introduction   353

Gravity and Cantilever Walls   355

8.2   Proportioning Retaining Walls   355
8.3   Application of Lateral Earth Pressure Theories to Design   356
8.4   Stability of Retaining Walls   358
8.5   Check for Overturning   359
8.6   Check for Sliding along the Base   361
8.7   Check for Bearing Capacity Failure   364
8.8   Construction Joints and Drainage from Backfill   374
8.9   Some Comments on Design of Retaining Walls   377

Mechanically Stabilized Retaining Walls   379

8.10   Soil Reinforcement   379
8.11   Considerations in Soil Reinforcement   380
8.12   General Design Considerations   382
8.13   Retaining Walls with Metallic Strip Reinforcement   383
8.14   Step-by-Step-Design Procedure Using Metallic Strip Reinforcement   390
8.15   Retaining Walls with Geotextile Reinforcement   395
8.16   Retaining Walls with Geogrid Reinforcement   399
8.17   General Comments   402
Problems   404
References   407

9   Sheet Pile Walls   **409**

9.1   Introduction   409
9.2   Construction Methods   413
9.3   Cantilever Sheet Pile Walls   414
9.4   Cantilever Sheet Piling Penetrating Sandy Soils   415
9.5   Special Cases for Cantilever Walls Penetrating a Sandy Soil   422
9.6   Cantilever Sheet Piling Penetrating Clay   423
9.7   Special Cases for Cantilever Walls Penetrating Clay   428
9.8   Anchored Sheet-Pile Walls   429
9.9   Free Earth Support Method for Penetration of Sandy Soil   430
9.10   Design Charts for Free Earth Support Method (Penetration into Sandy Soil)   435
9.11   Moment Reduction for Anchored Sheet-Pile Walls   440
9.12   Computational Pressure Diagram Method for Penetration into Sandy Soil   443
9.13   Field Observations of an Anchored Sheet Pile Wall   447
9.14   Free Earth Support Method for Penetration of Clay   448
9.15   Anchors   452
9.16   Holding Capacity of Anchor Plates in Sand   454

9.17   Holding Capacity of Anchor Plates in Clay ($\varphi = 0$ Condition)   460
9.18   Ultimate Resistance of Tiebacks   460
Problems   461
References   464

10   Braced Cuts   466

10.1   Introduction   466
10.2   Pressure Envelope for Braced-Cut Design   467
10.3   Pressure Envelope for Cuts in Layered Soil   471
10.4   Design of Various Components of a Braced Cut   472
10.5   Bottom Heave of a Cut in Clay   482
10.6   Stability of the Bottom of a Cut in Sand   485
10.7   Lateral Yielding of Sheet Piles and Ground Settlement   487
Problems   489
References   490

11   Pile Foundations   491

11.1   Introduction   491
11.2   Types of Piles and Their Structural Characteristics   493
11.3   Estimating Pile Length   502
11.4   Installation of Piles   504
11.5   Load Transfer Mechanism   508
11.6   Equations for Estimating Pile Capacity   509
11.7   Meyerhof's Method for Estimating Q_p 512
11.8   Vesic's Method for Estimating Q_p 515
11.9   Janbu's Method for Estimating Q_p 516
11.10   Coyle and Castello's Method for Estimating Q_p in Sand   520
11.11   Other Correlations for Calculating Q_p with SPT and CPT Results   521
11.12   Frictional Resistance (Q_s) in Sand   524
11.13   Frictional (Skin) Resistance in Clay   528
11.14   Point-Bearing Capacity of Piles Resting on Rock   531
11.15   Pile Load Tests   538
11.16   Comparison of Theory with Field Load Test Results   542
11.17   Elastic Settlement of Piles   543
11.18   Laterally Loaded Piles   546
11.19   Pile-Driving Formulas   562
11.20   Pile Capacity For Vibration-Driven Piles   568
11.21   Negative Skin Friction   570

Group Piles   573

11.22   Group Efficiency   573
11.23   Ultimate Capacity of Group Piles in Saturated Clay   576
11.24   Elastic Settlement of Group Piles   580

11.25 Consolidation Settlement of Group Piles   581
11.26 Piles in Rock   584
Problems   584
References   588

12

Drilled-Shaft Foundations   591

12.1 Introduction   591
12.2 Types of Drilled Shafts   592
12.3 Construction Procedures   593
12.4 Other Design Considerations   598
12.5 Load Transfer Mechanism   598
12.6 Estimation of Load-Bearing Capacity   599
12.7 Drilled Shafts in Granular Soil: Load-Bearing Capacity   602
12.8 Drilled Shafts in Clay: Load-Bearing Capacity   613
12.9 Settlement of Drilled Shafts at Working Load   620
12.10 Lateral Load-Carrying Capacity—Characteristic Load
and Moment Method   622
12.11 Drilled Shafts Extending into Rock   631
Problems   635
References   639

13

Foundations on Difficult Soils   640

13.1 Introduction   640

Collapsible Soil   640

13.2 Definition and Types of Collapsible Soil   640
13.3 Physical Parameters for Identification   641
13.4 Procedure for Calculating Collapse Settlement   645
13.5 Foundation Design in Soils Not Susceptible to Wetting   646
13.6 Foundation Design in Soils Susceptible to Wetting   648

Expansive Soils   649

13.7 General Nature of Expansive Soils   649
13.8 Laboratory Measurement of Swell   650
13.9 Classification of Expansive Soil on the Basis of Index Tests   655
13.10 Foundation Considerations for Expansive Soils   656
13.11 Construction on Expansive Soils   661

Sanitary Landfills   665

13.12 General Nature of Sanitary Landfills   665
13.13 Settlement of Sanitary Landfills   666
Problems   668
References   670

14   Soil Improvement and Ground Modification   **672**

14.1   Introduction   672
14.2   General Principles of Compaction   673
14.3   Correction for Compaction of Soils with Oversized Particles   676
14.4   Field Compaction   678
14.5   Compaction Control for Clay Hydraulic Barriers   681
14.6   Vibroflotation   682
14.7   Precompression   688
14.8   Sand Drains   696
14.9   Prefabricated Vertical Drains   706
14.10   Lime Stabilization   711
14.11   Cement Stabilization   714
14.12   Fly-Ash Stabilization   716
14.13   Stone Columns   717
14.14   Sand Compaction Piles   722
14.15   Dynamic Compaction   725
14.16   Jet Grouting   727
Problems   729
References   731

Appendix A   735

Answers to Selected Problems   740

Index   745

Preface

Soil mechanics and foundation engineering have rapidly developed during the last fifty years. Intensive research and observation in the field and the laboratory have refined and improved the science of foundation design. Originally published in 1984, this text on the principles of foundation engineering is now in the sixth edition. The use of this text throughout the world has increased greatly over the years; it also has been translated into several languages. New and improved materials that have been published in various geotechnical engineering journals and conference proceedings have been continuously incorporated into each edition of the text.

Principles of Foundation Engineering is intended primarily for undergraduate civil engineering students. The first chapter, on geotechnical properties of soil, reviews the topics covered in the introductory soil mechanics course, which is a prerequisite for the foundation engineering course. The text is comprised of fourteen chapters with examples and problems, one appendix, and an answer section for selected problems. The chapters are mostly devoted to the geotechnical aspects of foundation design. Systéme International (SI) units and English units are used in the text.

Because the text introduces civil engineering students to the application of the fundamental concepts of foundation analysis and design, the mathematical derivations are not always presented. Instead, just the final form of the equation is given. A list of references for further information and study is included at the end of each chapter.

Example problems that will help students understand the application of various equations and graphs are given in each chapter. A number of practice problems also are given at the end of each chapter. Answers to some of these problems are given at the end of the text.

Following is a brief overview of the changes from the fifth edition.

- Several tables were changed to graphical form to make interpolation convenient.
- Chapter 1 on geotechnical properties of soil has several additional empirical relationships for the compression index. Also included are the time factor and

average degree of consolidation relationships for initial pore water pressure increasing and having trapezoidal, sinusoidal, and triangular shapes.

- Chapter 2 on natural deposits and subsoil exploration has an expanded description on natural soil deposits. Also incorporated into this chapter are several recently developed correlations between:

  — the consistency index, standard penetration number, and unconfined compression strength
  — relative density, standard penetration number, and overconsolidation ratio for granular soils
  — cone penetration resistance, standard penetration number, and average grain size
  — peak and ultimate friction angles based on the results of dilatometer tests

- Chapter 3 on shallow foundations (ultimate bearing capacity) has added sections on Prakash and Saran's theory on the ultimate bearing capacity of eccentrically loaded foundations and bearing capacity of foundations due to eccentric and inclined loading.
- Chapter 4 on ultimate bearing capacity of shallow foundations (special cases) has new sections on the effect on ultimate bearing capacity for closely spaced foundations, bearing capacity of foundations on a slope, and ultimate uplift capacity.
- Chapter 5 on shallow foundations (allowable bearing capacity and settlement) has some more elaborate tables for elastic settlement calculation. Comparisons between predicted and measured settlements of shallow foundations on granular soils also are presented.
- Chapter 7 on lateral earth pressure has a new section on a generalized relationship for Rankine active pressure (retaining wall with inclined back face and inclined granular backfill). Relationships for active and passive earth pressures under earthquake conditions are presented. Also added in this chapter is the active earth-pressure theory for translations of a retaining wall with a granular backfill.
- Chapter 8 on retaining walls has several comparisons of predicted and observed lateral earth pressure on retaining walls and mechanically stabilized earth walls.
- New design charts for design of sheet-pile walls using the free earth support method has been incorporated into Chapter 9.
- Chapter 12 on drilled shaft foundations has some recently published results on the load bearing capacity based on settlement in gravelly sand.
- Chapter 13 on foundations on difficult soils has a new classification system of expansive soils based on free swell ratio.
- The fundamental concepts of jet grouting have been incorporated into Chapter 14 on soil improvement and ground modification.

Foundation analysis and design, as my colleagues in the geotechnical engineering area well know, is not just a matter of using theories, equations, and graphs from a textbook. Soil profiles found in nature are seldom homogeneous, elastic, and isotropic. The educated judgment needed to properly apply the theories, equations, and graphs to the evaluation of soils and foundation design cannot be be overemphasized or completely taught in the classroom. Field experience must supplement classroom work.

I am grateful to my wife for her continuous help during the past twenty-five years for the development of the original text and five subsequent revisions. Her apparent inexhaustible energy has been my primary source of inspiration.

Over 35 colleagues have reviewed the original text and revisions over the years. Their comments and suggestions have been invaluable. I am truly grateful for their critiques.

Thanks are also due to Chris Carson, General Manager, Engineering, and the Cengage Learning staff: Kamilah Reid-Burrell and Hilda Gowans, Developmental Editors, Susan Calvert, Editorial Content Manager, Renate McCloy, Production Manager, Vicki Gould, Permission Coordinator, Angela Cluer, Creative Director, and Andrew Adams, Cover Designer, for their interest and patience during the revision and production of the manuscript.

Braja M. Das

1

Geotechnical Properties of Soil

1.1 Introduction

The design of foundations of structures such as buildings, bridges, and dams generally requires a knowledge of such factors as (a) the load that will be transmitted by the superstructure to the foundation system, (b) the requirements of the local building code, (c) the behavior and stress-related deformability of soils that will support the foundation system, and (d) the geological conditions of the soil under consideration. To a foundation engineer, the last two factors are extremely important because they concern soil mechanics.

The geotechnical properties of a soil—such as its grain-size distribution, plasticity, compressibility, and shear strength—can be assessed by proper laboratory testing. In addition, recently emphasis has been placed on the *in situ* determination of strength and deformation properties of soil, because this process avoids disturbing samples during field exploration. However, under certain circumstances, not all of the needed parameters can be or are determined, because of economic or other reasons. In such cases, the engineer must make certain assumptions regarding the properties of the soil. To assess the accuracy of soil parameters—whether they were determined in the laboratory and the field or whether they were assumed—the engineer must have a good grasp of the basic principles of soil mechanics. At the same time, he or she must realize that the natural soil deposits on which foundations are constructed are not homogeneous in most cases. Thus, the engineer must have a thorough understanding of the geology of the area—that is, the origin and nature of soil stratification and also the groundwater conditions. Foundation engineering is a clever combination of soil mechanics, engineering geology, and proper judgment derived from past experience. To a certain extent, it may be called an art.

When determining which foundation is the most economical, the engineer must consider the superstructure load, the subsoil conditions, and the desired tolerable settlement. In general, foundations of buildings and bridges may be divided into two major categories: (1) *shallow foundations* and (2) *deep foundations. Spread footings, wall footings,* and *mat foundations* are all shallow foundations. In most shallow foundations, *the depth of embedment can be equal to or less than three to four times the width of the foundation.* Pile and *drilled shaft* foundations are deep foundations. They are used when top layers have poor load-bearing capacity and when the use of shallow foundations will cause considerable structural damage or instability. The problems relating to shallow foundations and mat foundations are considered in Chapters 3, 4, 5, and 6. Chapter 11 discusses pile foundations, and Chapter 12 examines drilled shafts.

This chapter serves primarily as a review of the basic geotechnical properties of soils. It includes topics such as grain-size distribution, plasticity, soil classification, effective stress, consolidation, and shear strength parameters. It is based on the assumption that you have already been exposed to these concepts in a basic soil mechanics course.

1.2   *Grain-Size Distribution*

In any soil mass, the sizes of the grains vary greatly. To classify a soil properly, you must know its *grain-size distribution*. The grain-size distribution of *coarse-grained* soil is generally determined by means of *sieve analysis*. For a *fine-grained* soil, the grain-size distribution can be obtained by means of *hydrometer analysis*. The fundamental features of these analyses are presented in this section. For detailed descriptions, see any soil mechanics laboratory manual (e.g., Das, 2002).

Sieve Analysis

A sieve analysis is conducted by taking a measured amount of dry, well-pulverized soil and passing it through a stack of progressively finer sieves with a pan at the bottom. The amount of soil retained on each sieve is measured, and the cumulative percentage of soil passing through each is determined. This percentage is generally referred to as *percent finer*. Table 1.1 contains a list of U.S. sieve numbers and the corresponding size of their openings. These sieves are commonly used for the analysis of soil for classification purposes.

Table 1.1   U.S. Standard Sieve Sizes

Sieve No.	Opening (mm)
4	4.750
6	3.350
8	2.360
10	2.000
16	1.180
20	0.850
30	0.600
40	0.425
50	0.300
60	0.250
80	0.180
100	0.150
140	0.106
170	0.088
200	0.075
270	0.053

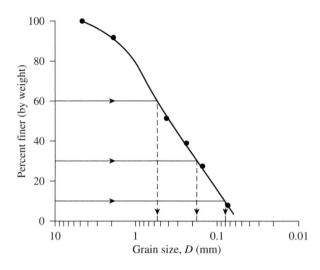

Figure 1.1 Grain-size distribution curve of a coarse-grained soil obtained from sieve analysis

The percent finer for each sieve, determined by a sieve analysis, is plotted on *semilogarithmic graph paper*, as shown in Figure 1.1. Note that the grain diameter, *D*, is plotted on the *logarithmic scale* and the percent finer is plotted on the *arithmetic scale*.

Two parameters can be determined from the grain-size distribution curves of coarse-grained soils: (1) the *uniformity coefficient* (C_u) and (2) the *coefficient of gradation,* or *coefficient of curvature* (C_c). These coefficients are

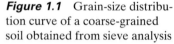

$$C_u = \frac{D_{60}}{D_{10}} \tag{1.1}$$

and

$$C_c = \frac{D_{30}^2}{(D_{60})(D_{10})} \tag{1.2}$$

where D_{10}, D_{30}, and D_{60} are the diameters corresponding to percents finer than 10, 30, and 60%, respectively.

For the grain-size distribution curve shown in Figure 1.1, $D_{10} = 0.08$ mm, $D_{30} = 0.17$ mm, and $D_{60} = 0.57$ mm. Thus, the values of C_u and C_c are

$$C_u = \frac{0.57}{0.08} = 7.13$$

and

$$C_c = \frac{0.17^2}{(0.57)(0.08)} = 0.63$$

Parameters C_u and C_c are used in the *Unified Soil Classification System,* which is described later in the chapter.

Hydrometer Analysis

Hydrometer analysis is based on the principle of sedimentation of soil particles in water. This test involves the use of 50 grams of dry, pulverized soil. A *deflocculating agent* is always added to the soil. The most common deflocculating agent used for hydrometer analysis is 125 cc of 4% solution of sodium hexametaphosphate. The soil is allowed to soak for at least 16 hours in the deflocculating agent. After the soaking period, distilled water is added, and the soil–deflocculating agent mixture is thoroughly agitated. The sample is then transferred to a 1000-ml glass cylinder. More distilled water is added to the cylinder to fill it to the 1000-ml mark, and then the mixture is again thoroughly agitated. A hydrometer is placed in the cylinder to measure the specific gravity of the soil–water suspension in the vicinity of the instrument's bulb (Figure 1.2), usually over a 24-hour period. Hydrometers are calibrated to show the amount of soil that is still in suspension at any given time t. The largest diameter of the soil particles still in suspension at time t can be determined by Stokes' law,

$$D = \sqrt{\frac{18\eta}{(G_s - 1)\gamma_w}}\sqrt{\frac{L}{t}} \qquad (1.3)$$

where

D = diameter of the soil particle
G_s = specific gravity of soil solids
η = viscosity of water

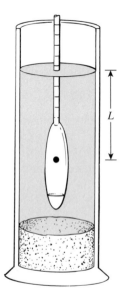

Figure 1.2 Hydrometer analysis

γ_w = unit weight of water

L = effective length (i.e., length measured from the water surface in the cylinder to the center of gravity of the hydrometer; see Figure 1.2)

t = time

Soil particles having diameters larger than those calculated by Eq. (1.3) would have settled beyond the zone of measurement. In this manner, with hydrometer readings taken at various times, the soil *percent finer* than a given diameter D can be calculated and a grain-size distribution plot prepared. The sieve and hydrometer techniques may be combined for a soil having both coarse-grained and fine-grained soil constituents.

1.3   *Size Limits for Soils*

Several organizations have attempted to develop the size limits for *gravel, sand, silt,* and *clay* on the basis of the grain sizes present in soils. Table 1.2 presents the size limits recommended by the American Association of State Highway and Transportation Officials (AASHTO) and the Unified Soil Classification systems (Corps of Engineers, Department of the Army, and Bureau of Reclamation). The table shows that soil particles smaller than 0.002 mm have been classified as *clay.* However, clays by nature are cohesive and can be rolled into a thread when moist. This property is caused by the presence of *clay minerals* such as *kaolinite, illite,* and *montmorillonite.* In contrast, some minerals, such as *quartz* and *feldspar,* may be present in a soil in particle sizes as small as clay minerals, but these particles will not have the cohesive property of clay minerals. Hence, they are called *clay-size particles,* not *clay particles.*

1.4   *Weight–Volume Relationships*

In nature, soils are three-phase systems consisting of solid soil particles, water, and air (or gas). To develop the *weight–volume relationships* for a soil, the three phases can be separated as shown in Figure 1.3a. Based on this separation, the volume relationships can then be defined.

The *void ratio, e,* is the ratio of the volume of voids to the volume of soil solids in a given soil mass, or

$$e = \frac{V_v}{V_s} \tag{1.4}$$

Table 1.2 Soil-Separate Size Limits

Classification system	Grain size (mm)
Unified	Gravel: 75 mm to 4.75 mm Sand: 4.75 mm to 0.075 mm Silt and clay (fines): <0.075 mm
AASHTO	Gravel: 75 mm to 2 mm Sand: 2 mm to 0.05 mm Silt: 0.05 mm to 0.002 mm Clay: <0.002 mm

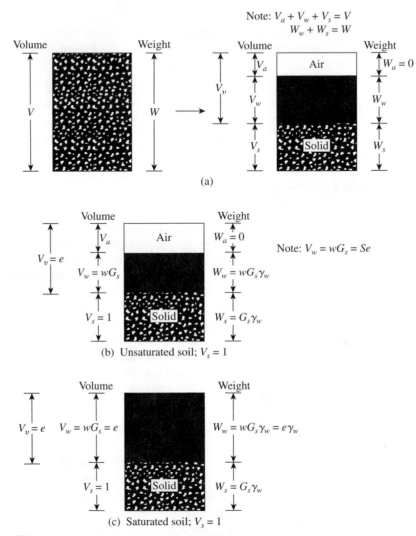

Note: $V_a + V_w + V_s = V$
$W_w + W_s = W$

(a)

(b) Unsaturated soil; $V_s = 1$

Note: $V_w = wG_s = Se$

(c) Saturated soil; $V_s = 1$

Figure 1.3   Weight–volume relationships

where

V_v = volume of voids
V_s = volume of soil solids

The *porosity, n,* is the ratio of the volume of voids to the volume of the soil specimen, or

$$n = \frac{V_v}{V} \tag{1.5}$$

where

V = total volume of soil

Moreover,

$$n = \frac{V_v}{V} = \frac{V_v}{V_s + V_v} = \frac{\dfrac{V_v}{V_s}}{\dfrac{V_s}{V_s} + \dfrac{V_v}{V_s}} = \frac{e}{1 + e} \tag{1.6}$$

The *degree of saturation, S,* is the ratio of the volume of water in the void spaces to the volume of voids, generally expressed as a percentage, or

$$S(\%) = \frac{V_w}{V_v} \times 100 \tag{1.7}$$

where

V_w = volume of water

Note that, for saturated soils, the degree of saturation is 100%.

The weight relationships are *moisture content, moist unit weight, dry unit weight,* and *saturated unit weight,* often defined as follows:

$$\text{Moisture content} = w(\%) = \frac{W_w}{W_s} \times 100 \tag{1.8}$$

where

W_s = weight of the soil solids
W_w = weight of water

$$\text{Moist unit weight} = \gamma = \frac{W}{V} \tag{1.9}$$

where

W = total weight of the soil specimen = $W_s + W_w$

The weight of air, W_a, in the soil mass is assumed to be negligible.

$$\text{Dry unit weight} = \gamma_d = \frac{W_s}{V} \tag{1.10}$$

When a soil mass is completely saturated (i.e., all the void volume is occupied by water), the moist unit weight of a soil [Eq. (1.9)] becomes equal to the *saturated unit weight* (γ_{sat}). So $\gamma = \gamma_{\text{sat}}$ if $V_v = V_w$.

More useful relations can now be developed by considering a representative soil specimen in which the volume of soil solids is equal to *unity,* as shown in Figure 1.3b. Note that if $V_s = 1$, then, from Eq. (1.4), $V_v = e$, and the weight of the soil solids is

$$W_s = G_s \gamma_w$$

where

G_s = specific gravity of soil solids
γ_w = unit weight of water (9.81 kN/m^3, or 62.4 lb/ft^3)

Also, from Eq. (1.8), the weight of water $W_w = wW_s$. Thus, for the soil specimen under consideration, $W_w = wW_s = wG_s\gamma_w$. Now, for the general relation for moist unit weight given in Eq. (1.9),

$$\gamma = \frac{W}{V} = \frac{W_s + W_w}{V_s + V_v} = \frac{G_s\gamma_w(1 + w)}{1 + e} \tag{1.11}$$

Similarly, the dry unit weight [Eq. (1.10)] is

$$\gamma_d = \frac{W_s}{V} = \frac{W_s}{V_s + V_v} = \frac{G_s\gamma_w}{1 + e} \tag{1.12}$$

From Eqs. (1.11) and (1.12), note that

$$\gamma_d = \frac{\gamma}{1 + w} \tag{1.13}$$

If a soil specimen is completely saturated, as shown in Figure 1.3c, then

$$V_v = e$$

Also, for this case,

$$V_v = \frac{W_w}{\gamma_w} = \frac{wG_s\gamma_w}{\gamma_w} = wG_s$$

Thus,

$$e = wG_s \quad \text{(for saturated soil } only) \tag{1.14}$$

The saturated unit weight of soil then becomes

$$\gamma_{\text{sat}} = \frac{W_s + W_w}{V_s + V_v} = \frac{G_s\gamma_w + e\gamma_w}{1 + e} \tag{1.15}$$

Relationships similar to Eqs. (1.11), (1.12), and (1.15) in terms of porosity can also be obtained by considering a representative soil specimen with a unit volume. These relationships are

$$\gamma = G_s\gamma_w(1 - n)(1 + w) \tag{1.16}$$

$$\gamma_d = (1 - n)G_s\gamma_w \tag{1.17}$$

and

$$\gamma_{\text{sat}} = [(1 - n)G_s + n]\gamma_w \tag{1.18}$$

Table 1.3 Specific Gravities of Some Soils

Type of Soil	G_s
Quartz sand	2.64–2.66
Silt	2.67–2.73
Clay	2.70–2.9
Chalk	2.60–2.75
Loess	2.65–2.73
Peat	1.30–1.9

Except for peat and highly organic soils, the general range of the values of specific gravity of soil solids (G_s) found in nature is rather small. Table 1.3 gives some representative values. For practical purposes, a reasonable value can be assumed in lieu of running a test.

1.5 *Relative Density*

In *granular soils*, the degree of compaction in the field can be measured according to the *relative density*, defined as

$$D_r(\%) = \frac{e_{max} - e}{e_{max} - e_{min}} \times 100 \qquad (1.19)$$

where

e_{max} = void ratio of the soil in the loosest state
e_{min} = void ratio in the densest state
e = *in situ* void ratio

The relative density can also be expressed in terms of dry unit weight, or

$$D_r(\%) = \left\{ \frac{\gamma_d - \gamma_{d(min)}}{\gamma_{d(max)} - \gamma_{d(min)}} \right\} \frac{\gamma_{d(max)}}{\gamma_d} \times 100 \qquad (1.20)$$

where

γ_d = *in situ* dry unit weight
$\gamma_{d(max)}$ = dry unit weight in the *densest* state; that is, when the void ratio is e_{min}
$\gamma_{d(min)}$ = dry unit weight in the *loosest* state; that is, when the void ratio is e_{max}

The denseness of a granular soil is sometimes related to the soil's relative density. Table 1.4 gives a general correlation of the denseness and D_r. For naturally occurring sands, the magnitudes of e_{max} and e_{min} [Eq. (1.19)] may vary widely. The main reasons for such wide variations are the uniformity coefficient, C_u, and the roundness of the particles.

Table 1.4   Denseness of a Granular Soil

Relative density, D_r(%)	Description
0–20	Very loose
20–40	Loose
40–60	Medium
60–80	Dense
80–100	Very dense

Example 1.1

A 0.25 ft³ moist soil weighs 30.8 lb. When dried in an oven, this soil weighs 28.2 lb. Given $G_s = 2.7$. Determine the

a. Moist unit weight, γ
b. Moisture content, w
c. Dry unit weight, γ_d
d. Void ratio, e
e. Porosity, n
f. Degree of saturation, S

Solution
Part a: Moist Unit Weight
From Eq. (1.9),

$$\gamma = \frac{W}{V} = \frac{30.8}{0.25} = \textbf{123.2 lb/ft}^3$$

Part b: Moisture Content
From Eq. (1.8),

$$w = \frac{W_w}{W_s} = \frac{30.8 - 28.2}{28.2} \times 100 = \textbf{9.2\%}$$

Part c: Dry Unit Weight
From Eq. (1.10),

$$\gamma_d = \frac{W_s}{V} = \frac{28.2}{0.25} = \textbf{112.8 lb/ft}^3$$

Part d: Void Ratio
From Eq. (1.4),

$$e = \frac{V_v}{V_s}$$

$$V_s = \frac{W_s}{G_s \gamma_w} = \frac{28.2}{(2.67)(62.4)} = 0.169 \text{ ft}^3$$

$$e = \frac{0.25 - 0.169}{0.169} = \mathbf{0.479}$$

Part e: Porosity
From Eq. (1.6),

$$n = \frac{e}{1 + e} = \frac{0.479}{1 + 0.479} = \mathbf{0.324}$$

Part f: Degree of Saturation
From Eq. (1.7),

$$S(\%) = \frac{V_w}{V_v} \times 100$$

$$V_v = V - V_s = 0.25 - 0.169 = 0.081 \text{ ft}^3$$

$$V_w = \frac{W_w}{\gamma_w} = \frac{30.8 - 28.2}{62.4} = 0.042 \text{ ft}^3$$

$$S = \frac{0.042}{0.081} \times 100 = \mathbf{51.9\%}$$ ∎

Example 1.2

A soil has a void ratio of 0.72, moisture content = 12%, and $G_s = 2.72$. Determine the

  a. Dry unit weight (kN/m³)
  b. Moist unit weight (kN/m³)
  c. Weight of water in kN/m³ to be added to make the soil saturated

Solution
Part a: Dry Unit Weight
From Eq. (1.12),

$$\gamma_d = \frac{G_s \gamma_w}{1 + e} = \frac{(2.72)(9.81)}{1 + 0.72} = \mathbf{15.51 \text{ kN/m}^3}$$

Part b: Moist Unit Weight
From Eq. (1.11),

$$\gamma = \frac{G_s\gamma_w(1 + w)}{1 + e} = \frac{(2.72)(9.81)(1 + 0.12)}{1 + 0.72} = \textbf{17.38 kN/m}^3$$

Part c: Weight of Water to be Added
From Eq. (1.15),

$$\gamma_{sat} = \frac{(G_s + e)\gamma_w}{1 + e} = \frac{(2.72 + 0.72)(9.81)}{1 + 0.72} = \textbf{19.62 kN/m}^3$$

Water to be added $= \gamma_{sat} - \gamma = 19.62 - 17.38 = \textbf{2.24 kN/m}^3$ ∎

Example 1.3

The maximum and minimum dry unit weights of a sand are 17.1 kN/m³ and 14.2 kN/m³, respectively. The sand in the field has a relative density of 70% with a moisture content of 8%. Determine the moist unit weight of the sand in the field.

Solution
From Eq. (1.20),

$$D_r = \left[\frac{\gamma_d - \gamma_{d(min)}}{\gamma_{d(max)} - \gamma_{d(min)}}\right]\left[\frac{\gamma_{d(max)}}{\gamma_d}\right]$$

$$0.7 = \left[\frac{\gamma_d - 14.2}{17.1 - 14.2}\right]\left[\frac{17.1}{\gamma_d}\right]$$

$$\gamma_d = 16.11 \text{ kN/m}^3$$

$$\gamma = \gamma_d(1 + w) = 16.11\left(1 + \frac{8}{100}\right) = \textbf{17.4 kN/m}^3$$ ∎

1.6   *Atterberg Limits*

When a clayey soil is mixed with an excessive amount of water, it may flow like a *semi-liquid*. If the soil is gradually dried, it will behave like a *plastic, semisolid,* or *solid* material, depending on its moisture content. The moisture content, in percent, at which the soil changes from a liquid to a plastic state is defined as the *liquid limit* (LL). Similarly, the

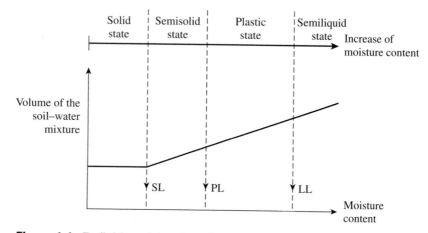

Figure 1.4 Definition of Atterberg limits

moisture content, in percent, at which the soil changes from a plastic to a semisolid state and from a semisolid to a solid state are defined as the *plastic limit* (PL) and the *shrinkage limit* (SL), respectively. These limits are referred to as *Atterberg limits* (Figure 1.4):

- The *liquid limit* of a soil is determined by Casagrande's liquid device (ASTM Test Designation D-4318) and is defined as the moisture content at which a groove closure of 12.7 mm (1/2 in.) occurs at 25 blows.
- The *plastic limit* is defined as the moisture content at which the soil crumbles when rolled into a thread of 3.18 mm (1/8 in.) in diameter (ASTM Test Designation D-4318).
- The *shrinkage limit* is defined as the moisture content at which the soil does not undergo any further change in volume with loss of moisture (ASTM Test Designation D-427).

The difference between the liquid limit and the plastic limit of a soil is defined as the *plasticity index* (PI), or

$$PI = LL - PL \tag{1.21}$$

1.7 *Soil Classification Systems*

Soil classification systems divide soils into groups and subgroups based on common engineering properties such as the *grain-size distribution, liquid limit,* and *plastic limit.* The two major classification systems presently in use are (1) the *American Association of State Highway and Transportation Officials (AASHTO) System* and (2) the *Unified Soil Classification System* (also *ASTM*). The AASHTO system is used mainly for the classification of highway subgrades. It is not used in foundation construction.

AASHTO System

The AASHTO Soil Classification System was originally proposed by the Highway Research Board's Committee on Classification of Materials for Subgrades and Granular Type Roads (1945). According to the present form of this system, soils can be classified according to eight major groups, A-1 through A-8, based on their grain-size distribution, liquid limit, and plasticity indices. Soils listed in groups A-1, A-2, and A-3 are coarse-grained materials, and those in groups A-4, A-5, A-6, and A-7 are fine-grained materials. Peat, muck, and other highly organic soils are classified under A-8. They are identified by visual inspection.

The AASHTO classification system (for soils A-1 through A-7) is presented in Table 1.5. Note that group A-7 includes two types of soil. For the A-7-5 type, the

Table 1.5 AASHTO Soil Classification System

General classification	Granular materials (35% or less of total sample passing No. 200 sieve)							
Group classification	A-1		A-3	A-2				
	A-1-a	A-1-b		A-2-4	A-2-5	A-2-6	A-2-7	
Sieve analysis (% passing)								
No. 10 sieve	50 max							
No. 40 sieve	30 max	50 max	51 min					
No. 200 sieve	15 max	25 max	10 max	35 max	35 max	35 max	35 max	
For fraction passing No. 40 sieve								
Liquid limit (LL)				40 max	41 min	40 max	41 min	
Plasticity index (PI)	6 max		Nonplastic	10 max	10 max	11 min	11 min	
Usual type of material	Stone fragments, gravel, and sand		Fine sand	Silty or clayey gravel and sand				
Subgrade rating				Excellent to good				

General classification	Silt–clay materials (More than 35% of total sample passing No. 200 sieve)			
Group classification	A-4	A-5	A-6	A-7
				A-7-5[a]
				A-7-6[b]
Sieve analysis (% passing)				
No. 10 sieve				
No. 40 sieve				
No. 200 sieve	36 min	36 min	36 min	36 min
For fraction passing No. 40 sieve				
Liquid limit (LL)	40 max	41 min	40 max	41 min
Plasticity index (PI)	10 max	10 max	11 min	11 min
Usual types of material	Mostly silty soils		Mostly clayey soils	
Subgrade rating		Fair to poor		

[a] If PI $\leq$ LL $-$ 30, the classification is A-7-5.
[b] If PI $>$ LL $-$ 30, the classification is A-7-6.

plasticity index of the soil is less than or equal to the liquid limit minus 30. For the A-7-6 type, the plasticity index is greater than the liquid limit minus 30.

For qualitative evaluation of the desirability of a soil as a highway subgrade material, a number referred to as the *group index* has also been developed. The higher the value of the group index for a given soil, the weaker will be the soil's performance as a subgrade. A group index of 20 or more indicates a very poor subgrade material. The formula for the group index is

$$GI = (F_{200} - 35)[0.2 + 0.005(LL - 40)] + 0.01(F_{200} - 15)(PI - 10) \qquad (1.22)$$

where

F_{200} = percent passing no. 200 sieve, expressed as a whole number
LL = liquid limit
PI = plasticity index

When calculating the group index for a soil belonging to group A-2-6 or A-2-7, use only the partial group-index equation relating to the plasticity index:

$$GI = 0.01(F_{200} - 15)(PI - 10) \qquad (1.23)$$

The group index is rounded to the nearest whole number and written next to the soil group in parentheses; for example, we have

A-4 (5)
| Group index
Soil group

Unified System

The Unified Soil Classification System was originally proposed by A. Casagrande in 1942 and was later revised and adopted by the United States Bureau of Reclamation and the U.S. Army Corps of Engineers. The system is currently used in practically all geotechnical work.

In the Unified System, the following symbols are used for identification:

Symbol	G	S	M	C	O	Pt	H	L	W	P
Description	Gravel	Sand	Silt	Clay	Organic silts and clay	Peat and highly organic soils	High plasticity	Low plasticity	Well graded	Poorly graded

The plasticity chart (Figure 1.5) and Table 1.6 show the procedure for determining the group symbols for various types of soil. When classifying a soil be sure

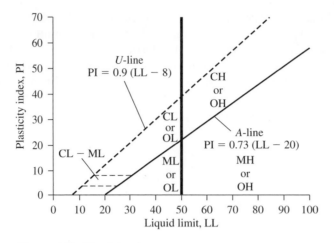

Figure 1.5   Plasticity chart

to provide the group name that generally describes the soil, along with the group symbol. Figures 1.6, 1.7, and 1.8 give flowcharts for obtaining the group names for coarse-grained soil, inorganic fine-grained soil, and organic fine-grained soil, respectively.

Example 1.4

Classify the following soil by the AASHTO classification system.

Percent passing No. 4 sieve = 92
Percent passing No. 10 sieve = 87
Percent passing No. 40 sieve = 65
Percent passing No. 200 sieve = 30
Liquid limit = 22
Plasticity index = 8

Solution

Table 1.5 shows that it is a granular material because less than 35% is passing a No. 200 sieve. With LL = 22 (that is, less than 40) and PI = 8 (that is, less than 10), the soil falls in group A-2-4. From Eq. (1.23),

$$GI = 0.01(F_{200} - 15)(PI - 10) = 0.01(30 - 15)(8 - 10)$$
$$= -0.3 \approx 0$$

The soil is **A-2-4(0)**.   ∎

Table 1.6 Unified Soil Classification Chart (after ASTM, 2005)

Criteria for assigning group symbols and group names using laboratory tests[a]				Soil classification	
				Group symbol	Group name[b]
Coarse-grained soils More than 50% retained on No. 200 sieve	Gravels More than 50% of coarse fraction retained on No. 4 sieve	Clean Gravels Less than 5% fines[c]	$C_u \geq 4$ and $1 \leq C_c \leq 3^e$	GW	Well-graded gravel[f]
			$C_u < 4$ and/or $1 > C_c > 3^e$	GP	Poorly graded gravel[f]
		Gravels with Fines More than 12% fines[c]	Fines classify as ML or MH	GM	Silty gravel[f,g,h]
			Fines classify as CL or CH	GC	Clayey gravel[f,g,h]
	Sands 50% or more of coarse fraction passes No. 4 sieve	Clean Sands Less than 5% fines[d]	$C_u \geq 6$ and $1 \leq C_c \leq 3^e$	SW	Well-graded sand[i]
			$C_u < 6$ and/or $1 > C_c > 3^e$	SP	Poorly graded sand[i]
		Sand with Fines More than 12% fines[d]	Fines classify as ML or MH	SM	Silty sand[g,h,i]
			Fines classify as CL or CH	SC	Clayey sand[g,h,i]
Fine-grained soils 50% or more passes the No. 200 sieve	Silts and Clays Liquid limit less than 50	Inorganic	PI > 7 and plots on or above "A" line[j]	CL	Lean clay[k,l,m]
			PI < 4 or plots below "A" line[j]	ML	Silt[k,l,m]
		Organic	$\dfrac{\text{Liquid limit—oven dried}}{\text{Liquid limit—not dried}} < 0.75$	OL	Organic clay[k,l,m,n] / Organic silt[k,l,m,o]
	Silts and Clays Liquid limit 50 or more	Inorganic	PI plots on or above "A" line	CH	Fat clay[k,l,m]
			PI plots below "A" line	MH	Elastic silt[k,l,m]
		Organic	$\dfrac{\text{Liquid limit—oven dried}}{\text{Liquid limit—not dried}} < 0.75$	OH	Organic clay[k,l,m,p] / Organic silt[k,l,m,q]
Highly organic soils	Primarily organic matter, dark in color, and organic odor			PT	Peat

[a] Based on the material passing the 75-mm. (3-in) sieve.

[b] If field sample contained cobbles or boulders, or both, add "with cobbles or boulders, or both" to group name.

[c] Gravels with 5 to 12% fines require dual symbols: GW-GM well-graded gravel with silt; GW-GC well-graded gravel with clay; GP-GM poorly graded gravel with silt; GP-GC poorly graded gravel with clay.

[d] Sands with 5 to 12% fines require dual symbols: SW-SM well-graded sand with silt; SW-SC well-graded sand with clay; SP-SM poorly graded sand with silt; SP-SC poorly graded sand with clay.

[e] $C_u = D_{60}/D_{10}$ $C_c = \dfrac{(D_{30})^2}{D_{10} \times D_{60}}$

[f] If soil contains ≥15% sand, add "with sand" to group name.

[g] If fines classify as CL-ML, use dual symbol GC-GM or SC-SM.

[h] If fines are organic, add "with organic fines" to group name.

[i] If soil contains ≥15% gravel, add "with gravel" to group name.

[j] If Atterberg limits plot in hatched area, soil is a CL-ML, silty clay.

[k] If soil contains 15 to 29% plus No. 200, add "with sand" or "with gravel," whichever is predominant.

[l] If soil contains ≥30% plus No. 200, predominantly sand, add "sandy" to group name.

[m] If soil contains ≥30% plus No. 200, predominantly gravel, add "gravelly" to group name.

[n] PI ≥ 4 and plots on or above "A" line.

[o] PI < 4 or plots below "A" line.

[p] PI plots on or above "A" line.

[q] PI plots below "A" line.

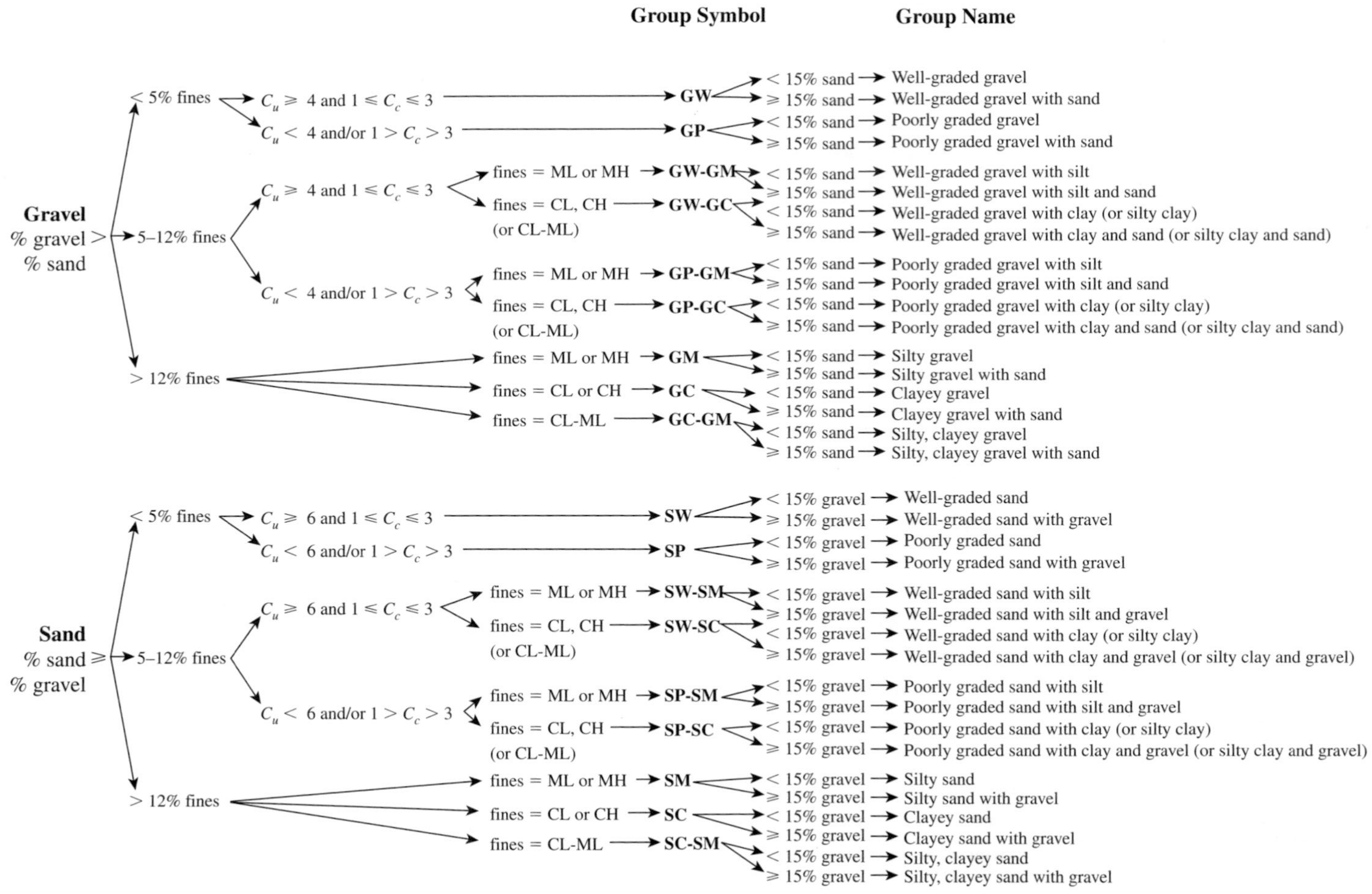

Group Symbol   **Group Name**

Gravel
% gravel > % sand

< 5% fines
- $C_u \geq 4$ and $1 \leq C_c \leq 3$ → **GW** → < 15% sand → Well-graded gravel
- → ≥ 15% sand → Well-graded gravel with sand
- $C_u < 4$ and/or $1 > C_c > 3$ → **GP** → < 15% sand → Poorly graded gravel
- → ≥ 15% sand → Poorly graded gravel with sand

5–12% fines
- fines = ML or MH → **GW-GM** → < 15% sand → Well-graded gravel with silt
- → ≥ 15% sand → Well-graded gravel with silt and sand
- fines = CL, CH (or CL-ML) → **GW-GC** → < 15% sand → Well-graded gravel with clay (or silty clay)
- → ≥ 15% sand → Well-graded gravel with clay and sand (or silty clay and sand)
- fines = ML or MH → **GP-GM** → < 15% sand → Poorly graded gravel with silt
- → ≥ 15% sand → Poorly graded gravel with silt and sand
- fines = CL, CH (or CL-ML) → **GP-GC** → < 15% sand → Poorly graded gravel with clay (or silty clay)
- → ≥ 15% sand → Poorly graded gravel with clay and sand (or silty clay and sand)

> 12% fines
- fines = ML or MH → **GM** → < 15% sand → Silty gravel
- → ≥ 15% sand → Silty gravel with sand
- fines = CL or CH → **GC** → < 15% sand → Clayey gravel
- → ≥ 15% sand → Clayey gravel with sand
- fines = CL-ML → **GC-GM** → < 15% sand → Silty, clayey gravel
- → ≥ 15% sand → Silty, clayey gravel with sand

Sand
% sand ≥ % gravel

< 5% fines
- $C_u \geq 6$ and $1 \leq C_c \leq 3$ → **SW** → < 15% gravel → Well-graded sand
- → ≥ 15% gravel → Well-graded sand with gravel
- $C_u < 6$ and/or $1 > C_c > 3$ → **SP** → < 15% gravel → Poorly graded sand
- → ≥ 15% gravel → Poorly graded sand with gravel

5–12% fines
- fines = ML or MH → **SW-SM** → < 15% gravel → Well-graded sand with silt
- → ≥ 15% gravel → Well-graded sand with silt and gravel
- fines = CL, CH (or CL-ML) → **SW-SC** → < 15% gravel → Well-graded sand with clay (or silty clay)
- → ≥ 15% gravel → Well-graded sand with clay and gravel (or silty clay and gravel)
- fines = ML or MH → **SP-SM** → < 15% gravel → Poorly graded sand with silt
- → ≥ 15% gravel → Poorly graded sand with silt and gravel
- fines = CL, CH (or CL-ML) → **SP-SC** → < 15% gravel → Poorly graded sand with clay (or silty clay)
- → ≥ 15% gravel → Poorly graded sand with clay and gravel (or silty clay and gravel)

> 12% fines
- fines = ML or MH → **SM** → < 15% gravel → Silty sand
- → ≥ 15% gravel → Silty sand with gravel
- fines = CL or CH → **SC** → < 15% gravel → Clayey sand
- → ≥ 15% gravel → Clayey sand with gravel
- fines = CL-ML → **SC-SM** → < 15% gravel → Silty, clayey sand
- → ≥ 15% gravel → Silty, clayey sand with gravel

Figure 1.6 Flowchart for classifying coarse-grained soils (more than 50% retained on No. 200 Sieve) (After ASTM, 2005)

Group Symbol

Group Name

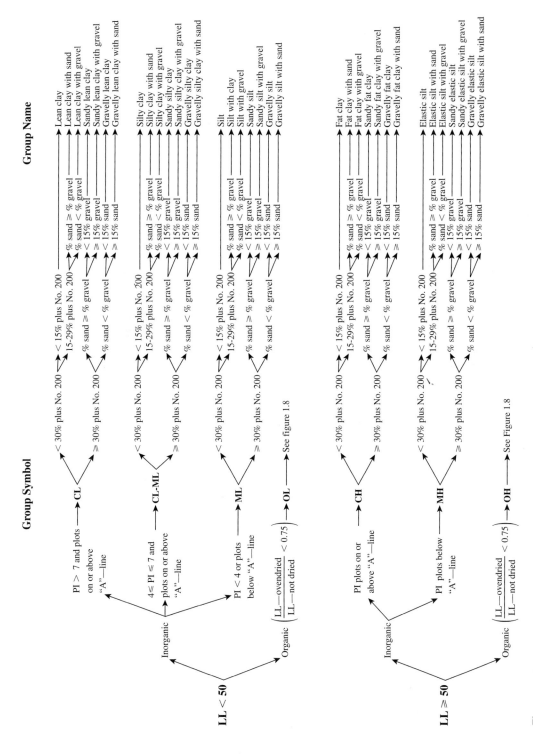

Figure 1.7 Flowchart for classifying fine-grained soil (50% or more passes No. 200 Sieve) (After ASTM, 2005)

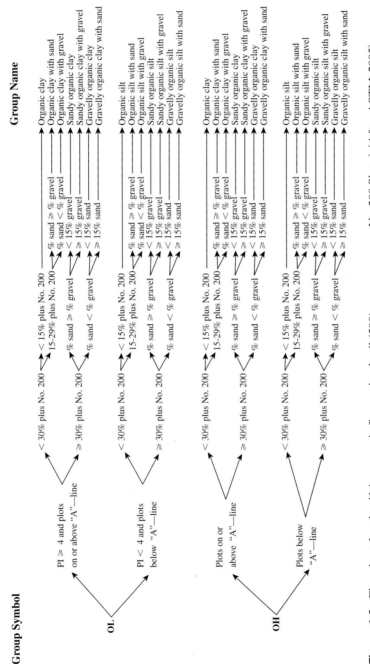

Figure 1.8 Flowchart for classifying organic fine-grained soil (50% or more passes No. 200 Sieve) (After ASTM, 2005)

Example 1.5

Classify the following soil by the Unified Soil Classification System:

Percent passing No. 4 sieve = 82
Percent passing No. 10 sieve = 71
Percent passing No. 40 sieve = 64
Percent passing No. 200 sieve = 41
Liquid limit = 31
Plasticity index = 12

Solution

We are given that $F_{200} = 41$, LL = 31, and PI = 12. Since 59% of the sample is retained on a No. 200 sieve, the soil is a coarse-grained material. The percentage passing a No. 4 sieve is 82, so 18% is retained on No. 4 sieve (gravel fraction). The coarse fraction passing a No. 4 sieve (sand fraction) is $59 - 18 = 41\%$ (which is more than 50% of the total coarse fraction). Hence, the specimen is a sandy soil.

Now, using Table 1.6 and Figure 1.5, we identify the group symbol of the soil as **SC.**

Again from Figure 1.6, since the gravel fraction is greater than 15%, the group name is **clayey sand with gravel.**   ∎

1.8   *Hydraulic Conductivity of Soil*

The void spaces, or pores, between soil grains allow water to flow through them. In soil mechanics and foundation engineering, you must know how much water is flowing through a soil per unit time. This knowledge is required to design earth dams, determine the quantity of seepage under hydraulic structures, and dewater foundations before and during their construction. Darcy (1856) proposed the following equation (Figure 1.9) for calculating the velocity of flow of water through a soil:

$$v = ki \tag{1.24}$$

In this equation,

v = Darcy velocity (unit: cm/sec)
k = hydraulic conductivity of soil (unit: cm/sec)
i = hydraulic gradient

The hydraulic gradient is defined as

$$i = \frac{\Delta h}{L} \tag{1.25}$$

where

Δh = piezometric head difference between the sections at AA and BB
L = distance between the sections at AA and BB

(*Note:* Sections AA and BB are perpendicular to the direction of flow.)

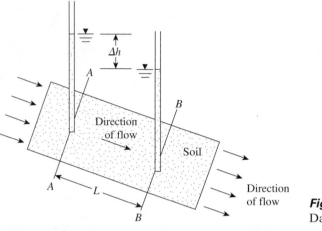

Figure 1.9 Definition of Darcy's law

Table 1.7 Range of the Hydraulic Conductivity for Various Soils

Type of soil	Hydraulic conductivity, k (cm/sec)
Medium to coarse gravel	Greater than 10^{-1}
Coarse to fine sand	10^{-1} to 10^{-3}
Fine sand, silty sand	10^{-3} to 10^{-5}
Silt, clayey silt, silty clay	10^{-4} to 10^{-6}
Clays	10^{-7} or less

Darcy's law [Eq. (1.24)] is valid for a wide range of soils. However, with materials like clean gravel and open-graded rockfills, the law breaks down because of the turbulent nature of flow through them.

The value of the hydraulic conductivity of soils varies greatly. In the laboratory, it can be determined by means of *constant-head* or *falling-head* permeability tests. The constant-head test is more suitable for granular soils. Table 1.7 provides the general range for the values of k for various soils. In granular soils, the value depends primarily on the void ratio. In the past, several equations have been proposed to relate the value of k to the void ratio in granular soil.

However the author recommends the following equation for use (also see Carrier, 2003):

$$k \propto \frac{e^3}{1 + e} \tag{1.26}$$

where

k = hydraulic conductivity
e = void ratio

More recently, Chapuis (2004) proposed an empirical relationship for k in conjunction with Eq. (1.26) as

$$k\,(\text{cm/s}) = 2.4622 \left[D_{10}^2 \frac{e^3}{(1 + e)} \right]^{0.7825} \tag{1.27}$$

where D = effective size (mm).

The preceding equation is valid for natural, uniform sand and gravel to predict k that is in the range of 10^{-1} to 10^{-3} cm/s. This can be extended to natural, silty sands without plasticity. It is not valid for crushed materials or silty soils with some plasticity.

According to their experimental observations, Samarasinghe, Huang, and Drnevich (1982) suggested that the hydraulic conductivity of normally consolidated clays could be given by the equation

$$k = C\frac{e^n}{1 + e} \tag{1.28}$$

where C and n are constants to be determined experimentally.

1.9 *Steady-State Seepage*

For most cases of seepage under hydraulic structures, the flow path changes direction and is not uniform over the entire area. In such cases, one of the ways of determining the rate of seepage is by a graphical construction referred to as the *flow net*, a concept based on Laplace's theory of continuity. According to this theory, for a steady flow condition, the flow at any point A (Figure 1.10) can be represented by the equation

$$k_x\frac{\partial^2 h}{\partial x^2} + k_y\frac{\partial^2 h}{\partial y^2} + k_z\frac{\partial^2 h}{\partial z^2} = 0 \tag{1.29}$$

where

k_x, k_y, k_z = hydraulic conductivity of the soil in the x, y, and z directions, respectively

h = hydraulic head at point A (i.e., the head of water that a piezometer placed at A would show with the *downstream water level* as *datum*, as shown in Figure 1.10)

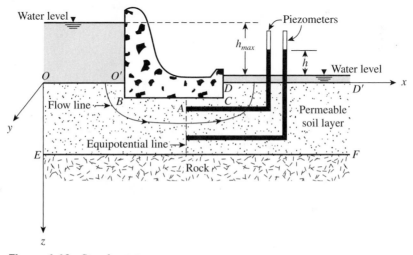

Figure 1.10 Steady-state seepage

For a two-dimensional flow condition, as shown in Figure 1.10,

$$\frac{\partial^2 h}{\partial^2 y} = 0$$

so Eq. (1.29) takes the form

$$k_x \frac{\partial^2 h}{\partial x^2} + k_z \frac{\partial^2 h}{\partial z^2} = 0 \qquad (1.30)$$

If the soil is isotropic with respect to hydraulic conductivity, $k_x = k_z = k$, and

$$\frac{\partial^2 h}{\partial x^2} + \frac{\partial^2 h}{\partial z^2} = 0 \qquad (1.31)$$

Equation (1.31), which is referred to as Laplace's equation and is valid for confined flow, represents two orthogonal sets of curves known as *flow lines* and *equipotential lines*. A flow net is a combination of numerous equipotential lines and flow lines. A flow line is a path that a water particle would follow in traveling from the upstream side to the downstream side. An equipotential line is a line along which water, in piezometers, would rise to the same elevation. (See Figure 1.10.)

In drawing a flow net, you need to establish the *boundary conditions*. For example, in Figure 1.10, the ground surfaces on the upstream (OO') and downstream (DD') sides are equipotential lines. The base of the dam below the ground surface, $O'BCD$, is a flow line. The top of the rock surface, EF, is also a flow line. Once the boundary conditions are established, a number of flow lines and equipotential lines are drawn by trial and error so that all the flow elements in the net have the same length-to-width ratio (L/B). In most cases, L/B is held to unity, that is, the flow elements are drawn as curvilinear "squares." This method is illustrated by the flow net shown in Figure 1.11. Note that all flow lines must intersect all equipotential lines at *right angles*.

Once the flow net is drawn, the seepage (in unit time per unit length of the structure) can be calculated as

$$q = k h_{max} \frac{N_f}{N_d} n \qquad (1.32)$$

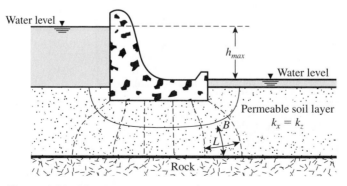

Figure 1.11 Flow net

where

N_f = number of flow channels
N_d = number of drops
n = width-to-length ratio of the flow elements in the flow net (B/L)
h_{max} = difference in water level between the upstream and downstream sides

The space between two consecutive flow lines is defined as a *flow channel,* and the space between two consecutive equipotential lines is called a *drop.* In Figure 1.11, $N_f = 2$, $N_d = 7$, and $n = 1$. When square elements are drawn in a flow net,

$$q = kh_{max}\frac{N_f}{N_d} \tag{1.33}$$

1.10 *Effective Stress*

The *total* stress at a given point in a soil mass can be expressed as

$$\sigma = \sigma' + u \tag{1.34}$$

where

σ = total stress
σ' = effective stress
u = pore water pressure

The effective stress, σ', is the vertical component of forces at solid-to-solid contact points over a unit cross-sectional area. Referring to Figure 1.12a, at point A

$$\sigma = \gamma h_1 + \gamma_{sat} h_2$$
$$u = h_2 \gamma_w$$

where

γ_w = unit weight of water
γ_{sat} = saturated unit weight of soil

So

$$\begin{aligned}\sigma' &= (\gamma h_1 + \gamma_{sat} h_2) - (h_2 \gamma_w) \\ &= \gamma h_1 + h_2(\gamma_{sat} - \gamma_w) \\ &= \gamma h_1 + \gamma' h_2\end{aligned} \tag{1.35}$$

where γ' = effective or submerged unit weight of soil.

For the problem in Figure 1.12a, there was *no seepage of water* in the soil. Figure 1.12b shows a simple condition in a soil profile in which there is upward seepage. For this case, at point A,

$$\sigma = h_1 \gamma_w + h_2 \gamma_{sat}$$

and

$$u = (h_1 + h_2 + h)\gamma_w$$

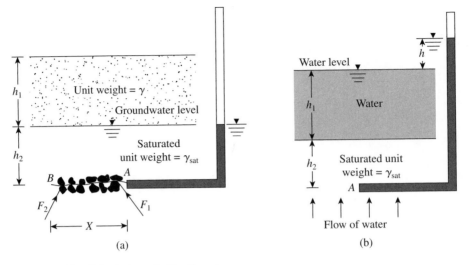

Figure 1.12　Calculation of effective stress

Thus, from Eq. (1.34),

$$\sigma' = \sigma - u = (h_1\gamma_w + h_2\gamma_{sat}) - (h_1 + h_2 + h)\gamma_w$$
$$= h_2(\gamma_{sat} - \gamma_w) - h\gamma_w = h_2\gamma' - h\gamma_w$$

or

$$\sigma' = h_2\left(\gamma' - \frac{h}{h_2}\gamma_w\right) = h_2(\gamma' - i\gamma_w) \qquad (1.36)$$

Note in Eq. (1.36) that h/h_2 is the hydraulic gradient i. If the hydraulic gradient is very high, so that $\gamma' - i\gamma_w$ becomes zero, *the effective stress will become zero*. In other words, there is no contact stress between the soil particles, and the soil will break up. This situation is referred to as the *quick condition*, or *failure by heave*. So, for heave,

$$i = i_{cr} = \frac{\gamma'}{\gamma_w} = \frac{G_s - 1}{1 + e} \qquad (1.37)$$

where i_{cr} = critical hydraulic gradient.

For most sandy soils, i_{cr} ranges from 0.9 to 1.1, with an average of about unity.

Example 1.6

For the soil profile shown in Figure 1.13, determine the total vertical stress, pore water pressure, and effective vertical stress at A, B, and C.

Solution
The following table can now be prepared.

Point	σ (kN/m^2)	u(kN/m^2)	$\sigma' = \sigma - u$ (kN/m^2)
A	0	0	0
B	$(4)(\gamma_d) = (4)(14.5) = 58$	0	58
C	$58 + (\gamma_{sat})(5) = 58 + (17.2)(5) = 144$	$(5)(\gamma_w) = (5)(9.81) = 49.05$	94.95

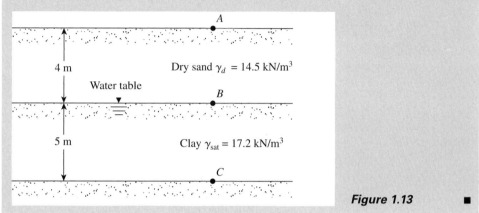

Figure 1.13 ∎

1.11 Consolidation

In the field, when the stress on a saturated clay layer is increased—for example, by the construction of a foundation—the pore water pressure in the clay will increase. Because the hydraulic conductivity of clays is very small, some time will be required for the excess pore water pressure to dissipate and the increase in stress to be transferred to the soil skeleton. According to Figure 1.14, if $\Delta\sigma$ is a surcharge at the ground surface over a very large area, the increase in total stress at any depth of the clay layer will be equal to $\Delta\sigma$.

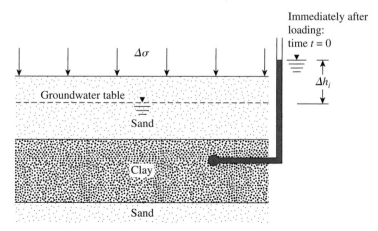

Figure 1.14 Principles of consolidation

However, at time $t = 0$ (i.e., immediately after the stress is applied), the excess pore water pressure at any depth Δu will equal $\Delta \sigma$, or

$$\Delta u = \Delta h_i \gamma_w = \Delta \sigma \text{ (at time } t = 0\text{)}$$

Hence, the increase in effective stress at time $t = 0$ will be

$$\Delta \sigma' = \Delta \sigma - \Delta u = 0$$

Theoretically, at time $t = \infty$, when all the excess pore water pressure in the clay layer has dissipated as a result of drainage into the sand layers,

$$\Delta u = 0 \quad \text{(at time } t = \infty\text{)}$$

Then the increase in effective stress in the clay layer is

$$\Delta \sigma' = \Delta \sigma - \Delta u = \Delta \sigma - 0 = \Delta \sigma$$

This gradual increase in the effective stress in the clay layer will cause settlement over a period of time and is referred to as *consolidation*.

Laboratory tests on undisturbed saturated clay specimens can be conducted (ASTM Test Designation D-2435) to determine the consolidation settlement caused by various incremental loadings. The test specimens are usually 63.5 mm (2.5 in.) in diameter and 25.4 mm (1 in.) in height. Specimens are placed inside a ring, with one porous stone at the top and one at the bottom of the specimen (Figure 1.15a). A load on the specimen is then applied so that the total vertical stress is equal to σ. Settlement readings for the specimen are taken periodically for 24 hours. After that, the load on the specimen is doubled and more settlement readings are taken. At all times during the test, the specimen is kept under water. The procedure is continued until the desired limit of stress on the clay specimen is reached.

Based on the laboratory tests, a graph can be plotted showing the variation of the void ratio e at the *end* of consolidation against the corresponding vertical effective stress σ'. (On a semilogarithmic graph, e is plotted on the arithmetic scale and σ' on the log scale.) The nature of the variation of e against log σ' for a clay specimen is shown in Figure 1.15b. After the desired consolidation pressure has been reached, the specimen gradually can be unloaded, which will result in the swelling of the specimen. The figure also shows the variation of the void ratio during the unloading period.

From the e–log σ' curve shown in Figure 1.15b, three parameters necessary for calculating settlement in the field can be determined:

1. The *preconsolidation pressure*, σ'_c, is the *maximum past effective overburden pressure* to which the soil specimen has been subjected. It can be determined by using a simple graphical procedure proposed by Casagrande (1936). The procedure involves five steps (see Figure 1.15b):
   a. Determine the point O on the e–log σ' curve that has the sharpest curvature (i.e., the smallest radius of curvature).
   b. Draw a horizontal line OA.
   c. Draw a line OB that is tangent to the e–log σ' curve at O.
   d. Draw a line OC that bisects the angle AOB.
   e. Produce the straight-line portion of the e–log σ' curve backwards to intersect OC. This is point D. The pressure that corresponds to point D is the preconsolidation pressure σ'_c.

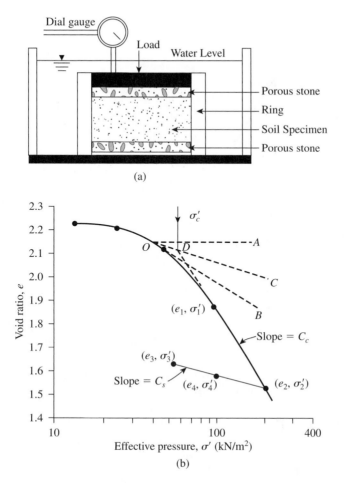

(a)

(b)

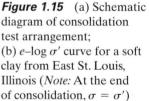

Figure 1.15 (a) Schematic diagram of consolidation test arrangement; (b) e–log σ' curve for a soft clay from East St. Louis, Illinois (*Note:* At the end of consolidation, $\sigma = \sigma'$)

Natural soil deposits can be *normally consolidated* or *overconsolidated* (or *preconsolidated*). If the present effective overburden pressure $\sigma' = \sigma'_o$ is equal to the preconsolidated pressure σ'_c the soil is *normally consolidated.* However, if $\sigma'_o < \sigma'_c$, the soil is *overconsolidated.*

2. The *compression index*, C_c, is the slope of the straight-line portion (the latter part) of the loading curve, or

$$C_c = \frac{e_1 - e_2}{\log \sigma'_2 - \log \sigma'_1} = \frac{e_1 - e_2}{\log \left(\dfrac{\sigma'_2}{\sigma'_1} \right)} \tag{1.38}$$

where e_1 and e_2 are the void ratios at the end of consolidation under effective stresses σ'_1 and σ'_2, respectively.

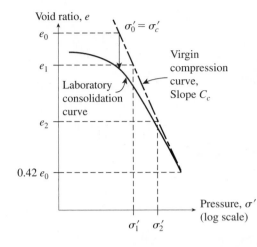

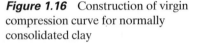

Figure 1.16   Construction of virgin compression curve for normally consolidated clay

The *compression index,* as determined from the laboratory e–log σ' curve, will be somewhat different from that encountered in the field. The primary reason is that the soil remolds itself to some degree during the field exploration. The nature of variation of the e–log σ' curve in the field for a normally consolidated clay is shown in Figure 1.16. The curve, generally referred to as the *virgin compression curve,* approximately intersects the laboratory curve at a void ratio of $0.42e_o$ (Terzaghi and Peck, 1967). Note that e_o is the void ratio of the clay in the field. Knowing the values of e_o and σ'_c, you can easily construct the virgin curve and calculate its compression index by using Eq. (1.38).

The value of C_c can vary widely, depending on the soil. Skempton (1944) gave an empirical correlation for the compression index in which

$$C_c = 0.009(\text{LL} - 10) \tag{1.39}$$

where LL = liquid limit.

Besides Skempton, several other investigators also have proposed correlations for the compression index. Some of those are given here:

Rendon-Herrero (1983):

$$C_c = 0.141G_s^{1.2}\left(\frac{1 + e_o}{G_s}\right)^{2.38} \tag{1.40}$$

Nagaraj and Murty (1985):

$$C_c = 0.2343\left[\frac{\text{LL}(\%)}{100}\right]G_s \tag{1.41}$$

Park and Koumoto (2004):

$$C_c = \frac{n_o}{371.747 - 4.275n_o} \tag{1.42}$$

where n_o = *in situ* porosity of soil.

Worth and Wood (1978):

$$C_c = 0.5G_s\left(\frac{PI(\%)}{100}\right) \qquad (1.43)$$

3. The *swelling index, C_s,* is the slope of the unloading portion of the e–$\log \sigma'$ curve. In Figure 1.15b, it is defined as

$$C_s = \frac{e_3 - e_4}{\log\left(\dfrac{\sigma_4'}{\sigma_3'}\right)} \qquad (1.44)$$

In most cases, the value of the swelling index is $\frac{1}{4}$ to $\frac{1}{5}$ of the compression index. Following are some representative values of C_s/C_c for natural soil deposits:

Description of soil	C_s/C_c
Boston Blue clay	0.24–0.33
Chicago clay	0.15–0.3
New Orleans clay	0.15–0.28
St. Lawrence clay	0.05–0.1

The swelling index is also referred to as the *recompression index.*

The determination of the swelling index is important in the estimation of consolidation settlement of *overconsolidated clays.* In the field, depending on the pressure increase, an overconsolidated clay will follow an e–$\log \sigma'$ path *abc,* as shown in Figure 1.17. Note that point *a,* with coordinates σ_o' and e_o, corresponds to the field conditions before any increase in pressure. Point *b* corresponds to the preconsolidation pressure (σ_c') of the clay. Line *ab* is approximately parallel to the laboratory unloading curve *cd* (Schmertmann, 1953). Hence, if you know $e_o, \sigma_o', \sigma_c', C_c,$ and C_s, you can easily construct the field consolidation curve.

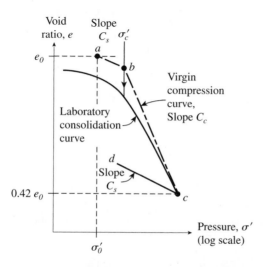

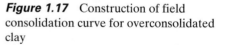

Figure 1.17 Construction of field consolidation curve for overconsolidated clay

1.12   *Calculation of Primary Consolidation Settlement*

The one-dimensional primary consolidation settlement (caused by an additional load) of a clay layer (Figure 1.18) having a thickness H_c may be calculated as

$$S_c = \frac{\Delta e}{1 + e_o} H_c \tag{1.45}$$

where

S_c = primary consolidation settlement
Δe = total change of void ratio caused by the additional load application
e_o = void ratio of the clay before the application of load

For normally consolidated clay (that is, $\sigma_o' = \sigma_c'$)

$$\Delta e = C_c \log \frac{\sigma_o' + \Delta \sigma'}{\sigma_o'} \tag{1.46}$$

where

σ_o' = average effective vertical stress on the clay layer
$\Delta \sigma'$ = $\Delta \sigma$ (that is, added pressure)

Now, combining Eqs. (1.45) and (1.46) yields

$$S_c = \frac{C_c H_c}{1 + e_o} \log \frac{\sigma_o' + \Delta \sigma'}{\sigma_o'} \tag{1.47}$$

For overconsolidated clay with $\sigma_o' + \Delta \sigma' \le \sigma_c'$,

$$\Delta e = C_s \log \frac{\sigma_o' + \Delta \sigma'}{\sigma_o'} \tag{1.48}$$

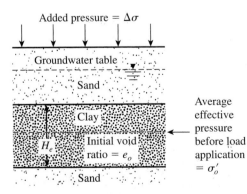

Figure 1.18   One-dimensional settlement calculation

Combining Eqs. (1.45) and (1.48) gives

$$S_c = \frac{H_c C_s}{1 + e_o} \log \frac{\sigma'_o + \Delta\sigma'}{\sigma'_o} \tag{1.49}$$

For overconsolidated clay, if $\sigma'_o < \sigma'_c < \sigma'_o + \Delta\sigma'$, then

$$\Delta e = \Delta e_1 + \Delta e_2 = C_s \log \frac{\sigma'_c}{\sigma'_o} + C_c \log \frac{\sigma'_o + \Delta\sigma'}{\sigma'_c} \tag{1.50}$$

Now, combining Eqs. (1.45) and (1.50) yields

$$S_c = \frac{C_s H_c}{1 + e_o} \log \frac{\sigma'_c}{\sigma'_o} + \frac{C_c H_c}{1 + e_o} \log \frac{\sigma'_o + \Delta\sigma'}{\sigma'_c} \tag{1.51}$$

1.13   *Time Rate of Consolidation*

In Section 1.11 (see Figure 1.14), we showed that consolidation is the result of the gradual dissipation of the excess pore water pressure from a clay layer. The dissipation of pore water pressure, in turn, increases the effective stress, which induces settlement. Hence, to estimate the degree of consolidation of a clay layer at some time t after the load is applied, you need to know the rate of dissipation of the excess pore water pressure.

Figure 1.19 shows a clay layer of thickness H_c that has highly permeable sand layers at its top and bottom. Here, the excess pore water pressure at any point A at any time t after the load is applied is $\Delta u = (\Delta h)\gamma_w$. For a vertical drainage condition (that is, in the direction of z only) from the clay layer, Terzaghi derived the differential equation

$$\frac{\partial(\Delta u)}{\partial t} = C_v \frac{\partial^2(\Delta u)}{\partial z^2} \tag{1.52}$$

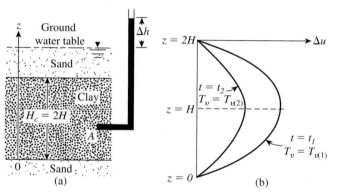

Figure 1.19   (a) Derivation of Eq. (1.54); (b) nature of variation of Δu with time

where C_v = coefficient of consolidation, defined by

$$C_v = \frac{k}{m_v \gamma_w} = \frac{k}{\dfrac{\Delta e}{\Delta \sigma'(1 + e_{av})} \gamma_w} \tag{1.53}$$

in which

k = hydraulic conductivity of the clay
Δe = total change of void ratio caused by an effective stress increase of $\Delta \sigma'$
e_{av} = average void ratio during consolidation
m_v = volume coefficient of compressibility = $\Delta e / [\Delta \sigma'(1 + e_{av})]$

Equation (1.52) can be solved to obtain Δu as a function of time t with the following boundary conditions:

1. Because highly permeable sand layers are located at $z = 0$ and $z = H_c$, the excess pore water pressure developed in the clay at those points will be immediately dissipated. Hence,

$$\Delta u = 0 \quad \text{at} \quad z = 0$$

and

$$\Delta u = 0 \quad \text{at} \quad z = H_c = 2H$$

where H = length of maximum drainage path (due to two-way drainage condition—that is, at the top and bottom of the clay).

2. At time $t = 0$, $\Delta u = \Delta u_0$ = initial excess pore water pressure after the load is applied. With the preceding boundary conditions, Eq. (1.53) yields

$$\Delta u = \sum_{m=0}^{m=\infty} \left[\frac{2(\Delta u_0)}{M} \sin\left(\frac{Mz}{H}\right) \right] e^{-M^2 T_v} \tag{1.54}$$

where

$M = [(2m + 1)\pi]/2$
m = an integer = $1, 2, \ldots$
T_v = nondimensional time factor = $(C_v t)/H^2$ $\qquad$ (1.55)

The value of Δu for various depths (i.e., $z = 0$ to $z = 2H$) at any given time t (and thus T_v) can be calculated from Eq. (1.52). The nature of this variation of Δu is shown in Figures 1.20a and b. Figure 1.20c shows the variation of $\Delta u / \Delta u_0$ with T_v and H/H_c using Eqs.(1.54) and (1.55).

The *average degree of consolidation* of the clay layer can be defined as

$$U = \frac{S_{c(t)}}{S_{c(\text{max})}} \tag{1.56}$$

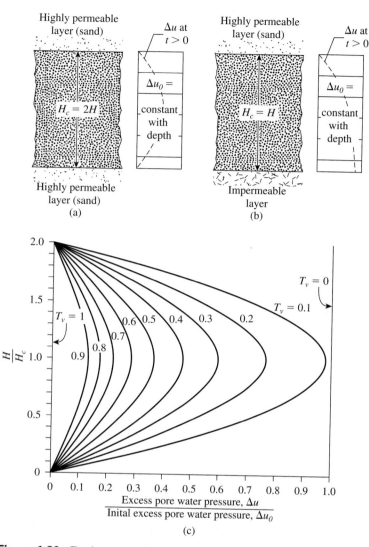

Figure 1.20 Drainage condition for consolidation: (a) two-way drainage; (b) one-way drainage; (c) plot of $\Delta u / \Delta u_0$ with T_v and H/H_c

where

$S_{c(t)}$ = settlement of a clay layer at time t after the load is applied
$S_{c(max)}$ = maximum consolidation settlement that the clay will undergo under a given loading

If the initial pore water pressure (Δu_0) distribution is constant with depth, as shown in Figure 1.20a, the average degree of consolidation also can be expressed as

$$U = \frac{S_{c(t)}}{S_{c(max)}} = \frac{\int_0^{2H} (\Delta u_0)dz - \int_0^{2H} (\Delta u)dz}{\int_0^{2H} (\Delta u_0)dz} \qquad (1.57)$$

or

$$U = \frac{(\Delta u_0)2H - \int_0^{2H} (\Delta u)\,dz}{(\Delta u_0)2H} = 1 - \frac{\int_0^{2H} (\Delta u)\,dz}{2H(\Delta u_0)} \tag{1.58}$$

Now, combining Eqs. (1.54) and (1.58), we obtain

$$U = \frac{S_{c(t)}}{S_{c(max)}} = 1 - \sum_{m=0}^{m=\infty} \left(\frac{2}{M^2}\right) e^{-M^2 T_v} \tag{1.59}$$

The variation of U with T_v can be calculated from Eq. (1.59) and is plotted in Figure 1.21. Note that Eq. (1.59) and thus Figure 1.21 are also valid when an impermeable layer is located at the bottom of the clay layer (Figure 1.20). In that case, the dissipation of excess pore water pressure can take place in one direction only. The length of the *maximum drainage path* is then equal to $H = H_c$.

The variation of T_v with U shown in Figure 1.21 can also be approximated by

$$T_v = \frac{\pi}{4}\left(\frac{U\%}{100}\right)^2 \quad \text{(for } U = 0 \text{ to } 60\%) \tag{1.60}$$

and

$$T_v = 1.781 - 0.933 \log(100 - U\%) \quad \text{(for } U > 60\%) \tag{1.61}$$

Trapezoidal Variation Figure 1.22 shows a trapezoidal variation of initial excess pore water pressure with *two-way drainage*. For this case the variation of T_v with U will be the same as shown in Figure 1.21.

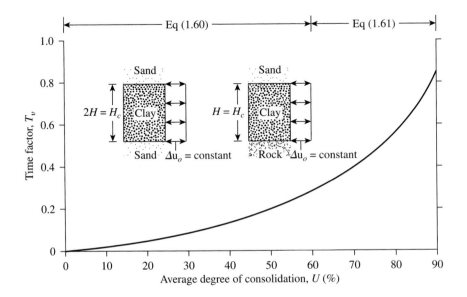

Figure 1.21 Plot of time factor against average degree of consolidation (Δu_0 = constant)

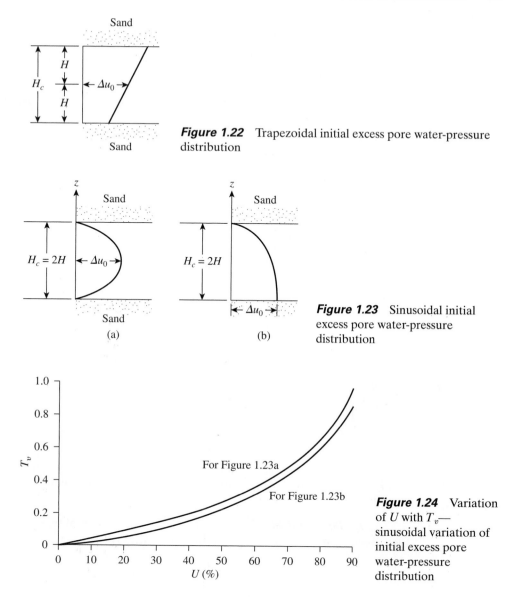

Figure 1.22 Trapezoidal initial excess pore water-pressure distribution

Figure 1.23   Sinusoidal initial excess pore water-pressure distribution

Figure 1.24   Variation of U with T_v—sinusoidal variation of initial excess pore water-pressure distribution

Sinusoidal Variation This variation is shown in Figures 1.23a and 1.23b. For the initial excess pore water-pressure variation shown in Figure 1.23a,

$$\Delta u = \Delta u_0 \sin \frac{\pi z}{2H} \tag{1.62}$$

Similarly, for the case shown in Figure 1.23b,

$$\Delta u = \Delta u_0 \cos \frac{\pi z}{4H} \tag{1.63}$$

The variations of T_v with U for these two cases are shown in Figure 1.24.

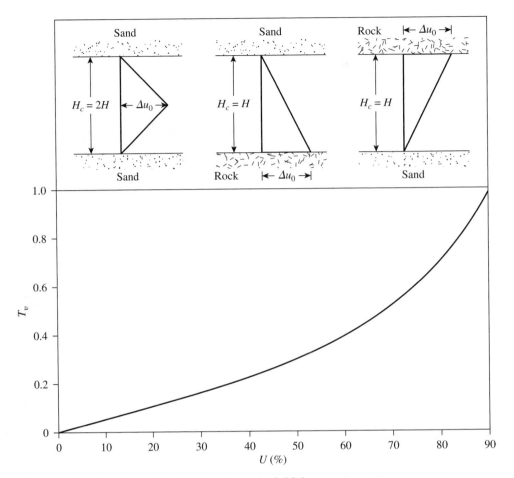

Figure 1.25   Variation of U with T_v—triangular initial excess pore water-pressure distribution

Triangular Variation   Figures 1.25 and 1.26 show several types of initial pore water-pressure variation and the variations of T_v with the average degree of consolidation.

Example 1.7

A laboratory consolidation test on a normally consolidated clay showed the following results:

Load, $\Delta\sigma'$ (kN/m²)	Void ratio at the end of consolidation, e
140	0.92
212	0.86

The specimen tested was 25.4 mm in thickness and drained on both sides. The time required for the specimen to reach 50% consolidation was 4.5 min.

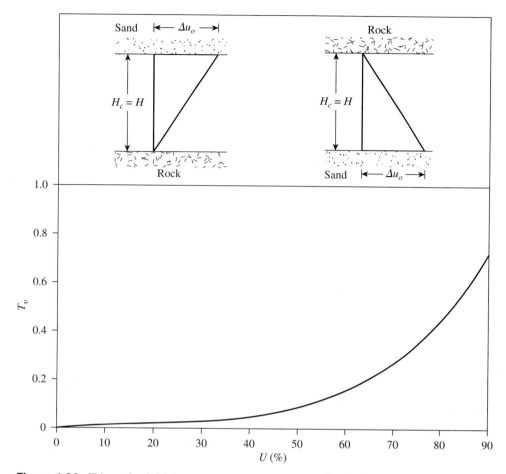

Figure 1.26   Triangular initial excess pore water-pressure distribution—variation of U with T_v

A similar clay layer in the field 2.8 m thick and drained on both sides, is subjected to a similar increase in average effective pressure (i.e., $\sigma_0' = 140 \text{ kN/m}^2$ and $\sigma_0' + \Delta\sigma' = 212 \text{ kN/m}^2$). Determine

a. the expected maximum primary consolidation settlement in the field.
b. the length of time required for the total settlement in the field to reach 40 mm. (Assume a uniform initial increase in excess pore water pressure with depth.)

Solution

Part a
For normally consolidated clay [Eq. (1.38)],

$$C_c = \frac{e_1 - e_2}{\log\left(\dfrac{\sigma_2'}{\sigma_1'}\right)} = \frac{0.92 - 0.86}{\log\left(\dfrac{212}{140}\right)} = 0.333$$

From Eq. (1.47),

$$S_c = \frac{C_c H_c}{1 + e_0} \log \frac{\sigma'_0 + \Delta \sigma'}{\sigma'_0} = \frac{(0.333)(2.8)}{1 + 0.92} \log \frac{212}{140} = 0.0875 \text{ m} = \textbf{87.5 mm}$$

Part b

From Eq. (1.56), the average degree of consolidation is

$$U = \frac{S_{c(t)}}{S_{c(max)}} = \frac{40}{87.5}(100) = 45.7\%$$

The coefficient of consolidation, C_v, can be calculated from the laboratory test. From Eq. (1.55),

$$T_v = \frac{C_v t}{H^2}$$

For 50% consolidation (Figure 1.21), $T_v = 0.197$, $t = 4.5$ min, and $H = H_c/2 = 12.7$ mm, so

$$C_v = T_{50} \frac{H^2}{t} = \frac{(0.197)(12.7)^2}{4.5} = 7.061 \text{ mm}^2/\text{min}$$

Again, for field consolidation, $U = 45.7\%$. From Eq. (1.60)

$$T_v = \frac{\pi}{4} \left(\frac{U\%}{100} \right)^2 = \frac{\pi}{4} \left(\frac{45.7}{100} \right)^2 = 0.164$$

But

$$T_v = \frac{C_v t}{H^2}$$

or

$$t = \frac{T_v H^2}{C_v} = \frac{0.164 \left(\dfrac{2.8 \times 1000}{2} \right)^2}{7.061} = 45{,}523 \text{ min} = \textbf{31.6 days} \quad \blacksquare$$

1.14   *Degree of Consolidation Under Ramp Loading*

The relationships derived for the average degree of consolidation in Section 1.13 assume that the surcharge load per unit area ($\Delta \sigma$) is applied instantly at time $t = 0$. However, in most practical situations, $\Delta \sigma$ increases gradually with time to a maximum value and remains constant thereafter. Figure 1.27 shows $\Delta \sigma$ increasing linearly with time (t) up to a maximum at time t_c (a condition called ramp loading).

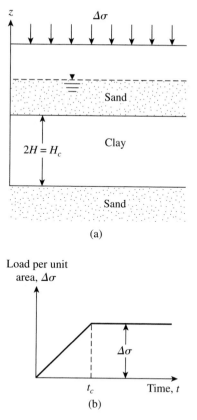

Figure 1.27 One-dimensional consolidation due to single ramp loading

For $t \geq t_c$, the magnitude of $\Delta\sigma$ remains constant. Olson (1977) considered this phenomenon and presented the average degree of consolidation, U, in the following form:

For $T_v \leq T_c$,

$$U = \frac{T_v}{T_c}\left\{1 - \frac{2}{T_v}\sum_{m=0}^{m=\infty}\frac{1}{M^4}[1 - \exp(-M^2T_v)]\right\} \qquad (1.64)$$

and for $T_v \geq T_c$,

$$U = 1 - \frac{2}{T_c}\sum_{m=0}^{m=\infty}\frac{1}{M^4}[\exp(M^2T_c) - 1]\exp(-M^2T_c) \qquad (1.65)$$

where m, M, and T_v have the same definition as in Eq. (1.54) and where

$$T_c = \frac{C_v t_c}{H^2} \qquad (1.66)$$

Figure 1.28 shows the variation of U with T_v for various values of T_c, based on the solution given by Eqs. (1.64) and (1.65).

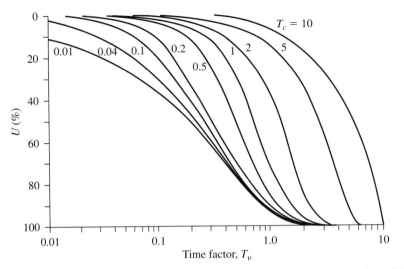

Figure 1.28 Olson's ramp-loading solution: plot of U vs. T_v (Eqs. 1.64 and 1.65)

Example 1.8

In Example 1.7, Part (b), if the increase in $\Delta\sigma$ would have been in the manner shown in Figure 1.29, calculate the settlement of the clay layer at time $t = 31.6$ days after the beginning of the surcharge.

Solution
From Part (b) of Example 1.7, $C_v = 7.061$ mm^2/min. From Eq. (1.66),

$$T_c = \frac{C_v t_c}{H^2} = \frac{(7.061 \text{ mm}^2/\text{min})(15 \times 24 \times 60 \text{ min})}{\left(\dfrac{2.8}{2} \times 1000 \text{ mm}\right)^2} = 0.0778$$

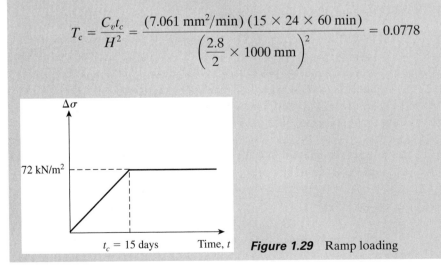

Figure 1.29 Ramp loading

Also,

$$T_v = \frac{C_v t}{H^2} = \frac{(7.061 \text{ mm}^2/\text{min})(31.6 \times 24 \times 60 \text{ min})}{\left(\dfrac{2.8}{2} \times 1000 \text{ mm}\right)^2} = 0.164$$

From Figure 1.28, for $T_v = 0.164$ and $T_c = 0.0778$, the value of U is about 36%. Thus,

$$S_{c(t=31.6\text{ days})} = S_{c(\text{max})}(0.36) = (87.5)(0.36) = \textbf{31.5 mm}$$ ∎

1.15 *Shear Strength*

The shear strength of a soil, defined in terms of effective stress, is

$$s = c' + \sigma' \tan \phi' \tag{1.67}$$

where

σ' = effective normal stress on plane of shearing
c' = cohesion, or apparent cohesion
ϕ' = effective stress angle of friction

Equation (1.67) is referred to as the *Mohr–Coulomb failure criterion*. The value of c' for sands and normally consolidated clays is equal to zero. For overconsolidated clays, $c' > 0$.

For most day-to-day work, the shear strength parameters of a soil (i.e., c' and ϕ') are determined by two standard laboratory tests: the *direct shear test* and the *triaxial test*.

Direct Shear Test

Dry sand can be conveniently tested by direct shear tests. The sand is placed in a shear box that is split into two halves (Figure 1.30a). First a normal load is applied to the specimen. Then a shear force is applied to the top half of the shear box to cause failure in the sand. The normal and shear stresses at failure are

$$\sigma' = \frac{N}{A}$$

and

$$s = \frac{R}{A}$$

where A = area of the failure plane in soil—that is, the cross-sectional area of the shear box.

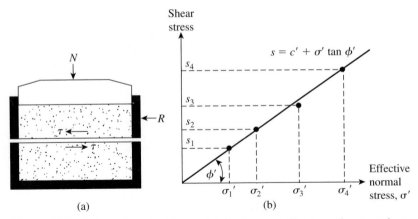

Figure 1.30   Direct shear test in sand: (a) schematic diagram of test equipment; (b) plot of test results to obtain the friction angle ϕ'

Table 1.8   Relationship between Relative Density and Angle of Friction of Cohesionless Soils

State of packing	Relative density (%)	Angle of friction, ϕ' (deg.)
Very loose	<20	<30
Loose	20–40	30–35
Compact	40–60	35–40
Dense	60–80	40–45
Very dense	>80	>45

Several tests of this type can be conducted by varying the normal load. The angle of friction of the sand can be determined by plotting a graph of s against σ' ($= \sigma$ for dry sand), as shown in Figure 1.30b, or

$$\phi' = \tan^{-1}\left(\frac{s}{\sigma'}\right) \tag{1.68}$$

For sands, the angle of friction usually ranges from 26° to 45°, increasing with the relative density of compaction. A general range of the friction angle, ϕ', for sands is given in Table 1.8.

Triaxial Tests

Triaxial compression tests can be conducted on sands and clays Figure 1.31a shows a schematic diagram of the triaxial test arrangement. Essentially, the test consists of placing a soil specimen confined by a rubber membrane into a lucite chamber and then applying an all-around confining pressure (σ_3) to the specimen by means of the chamber fluid (generally, water or glycerin). An added stress ($\Delta\sigma$) can also be

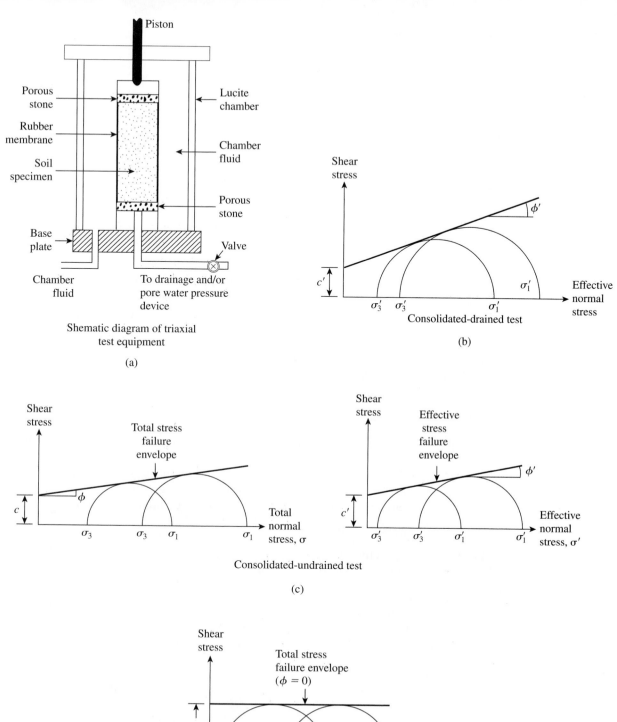

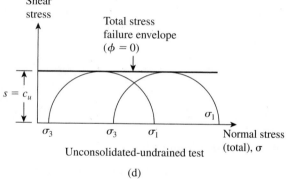

Figure 1.31 Triaxial test

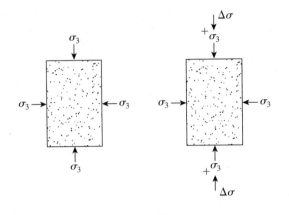

Figure 1.32 Sequence of stress application in triaxial test

applied to the specimen in the axial direction to cause failure ($\Delta\sigma = \Delta\sigma_f$ at failure). Drainage from the specimen can be allowed or stopped, depending on the condition being tested. For clays, three main types of tests can be conducted with triaxial equipment (see Figure 1.32):

1. Consolidated-drained test (CD test)
2. Consolidated-undrained test (CU test)
3. Unconsolidated-undrained test (UU test)

Consolidated-Drained Tests:

Step 1. Apply chamber pressure σ_3. Allow complete drainage, so that the pore water pressure ($u = u_0$) developed is zero.

Step 2. Apply a deviator stress $\Delta\sigma$ slowly. Allow drainage, so that the pore water pressure ($u = u_d$) developed through the application of $\Delta\sigma$ is zero. At failure, $\Delta\sigma = \Delta\sigma_f$; the total pore water pressure $u_f = u_0 + u_d = 0$.

So for *consolidated-drained tests,* at failure,

Major principal effective stress $= \sigma_3 + \Delta\sigma_f = \sigma_1 = \sigma_1'$
Minor principal effective stress $= \sigma_3 = \sigma_3'$

Changing σ_3 allows several tests of this type to be conducted on various clay specimens. The shear strength parameters (c' and ϕ') can now be determined by plotting Mohr's circle at failure, as shown in Figure 1.31b, and drawing a common tangent to the Mohr's circles. This is the *Mohr–Coulomb failure envelope.* (*Note:* For normally consolidated clay, $c' \approx 0$.) At failure,

$$\sigma_1' = \sigma_3' \tan^2\left(45 + \frac{\phi'}{2}\right) + 2c' \tan\left(45 + \frac{\phi'}{2}\right) \qquad (1.69)$$

Consolidated-Undrained Tests:

Step 1. Apply chamber pressure σ_3. Allow complete drainage, so that the pore water pressure ($u = u_0$) developed is zero.

Step 2.   Apply a deviator stress $\Delta\sigma$. Do not allow drainage, so that the pore water pressure $u = u_d \neq 0$. At failure, $\Delta\sigma = \Delta\sigma_f$; the pore water pressure $u_f = u_0 + u_d = 0 + u_{d(f)}$.

Hence, at failure,

Major principal total stress $= \sigma_3 + \Delta\sigma_f = \sigma_1$
Minor principal total stress $= \sigma_3$
Major principal effective stress $= (\sigma_3 + \Delta\sigma_f) - u_f = \sigma_1'$
Minor principal effective stress $= \sigma_3 - u_f = \sigma_3'$

Changing σ_3 permits multiple tests of this type to be conducted on several soil specimens. The total stress Mohr's circles at failure can now be plotted, as shown in Figure 1.31c, and then a common tangent can be drawn to define the *failure envelope*. This *total stress failure envelope* is defined by the equation

$$s = c + \sigma \tan \phi \tag{1.70}$$

where c and ϕ are the *consolidated-undrained cohesion* and *angle of friction*, respectively. (*Note:* $c \approx 0$ for normally consolidated clays.)

Similarly, effective stress Mohr's circles at failure can be drawn to determine the *effective stress failure envelope* (Figure 1.31c), which satisfy the relation expressed in Eq. (1.67).

Unconsolidated-Undrained Tests:

Step 1.   Apply chamber pressure σ_3. Do not allow drainage, so that the pore water pressure $(u = u_0)$ developed through the application of σ_3 is not zero.
Step 2.   Apply a deviator stress $\Delta\sigma$. Do not allow drainage $(u = u_d \neq 0)$. At failure, $\Delta\sigma = \Delta\sigma_f$; the pore water pressure $u_f = u_0 + u_{d(f)}$

For *unconsolidated-undrained* triaxial tests,

Major principal total stress $= \sigma_3 + \Delta\sigma_f = \sigma_1$
Minor principal total stress $= \sigma_3$

The total stress Mohr's circle at failure can now be drawn, as shown in Figure 1.31d. For saturated clays, the value of $\sigma_1 - \sigma_3 = \Delta\sigma_f$ is a constant, irrespective of the chamber confining pressure σ_3 (also shown in Figure 1.31d). The tangent to these Mohr's circles will be a horizontal line, called the $\phi = 0$ condition. The shear strength for this condition is

$$s = c_u = \frac{\Delta\sigma_f}{2} \tag{1.71}$$

where $c_u = $ undrained cohesion (or undrained shear strength).

The pore pressure developed in the soil specimen during the unconsolidated-undrained triaxial test is

$$u = u_0 + u_d \tag{1.72}$$

The pore pressure u_0 is the contribution of the hydrostatic chamber pressure σ_3. Hence,

$$u_0 = B\sigma_3 \tag{1.73}$$

where B = Skempton's pore pressure parameter.
Similarly, the pore parameter u_d is the result of the added axial stress $\Delta\sigma$, so

$$u_d = A\Delta\sigma \tag{1.74}$$

where A = Skempton's pore pressure parameter.
However,

$$\Delta\sigma = \sigma_1 - \sigma_3 \tag{1.75}$$

Combining Eqs. (1.72), (1.73), (1.74), and (1.75) gives

$$u = u_0 + u_d = B\sigma_3 + A(\sigma_1 - \sigma_3) \tag{1.76}$$

The pore water pressure parameter B in soft saturated soils is approximately 1, so

$$u = \sigma_3 + A(\sigma_1 - \sigma_3) \tag{1.77}$$

The value of the pore water pressure parameter A at failure will vary with the type of soil. Following is a general range of the values of A at failure for various types of clayey soil encountered in nature:

Type of soil	A at failure
Sandy clays	0.5–0.7
Normally consolidated clays	0.5–1
Overconsolidated clays	−0.5–0

1.16   *Unconfined Compression Test*

The *unconfined compression test* (Figure 1.33a) is a special type of unconsolidated-undrained triaxial test in which the confining pressure $\sigma_3 = 0$, as shown in Figure 1.33b. In this test, an axial stress $\Delta\sigma$ is applied to the specimen to cause failure (i.e., $\Delta\sigma = \Delta\sigma_f$). The corresponding Mohr's circle is shown in Figure 1.33b. Note that, for this case,

Major principal total stress = $\Delta\sigma_f = q_u$
Minor principal total stress = 0

The axial stress at failure, $\Delta\sigma_f = q_u$, is generally referred to as the *unconfined compression strength*. The shear strength of saturated clays under this condition ($\phi = 0$), from Eq. (1.67), is

$$s = c_u = \frac{q_u}{2} \tag{1.78}$$

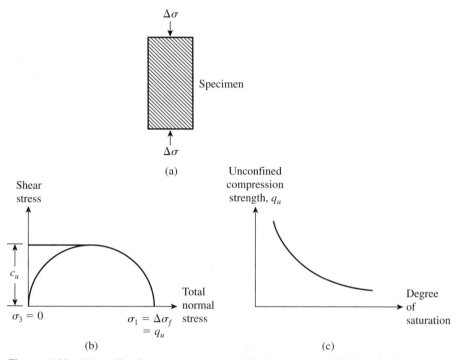

Figure 1.33 Unconfined compression test: (a) soil specimen; (b) Mohr's circle for the test; (c) variation of q_u with the degree of saturation

The unconfined compression strength can be used as an indicator of the consistency of clays.

Unconfined compression tests are sometimes conducted on unsaturated soils. With the void ratio of a soil specimen remaining constant, the unconfined compression strength rapidly decreases with the degree of saturation (Figure 1.33c).

1.17 Comments on Friction Angle, ϕ'

Effective Stress Friction Angle of Granular Soils

In general, the direct shear test yields a higher angle of friction compared with that obtained by the triaxial test. Also, note that the failure envelope for a given soil is actually curved. The Mohr–Coulomb failure criterion defined by Eq. (1.67) is only an approximation. Because of the curved nature of the failure envelope, a soil tested at higher normal stress will yield a lower value of ϕ'. An example of this relationship is shown in Figure 1.34, which is a plot of ϕ' versus the void ratio e for Chattachoochee River sand near Atlanta, Georgia (Vesic, 1963). The friction angles shown were obtained from triaxial tests. Note that, for a given value of e, the magnitude of ϕ' is about 4° to 5° smaller when the confining pressure σ'_3 is greater than about 70 kN/m² (10 lb/in²), compared with that when $\sigma'_3 < 70$ kN/m²(≈ 10 lb/in²).

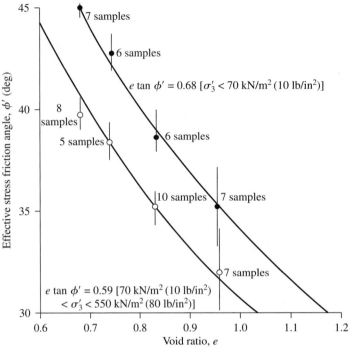

Figure 1.34 Variation of friction angle ϕ' with void ratio for Chattachoochee River sand (After Vesic, 1963)

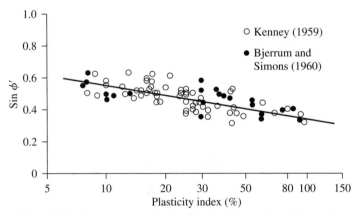

Figure 1.35 Variation of sin ϕ' with plasticity index (PI) for several normally consolidated clays

Effective Stress Friction Angle of Cohesive Soils

Figure 1.35 shows the variation of effective stress friction angle, ϕ', for several normally consolidated clays (Bejerrum and Simons, 1960; Kenney, 1959). It can be seen from the figure that, in general, the friction angle ϕ' decreases with the increase in plasticity index. The value of ϕ' generally decreases from about 37 to 38° with a plasticity index of about 10 to about 25° or less with a plasticity index of about 100. The consolidated

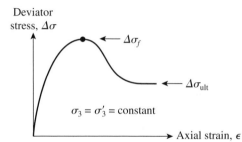

Figure 1.36   Plot of deviator stress versus axial strain–drained triaxial test

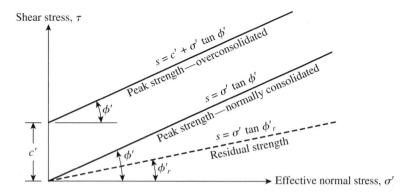

Figure 1.37   Peak- and residual-strength envelopes for clay

undrained friction angle (ϕ) of normally consolidated saturated clays generally ranges from 5 to 20°.

The consolidated drained triaxial test was described in Section 1.15. Figure 1.36 shows a schematic diagram of a plot of $\Delta\sigma$ versus axial strain in a drained triaxial test for a clay. At failure, for this test, $\Delta\sigma = \Delta\sigma_f$. However, at large axial strain (i.e., the ultimate strength condition), we have the following relationships:

Major principal stress: $\sigma'_{1(\text{ult})} = \sigma_3 + \Delta\sigma_{\text{ult}}$
Minor principal stress: $\sigma'_{3(\text{ult})} = \sigma_3$

At failure (i.e., peak strength), the relationship between σ'_1 and σ'_3 is given by Eq. (1.69). However, for ultimate strength, it can be shown that

$$\sigma'_{1(\text{ult})} = \sigma'_3 \tan^2\left(45 + \frac{\phi'_r}{2}\right) \tag{1.79}$$

where ϕ'_r = residual effective stress friction angle.

Figure 1.37 shows the general nature of the failure envelopes at peak strength and ultimate strength (or *residual strength*). The residual shear strength of clays is important in the evaluation of the long-term stability of new and existing slopes and the design of remedial measures. The effective stress residual friction angles ϕ'_r

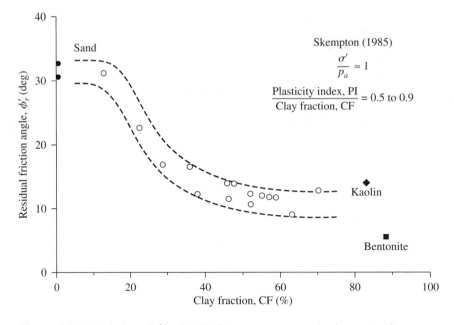

Figure 1.38   Variation of ϕ'_r with CF (*Note:* p_a = atmospheric pressure)

of clays may be substantially smaller than the effective stress peak friction angle ϕ'. Past research has shown that the clay fraction (i.e., the percent finer than 2 microns) present in a given soil, CF, and the clay mineralogy are the two primary factors that control ϕ'_r. The following is a summary of the effects of CF on ϕ'_r.

1. If CF is less than about 15%, then ϕ'_r is greater than about 25°.
2. For CF > about 50%, ϕ'_r is entirely governed by the sliding of clay minerals and may be in the range of about 10 to 15°.
3. For kaolinite, illite, and montmorillonite, ϕ'_r is about 15°, 10°, and 5°, respectively.

Illustrating these facts, Figure 1.38 shows the variation of ϕ'_r with CF for several soils (Skempton, 1985).

1.18   *Correlations for Undrained Shear Strength, c_u*

The undrained shear strength, c_u, is an important parameter in the design of foundations. For normally consolidated clay deposits (Figure 1.39), the magnitude of c_u increases almost linearly with the increase in effective overburden pressure.

For normally consolidated clays, Skempton (1957) has given the following correlation for the undrained shear strength:

$$\frac{c_{u(\text{VST})}}{\sigma'_0} = 0.11 + 0.0037\text{PI} \tag{1.80}$$

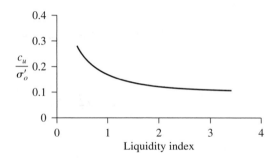

Figure 1.39 Variation of c_u/σ'_o with liquidity index (Based on Bjerrum and Simons 1960)

where

PI = plasticity index (%)

$c_{u(\text{VST})}$ = undrained shear strength from vane shear test (See Chapter 2 for details of vane shear test.)

For overconsolidated clays (Ladd et al., 1977)

$$\frac{\left(\dfrac{c_u}{\sigma'_0}\right)_{\text{overconsolidated}}}{\left(\dfrac{c_u}{\sigma'_0}\right)_{\text{normally consolidated}}} = (\text{OCR})^{0.8} \tag{1.81}$$

where OCR = overconsolidation ratio = $\dfrac{\sigma'_c}{\sigma'_0}$.

1.19 *Sensitivity*

For many naturally deposited clay soils, the unconfined compression strength is much less when the soils are tested after remolding without any change in the moisture content. This property of clay soil is called *sensitivity*. The degree of sensitivity is the ratio of the unconfined compression strength in an undisturbed state to that in a remolded state, or

$$S_t = \frac{q_{u(\text{undisturbed})}}{q_{u(\text{remolded})}} \tag{1.82}$$

The sensitivity ratio of most clays ranges from about 1 to 8; however, highly flocculent marine clay deposits may have sensitivity ratios ranging from about 10 to 80. Some clays turn to viscous liquids upon remolding, and these clays are referred to as "quick" clays. The loss of strength of clay soils from remolding is caused primarily by the destruction of the clay particle structure that was developed during the original process of sedimentation.

Problems

(handwritten annotations: ASTM; The Massachusetts State Building Code; Providence Engineering Society; The RI Society of Professional Eng's)

1.1   A soil specimen has a volume of 0.05 m³ and a mass of 87.5 kg. Given: $w = 15\%$, $G_s = 2.68$. Determine
   a. Void ratio
   b. Porosity
   c. Dry unit weight
   d. Moist unit weight
   e. Degree of saturation

1.2   The saturated unit weight of a soil is 20.1 kN/m³ at a moisture content of 22%. Determine (a) the dry unit weight and (b) the specific gravity of soil solids, G_s.

1.3   For a soil, given: void ratio = 0.81, moisture content = 21%, and $G_s = 2.68$. Calculate the following:
   a. Porosity
   b. Degree of saturation
   c. Moist unit weight in kN/m³
   d. Dry unit weight in kN/m³

1.4   For a given soil, the following are given: moist unit weight = 122 lb/ft³, moisture content = 14.7%, and $G_s = 2.68$. Calculate the following:
   a. Void ratio
   b. Porosity
   c. Degree of saturation
   d. Dry unit weight

1.5   For the soil described in Problem 1.4:
   a. What would be the saturated unit weight in lb/ft³?
   b. How much water, in lb/ft³, needs to be added to the soil for complete saturation?
   c. What would be the moist unit weight in lb/ft³ when the degree of saturation is 80%?

1.6   For a granular soil, given: $\gamma = 116.64$ lb/ft³, $D_r = 82\%$, $w = 8\%$, and $G_s = 2.65$. For this soil, if $e_{min} = 0.44$, what would be e_{max}? What would be the dry unit weight in the loosest state?

1.7   The laboratory test results of six soils are given in the following table. Classify the soils by the AASHTO Soil Classification System and give the group indices.

Sieve Analysis—Percent Passing

Sieve No.	Soil					
	A	B	C	D	E	F
4	100	100	95	95	100	100
10	95	80	80	90	94	94
40	82	61	54	79	76	86
200	65	55	8	64	33	76
Liquid limit	42	38	NP*	35	38	52
Plastic limit	26	25	NP	26	25	28

*NP = nonplastic

1.8 Classify the soils given in Problem 1.7 by the Unified Soil Classification System and determine the group symbols and group names.

1.9 For a sandy soil, given: void ratio, $e = 0.63$; hydraulic conductivity, $k = 0.22$ cm/sec; specific gravity of soil solids, $G_s = 2.68$. Estimate the hydraulic conductivity of the sand (cm/sec) when the dry unit weight of compaction is 117 lb/ft^3. Use Eq. (1.26).

1.10 A sand has a hydraulic conductivity of 0.25 cm/sec at a void ratio of 0.7. Estimate the void ratio at which its hydraulic conductivity would be 0.115 cm/sec. Use Eq. (1.26).

1.11 The *in situ* hydraulic conductivity of a clay is 5.4×10^{-6} cm/sec at a void ratio of 0.92. What could be its hydraulic conductivity at a void ratio of 0.72? Use Eq. (1.28), and $n = 5.1$.

1.12 Refer to the soil profile shown in Figure P1.12. Determine the total stress, pore water pressure, and effective stress at A, B, C, and D.

1.13 Refer to Problem 1.12. If the ground water table rises to 4 ft below the ground surface, what will be the change in the effective stress at D?

1.14 A sandy soil ($G_s = 2.65$), in its densest and loosest states, has void ratios of 0.42 and 0.85, respectively. Estimate the range of the critical hydraulic gradient in this soil at which a quicksand condition might occur.

1.15 For a normally consolidated clay layer, the following are given:
Thickness = 3.7 m
Void ratio = 0.82
Liquid limit = 42
Average effective stress on the clay layer = 110 kN/m^2
How much consolidation settlement would the clay undergo if the average effective stress on the clay layer is increased to 155 kN/m^2 as the result of the construction of a foundation?

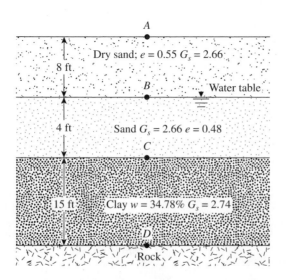

Figure P1.12

1.16 Refer to Problem 1.15. Assume that the clay layer is preconsolidated, $\sigma_c' = 128$ kN/m² and $C_s = \frac{1}{5}C_c$. Estimate the consolidation settlement.

1.17 Refer to the soil profile shown in Figure P1.12. The clay is normally consolidated. A laboratory consolidation test on the clay gave the following results:

Pressure (lb/in.²)	Void ratio
21	0.91
42	0.792

If the average effecive stress on the clay layer increases by 1000 lb/ft²,
a. What would be the total consolidation settlement?
b. If $C_v = 1.45 \times 10^{-4}$ in²/sec, how long will it take for half the consolidation settlement to take place?

1.18 For a normally consolidated soil, the following is given:

Pressure (kN/m²)	Void ratio
120	0.82
360	0.64

Determine the following:
a. The compression index, C_c.
b. The void ratio corresponding to pressure of 200 kN/m².

1.19 A clay soil specimen, 1.5 in. thick (drained on top only) was tested in the laboratory. For a given load increment, the time for 60% consolidation was 8 min 10 sec. How long will it take for 50% consolidation for a similar clay layer in the field that is 10-ft thick and drained on both sides?

1.20 Refer to Figure P1.20. A total of 60 mm consolidation settlement is expected in the two clay layers due to a surcharge of $\Delta\sigma$. Find the duration of surcharge application at which 30 mm of total settlement would take place.

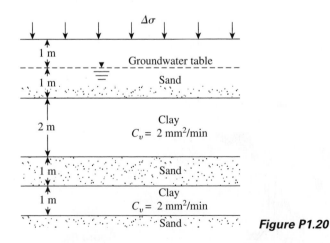

Figure P1.20

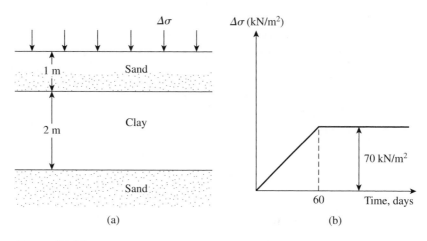

Figure P1.21

1.21 The coefficient of consolidation of a clay for a given pressure range was obtained as 8×10^{-3} mm²/sec on the basis of one-dimensional consolidation test results. In the field, there is a 2-m-thick layer of the same clay (Figure P1.21a). Based on the assumption that a uniform surcharge of 70 kN/m² was to be applied instantaneously, the total consolidation settlement was estimated to be 150 mm. However, during construction, the loading was gradual; the resulting surcharge can be approximated as shown in Figure P1.21b. Estimate the settlement at $t = 30$ and $t = 120$ days after the beginning of construction.

1.22 A direct shear test was conducted on dry sand. The results were as follows:
Area of the specimen = 2 in. × 2 in.

Normal force (lb)	Shear force at failure (lb)
50	43.5
110	95.5
150	132.0

Graph the shear stress at failure against normal stress and determine the soil friction angle, ϕ'.

1.23 A consolidated-drained triaxial test on a sand yielded the following results:
All-around confining pressure = σ_3 = 30 lb/in²
Added axial stress at failure = $\Delta\sigma$ = 96 lb/in²
Determine the shear stress parameters (i.e., ϕ' and c')

1.24 Repeat Problem 1.23 with the following:
All-around confining pressure = σ_3 = 20 lb/in²
Added axial stress at failure = $\Delta\sigma$ = 40 lb/in²

1.25   A consolidated-drained triaxial test on a normally consolidated clay yielded a friction angle, ϕ', of 28°. If the all-around confining pressure during the test was 140 kN/m², what was the major principal stress at failure?

1.26   Following are the results of two consolidated-drained triaxial tests on a clay:

Test I: $\sigma_3 = 140$ kN/m²; $\sigma_{1(\text{failure})} = 368$ kN/m²
Test II: $\sigma_3 = 280$ kN/m²; $\sigma_{1(\text{failure})} = 701$ kN/m²

Determine the shear strength parameters; that is, c' and ϕ'.

1.27   A consolidated-undrained triaxial test was conducted on a saturated normally consolidated clay. Following are the test results:

$\sigma_3 = 13$ lb/in²
$\sigma_{1(\text{failure})} = 32$ lb/in²
Pore pressure at failure $= u = 5.5$ lb/in²

Determine c, ϕ, c', and ϕ'.

1.28   For a normally consolidated clay, given $\phi' = 28°$ and $\phi = 20°$. If a consolidated-undrained triaxial test is conducted on the same clay with $\sigma_3 = 150$ kN/m², what would be the pore water pressure at failure?

1.29   A saturated clay layer has

Saturated unit weight $\gamma_{\text{sat}} = 19.6$ kN/m³
Plasticity index $= 21$

The water table coincides with the ground surface. If the clay is normally consolidated, estimate the magnitude of c_u (kN/m²) that can be obtained from a vane shear test at a depth of 8 m from the ground surface. Use the Skempton relationship given in Eq. (1.80).

References

American Society for Testing and Materials (2005). *Annual Book of ASTM Standards,* Vol. 04.08, Conshohocken, PA.

Bjerrum, L., and Simons, N. E. (1960). "Comparison of Shear Strength Characteristics of Normally Consolidated Clay," *Proceedings, Research Conference on Shear Strength of Cohesive Soils,* ASCE, 711–726.

Carrier III, W. D. (2003). "Goodbye, Hazen; Hello, Kozeny-Carman," *Journal of Geotechnical and Geoenvironmental Engineering,* ASCE, Vol. 129, No. 11, pp. 1054–1056.

Casagrande, A. (1936). "Determination of the Preconsolidation Load and Its Practical Significance," *Proceedings, First International Conference on Soil Mechanics and Foundation Engineering,* Cambridge, MA, Vol. 3, pp. 60–64.

Chapuis, R. P. (2004). "Predicting the Saturated Hydraulic Conductivity of Sand and Gravel Using Effective Diameter and Void Ratio," *Canadian Geotechnical Journal,* Vol. 41, No. 5, pp. 787–795.

Darcy, H. (1856). *Les Fontaines Publiques de la Ville de Dijon,* Paris.

Das, B. M. (2002). *Soil Mechanics Laboratory Manual,* 6th ed., Oxford University Press, New York.

Highway Research Board (1945). *Report of the Committee on Classification of Materials for Subgrades and Granular Type Roads,* Vol. 25, pp. 375–388.

Kenney, T. C. (1959). "Discussion," *Journal of the Soil Mechanics and Foundations Division,* American Society of Civil Engineers, Vol. 85, No. SM3, pp. 67–69.

Ladd, C. C., Foote, R., Ishihara, K., Schlosser, F., and Poulos, H. G. (1977). "Stress Deformation and Strength Characteristics," *Proceedings, Ninth International Conference on Soil Mechanics and Foundation Engineering,* Tokyo, Vol. 2, 421–494.

Nagaraj, T. S., and Murthy, B. R. S. (1985). "Prediction of the Preconsolidation Pressure and Recompression Index of Soils," *Geotechnical Testing Journal,* American Society for Testing and Materials, Vol. 8, No. 4, pp. 199–202.

Olson, R. E. (1977). "Consolidation Under Time-Dependent Loading," *Journal of Geotechnical Engineering,* ASCE, Vol. 103, No. GT1, pp. 55–60.

Park, J. H., and Koumoto, T. (2004). "New Compression Index Equation," *Journal of Geotechnical and Geoenvironmental Engineering*, ASCE, Vol. 130, No. 2, pp. 223–226.

Rendon-Herrero, O. (1980). "Universal Compression Index Equation," *Journal of the Geotechnical Engineering Division*, American Society of Civil Engineers, Vol. 106, No. GT11, pp. 1178–1200.

Samarasinghe, A. M., Huang, Y. H., and Drnevich, V. P. (1982). "Permeability and Consolidation of Normally Consolidated Soils," *Journal of the Geotechnical Engineering Division,* ASCE, Vol. 108, No. GT6, 835–850.

Schmertmann, J. H. (1953). "Undisturbed Consolidation Behavior of Clay," *Transactions,* American Society of Civil Engineers, Vol. 120, p. 1201.

Skempton, A. W. (1944). "Notes on the Compressibility of Clays," *Quarterly Journal of Geological Society,* London, Vol. C, pp. 119–135.

Skempton, A. W. (1957). "The Planning and Design of New Hong Kong Airport," *Proceedings, The Institute of Civil Engineers,* London, Vol. 7, pp. 305–307.

Skempton, A. W. (1985). "Residual Strength of Clays in Landslides, Folded Strata, and the Laboratory," *Geotechnique,* Vol. 35, No. 1, pp. 3–18.

Terzaghi, K., and Peck, R. B. (1967). *Soil Mechanics in Engineering Practice,* Wiley, New York.

Vesic, A. S. (1963). "Bearing Capacity of Deep Foundations in Sand," *Highway Research Record No. 39,* National Academy of Sciences, Washington, DC., pp. 112–154.

2

Natural Soil Deposits and Subsoil Exploration

2.1 Introduction

To design a foundation that will support a structure, an engineer must understand the types of soil deposits that will support the foundation. Moreover, foundation engineers must remember that soil at any site frequently is nonhomogeneous; that is, the soil profile may vary. Soil mechanics theories involve idealized conditions, so the application of the theories to foundation engineering problems involves a judicious evaluation of site conditions and soil parameters. To do this requires some knowledge of the geological process by which the soil deposit at the site was formed, supplemented by subsurface exploration. Good professional judgment constitutes an essential part of geotechnical engineering—and it comes only with practice.

This chapter is divided into two parts. The first is a general overview of natural soil deposits generally encountered, and the second describes the general principles of subsoil exploration.

Natural Soil Deposits

2.2 Soil Origin

Most of the soils that cover the earth are formed by the weathering of various rocks. There are two general types of weathering: (1) mechanical weathering and (2) chemical weathering.

Mechanical weathering is the process by which rocks are broken into smaller and smaller pieces by physical forces, including running water, wind, ocean waves, glacier ice, frost, and expansion and contraction caused by the gain and loss of heat.

Chemical weathering is the process of chemical decomposition of the original rock. In the case of mechanical weathering, the rock breaks into smaller pieces without a change in its chemical composition. However, in chemical weathering, the original material may be changed to something entirely different. For example, the chemical weathering of feldspar can produce clay minerals. Most rock weathering is a combination of mechanical and chemical weathering.

Soil produced by the weathering of rocks can be transported by physical processes to other places. The resulting soil deposits are called *transported soils*. In contrast, some soils stay where they were formed and cover the rock surface from which they derive. These soils are referred to as *residual soils*.

Transported soils can be subdivided into five major categories based on the *transporting agent:*

1. *Gravity transported* soil
2. *Lacustrine* (lake) deposits
3. *Alluvial* or *fluvial* soil deposited by running water
4. *Glacial* deposited by glaciers
5. *Aeolian* deposited by the wind

In addition to transported and residual soils, there are *peats* and *organic soils,* which derive from the decomposition of organic materials.

2.3 *Residual Soil*

Residual soils are found in areas where the rate of weathering is more than the rate at which the weathered materials are carried away by transporting agents. The rate of weathering is higher in warm and humid regions compared to cooler and drier regions and, depending on the climatic conditions, the effect of weathering may vary widely.

Residual soil deposits are common in the tropics, on islands such as the Hawaiian islands, and in the southeastern United States. The nature of a residual soil deposit will generally depend on the parent rock. When hard rocks such as granite and gneiss undergo weathering, most of the materials are likely to remain in place. These soil deposits generally have a top layer of clayey or silty clay material, below which are silty or sandy soil layers. These layers in turn are generally underlain by a partially weathered rock and then sound bedrock. The depth of the sound bedrock may vary widely, even within a distance of a few meters. Figure 2.1 shows the boring log of a residual soil deposit derived from the weathering of granite.

In contrast to hard rocks, there are some chemical rocks, such as limestone, that are chiefly made up of calcite ($CaCo_3$) mineral. Chalk and dolomite have large concentrations of dolomite minerals [$Ca Mg(Co_3)_2$]. These rocks have large amounts of soluble materials, some of which are removed by groundwater, leaving behind the insoluble fraction of the rock. Residual soils that derive from chemical rocks do not possess a gradual transition zone to the bedrock, as seen in Figure 2.1. The residual soils derived from the weathering of limestone-like rocks are mostly red in color. Although uniform in kind, the depth of weathering may vary greatly. The residual soils immediately above the bedrock may be normally consolidated. Large foundations with heavy loads may be susceptible to large consolidation settlements on these soils.

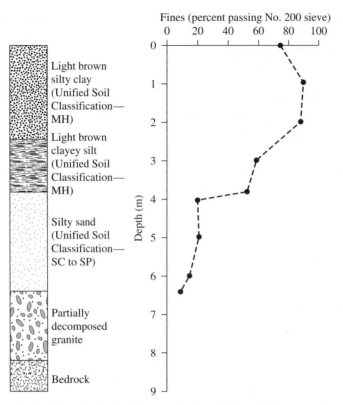

Figure 2.1 Boring log for a residual soil derived from granite

Gravity Transported Soil

Residual soils on a steep natural slope can move slowly downward, and this is usually referred to as *creep*. When the downward soil movement is sudden and rapid, it is called as a *landslide*. The soil deposits formed by landslides are *colluvium*. *Mud flows* are one type of gravity transported soil. In this case, highly saturated loose sandy residual soils on relatively flat slopes move downward like a viscous liquid and come to rest in a more dense condition. The soil deposits derived from past mud flows are highly heterogeneous in composition.

Alluvial Deposits

Alluvial soil deposits derive from the action of streams and rivers and can be divided into two major categories: (1) *braided-stream deposits* and (2) deposits caused by the *meandering belt of streams.*

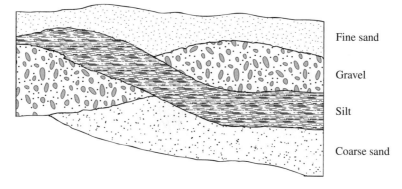

Fine sand

Gravel

Silt

Coarse sand

Figure 2.2 Cross section of a braided-stream deposit

Deposits from Braided Streams

Braided streams are high-gradient, rapidly flowing streams that are highly erosive and carry large amounts of sediment. Because of the high bed load, a minor change in the velocity of flow will cause sediments to deposit. By this process, these streams may build up a complex tangle of converging and diverging channels separated by sandbars and islands.

The deposits formed from braided streams are highly irregular in stratification and have a wide range of grain sizes. Figure 2.2 shows a cross section of such a deposit. These deposits share several characteristics:

1. The grain sizes usually range from gravel to silt. Clay-sized particles are generally *not* found in deposits from braided streams.
2. Although grain size varies widely, the soil in a given pocket or lens is rather uniform.
3. At any given depth, the void ratio and unit weight may vary over a wide range within a lateral distance of only a few meters. This variation can be observed during soil exploration for the construction of a foundation for a structure. The standard penetration resistance at a given depth obtained from various boreholes will be highly irregular and variable.

Alluvial deposits are present in several parts of the western United States, such as Southern California, Utah, and the basin and range sections of Nevada. Also, a large amount of sediment originally derived from the Rocky Mountain range was carried eastward to form the alluvial deposits of the Great Plains. On a smaller scale, this type of natural soil deposit, left by braided streams, can be encountered locally.

Meander Belt Deposits

The term *meander* is derived from the Greek word *maiandros,* after the Maiandros (now Menderes) River in Asia, famous for its winding course. Mature streams in a valley curve back and forth. The valley floor in which a river meanders is referred to as the *meander belt.* In a meandering river, the soil from the bank is continually eroded from the points where it is concave in shape and is deposited at points where the bank is convex in shape, as shown in Figure 2.3. These deposits are called *point bar deposits,*

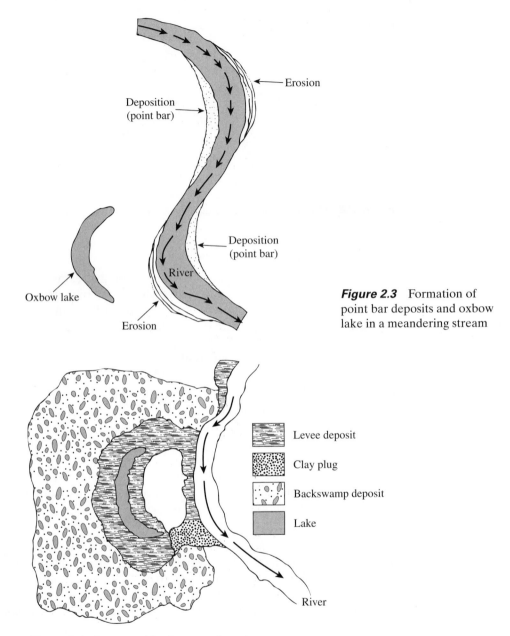

Figure 2.3 Formation of point bar deposits and oxbow lake in a meandering stream

Figure 2.4   Levee and backswamp deposit

and they usually consist of sand and silt-size particles. Sometimes, during the process of erosion and deposition, the river abandons a meander and cuts a shorter path. The abandoned meander, when filled with water, is called an *oxbow lake*. (See Figure 2.3.)

During floods, rivers overflow low-lying areas. The sand and silt-size particles carried by the river are deposited along the banks to form ridges known as *natural levees* (Figure 2.4). Finer soil particles consisting of silts and clays are carried by the

Table 2.1 Properties of Deposits within the Mississippi Alluvial Valley

Environment	Soil texture	Natural water content (%)	Liquid limit	Plasticity index
Natural levee	Clay (CL)	25–35	35–45	15–25
	Silt (ML)	15–35	NP–35	NP–5
Point bar	Silt (ML) and silty sand (SM)	25–45	30–55	10–25
Abandoned channel	Clay (CL, CH)	30–95	30–100	10–65
Backswamps	Clay (CH)	25–70	40–115	25–100
Swamp	Organic clay (OH)	100–265	135–300	100–165

(*Note:* NP—Nonplastic)

water farther onto the floodplains. These particles settle at different rates to form what is referred to as *backswamp deposits* (Figure 2.4), often highly plastic clays.

Table 2.1 gives some properties of soil deposits found in natural levees, point bars, abandoned channels, backswamps and swamps within the alluvial Mississippi Valley (Kolb and Shockley, 1959).

2.6 Lacustrine Deposits

Water from rivers and springs flows into lakes. In arid regions, streams carry large amounts of suspended solids. Where the stream enters the lake, granular particles are deposited in the area forming a delta. Some coarser particles and the finer particles (that is, silt and clay) that are carried into the lake are deposited onto the lake bottom in alternate layers of coarse-grained and fine-grained particles. The deltas formed in humid regions usually have finer grained soil deposits compared to those in arid regions.

2.7 Glacial Deposits

During the Pleistocene Ice Age, glaciers covered large areas of the earth. The glaciers advanced and retreated with time. During their advance, the glaciers carried large amounts of sand, silt, clay, gravel, and boulders. *Drift* is a general term usually applied to the deposits laid down by glaciers. Unstratified deposits laid down by melting glaciers are referred to as *till*. The physical characteristics of till may vary from glacier to glacier.

The landforms that developed from the deposits of till are called *moraines*. A *terminal moraine* (Figure 2.5) is a ridge of till that marks the maximum limit of a

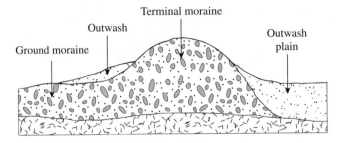

Figure 2.5   Terminal moraine, ground moraine, and outwash plain

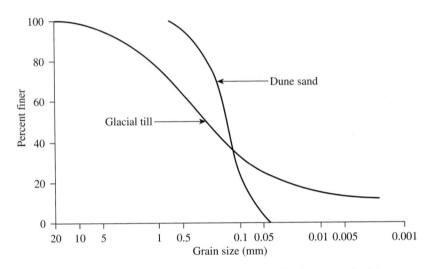

Figure 2.6   Comparison of the grain-size distribution between glacial till and dune sand

glacier's advance. *Recessional moraines* are ridges of till developed behind the terminal moraine at varying distances apart. They are the result of temporary stabilization of the glacier during the recessional period. The till deposited by the glacier between the moraines is referred to as *ground moraine* (Figure 2.5). Ground moraines constitute large areas of the central United States and are called *till plains.*

The sand, silt, and gravel that are carried by the melting water from the front of a glacier are called *outwash.* In a pattern similar to that of braided-stream deposits, the melted water deposits the outwash, forming *outwash plains* (Figure 2.5), also called *glaciofluvial deposits.*

The range of grain sizes present in a given till varies greatly. Figure 2.6 compares the grain-size distribution of *glacial-till* and *dune sand* (see Section 2.8). The amount of clay-size fractions present and the plasticity indices of tills also vary widely. During field-exploration programs, erratic values of standard penetration resistances also may be expected.

2.8   *Aeolian Soil Deposits*

Wind is also a major transporting agent leading to the formation of soil deposits. When large areas of sand lie exposed, wind can blow the sand away and redeposit it elsewhere. Deposits of windblown sand generally take the shape of *dunes* (Figure 2.7). As dunes are formed, the sand is blown over the crest by the wind. Beyond the crest, the sand particles roll down the slope. The process tends to form a *compact sand deposit* on the *windward side,* and a rather *loose deposit* on the *leeward side,* of the dune.

Dunes exist along the southern and eastern shores of Lake Michigan, the Atlantic Coast, the southern coast of California, and at various places along the coasts of Oregon and Washington. Sand dunes can also be found in the alluvial and rocky plains of the western United States. Following are some of the typical properties of *dune sand:*

1. The grain-size distribution of the sand at any particular location is surprisingly uniform. This uniformity can be attributed to the sorting action of the wind.
2. The general grain size decreases with distance from the source, because the wind carries the small particles farther than the large ones.
3. The relative density of sand deposited on the windward side of dunes may be as high as 50 to 65%, decreasing to about 0 to 15% on the leeward side.

Loess is an aeolian deposit consisting of silt and silt-sized particles. The grain-size distribution of loess is rather uniform. The cohesion of loess is generally derived from a clay coating over the silt-sized particles, which contributes to a stable soil structure in an unsaturated state. The cohesion may also be the result of the precipitation of chemicals leached by rainwater. Loess is a *collapsing* soil, because when the soil becomes saturated, it loses its binding strength between particles. Special precautions need to be taken for the construction of foundations over loessial deposits. There are extensive deposits of loess in the United States, mostly in the midwestern states of Iowa, Missouri, Illinois, and Nebraska and for some distance along the Mississippi River in Tennessee and Mississippi.

Volcanic ash (with grain sizes between 0.25 to 4 mm) and volcanic dust (with grain sizes less than 0.25 mm) may be classified as wind-transported soil. Volcanic ash is a lightweight sand or sandy gravel. Decomposition of volcanic ash results in highly plastic and compressible clays.

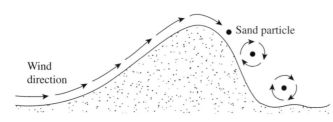

Figure 2.7   Sand dune

2.9   *Organic Soil*

Organic soils are usually found in low-lying areas where the water table is near or above the ground surface. The presence of a high water table helps in the growth of aquatic plants that, when decomposed, form organic soil. This type of soil deposit is usually encountered in coastal areas and in glaciated regions. Organic soils show the following characteristics:

1. Their natural moisture content may range from 200 to 300%.
2. They are highly compressible.
3. Laboratory tests have shown that, under loads, a large amount of settlement is derived from secondary consolidation.

Subsurface Exploration

2.10   *Purpose of Subsurface Exploration*

The process of identifying the layers of deposits that underlie a proposed structure and their physical characteristics is generally referred to as *subsurface exploration*. The purpose of subsurface exploration is to obtain information that will aid the geotechnical engineer in

1. Selecting the type and depth of foundation suitable for a given structure.
2. Evaluating the load-bearing capacity of the foundation.
3. Estimating the probable settlement of a structure.
4. Determining potential foundation problems (e.g., expansive soil, collapsible soil, sanitary landfill, and so on).
5. Determining the location of the water table.
6. Predicting the lateral earth pressure for structures such as retaining walls, sheet pile bulkheads, and braced cuts.
7. Establishing construction methods for changing subsoil conditions.

Subsurface exploration may also be necessary when additions and alterations to existing structures are contemplated.

2.11   *Subsurface Exploration Program*

Subsurface exploration comprises several steps, including the collection of preliminary information, reconnaissance, and site investigation.

Collection of Preliminary Information
This step involves obtaining information regarding the type of structure to be built and its general use. For the construction of buildings, the approximate column loads and their spacing and the local building-code and basement requirements should be

known. The construction of bridges requires determining the lengths of their spans and the loading on piers and abutments.

A general idea of the topography and the type of soil to be encountered near and around the proposed site can be obtained from the following sources:

1. United States Geological Survey maps.
2. State government geological survey maps.
3. United States Department of Agriculture's Soil Conservation Service county soil reports.
4. Agronomy maps published by the agriculture departments of various states.
5. Hydrological information published by the United States Corps of Engineers, including records of stream flow, information on high flood levels, tidal records, and so on.
6. Highway department soil manuals published by several states.

The information collected from these sources can be extremely helpful in planning a site investigation. In some cases, substantial savings may be realized by anticipating problems that may be encountered later in the exploration program.

Reconnaissance

The engineer should always make a visual inspection of the site to obtain information about

1. The general topography of the site, the possible existence of drainage ditches, abandoned dumps of debris, and other materials present at the site. Also, evidence of creep of slopes and deep, wide shrinkage cracks at regularly spaced intervals may be indicative of expansive soils.
2. Soil stratification from deep cuts, such as those made for the construction of nearby highways and railroads.
3. The type of vegetation at the site, which may indicate the nature of the soil. For example, a mesquite cover in central Texas may indicate the existence of expansive clays that can cause foundation problems.
4. High-water marks on nearby buildings and bridge abutments.
5. Groundwater levels, which can be determined by checking nearby wells.
6. The types of construction nearby and the existence of any cracks in walls or other problems.

The nature of the stratification and physical properties of the soil nearby also can be obtained from any available soil-exploration reports on existing structures.

Site Investigation

The site investigation phase of the exploration program consists of planning, making test boreholes, and collecting soil samples at desired intervals for subsequent observation and laboratory tests. The approximate required minimum depth of the borings should be predetermined. The depth can be changed during the drilling operation, depending on the subsoil encountered. To determine the approximate minimum depth of boring, engineers may use the rules established by the American Society of Civil Engineers (1972):

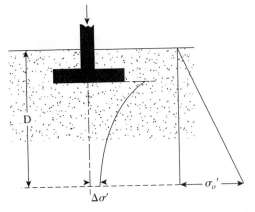

Figure 2.8   Determination of the minimum depth of boring

1. Determine the net increase in the effective stress, $\Delta\sigma'$, under a foundation with depth as shown in Figure 2.8. (The general equations for estimating increases in stress are given in Chapter 5.)
2. Estimate the variation of the vertical effective stress, σ'_o, with depth.
3. Determine the depth, $D = D_1$, at which the effective stress increase $\Delta\sigma'$ is equal to $\left(\frac{1}{10}\right)q$ (q = estimated net stress on the foundation).
4. Determine the depth, $D = D_2$, at which $\Delta\sigma'/\sigma'_o = 0.05$.
5. Choose the smaller of the two depths, D_1 and D_2, just determined as the approximate minimum depth of boring required, unless bedrock is encountered.

If the preceding rules are used, the depths of boring for a building with a width of 30 m (100 ft) will be approximately the following, according to Sowers and Sowers (1970):

No. of stories	Boring depth	
1	3.5 m	(11 ft)
2	6 m	(20 ft)
3	10 m	(33 ft)
4	16 m	(53 ft)
5	24 m	(79 ft)

To determine the boring depth for hospitals and office buildings, Sowers and Sowers also use the rule

$$D_b = 3S^{0.7} \quad \text{(for light steel or narrow concrete buildings)} \tag{2.1}$$

and

$$D_b = 6S^{0.7} \quad \text{(for heavy steel or wide concrete buildings)} \tag{2.2}$$

Table 2.2 Approximate Spacing of Boreholes

Type of project	Spacing (m)	Spacing (ft)
Multistory building	10–30	30–100
One-story industrial plants	20–60	60–200
Highways	250–500	800–1600
Residential subdivision	250–500	800–1600
Dams and dikes	40–80	130–260

where

D_b = depth of boring, in meters
S = number of stories

In English units, the preceding equations take the form

$$D_b \text{ (ft)} = 10S^{0.7} \quad \text{(for light steel or narrow concrete buildings)} \qquad (2.3)$$

and

$$D_b \text{ (ft)} = 20S^{0.7} \quad \text{(for heavy steel or wide concrete buildings)} \qquad (2.4)$$

When deep excavations are anticipated, the depth of boring should be at least 1.5 times the depth of excavation.

Sometimes, subsoil conditions require that the foundation load be transmitted to bedrock. The minimum depth of core boring into the bedrock is about 3 m (10 ft). If the bedrock is irregular or weathered, the core borings may have to be deeper.

There are no hard-and-fast rules for borehole spacing. Table 2.2 gives some general guidelines. Spacing can be increased or decreased, depending on the condition of the subsoil. If various soil strata are more or less uniform and predictable, fewer boreholes are needed than in nonhomogeneous soil strata.

The engineer should also take into account the ultimate cost of the structure when making decisions regarding the extent of field exploration. The exploration cost generally should be 0.1 to 0.5% of the cost of the structure. Soil borings can be made by several methods, including auger boring, wash boring, percussion drilling, and rotary drilling.

2.12 Exploratory Borings in the Field

Auger boring is the simplest method of making exploratory boreholes. Figure 2.9 shows two types of hand auger: the *posthole auger* and the *helical auger*. Hand augers cannot be used for advancing holes to depths exceeding 3 to 5 m (10 to 16 ft). However, they can be used for soil exploration work on some highways and small structures. *Portable power-driven helical augers* (76 mm to 305 mm in diameter) are available for making deeper boreholes. The soil samples obtained from such borings are highly disturbed. In some noncohesive soils or soils having low cohesion, the

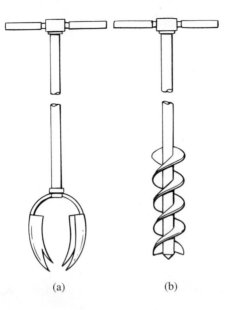

(a)   (b)

Figure 2.9   Hand tools: (a) posthole auger;
(b) helical auger

walls of the boreholes will not stand unsupported. In such circumstances, a metal
pipe is used as a *casing* to prevent the soil from caving in.

When power is available, *continuous-flight augers* are probably the most common method used for advancing a borehole. The power for drilling is delivered by
truck- or tractor-mounted drilling rigs. Boreholes up to about 60 to 70 m (200 to 230 ft)
can easily be made by this method. Continuous-flight augers are available in sections
of about 1 to 2 m (3 to 6 ft) with either a solid or hollow stem. Some of the commonly
used solid—stem augers have outside diameters of 66.68 mm ($2\frac{5}{8}$ in.), 82.55 mm
($3\frac{1}{4}$ in.), 101.6 mm (4 in.), and 114.3 mm ($4\frac{1}{2}$ in.). Common commercially available
hollow-stem augers have dimensions of 63.5 mm ID and 158.75 mm OD
(2.5 in. × 6.25 in.), 69.85 mm ID and 177.8 OD (2.75 in. × 7 in.), 76.2 mm ID and
203.2 OD (3 in. × 8 in.), and 82.55 mm ID and 228.6 mm OD (3.25 in. × 9 in.).

The tip of the auger is attached to a cutter head (Figure 2.10). During the
drilling operation (Figure 2.11), section after section of auger can be added and the
hole extended downward. The flights of the augers bring the loose soil from the bottom of the hole to the surface. The driller can detect changes in the type of soil by
noting changes in the speed and sound of drilling. When solid-stem augers are used,
the auger must be withdrawn at regular intervals to obtain soil samples and also to
conduct other operations such as standard penetration tests. Hollow-stem augers
have a distinct advantage over solid-stem augers in that they do not have to be removed frequently for sampling or other tests. As shown schematically in Figure 2.12,
the outside of the hollow-stem auger acts as a casing.

The hollow-stem auger system includes the following components:

Outer component: (a) hollow auger sections, (b) hollow auger cap, and
(c) drive cap

Inner component: (a) pilot assembly, (b) center rod column, and (c) rod-to-
cap adapter

Figure 2.10 Carbide-tipped cutting head on auger flight

The auger head contains replaceable carbide teeth. During drilling, if soil samples are to be collected at a certain depth, the pilot assembly and the center rod are removed. The soil sampler is then inserted through the hollow stem of the auger column.

Wash boring is another method of advancing boreholes. In this method, a casing about 2 to 3 m (6 to 10 ft) long is driven into the ground. The soil inside the casing is then removed by means of a chopping bit attached to a drilling rod. Water is forced through the drilling rod and exits at a very high velocity through the holes at the bottom of the chopping bit (Figure 2.13). The water and the chopped soil particles rise in the drill hole and overflow at the top of the casing through a T connection. The washwater is collected in a container. The casing can be extended with additional pieces as the borehole progresses; however, that is not required if the borehole will stay open and not cave in. Wash borings are rarely used now in the United States and other developed countries.

Rotary drilling is a procedure by which rapidly rotating drilling bits attached to the bottom of drilling rods cut and grind the soil and advance the borehole. There are several types of drilling bit. Rotary drilling can be used in sand, clay, and rocks (unless they are badly fissured). Water or *drilling mud* is forced down the drilling

Figure 2.11   Drilling with continuous-flight augers

rods to the bits, and the return flow forces the cuttings to the surface. Boreholes with diameters of 50 to 203 mm (2 to 8 in.) can easily be made by this technique. The drilling mud is a slurry of water and bentonite. Generally, it is used when the soil that is encountered is likely to cave in. When soil samples are needed, the drilling rod is raised and the drilling bit is replaced by a sampler. With the environmental drilling applications, rotary drilling with air is becoming more common.

Percussion drilling is an alternative method of advancing a borehole, particularly through hard soil and rock. A heavy drilling bit is raised and lowered to chop the hard soil. The chopped soil particles are brought up by the circulation of water. Percussion drilling may require casing.

2.13   *Procedures for Sampling Soil*

Two types of soil samples can be obtained during subsurface exploration: *disturbed* and *undisturbed*. Disturbed, but representative, samples can generally be used for the following types of laboratory test:

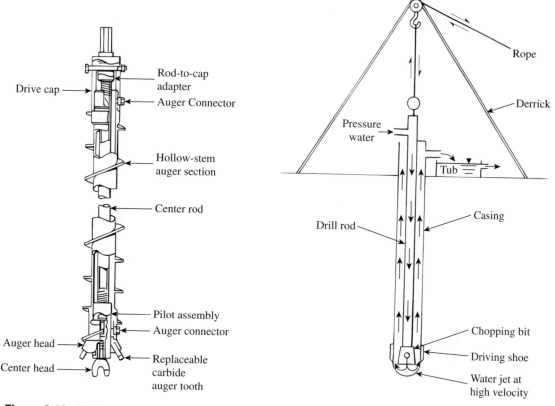

Figure 2.12   Hollow-stem auger components (After ASTM, 2001)

Figure 2.13   Wash boring

1. Grain-size analysis
2. Determination of liquid and plastic limits
3. Specific gravity of soil solids
4. Determination of organic content
5. Classification of soil

Disturbed soil samples, however, cannot be used for consolidation, hydraulic conductivity, or shear strength tests. Undisturbed soil samples must be obtained for these types of laboratory tests.

Split-Spoon Sampling

Split-spoon samplers can be used in the field to obtain soil samples that are generally disturbed, but still representative. A section of a *standard split-spoon sampler* is shown in Figure 2.14a. The tool consists of a steel driving shoe, a steel tube that is split longitudinally in half, and a coupling at the top. The coupling connects the

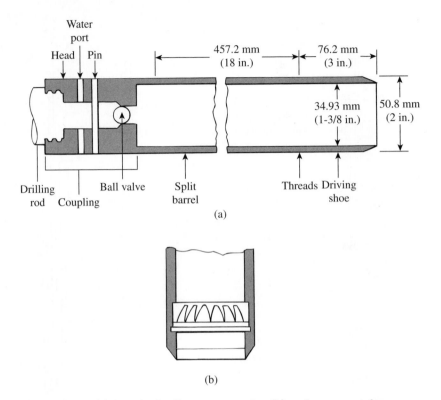

Figure 2.14   (a) Standard split-spoon sampler; (b) spring core catcher

sampler to the drill rod. The standard split tube has an inside diameter of 34.93 mm ($1\frac{3}{8}$ in.) and an outside diameter of 50.8 mm (2 in.); however, samplers having inside and outside diameters up to 63.5 mm ($2\frac{1}{2}$ in.) and 76.2 mm (3 in.), respectively, are also available. When a borehole is extended to a predetermined depth, the drill tools are removed and the sampler is lowered to the bottom of the hole. The sampler is driven into the soil by hammer blows to the top of the drill rod. The standard weight of the hammer is 622.72 N (140 lb), and for each blow, the hammer drops a distance of 0.762 m (30 in.). The number of blows required for a spoon penetration of three 152.4-mm (6-in.) intervals are recorded. The number of blows required for the last two intervals are added to give the *standard penetration number, N,* at that depth. This number is generally referred to as the *N value* (American Society for Testing and Materials, 2001, Designation D-1586-99). The sampler is then withdrawn, and the shoe and coupling are removed. Finally, the soil sample recovered from the tube is placed in a glass bottle and transported to the laboratory. This field test is called the standard penetration test (SPT).

The degree of disturbance for a soil sample is usually expressed as

$$A_R(\%) = \frac{D_o^2 - D_i^2}{D_i^2}(100) \tag{2.5}$$

where

A_R = area ratio (ratio of disturbed area to total area of soil)
D_o = outside diameter of the sampling tube
D_i = inside diameter of the sampling tube

When the area ratio is 10% or less, the sample generally is considered to be undisturbed. For a standard split-spoon sampler,

$$A_R(\%) = \frac{(50.8)^2 - (34.93)^2}{(34.93)^2}(100) = 111.5\%$$

Hence, these samples are highly disturbed. Split-spoon samples generally are taken at intervals of about 1.5 m (5 ft). When the material encountered in the field is sand (particularly fine sand below the water table), recovery of the sample by a split-spoon sampler may be difficult. In that case, a device such as a *spring core catcher* may have to be placed inside the split spoon (Figure 2.14b).

At this juncture, it is important to point out that several factors contribute to the variation of the standard penetration number N at a given depth for similar soil profiles. Among these factors are the SPT hammer efficiency, borehole diameter, sampling method, and rod length factor (Skempton, 1986; Seed, et al., 1985). The SPT hammer energy efficiency can be expressed as

$$E_r(\%) = \frac{\text{actual hammer energy to the sampler}}{\text{input energy}} \times 100 \qquad (2.6)$$

$$\text{Theoretical input energy} = Wh \qquad (2.7)$$

where

W = weight of the hammer $\approx$ 0.623 kN (140 lb)
h = height of drop $\approx$ 0.76 mm (30 in.)

So,

$$Wh = (0.623)(0.76) = 0.474 \text{ kN-m } (4200 \text{ in.-lb})$$

In the field, the magnitude of E_r can vary from 30 to 90%. The standard practice now in the U.S. is to express the N-value to an average energy ratio of 60% ($\approx N_{60}$). Thus, correcting for field procedures and on the basis of field observations, it appears reasonable to standardize the field penetration number as a function of the input driving energy and its dissipation around the sampler into the surrounding soil, or

$$N_{60} = \frac{N\eta_H \eta_B \eta_S \eta_R}{60} \qquad (2.8)$$

where

N_{60} = standard penetration number, corrected for field conditions
N = measured penetration number
η_H = hammer efficiency (%)

η_B = correction for borehole diameter

η_S = sampler correction

η_R = correction for rod length

Variations of η_H, η_B, η_S, and η_R, based on recommendations by Seed et al. (1985) and Skempton (1986), are summarized in Table 2.3.

Besides compelling the geotechnical engineer to obtain soil samples, standard penetration tests provide several useful correlations. For example, the consistency of clay soils can be estimated from the standard penetration number, N_{60}. In order to achieve that, Szechy and Vargi (1978) calculated the *consistency index* (CI) as

$$CI = \frac{LL - w}{LL - PL} \tag{2.9}$$

where

w = natural moisture content

LL = liquid limit

PL = plastic limit

The approximate correlation between CI, N_{60}, and the unconfined compression strength (q_u) is given in Table 2.4.

The literature contains many correlations between the standard penetration number and the undrained shear strength of clay, c_u. On the basis of results of undrained triaxial tests conducted on insensitive clays, Stroud (1974) suggested that

$$c_u = KN_{60} \tag{2.10}$$

Table 2.3 Variations of η_H, η_B, η_S, and η_R [Eq. (2.8)]

1. Variation of η_H

Country	Hammer type	Hammer release	η_H (%)
Japan	Donut	Free fall	78
	Donut	Rope and pulley	67
United States	Safety	Rope and pulley	60
	Donut	Rope and pulley	45
Argentina	Donut	Rope and pulley	45
China	Donut	Free fall	60
	Donut	Rope and pulley	50

2. Variation of η_B

Diameter		η_B
mm	in.	
60–120	2.4–4.7	1
150	6	1.05
200	8	1.15

3. Variation of η_S

Variable	η_S
Standard sampler	1.0
With liner for dense sand and clay	0.8
With liner for loose sand	0.9

4. Variation of η_R

Rod length		η_R
m	ft	
>10	>30	1.0
6–10	20–30	0.95
4–6	12–20	0.85
0–4	0–12	0.75

Table 2.4 Approximate Correlation between CI, N_{60}, and q_u

Standard penetration number, N_{60}	Consistency	CI	Unconfined compression strength, q_u	
			(kN/m²)	(lb/ft²)
<2	Very soft	<0.5	<25	500
2–8	Soft to medium	0.5–0.75	25–80	500–1700
8–15	Stiff	0.75–1.0	80–150	1700–3100
15–30	Very stiff	1.0–1.5	150–400	3100–8400
>30	Hard	>1.5	>400	8400

where

K = constant = 3.5–6.5 kN/m² (0.507–0.942 lb/in²)
N_{60} = standard penetration number obtained from the field

The average value of K is about 4.4 kN/m² (0.638 lb/in²).
   Hara, et al. (1971) also suggested that

$$c_u(\text{kN/m}^2) = 29N_{60}^{0.72} \tag{2.11}$$

   The overconsolidation ratio, OCR, of a natural clay deposit can also be correlated with the standard penetration number. On the basis of the regression analysis of 110 data points, Mayne and Kemper (1988) obtained the relationship

$$\text{OCR} = 0.193\left(\frac{N_{60}}{\sigma'_o}\right)^{0.689} \tag{2.12}$$

where σ'_o = effective vertical stress in MN/m².

   It is important to point out that any correlation between c_u and N_{60} is only approximate.
   In granular soils, the value of N is affected by the effective overburden pressure, σ'_o. For that reason, the value of N_{60} obtained from field exploration under different effective overburden pressures should be changed to correspond to a standard value of σ'_o. That is,

$$(N_1)_{60} = C_N N_{60} \tag{2.13}$$

where

$(N_1)_{60}$ = value of N_{60} corrected to a standard value of σ'_o $[100 \text{ kN/m}^2 \ (2000 \text{ lb/ft}^2)]$
C_N = correction factor
N_{60} = value of N obtained from field exploration [Eq. (2.8)]

In the past, a number of empirical relations were proposed for C_N. Some of the relationships are given next. The most commonly cited relationships are those of Liao and Whitman (1986) and Skempton (1986).

In the following relationships for C_N, note that σ'_o is the effective overburden pressure and p_a = atmospheric pressure $(\approx 100 \text{ kN/m}^2, \text{ or } \approx 2000 \text{ lb/ft}^2)$
Liao and Whitman's relationship (1986):

$$C_N = \left[\frac{1}{\left(\dfrac{\sigma'_o}{p_a} \right)} \right]^{0.5} \tag{2.14}$$

Skempton's relationship (1986):

$$C_N = \frac{2}{1 + \left(\dfrac{\sigma'_o}{p_a} \right)} \tag{2.15}$$

Seed et al.'s relationship (1975):

$$C_N = 1 - 1.25 \log\left(\frac{\sigma'_o}{p_a} \right) \tag{2.16}$$

Peck et al.'s relationship (1974):

$$C_N = 0.77 \log\left[\frac{20}{\left(\dfrac{\sigma'_o}{p_a} \right)} \right] \tag{2.17}$$

for $\sigma'_o \geq 25 \text{ kN//m}^2 \ (\approx 500 \text{ lb/ft}^2)$

An approximate relationship between the corrected standard penetration number and the relative density of sand is given in Table 2.5. The values are approximate primarily because the effective overburden pressure and the stress history of

Table 2.5 Relation between the Corrected $(N_1)_{60}$ Values and the Relative Density in Sands

Standard penetration number, $(N_1)_{60}$	Approximate relative density, D_r, (%)
0–5	0–5
5–10	5–30
10–30	30–60
30–50	60–95

the soil significantly influence the N_{60} values of sand. An extensive study conducted by Marcuson and Bieganousky (1977) produced the empirical relationship

$$D_r(\%) = 11.7 + 0.76\left(222N_{60} + 1600 - 53\sigma'_o - 50C_u^2\right)^{0.5} \tag{2.18a}$$

where

D_r = relative density
N_{60} = standard penetration number in the field
σ'_o = effective overburden pressure (lb/in²)
C_u = uniformity coefficient of the sand

Kulhawy and Mayne (1990) modified the preceding relationship to consider the stress history (i.e., OCR) of soil, which can be expressed as

$$D_r(\%) = 12.2 + 0.75\left[222N_{60} + 2311 - 711\text{OCR} - 779\left(\frac{\sigma'_o}{p_a}\right) - 50C_u^2\right]^{0.5} \tag{2.18b}$$

where

$$\text{OCR} = \frac{\text{preconsolidation pressure}, \sigma'_c}{\text{effective overburden pressure}, \sigma'_o}$$

p_a = atmospheric pressure

Cubrinovski and Ishihara (1999) also proposed a correlation between N_{60} and the relative density of sand (D_r) that can be expressed as

$$D_r(\%) = \left[\frac{N_{60}\left(0.23 + \dfrac{0.06}{D_{50}}\right)^{1.7}}{9}\left(\frac{1}{\dfrac{\sigma'_o}{p_a}}\right)\right]^{0.5}(100) \tag{2.19}$$

where

p_a = atmospheric pressure (≈ 100 kN/m², or ≈ 2000 lb/ft²)
D_{50} = sieve size through which 50% of the soil will pass (mm)

Kulhawy and Mayne (1990) correlated the corrected standard penetration number and the relative density of sand in the form

$$D_r(\%) = \left[\frac{(N_1)_{60}}{C_p C_A C_{\text{OCR}}}\right]^{0.5}(100) \tag{2.20}$$

where

$$C_P = \text{grain-size correlations factor} = 60 + 25 \log D_{50} \tag{2.21}$$

$$C_A = \text{correlation factor for aging} = 1.2 + 0.05 \log\left(\frac{t}{100}\right) \tag{2.22}$$

$$C_{\text{OCR}} = \text{correlation factor for overconsolidation} = \text{OCR}^{0.18} \tag{2.23}$$

D_{50} = diameter through which 50% soil will pass through (mm)
  t = age of soil since deposition (years)
OCR = overconsolidation ratio

The peak friction angle, ϕ', of granular soil has also been correlated with N_{60} or $(N_1)_{60}$ by several investigators. Some of these correlations are as follows:

1. Peck, Hanson, and Thornburn (1974) give a correlation between N_{60} and ϕ' in a graphical form, which can be approximated as (Wolff, 1989)

$$\phi'(\text{deg}) = 27.1 + 0.3N_{60} - 0.00054[N_{60}]^2 \tag{2.24}$$

2. Schmertmann (1975) provided the correlation between N_{60}, σ_o', and ϕ'. Mathematically, the correlation can be approximated as (Kulhawy and Mayne, 1990)

$$\phi' = \tan^{-1}\left[\frac{N_{60}}{12.2 + 20.3\left(\dfrac{\sigma_o'}{p_a}\right)}\right]^{0.34} \tag{2.25}$$

where

N_{60} = field standard penetration number
σ_o' = effective overburden pressure
p_a = atmospheric pressure in the same unit as σ_o'
ϕ' = soil friction angle

3. Hatanaka and Uchida (1996) provided a simple correlation between ϕ' and $(N_1)_{60}$ that can be expressed as

$$\phi' = \sqrt{20(N_1)_{60}} + 20 \tag{2.26}$$

The following qualifications should be noted when standard penetration resistance values are used in the preceding correlations to estimate soil parameters:

1. The equations are approximate.
2. Because the soil is not homogeneous, the values of N_{60} obtained from a given borehole vary widely.
3. In soil deposits that contain large boulders and gravel, standard penetration numbers may be erratic and unreliable.

Although approximate, with correct interpretation the standard penetration test provides a good evaluation of soil properties. The primary sources of error in standard penetration tests are inadequate cleaning of the borehole, careless measurement of the blow count, eccentric hammer strikes on the drill rod, and inadequate maintenance of water head in the borehole.

The modulus of elasticity of granular soils (E_s) is an important parameter in estimating the elastic settlement of foundations. A first order estimation for E_s was given by Kulhawy and Mayne (1990) as

$$\frac{E_s}{p_a} = \alpha N_{60} \tag{2.27}$$

where

p_a = atmospheric pressure (same unit as E_s)

$\alpha = \begin{cases} 5 \text{ for sands with fines} \\ 10 \text{ for clean normally consolidated sand} \\ 15 \text{ for clean overconsolidated sand} \end{cases}$

Scraper Bucket

When the soil deposits are sand mixed with pebbles, obtaining samples by split spoon with a spring core catcher may not be possible because the pebbles may prevent the springs from closing. In such cases, a scraper bucket may be used to obtain disturbed representative samples (Figure 2.15a). The scraper bucket has a driving point and can be attached to a drilling rod. The sampler is driven down into the soil and rotated, and the scrapings from the side fall into the bucket.

Thin-Walled Tube

Thin-walled tubes are sometimes referred to as *Shelby tubes*. They are made of seamless steel and are frequently used to obtain undisturbed clayey soils. The most common thin-walled tube samplers have outside diameters of 50.8 mm (2 in.) and 76.2 mm (3 in.). The bottom end of the tube is sharpened. The tubes can be attached

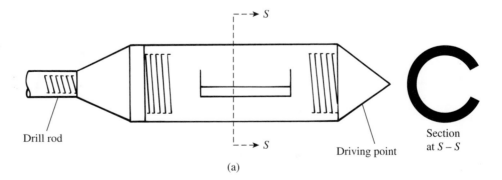

(a)

Figure 2.15 Sampling devices: (a) scraper bucket

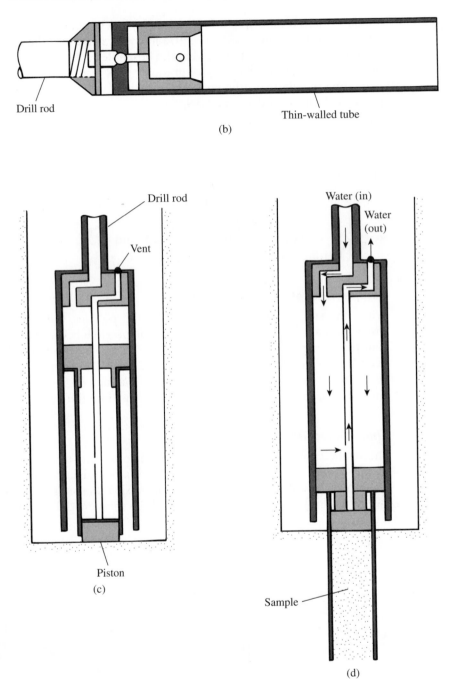

Drill rod

Thin-walled tube

(b)

Drill rod

Vent

Piston

(c)

Water (in)

Water (out)

Sample

(d)

Figure 2.15 *(Continued)* (b) thin-walled tube; (c) and (d) piston sampler

to drill rods (Figure 2.15b). The drill rod with the sampler attached is lowered to the bottom of the borehole, and the sampler is pushed into the soil. The soil sample inside the tube is then pulled out. The two ends are sealed, and the sampler is sent to the laboratory for testing.

Samples obtained in this manner may be used for consolidation or shear tests. A thin-walled tube with a 50.8-mm (2-in.) outside diameter has an inside diameter of about 47.63 mm ($1\frac{7}{8}$ in.). The area ratio is

$$A_R(\%) = \frac{D_o^2 - D_i^2}{D_i^2}(100) = \frac{(50.8)^2 - (47.63)^2}{(47.63)^2}(100) = 13.75\%$$

Increasing the diameters of samples increases the cost of obtaining them.

Piston Sampler

When undisturbed soil samples are very soft or larger than 76.2 mm (3 in.) in diameter, they tend to fall out of the sampler. Piston samplers are particularly useful under such conditions. There are several types of piston sampler; however, the sampler proposed by Osterberg (1952) is the most useful. (see Figures 2.15c and 2.15d). It consists of a thin-walled tube with a piston. Initially, the piston closes the end of the tube. The sampler is lowered to the bottom of the borehole (Figure 2.15c), and the tube is pushed into the soil hydraulically, past the piston. Then the pressure is released through a hole in the piston rod (Figure 2.15d). To a large extent, the presence of the piston prevents distortion in the sample by not letting the soil squeeze into the sampling tube very fast and by not admitting excess soil. Consequently, samples obtained in this manner are less disturbed than those obtained by Shelby tubes.

2.14   *Observation of Water Tables*

The presence of a water table near a foundation significantly affects the foundation's load-bearing capacity and settlement, among other things. The water level will change seasonally. In many cases, establishing the highest and lowest possible levels of water during the life of a project may become necessary.

If water is encountered in a borehole during a field exploration, that fact should be recorded. In soils with high hydraulic conductivity, the level of water in a borehole will stabilize about 24 hours after completion of the boring. The depth of the water table can then be recorded by lowering a chain or tape into the borehole.

In highly impermeable layers, the water level in a borehole may not stabilize for several weeks. In such cases, if accurate water-level measurements are required, a *piezometer* can be used. A piezometer basically consists of a porous stone or a perforated pipe with a plastic standpipe attached to it. Figure 2.16 shows the general placement of a piezometer in a borehole (also see Figure A.1 in Appendix A). This procedure will allow periodic checking until the water level stabilizes.

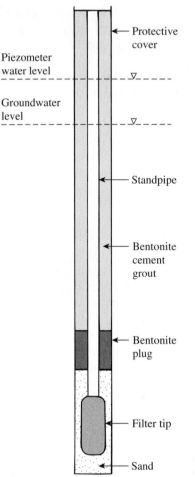

Piezometer
water level

Groundwater
level

Protective
cover

Standpipe

Bentonite
cement
grout

Bentonite
plug

Filter tip

Sand

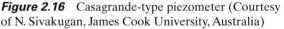

Figure 2.16 Casagrande-type piezometer (Courtesy of N. Sivakugan, James Cook University, Australia)

2.15   *Vane Shear Test*

The *vane shear test* (ASTM D-2573) may be used during the drilling operation to determine the *in situ* undrained shear strength (c_u) of clay soils—particularly soft clays. The vane shear apparatus consists of four blades on the end of a rod, as shown in Figure 2.17. The height, H, of the vane is twice the diameter, D. The vane can be either rectangular or tapered (see Figure 2.17). The dimensions of vanes used in the field are given in Table 2.6. The vanes of the apparatus are pushed into the soil at the bottom of a borehole without disturbing the soil appreciably. Torque is applied at the top of the rod to rotate the vanes at a standard rate of $0.1°/\text{sec}$. This rotation will induce failure in a soil of cylindrical shape surrounding the vanes. The maximum torque, T, applied to cause failure is measured. Note that

$$T = f(c_u, H, \text{ and } D) \qquad (2.28)$$

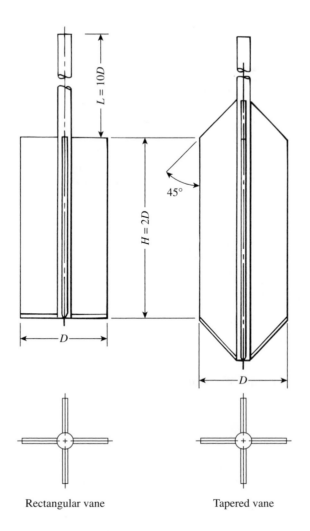

Rectangular vane   Tapered vane

Figure 2.17 Geometry of field vane (After ASTM, 2001)

or

$$c_u = \frac{T}{K} \tag{2.29}$$

where

T is in N·m, c_u is in kN/m², and
K = a constant with a magnitude depending on the dimension and shape
    of the vane

The constant

$$K = \left(\frac{\pi}{10^6}\right)\left(\frac{D^2 H}{2}\right)\left(1 + \frac{D}{3H}\right) \tag{2.30}$$

Table 2.6   ASTM Recommended Dimensions of Field Vanes[a]

Casing size	Diameter, D mm (in.)	Height, H mm (in.)	Thickness of blade mm (in.)	Diameter of rod mm (in.)
AX	38.1 ($1\frac{1}{2}$)	76.2 (3)	1.6 ($\frac{1}{16}$)	12.7 ($\frac{1}{2}$)
BX	50.8 (2)	101.6 (4)	1.6 ($\frac{1}{16}$)	12.7 ($\frac{1}{2}$)
NX	63.5 ($2\frac{1}{2}$)	127.0 (5)	3.2 ($\frac{1}{8}$)	12.7 ($\frac{1}{2}$)
101.6 mm (4 in.)[b]	92.1 ($3\frac{5}{8}$)	184.1 ($7\frac{1}{4}$)	3.2 ($\frac{1}{8}$)	12.7 ($\frac{1}{2}$)

[a]The selection of a vane size is directly related to the consistency of the soil being tested; that is, the softer the soil, the larger the vane diameter should be.
[b]Inside diameter.

where

D = diameter of vane in cm
H = measured height of vane in cm

If $H/D = 2$, Eq. (2.30) yields

$$K = 366 \times 10^{-8} D^3 \qquad (2.31)$$
$$\uparrow$$
$$(\text{cm})$$

In English units, if c_u and T in Eq. (2.29) are expressed in lb/ft² and lb-ft, respectively, then

$$K = \left(\frac{\pi}{1728}\right)\left(\frac{D^2 H}{2}\right)\left(1 + \frac{D}{3H}\right) \qquad (2.32)$$

If $H/D = 2$, Eq. (2.32) yields

$$K = 0.0021 D^3 \qquad (2.33)$$
$$\uparrow$$
$$(\text{in.})$$

Field vane shear tests are moderately rapid and economical and are used extensively in field soil-exploration programs. The test gives good results in soft and medium-stiff clays and gives excellent results in determining the properties of sensitive clays.

Sources of significant error in the field vane shear test are poor calibration of torque measurement and damaged vanes. Other errors may be introduced if the rate of rotation of the vane is not properly controlled.

For actual design purposes, the undrained shear strength values obtained from field vane shear tests $[c_{u(\text{VST})}]$ are too high, and it is recommended that they be corrected according to the equation

$$c_{u(\text{corrected})} = \lambda c_{u(\text{VST})} \qquad (2.34)$$

where λ = correction factor.

Several correlations have been given previously for the correction factor λ. The most commonly used correlation for λ is that given by Bjerrum (1972), which can be expressed as

$$\lambda = 1.7 - 0.54 \log[\text{PI}(\%)] \tag{2.35}$$

The field vane shear strength can be correlated with the preconsolidation pressure and the overconsolidation ratio of the clay. Using 343 data points, Mayne and Mitchell (1988) derived the following empirical relationship for estimating the preconsolidation pressure of a natural clay deposit:

$$\sigma'_c = 7.04[c_{u(\text{field})}]^{0.83} \tag{2.36}$$

Here,

σ'_c = preconsolidation pressure (kN/m^2)

$c_{u(\text{field})}$ = field vane shear strength (kN/m^2)

The overconsolidation ratio, OCR, also can be correlated to $c_{u(\text{field})}$ according to the equation

$$\text{OCR} = \beta \frac{c_{u(\text{field})}}{\sigma'_o} \tag{2.37}$$

where σ'_o = effective overburden pressure.

The magnitudes of β developed by various investigators are given below.

- Mayne and Mitchell (1988):

$$\beta = 22[\text{PI}(\%)]^{-0.48} \tag{2.38}$$

- Hansbo (1957):

$$\beta = \frac{222}{w(\%)} \tag{2.39}$$

- Larsson (1980):

$$\beta = \frac{1}{0.08 + 0.0055(\text{PI})} \tag{2.40}$$

2.16   *Cone Penetration Test*

The cone penetration test (CPT), originally known as the Dutch cone penetration test, is a versatile sounding method that can be used to determine the materials in a soil profile and estimate their engineering properties. The test is also called the *static penetration test,* and no boreholes are necessary to perform it. In the original version, a 60° cone with a base area of 10 cm² (1.55 in.²) was pushed into the ground at a steady rate of about 20 mm/sec (≈0.8 in./sec), and the resistance to penetration (called the point resistance) was measured.

The cone penetrometers in use at present measure (a) the *cone resistance* (q_c) to penetration developed by the cone, which is equal to the vertical force applied to the cone, divided by its horizontally projected area; and (b) the *frictional resistance* (f_c), which is the resistance measured by a sleeve located above the cone with the local soil surrounding it. The frictional resistance is equal to the vertical force applied to the sleeve, divided by its surface area—actually, the sum of friction and adhesion.

Generally, two types of penetrometers are used to measure q_c and f_c:

1. *Mechanical friction-cone penetrometer* (Figure 2.18). The tip of this penetrometer is connected to an inner set of rods. The tip is first advanced about 40 mm, giving the cone resistance. With further thrusting, the tip engages the friction sleeve. As the inner rod advances, the rod force is equal to the sum of the vertical force on the cone and sleeve. Subtracting the force on the cone gives the side resistance.
2. *Electric friction-cone penetrometer* (Figure 2.19). The tip of this penetrometer is attached to a string of steel rods. The tip is pushed into the ground at the rate of 20 mm/sec. Wires from the transducers are threaded through the center of the rods and continuously measure the cone and side resistances.

Figure 2.20 shows the results of penetrometer test in a soil profile with friction measurement by an electric friction-cone penetrometer.

Several correlations that are useful in estimating the properties of soils encountered during an exploration program have been developed for the point resistance (q_c) and the friction ratio (F_r) obtained from the cone penetration tests. The friction ratio is defined as

$$F_r = \frac{\text{frictional resistance}}{\text{cone resistance}} = \frac{f_c}{q_c} \tag{2.41}$$

In a more recent study on several soils in Greece, Anagnostopoulos et al. (2003) expressed F_s as

$$F_s(\%) = 1.45 - 1.36 \log D_{50} \text{ (electric cone)} \tag{2.42}$$

and

$$F_s(\%) = 0.7811 - 1.611 \log D_{50} \text{ (mechanical cone)} \tag{2.43}$$

where D_{50} = size through which 50% of soil will pass through (mm).

The D_{50} for soils based on which Eqs. (2.42) and (2.43) have been developed ranged from 0.001 mm to about 10 mm.

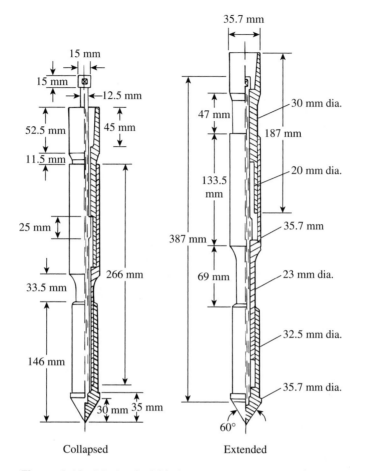

Figure 2.18 Mechanical friction-cone penetrometer (After ASTM, 2001)

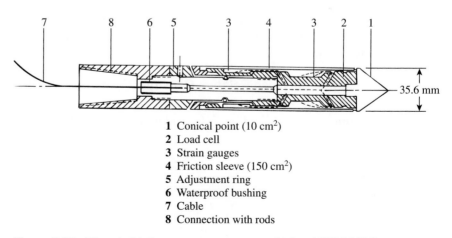

1 Conical point (10 cm^2)
2 Load cell
3 Strain gauges
4 Friction sleeve (150 cm^2)
5 Adjustment ring
6 Waterproof bushing
7 Cable
8 Connection with rods

Figure 2.19 Electric friction-cone penetrometer (After ASTM, 2001)

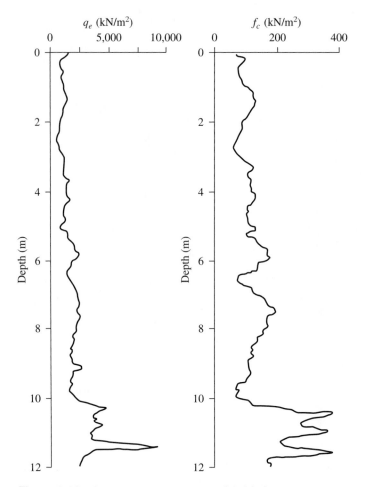

Figure 2.20 Cone penetrometer test with friction measurement

As in the case of standard penetration tests, several correlations have been developed between q_c and other soil properties. Some of these correlations are presented next.

Correlation between Relative Density (D_r) and q_c for Sand

Lancellotta (1983) and Jamiolkowski et al. (1985) showed that the relative density of *normally consolidated sand*, D_r, and q_c can be correlated according to the formula

$$D_r(\%) = A + B \log_{10}\left(\frac{q_c}{\sqrt{\sigma'_o}}\right) \qquad (2.44)$$

The preceding relationship can be rewritten as (Kulhawy and Mayne, 1990)

$$D_r(\%) = 68\left[\log\left(\frac{q_c}{\sqrt{p_a \cdot \sigma_0'}}\right) - 1\right] \qquad (2.45)$$

where

p_a = atmospheric pressure
σ_o' = vertical effective stress

Baldi et al. (1982), and Robertson and Campanella (1983) recommended the empirical relationship shown in Figure 2.21 between vertical effective stress (σ_o'), relative density (D_r), and for *normally consolidated sand* (q_c).

Kulhawy and Mayne (1990) proposed the following relationship to correlate D_r, q_c, and the vertical effective stress σ_o':

$$D_r = \sqrt{\left[\frac{1}{305Q_c OCR^{1.8}}\right]\left[\frac{\dfrac{q_c}{p_a}}{\left(\dfrac{\sigma_o'}{p_a}\right)^{0.5}}\right]} \qquad (2.46)$$

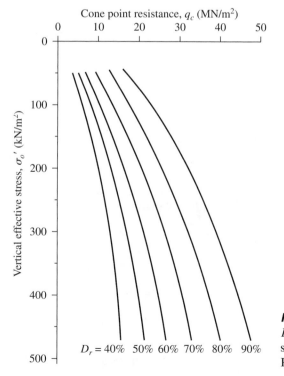

Figure 2.21 Variation of q_c, σ_o', and D_r for normally consolidated quartz sand (Based on Baldi et al., 1982, and Robertson and Campanella, 1983)

In this equation,

OCR = overconsolidation ratio
p_a = atmospheric pressure
Q_c = compressibility factor

The recommended values of Q_c are as follows:

Highly compressible sand = 0.91
Moderately compressible sand = 1.0
Low compressible sand = 1.09

Correlation between q_c and Drained Friction Angle (ϕ') for Sand

On the basis of experimental results, Robertson and Campanella (1983) suggested the variation of D_r, σ'_o, and ϕ' for normally consolidated quartz sand. This relationship can be expressed as (Kulhawy and Mayne, 1990)

$$\phi' = \tan^{-1}\left[0.1 + 0.38 \log\left(\frac{q_c}{\sigma'_o}\right)\right] \qquad (2.47)$$

Based on the cone penetration tests on the soils in the Venice Lagoon (Italy), Ricceri et al. (2002) proposed a similar relationship for soil with classifications of ML and SP-SM as

$$\phi' = \tan^{-1}\left[0.38 + 0.27 \log\left(\frac{q_c}{\sigma'_o}\right)\right] \qquad (2.48)$$

In a more recent study, Lee et al. (2004) developed a correlation between ϕ', q_c, and the horizontal effective stress (σ'_h) in the form

$$\phi' = 15.575\left(\frac{q_c}{\sigma'_h}\right)^{0.1714} \qquad (2.49)$$

Correlation between q_c and N_{60}

Figure 2.22 shows a plot of q_c (kN/m²)/N_{60} (N_{60} = standard penetration resistance) against the mean grain size (D_{50} in mm) for various types of soil. This was developed from field test results by Robertson and Campanella (1983).

Anagnostopoulos et al. (2003) provided a similar relationship correlating q_c, N_{60}, and D_{50}. Or

$$\frac{\left(\dfrac{q_c}{p_a}\right)}{N_{60}} = 7.6429 D_{50}^{0.26} \qquad (2.50)$$

where p_a = atmospheric pressure (same unit as q_c).

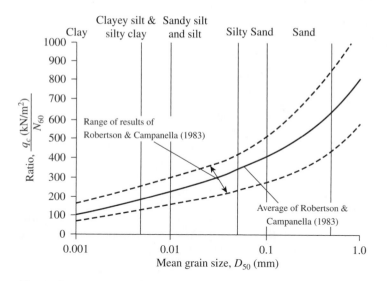

Figure 2.22 General range of variation of q_c/N_{60} for various types of soil

Correlations of Soil Types

Robertson and Campanella (1983) provided the correlations shown in Figure 2.23 between q_c and the friction ratio [Eq. (2.41)] to identify various types of soil encountered in the field.

Correlations for Undrained Shear Strength (c_u), Preconsolidation Pressure (σ_c'), and Overconsolidation Ratio (OCR) for Clays

The undrained shear strength, c_u, can be expressed as

$$c_u = \frac{q_c - \sigma_o}{N_K} \tag{2.51}$$

where

σ_o = total vertical stress
N_K = bearing capacity factor

The bearing capacity factor, N_K, may vary from 11 to 19 for normally consolidated clays and may approach 25 for overconsolidated clay. According to Mayne and Kemper (1988)

$$N_K = 15 \text{ (for electric cone)}$$

and

$$N_K = 20 \text{ (for mechanical cone)}$$

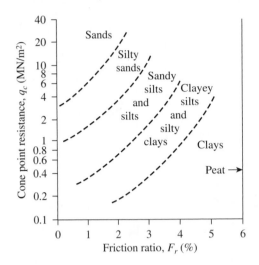

Figure 2.23 Robertson and Campanella's correlation (1983) between q_c, F_r, and the type of soil

Based on tests in Greece, Anagnostopoulos et al. (2003) determined

$$N_K = 17.2 \text{ (for electric cone)}$$

and

$$N_K = 18.9 \text{ (for mechanical cone)}$$

These field tests also showed that

$$c_u = \frac{f_c}{1.26} \text{ (for mechanical cones)} \tag{2.52}$$

and

$$c_u = f_c \text{ (for electrical cones)} \tag{2.53}$$

Mayne and Kemper (1988) provided correlations for preconsolidation pressure (σ_c') and overconsolidation ratio (OCR) as

$$\begin{array}{cc} \sigma_c' = 0.243\,(q_c)^{0.96} \\ \uparrow \qquad\quad \uparrow \\ \mathrm{MN/m^2} \quad \mathrm{MN/m^2} \end{array} \tag{2.54}$$

and

$$\mathrm{OCR} = 0.37\left(\frac{q_c - \sigma_o}{\sigma_o'}\right)^{1.01} \tag{2.55}$$

where σ_o and σ_o' = total and effective stress, respectively.

2.17   *Pressuremeter Test (PMT)*

The pressuremeter test is an *in situ* test conducted in a borehole. It was originally developed by Menard (1956) to measure the strength and deformability of soil. It has also been adopted by ASTM as Test Designation 4719. The Menard-type PMT consists essentially of a probe with three cells. The top and bottom ones are *guard cells* and the middle one is the *measuring cell,* as shown schematically in Figure 2.24a. The test is conducted in a prebored hole with a diameter that is between 1.03 and 1.2 times the nominal diameter of the probe. The probe that is most commonly used has a diameter of 58 mm and a length of 420 mm. The probe cells can be expanded by either liquid or gas. The guard cells are expanded to reduce the end-condition effect on the measuring cell, which has a volume (V_o) of 535 cm³. Following are the dimensions for the probe diameter and the diameter of the borehole, as recommended by ASTM:

Probe diameter (mm)	Borehole diameter	
	Nominal (mm)	Maximum (mm)
44	45	53
58	60	70
74	76	89

In order to conduct a test, the measuring cell volume, V_o, is measured and the probe is inserted into the borehole. Pressure is applied in increments and the new volume of the cell is measured. The process is continued until the soil fails or until the pressure limit of the device is reached. The soil is considered to have failed when

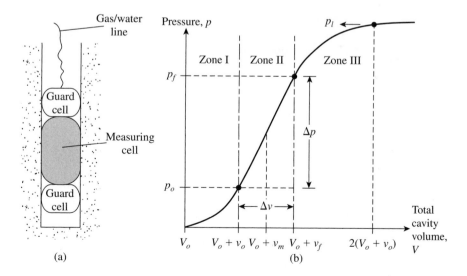

Figure 2.24 (a) Pressuremeter; (b) plot of pressure versus total cavity volume

the total volume of the expanded cavity (V) is about twice the volume of the original cavity. After the completion of the test, the probe is deflated and advanced for
testing at another depth.

The results of the pressuremeter test are expressed in the graphical form of
pressure versus volume, as shown in Figure 2.24b. In the figure, Zone I represents the
reloading portion during which the soil around the borehole is pushed back into the
initial state (i.e., the state it was in before drilling). The pressure p_o represents the *in
situ* total horizontal stress. Zone II represents a pseudoelastic zone in which the cell
volume versus cell pressure is practically linear. The pressure p_f represents the creep,
or yield, pressure. The zone marked III is the plastic zone. The pressure p_l represents
the limit pressure.

The pressuremeter modulus, E_p, of the soil is determined with the use of the
theory of expansion of an infinitely thick cylinder. Thus,

$$E_p = 2(1 + \mu_s) (V_o + v_m)\left(\frac{\Delta p}{\Delta v}\right) \tag{2.56}$$

where

$$v_m = \frac{v_o + v_f}{2}$$
$$\Delta p = p_f - p_o$$
$$\Delta v = v_f - v_o$$

μ_s = Poisson's ratio (which may be assumed to be 0.33)

The limit pressure p_l is usually obtained by extrapolation and not by direct measurement.

In order to overcome the difficulty of preparing the borehole to the proper
size, self-boring pressuremeters (SBPMTs) have also been developed. The details
concerning SBPMTs can be found in the work of Baguelin et al. (1978).

Correlations between various soil parameters and the results obtained from
the pressuremeter tests have been developed by various investigators. Kulhawy and
Mayne (1990) proposed that

$$\sigma'_c = 0.45 p_l \tag{2.57}$$

where σ'_c = preconsolidation pressure.

On the basis of the cavity expansion theory, Baguelin et al. (1978) proposed that

$$c_u = \frac{(p_l - p_o)}{N_p} \tag{2.58}$$

where

c_u = undrained shear strength of a clay

$$N_p = 1 + \ln\left(\frac{E_p}{3c_u}\right)$$

Typical values of N_p vary between 5 and 12, with an average of about 8.5. Ohya et al. (1982) (see also Kulhawy and Mayne, 1990) correlated E_p with field standard penetration numbers (N_{60}) for sand and clay as follows:

Clay: $E_p(\text{kN/m}^2) = 1930N_{60}^{0.63}$ (2.59)

Sand: $E_p(\text{kN/m}^2) = 908N_{60}^{0.66}$ (2.60)

2.18   *Dilatometer Test*

The use of the flat-plate dilatometer test (DMT) is relatively recent (Marchetti, 1980; Schmertmann, 1986). The equipment essentially consists of a flat plate measuring 220 mm (length) × 95 mm (width) × 14 mm (thickness) (8.66 in. × 3.74 in. × 0.55 in.). A thin, flat, circular, expandable steel membrane having a diameter of 60 mm (2.36 in.) is located flush at the center on one side of the plate (Figure 2.25a). The dilatometer probe is inserted into the ground with a cone penetrometer testing rig (Figure 2.25b). Gas and electric lines extend from the surface control box, through the penetrometer rod, and into the blade. At the required depth, high-pressure nitrogen gas is used to inflate the membrane. Two pressure readings are taken:

1. The pressure A required to "lift off" the membrane.
2. The pressure B at which the membrane expands 1.1 mm (0.4 in.) into the surrounding soil

The A and B readings are corrected as follows (Schmertmann, 1986):

$$\text{Contact stress, } p_o = 1.05(A + \Delta A - Z_m) - 0.05(B - \Delta B - Z_m) \quad (2.61)$$

$$\text{Expansion stress, } p_1 = B - Z_m - \Delta B \quad (2.62)$$

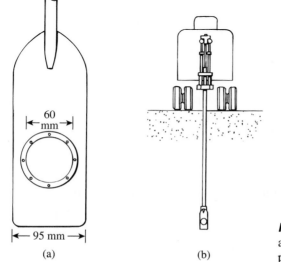

Figure 2.25 (a) Schematic diagram of a flat-plate dilatometer; (b) dilatometer probe inserted into ground

(a)

(b)

where

ΔA = vacuum pressure required to keep the membrane in contact with its seating
ΔB = air pressure required inside the membrane to deflect it outward to a center expansion of 1.1 mm
Z_m = gauge pressure deviation from zero when vented to atmospheric pressure

The test is normally conducted at depths 200 to 300 mm apart. The result of a given test is used to determine three parameters:

1. Material index, $I_D = \dfrac{p_1 - p_o}{p_o - u_o}$

2. Horizontal stress index, $K_D = \dfrac{p_o - u_o}{\sigma'_o}$

3. Dilatometer modulus, $E_D(\text{kN/m}^2) = 34.7(p_1 \text{ kN/m}^2 - p_o \text{ kN/m}^2)$

where

u_o = pore water pressure
σ'_o = *in situ* vertical effective stress

Figure 2.26 shows the results of a dilatometer test conducted in Bangkok soft clay and reported by Shibuya and Hanh (2001). Based on his initial tests, Marchetti (1980) provided the following correlations.

$$K_o = \left(\frac{K_D}{1.5}\right)^{0.47} - 0.6 \tag{2.63}$$

$$\text{OCR} = (0.5K_D)^{1.56} \tag{2.64}$$

$$\frac{c_u}{\sigma'_o} = 0.22 \qquad \text{(for normally consolidated clay)} \tag{2.65}$$

$$\left(\frac{c_u}{\sigma'_o}\right)_{\text{OC}} = \left(\frac{c_u}{\sigma'_o}\right)_{\text{NC}} (0.5K_D)^{1.25} \tag{2.66}$$

$$E_s = (1 - \mu_s^2)E_D \tag{2.67}$$

where

K_o = coefficient of at-rest earth pressure
OCR = overconsolidation ratio
OC = overconsolidated soil
NC = normally consolidated soil
E_s = modulus of elasticity

Other relevant correlations using the results of dilatometer tests are as follows:

- For undrained cohesion in clay (Kamei and Iwasaki, 1995):

$$c_u = 0.35\,\sigma'_0\,(0.47K_D)^{1.14} \tag{2.68}$$

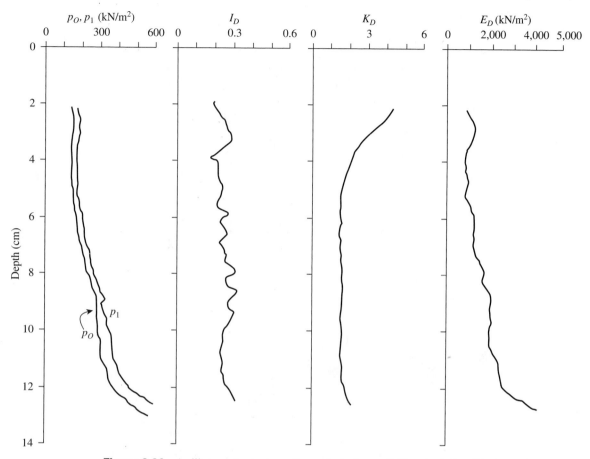

Figure 2.26 A dilatometer test result conducted on soft Bangkok clay (Redrawn from Shibuya and Hanh, 2001)

- For soil friction angle (ML and SP-SM soils) (Ricceri et al., 2002):

$$\phi' = 31 + \frac{K_D}{0.236 + 0.066K_D} \tag{2.69a}$$

$$\phi'_{ult} = 28 + 14.6 \log K_D - 2.1(\log K_D)^2 \tag{2.69b}$$

For definition of ϕ'_{ult}, see Fig. 1.36.

Schmertmann (1986) also provided a correlation between the material index (I_D) and the dilatometer modulus (E_D) for a determination of the nature of the soil and its unit weight (γ). This relationship is shown in Figure 2.27.

2.19 *Coring of Rocks*

When a rock layer is encountered during a drilling operation, rock coring may be necessary. To core rocks, a *core barrel* is attached to a drilling rod. A *coring bit* is attached to the bottom of the barrel (Fig. 2.28). The cutting elements may be diamond,

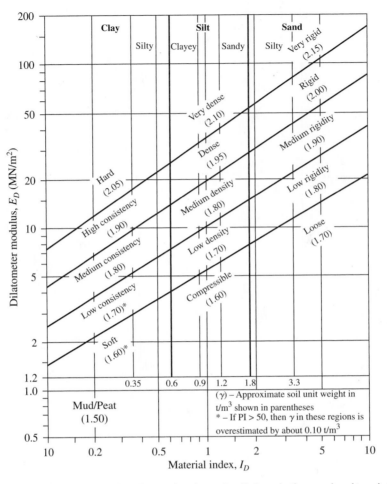

Figure 2.27 Chart for determination of soil description and unit weight (After Schmertmann, 1986) (*Note:* 1 t/m³ = 9.81 kN/m³)

tungsten, carbide, and so on. Table 2.7 summarizes the various types of core barrel and their sizes, as well as the compatible drill rods commonly used for exploring foundations. Figure 2.29 shows the photographs of a diamond coring bit. The coring is advanced by rotary drilling. Water is circulated through the drilling rod during coring, and the cutting is washed out.

Two types of core barrel are available: the *single-tube core barrel* (Figure 2.28a) and the *double-tube core barrel* (Figure 2.28b). Rock cores obtained by single-tube core barrels can be highly disturbed and fractured because of torsion. Rock cores smaller than the BX size tend to fracture during the coring process.

When the core samples are recovered, the depth of recovery should be properly recorded for further evaluation in the laboratory. Based on the length of the

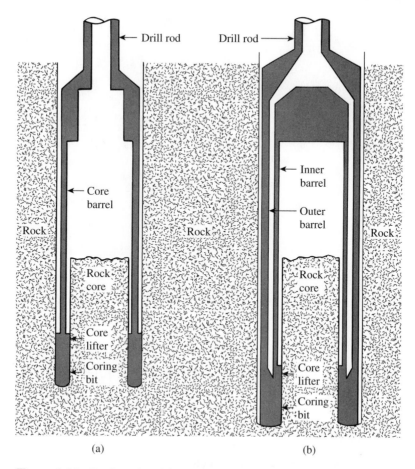

Figure 2.28 Rock coring: (a) single-tube core barrel; (b) double-tube core barrel

Table 2.7 Standard Size and Designation of Casing, Core Barrel, and Compatible Drill Rod

Casing and core barrel designation	Outside diameter of core barrel bit		Drill rod designation	Outside diameter of drill rod		Diameter of borehole		Diameter of core sample	
	(mm)	(in.)		(mm)	(in.)	(mm)	(in.)	(mm)	(in.)
EX	36.51	$1\frac{7}{16}$	E	33.34	$1\frac{5}{16}$	38.1	$1\frac{1}{2}$	22.23	$\frac{7}{8}$
AX	47.63	$1\frac{7}{8}$	A	41.28	$1\frac{5}{8}$	50.8	2	28.58	$1\frac{1}{8}$
BX	58.74	$2\frac{5}{16}$	B	47.63	$1\frac{7}{8}$	63.5	$2\frac{1}{2}$	41.28	$1\frac{5}{8}$
NX	74.61	$2\frac{15}{16}$	N	60.33	$2\frac{3}{8}$	76.2	3	53.98	$2\frac{1}{8}$

Figure 2.29 Diamond coring bit

rock core recovered from each run, the following quantities may be calculated for a general evaluation of the rock quality encountered:

$$\text{Recovery ratio} = \frac{\text{length of core recovered}}{\text{theoretical length of rock cored}} \qquad (2.70)$$

Rock quality designation (RQD)

$$= \frac{\Sigma \text{ length of recovered pieces equal to or larger than 101.6 m (4 in.)}}{\text{theoretical length of rock cored}} \qquad (2.71)$$

A recovery ratio of unity indicates the presence of intact rock; for highly fractured rocks, the recovery ratio may be 0.5 or smaller. Table 2.8 presents the general relationship (Deere, 1963) between the RQD and the *in situ* rock quality.

Table 2.8 Relation between *in situ* Rock Quality and RQD

RQD	Rock quality
0–0.25	Very poor
0.25–0.5	Poor
0.5–0.75	Fair
0.75–0.9	Good
0.9–1	Excellent

2.20   *Preparation of Boring Logs*

The detailed information gathered from each borehole is presented in a graphical form called the *boring log*. As a borehole is advanced downward, the driller generally should record the following information in a standard log:

1. Name and address of the drilling company
2. Driller's name
3. Job description and number
4. Number, type, and location of boring
5. Date of boring
6. Subsurface stratification, which can be obtained by visual observation of the soil brought out by auger, split-spoon sampler, and thin-walled Shelby tube sampler
7. Elevation of water table and date observed, use of casing and mud losses, and so on
8. Standard penetration resistance and the depth of SPT
9. Number, type, and depth of soil sample collected
10. In case of rock coring, type of core barrel used and, for each run, the actual length of coring, length of core recovery, and RQD

This information should never be left to memory, because doing so often results in erroneous boring logs.

After completion of the necessary laboratory tests, the geotechnical engineer prepares a finished log that includes notes from the driller's field log and the results of tests conducted in the laboratory. Figure 2.30 shows a typical boring log. These logs have to be attached to the final soil-exploration report submitted to the client. The figure also lists the classifications of the soils in the left-hand column, along with the description of each soil (based on the Unified Soil Classification System).

2.21   *Geophysical Exploration*

Several types of geophysical exploration techniques permit a rapid evaluation of subsoil characteristics. These methods also allow rapid coverage of large areas and are less expensive than conventional exploration by drilling. However, in many cases, definitive interpretation of the results is difficult. For that reason, such techniques should be used for preliminary work only. Here, we discuss three types of geophysical exploration technique: the seismic refraction survey, cross-hole seismic survey, and resistivity survey.

Seismic Refraction Survey

Seismic refraction surveys are useful in obtaining preliminary information about the thickness of the layering of various soils and the depth to rock or hard soil at a site. Refraction surveys are conducted by impacting the surface, such as at point *A* in Figure 2.31a, and observing the first arrival of the disturbance (stress waves) at several other points (e.g., *B, C, D, . . .*). The impact can be created by a hammer blow or by a small explosive charge. The first arrival of disturbance waves at various points can be recorded by geophones.

Boring Log

Name of the Project Two-story apartment building

Location Johnson & Olive St. Date of Boring March 2, 2005

Boring No. 3 Type of Hollow-stem auger Ground   60.8 m
                    Boring   Elevation

Soil description	Depth (m)	Soil sample type and number	N_{60}	w_n (%)	Comments
Light brown clay (fill)					
Silty sand (SM)	1 2	SS-1	9	8.2	
°G.W.T. 3.5 m	3 4	SS-2	12	17.6	LL = 38 PI = 11
Light gray clayey silt (ML)	5	ST-1		20.4	LL = 36 $q_u = 112$ kN/m²
	6	SS-3	11	20.6	
Sand with some gravel (SP) End of boring @ 8 m	7 8	SS-4	27	9	

N_{60} = standard penetration number   °Groundwater table
w_n = natural moisture content   observed after one
LL = liquid limit; PI = plasticity index   week of drilling
q_u = unconfined compression strength
SS = split-spoon sample; ST = Shelby tube sample

Figure 2.30   A typical boring log

The impact on the ground surface creates two types of *stress wave: P waves* (or *plane waves*) and *S waves* (or *shear waves*). *P* waves travel faster than *S* waves; hence, the first arrival of disturbance waves will be related to the velocities of the *P* waves in various layers. The velocity of *P* waves in a medium is

$$v = \sqrt{\frac{E_s}{\left(\dfrac{\gamma}{g}\right)} \frac{(1 - \mu_s)}{(1 - 2\mu_s)(1 + \mu_s)}} \tag{2.72}$$

where

E_s = modulus of elasticity of the medium
γ = unit weight of the medium
g = acceleration due to gravity
μ_s = Poisson's ratio

To determine the velocity v of *P* waves in various layers and the thicknesses of those layers, we use the following procedure:

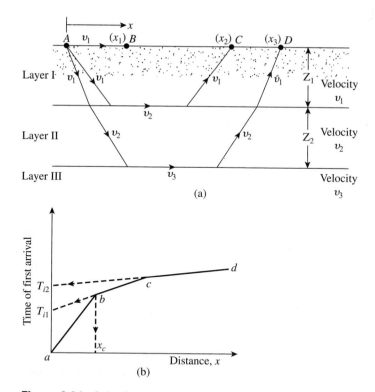

Figure 2.31 Seismic refraction survey

Step 1. Obtain the times of first arrival, $t_1, t_2, t_3, \ldots$, at various distances $x_1, x_2, x_3, \ldots$ from the point of impact.

Step 2. Plot a graph of time t against distance x. The graph will look like the one shown in Figure 2.31b.

Step 3. Determine the slopes of the lines ab, bc, cd, $\ldots$:

$$\text{Slope of } ab = \frac{1}{v_1}$$

$$\text{Slope of } bc = \frac{1}{v_2}$$

$$\text{Slope of } cd = \frac{1}{v_3}$$

Here, $v_1, v_2, v_3, \ldots$ are the *P*-wave velocities in layers I, II, III, $\ldots$, respectively (Figure 2.31a).

Step 4. Determine the thickness of the top layer:

$$Z_1 = \frac{1}{2}\sqrt{\frac{v_2 - v_1}{v_2 + v_1}}\, x_c \tag{2.73}$$

The value of x_c can be obtained from the plot, as shown in Figure 2.31b.

Step 5. Determine the thickness of the second layer:

$$Z_2 = \frac{1}{2}\left[T_{i2} - 2Z_1 \frac{\sqrt{v_3^2 - v_1^2}}{v_3 v_1} \right] \frac{v_3 v_2}{\sqrt{v_3^2 - v_2^2}} \tag{2.74}$$

Here, T_{i2} is the time intercept of the line *cd* in Figure 2.31b, extended backwards.

(For detailed derivatives of these equations and other related information, see Dobrin, 1960, and Das, 1992).

The velocities of *P* waves in various layers indicate the types of soil or rock that are present below the ground surface. The range of the *P*-wave velocity that is generally encountered in different types of soil and rock at shallow depths is given in Table 2.9.

In analyzing the results of a refraction survey, two limitations need to be kept in mind:

1. The basic equations for the survey—that is, Eqs. (2.73) and (2.74)—are based on the assumption that the *P*-wave velocity $v_1 < v_2 < v_3 < \cdots$.
2. When a soil is saturated below the water table, the *P*-wave velocity may be deceptive. *P* waves can travel with a velocity of about 1500 m/sec (5000 ft/sec) through water. For dry, loose soils, the velocity may be well below 1500 m/sec. However, in a saturated condition, the waves will travel through water that is present in the void spaces with a velocity of about 1500 m/sec. If the presence of groundwater has not been detected, the *P*-wave velocity may be erroneously interpreted to indicate a stronger material (e.g., sandstone) than is actually present *in situ*. In general, geophysical interpretations should always be verified by the results obtained from borings.

Table 2.9 Range of *P*-Wave Velocity in Various Soils and Rocks

Type of soil or rock	P-wave velocity	
	m/sec	ft/sec
Soil		
Sand, dry silt, and fine-grained topsoil	200–1000	650–3300
Alluvium	500–2000	1650–6600
Compacted clays, clayey gravel, and	1000–2500	3300–8200
dense clayey sand		
Loess	250–750	800–2450
Rock		
Slate and shale	2500–5000	8200–16,400
Sandstone	1500–5000	4900–16,400
Granite	4000–6000	13,100–19,700
Sound limestone	5000–10,000	16,400–32,800

Example 2.1

The results of a refraction survey at a site are given in the following table:

Distance of geophone from the source of disturbance (m)	Time of first arrival (sec × 10³)
2.5	11.2
5	23.3
7.5	33.5
10	42.4
15	50.9
20	57.2
25	64.4
30	68.6
35	71.1
40	72.1
50	75.5

Determine the *P*-wave velocities and the thickness of the material encountered.

Solution

Velocity
In Figure 2.32, the times of first arrival of the *P* waves are plotted against the distance of the geophone from the source of disturbance. The plot has three straight-line segments. The velocity of the top three layers can now be calculated as follows:

$$\text{Slope of segment } 0a = \frac{1}{v_1} = \frac{\text{time}}{\text{distance}} = \frac{23 \times 10^{-3}}{5.25}$$

or

$$v_1 = \frac{5.25 \times 10^3}{23} = \textbf{228 m/sec (top layer)}$$

$$\text{Slope of segment } ab = \frac{1}{v_2} = \frac{13.5 \times 10^{-3}}{11}$$

or

$$v_2 = \frac{11 \times 10^3}{13.5} = \textbf{814.8 m/sec (middle layer)}$$

$$\text{Slope of segment } bc = \frac{1}{v_3} = \frac{3.5 \times 10^{-3}}{14.75}$$

or

$$v_3 = \textbf{4214 m/sec (third layer)}$$

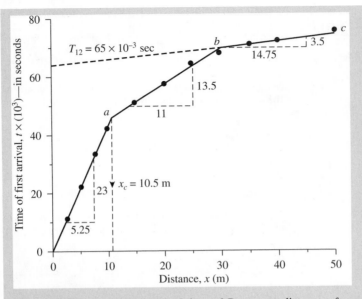

Figure 2.32 Plot of first arrival time of *P* wave vs. distance of geophone from source of disturbance

Comparing the velocities obtained here with those given in Table 2.9 indicates that the third layer is a *rock layer*.

Thickness of Layers
From Figure 2.32, $x_c = 10.5$ m, so

$$Z_1 = \frac{1}{2}\sqrt{\frac{v_2 - v_1}{v_2 + v_1}}\, x_c \qquad\qquad [\text{Eq. (2.73)}]$$

Thus,

$$Z_1 = \frac{1}{2}\sqrt{\frac{814.8 - 228}{814.8 + 228}} \times 10.5 = \textbf{3.94 m}$$

Again, from Eq. (2.74)

$$Z_2 = \frac{1}{2}\left[T_{i2} - \frac{2Z_1\sqrt{v_3^2 - v_1^2}}{(v_3 v_1)} \right] \frac{(v_3)(v_2)}{\sqrt{v_3^2 - v_2^2}}$$

The value of T_{i2} (from Figure 2.32) is 65×10^{-3} sec. Hence,

$$Z_2 = \frac{1}{2}\left[65 \times 10^{-3} - \frac{2(3.94)\sqrt{(4214)^2 - (228)^2}}{(4214)(228)} \right] \frac{(4214)(814.8)}{\sqrt{(4214)^2 - (814.8)^2}}$$

$$= \frac{1}{2}(0.065 - 0.0345)830.47 = \textbf{12.66 m}$$

Thus, the rock layer lies at a depth of $Z_1 + Z_2 = 3.94 + 12.66 = $ **16.60 m from the surface of the ground.**

∎

Cross-Hole Seismic Survey

The velocity of shear waves created as the result of an impact to a given layer of soil can be effectively determined by the *cross-hole seismic survey* (Stokoe and Woods, 1972). The principle of this technique is illustrated in Figure 2.33, which shows two holes drilled into the ground a distance L apart. A vertical impulse is created at the bottom of one borehole by means of an impulse rod. The shear waves thus generated are recorded by a vertically sensitive transducer. The velocity of shear waves can be calculated as

$$v_s = \frac{L}{t} \tag{2.75}$$

where $t = $ travel time of the waves.

The shear modulus G_s of the soil at the depth at which the test is taken can be determined from the relation

$$v_s = \sqrt{\frac{G_s}{(\gamma/g)}}$$

or

$$G_s = \frac{v_s^2 \gamma}{g} \tag{2.76}$$

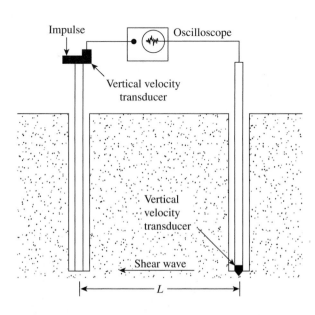

Figure 2.33   Cross-hole method of seismic survey

where

v_s = velocity of shear waves
γ = unit weight of soil
g = acceleration due to gravity

The shear modulus is useful in the design of foundations to support vibrating machinery and the like.

Resistivity Survey

Another geophysical method for subsoil exploration is the *electrical resistivity survey*. The electrical resistivity of any conducting material having a length L and an area of cross section A can be defined as

$$\rho = \frac{RA}{L} \tag{2.77}$$

where R = electrical resistance.

The unit of resistivity is the *ohm-centimeter* or *ohm-meter*. The resistivity of various soils depends primarily on their moisture content and also on the concentration of dissolved ions in them. Saturated clays have a very low resistivity; dry soils and rocks have a high resistivity. The range of resistivity generally encountered in various soils and rocks is given in Table 2.10.

The most common procedure for measuring the electrical resistivity of a soil profile makes use of four electrodes driven into the ground and spaced equally along a straight line. The procedure is generally referred to as the *Wenner method* (Figure 2.34a). The two outside electrodes are used to send an electrical current I (usually a dc current with nonpolarizing potential electrodes) into the ground. The current is typically in the range of 50–100 milliamperes. The voltage drop, V, is measured between the two inside electrodes. If the soil profile is homogeneous, its electrical resistivity is

$$\rho = \frac{2\pi dV}{I} \tag{2.78}$$

In most cases, the soil profile may consist of various layers with different resistivities, and Eq. (2.78) will yield the *apparent resistivity*. To obtain the *actual*

Table 2.10   Representative Values of Resistivity

Material	Resistivity (ohm · m)
Sand	500–1500
Clays, saturated silt	0–100
Clayey sand	200–500
Gravel	1500–4000
Weathered rock	1500–2500
Sound rock	>5000

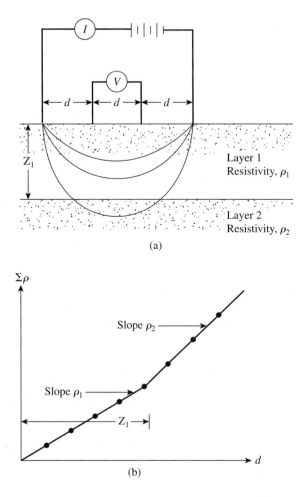

(a)

(b)

Figure 2.34 Electrical resistivity survey: (a) Wenner method; (b) empirical method for determining resistivity and thickness of each layer

resistivity of various layers and their thicknesses, one may use an empirical method that involves conducting tests at various electrode spacings (i.e., d is changed). The sum of the apparent resistivities, $\Sigma\rho$, is plotted against the spacing d, as shown in Figure 2.34b. The plot thus obtained has relatively straight segments, the slopes of which give the resistivity of individual layers. The thicknesses of various layers can be estimated as shown in Figure 2.34b.

The resistivity survey is particularly useful in locating gravel deposits within a fine-grained soil.

2.22   *Subsoil Exploration Report*

At the end of all soil exploration programs, the soil and rock specimens collected in the field are subject to visual observation and appropriate laboratory testing. (The basic soil tests were described in Chapter 1.) After all the required information has been compiled, a soil exploration report is prepared for use by the design office and for reference during future construction work. Although the details and sequence of information in such reports may vary to some degree, depending on the structure under consideration and the person compiling the report, each report should include the following items:

1. A description of the scope of the investigation
2. A description of the proposed structure for which the subsoil exploration has been conducted
3. A description of the location of the site, including any structures nearby, drainage conditions, the nature of vegetation on the site and surrounding it, and any other features unique to the site
4. A description of the geological setting of the site
5. Details of the field exploration—that is, number of borings, depths of borings, types of borings involved, and so on
6. A general description of the subsoil conditions, as determined from soil specimens and from related laboratory tests, standard penetration resistance and cone penetration resistance, and so on
7. A description of the water-table conditions
8. Recommendations regarding the foundation, including the type of foundation recommended, the allowable bearing pressure, and any special construction procedure that may be needed; alternative foundation design procedures should also be discussed in this portion of the report
9. Conclusions and limitations of the investigations

The following graphical presentations should be attached to the report:

1. A site location map
2. A plan view of the location of the borings with respect to the proposed structures and those nearby
3. Boring logs
4. Laboratory test results
5. Other special graphical presentations

The exploration reports should be well planned and documented, as they will help in answering questions and solving foundation problems that may arise later during design and construction.

Problems

2.1   For a Shelby tube, given: outside diameter $= 2$ in. and inside diameter $1\frac{7}{8}$ in.
   a. What is the area ratio of the tube?
   b. The outside diameter remaining the same, what should be the inside diameter of the tube to give an area ratio of 10%.

2.2   A soil profile is shown in Figure P2.2 along with the standard penetration numbers in the clay layer. Use Eqs. (2.11) and (2.12) to determine and plot the variation of c_u and OCR with depth.

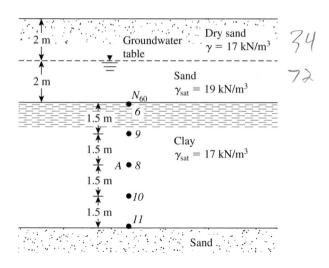

Figure P2.2

2.3 Following is the variation of the field standard penetration number (N_{60}) in a sand deposit:

Depth (m)	N_{60}
1.5	5
3	7
4.5	9
6	8
7.9	13
9	12

The groundwater table is located at a depth of 5.5 m. Given: the dry unit weight of sand from 0 to a depth of 5.5 m is 18.08 kN/m³, and the saturated unit weight of sand for depth 5.5 to 10.5 m is 19.34 kN/m³. Use the relationship of Liao and Whitman given in Eq. (2.14) to calculate the corrected penetration numbers.

2.4 For the soil profile described in Problem 2.3, estimate an average peak soil friction angle. Use Eq. (2.24) and (2.25).

2.5 The following table gives the variation of the field standard penetration number (N_{60}) in a sand deposit:

Depth (m)	N_{60}
1.5	6
3.0	8
4.5	9
6.0	8
7.5	13
9.0	14

The groundwater table is located at a depth of 6 m. The dry unit weight of sand from 0 to a depth of 6 m is 18 kN/m³, and the saturated unit weight of sand for a depth of 6 to 12 m is 20.2 kN/m³. Assume that the mean grain size (D_{50}) of the sand deposit to be about 0.6 mm. Estimate the variation of the relative density with depth for sand. Use Eq. (2.19).

2.6   Following are the standard penetration numbers determined from a sandy soil in the field:

Depth (ft)	Unit weight of soil (lb/ft³)	N_{60}
10	106	7
15	106	9
20	106	11
25	118	16
30	118	18
35	118	20
40	118	22

Using Eq. (2.25), determine the variation of the peak soil friction angle, ϕ'. Estimate an average value of ϕ' for the design of a shallow foundation. (*Note:* For depth greater than 20 ft, the unit weight of soil is 118 lb/ft³. Use $p_a = 14.7$ lb/in².)

2.7   Following are the details for a soil deposit in sand:

Depth (m)	Effective overburden pressure (lb/ft²)	Field standard penetration number, N_{60}
10	1150	9
15	1725	11
20	2030	12

Assume the uniformity coefficient (C_u) of the sand to be 3.2 and an overconsolidation ratio (OCR) of 3. Estimate the average relative density of the sand between the depth of 10 to 20 ft. Use Eq. (2.18b).

2.8   Refer to Figure P2.2. Vane shear tests were conducted in the clay layer. The vane dimensions were 63.5 mm (D) × 127 mm (H). For the test at A, the torque required to cause failure was 0.072 N·m. For the clay, given: liquid limit = 51 and plastic limit = 18. Estimate the undrained cohesion of the clay for use in the design by using Bjerrum's λ relationship [Eq. (2.35)].

2.9   **a.** A vane shear test was conducted in a saturated clay. The height and diameter of the vane were 4 in. and 2 in., respectively. During the test, the maximum torque applied was 12.4 lb-ft. Determine the undrained shear strength of the clay.

b. The clay soil described in part (a) has a liquid limit of 64 and a plastic limit of 29. What would be the corrected undrained shear strength of the clay for design purposes? Use Bjerrum's relationship for λ [Eq. (2.35)].

2.10   Refer to Problem 2.8. Determine the overconsolidation ratio for the clay. Use Eqs. (2.37) and (2.38).

2.11 In a deposit of normally consolidated dry sand, a cone penetration test was conducted. Following are the results:

Depth (m)	Point resistance of cone, q_c (MN/m²)
1.5	2.06
3.0	4.23
4.5	6.01
6.0	8.18
7.5	9.97
9.0	12.42

Assuming the dry unit weight of sand to be 15.5 kN/m³, estimate the average peak friction angle, ϕ', of the sand. Use Eq. (2.47).

2.12 Refer to Problem 2.11. Using Eq. (2.45), determine the variation of the relative density with depth.

2.13 In the soil profile shown in Figure P2.13, if the cone penetration resistance (q_c) at *A* (as determined by an electric friction-cone penetrometer) is 90 lb/in.², estimate
a. The undrained cohesion, c_u
b. The overconsolidation ratio, OCR

2.14 In a pressuremeter test in a soft saturated clay, the measuring cell volume $V_o = 535$ cm³, $p_o = 42.4$ kN/m², $p_f = 326.5$ kN/m², $v_o = 46$ cm³, and $v_f = 180$ cm³. Assuming Poisson's ratio (μ_s) to be 0.5 and using Figure 2.24, calculate the pressuremeter modulus (E_p).

2.15 A dilatometer test was conducted in a clay deposit. The groundwater table was located at a depth of 3 m below the surface. At a depth of 8 m below the surface, the contact pressure (p_o) was 280 kN/m² and the expansion stress (p_1) was 350 kN/m². Determine the following:

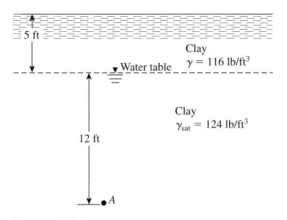

Figure P2.13

        a. Coefficient of at-rest earth pressure, K_o
        b. Overconsolidation ratio, OCR
        c. Modulus of elasticity, E_s
        Assume σ_o' at a depth of 8 m to be 95 kN/m^2 and $\mu_s = 0.35$.

2.16 A dilatometer test was conducted in a sand deposit at a depth of 5 m. The groundwater table was located at a depth of 2 m below the ground surface. Given, for the sand: $\gamma_d = 15$ kN/m^3 and $\gamma_{sat} = 19$ kN/m^3. The contact stress during the test was 260 kN/m^2. Estimate the soil friction angle, ϕ'.

2.17 During a field exploration, coring of rock was required. The core barrel was advanced 5 ft during the coring. The length of the core recovered was 3.2 ft. What was the recovery ratio?

2.18 The P-wave velocity in a soil is 1900 m/sec. Assuming Poisson's ratio to be 0.32, calculate the modulus of elasticity of the soil. Assume that the unit weight of soil is 18 kN/m^3.

2.19 The results of a refraction survey (Figure 2.31a) at a site are given in the following table. Determine the thickness and the P-wave velocity of the materials encountered.

Distance from the source of disturbance (m)	Time of first arrival of P-waves (sec $\times$ 10^3)
2.5	5.08
5.0	10.16
7.5	15.24
10.0	17.01
15.0	20.02
20.0	24.2
25.0	27.1
30.0	28.0
40.0	31.1
50.0	33.9

2.20 Repeat Problem 2.19 for the following data.

Distance from the source of disturbance (ft)	Time of first arrival of P-waves (sec $\times$ 10^3)
25	49.06
50	81.96
75	122.8
100	148.2
150	174.2
200	202.8
250	228.6
300	256.7

References

American Society for Testing and Materials (2001). *Annual Book of ASTM Standards,* Vol. 04.08, West Conshohocken, PA.

American Society of Civil Engineers (1972). "Subsurface Investigation for Design and Construction of Foundations of Buildings," *Journal of the Soil Mechanics and Foundations Division,* American Society of Civil Engineers, Vol. 98, No. SM5, pp. 481–490.

Anagnostopoulos, A., Koukis, G., Sabatakakis, N., and Tsiambaos, G. (2003). "Empirical Correlations of Soil Parameters Based on Cone Penetration Tests (CPT) for Greek Soils," *Geotechnical and Geological Engineering,* Vol. 21, No. 4, pp. 377–387.

Baguelin, F., Jézéquel, J. F., and Shields, D. H. (1978). *The Pressuremeter and Foundation Engineering,* Trans Tech Publications, Clausthal, Germany.

Baldi, G., Bellotti, R., Ghionna, V., and Jamiolkowski, M. (1982). "Design Parameters for Sands from CPT," *Proceedings, Second European Symposium on Penetration Testing,* Amsterdam, Vol. 2, pp. 425–438.

Bjerrum, L. (1972). "Embankments on Soft Ground," *Proceedings of the Specialty Conference,* American Society of Civil Engineers, Vol. 2, pp. 1–54.

Cubrinovski, M., and Ishihara, K. (1999). "Empirical Correlations between SPT N-Values and Relative Density for Sandy Soils," *Soils and Foundations,* Vol. 39, No. 5, pp. 61–92.

Das, B. M. (1992). *Principles of Soil Dynamics,* PWS Publishing Company, Boston.

Deere, D. U. (1963). "Technical Description of Rock Cores for Engineering Purposes," *Felsmechanik und Ingenieurgeologie,* Vol. 1, No. 1, pp. 16–22.

Dobrin, M. B. (1960). *Introduction to Geophysical Prospecting,* McGraw-Hill, New York.

Hansbo, S. (1957). *A New Approach to the Determination of the Shear Strength of Clay by the Fall Cone Test,* Swedish Geotechnical Institute, Report No. 114.

Hara, A., Ohata, T., and Niwa, M. (1971). "Shear Modulus and Shear Strength of Cohesive Soils," *Soils and Foundations,* Vol. 14, No. 3, pp. 1–12.

Hatanaka, M., and Uchida, A. (1996). "Empirical Correlation between Penetration Resistance and Internal Friction Angle of Sandy Soils," *Soils and Foundations,* Vol. 36, No. 4, pp. 1–10.

Jamiolkowski, M., Ladd, C. C., Germaine, J. T., and Lancellotta, R. (1985). "New Developments in Field and Laboratory Testing of Soils," *Proceedings, 11th International Conference on Soil Mechanics and Foundation Engineering,* Vol. 1, pp. 57–153.

Kamei, T., and Iwasaki, K. (1995). "Evaluation of Undrained Shear Strength of Cohesive Soils using a Flat Dilatometer," *Soils and Foundations,* Vol. 35, No. 2, pp. 111–116.

Kolb, C. R., and Shockley, W. G. (1959). "Mississippi Valley Geology: Its Engineering Significance" *Proceedings,* American Society of Civil Engineers, Vol. 124, pp. 633–656.

Kulhawy, F. H., and Mayne, P. W. (1990). *Manual on Estimating Soil Properties for Foundation Design,* Electric Power Research Institute, Palo Alto, California.

Lancellotta, R. (1983). *Analisi di Affidabilità in Ingegneria Geotecnica,* Atti Istituto Scienza Construzioni, No. 625, Politecnico di Torino.

Larsson, R. (1980). "Undrained Shear Strength in Stability Calculation of Embankments and Foundations on Clay," *Canadian Geotechnical Journal,* Vol. 17, pp. 591–602.

Lee, J., Salgado, R., and Carraro, A. H. (2004). "Stiffness Degradation and Shear Strength of Silty Sand," *Canadian Geotechnical Journal,* Vol. 41, No. 5, pp. 831–843.

Liao, S. S. C. and Whitman, R. V. (1986). "Overburden Correction Factors for SPT in Sand," *Journal of Geotechnical Engineering,* American Society of Civil Engineers, Vol. 112, No. 3, pp. 373–377.

Marchetti, S. (1980). "*In Situ* Test by Flat Dilatometer," *Journal of Geotechnical Engineering Division,* ASCE, Vol. 106, GT3, pp. 299–321.

Marcuson, W. F., III, and Bieganousky, W. A. (1977). "SPT and Relative Density in Coarse Sands," *Journal of Geotechnical Engineering Division,* American Society of Civil Engineers, Vol. 103, No. 11, pp. 1295–1309.

Mayne, P. W., and Kemper, J. B. (1988). "Profiling OCR in Stiff Clays by CPT and SPT," *Geotechnical Testing Journal,* ASTM, Vol. 11, No. 2, pp. 139–147.

Mayne, P. W., and Mitchell, J. K. (1988). "Profiling of Overconsolidation Ratio in Clays by Field Vane," *Canadian Geotechnical Journal,* Vol. 25, No. 1, pp. 150–158.

Menard, L. (1956). *An Apparatus for Measuring the Strength of Soils in Place,* master's thesis, University of Illinois, Urbana, Illinois.

Ohya, S., Imai, T., and Matsubara, M. (1982). "Relationships between N Value by SPT and LLT Pressuremeter Results," *Proceedings, 2nd European Symposium on Penetration Testing,* Vol. 1, Amsterdam, pp. 125–130.

Osterberg, J. O. (1952). "New Piston-Type Soil Sampler," *Engineering News-Record,* April 24.

Peck, R. B., Hanson, W. E., and Thornburn, T. H. (1974). *Foundation Engineering,* 2d ed., Wiley, New York.

Ricceri, G., Simonini, P., and Cola, S. (2002). "Applicability of Piezocone and Dilatometer to Characterize the Soils of the Venice Lagoon" *Geotechnical and Geological Engineering,* Vol. 20, No. 2, pp. 89–121.

Robertson, P. K., and Campanella, R. G. (1983). "Interpretation of Cone Penetration Tests. Part I: Sand," *Canadian Geotechnical Journal,* Vol. 20, No. 4, pp. 718–733.

Schmertmann, J. H. (1975). "Measurement of *In Situ* Shear Strength," *Proceedings, Specialty Conference on* In Situ *Measurement of Soil Properties,* ASCE, Vol. 2, pp. 57–138.

Schmertmann, J. H. (1986). "Suggested Method for Performing the Flat Dilatometer Test," *Geotechnical Testing Journal,* ASTM, Vol. 9, No. 2, pp. 93–101.

Seed, H. B., Arango, I., and Chan, C. K. (1975). *Evaluation of Soil Liquefaction Potential during Earthquakes,* Report No. EERC 75-28, Earthquake Engineering Research Center, University of California, Berkeley.

Seed, H. B., Tokimatsu, K., Harder, L. F., and Chung, R. M. (1985). "Influence of SPT Procedures in Soil Liquefaction Resistance Evaluations," *Journal of Geotechnical Engineering,* ASCE, Vol. 111, No. 12, pp. 1425–1445.

Shibuya, S., and Hanh, L. T. (2001). "Estimating Undrained Shear Strength of Soft Clay Ground Improved by Pre-Loading with PVD—Case History in Bangkok," *Soils and Foundations,* Vol. 41, No. 4, pp. 95–101.

Skempton, A. W. (1986). "Standard Penetration Test Procedures and the Effect in Sands of Overburden Pressure, Relative Density, Particle Size, Aging and Overconsolidation," *Geotechnique,* Vol. 36, No. 3, pp. 425–447.

Sowers, G. B., and Sowers, G. F. (1970). *Introductory Soil Mechanics and Foundations,* 3d ed., Macmillan, New York.

Stokoe, K. H., and Woods, R. D. (1972). "*In Situ* Shear Wave Velocity by Cross-Hole Method," *Journal of Soil Mechanics and Foundations Division,* American Society of Civil Engineers, Vol. 98, No. SM5, pp. 443–460.

Stroud, M. (1974). "SPT in Insensitive Clays," *Proceedings, European Symposium on Penetration Testing,* Vol. 2.2, pp. 367–375.

Szechy, K., and Varga, L. (1978). *Foundation Engineering—Soil Exploration and Spread Foundation,* Akademiai Kiado, Budapest, Hungary.

Wolff, T. F. (1989). "Pile Capacity Prediction Using Parameter Functions," in *Predicted and Observed Axial Behavior of Piles, Results of a Pile Prediction Symposium,* sponsored by the Geotechnical Engineering Division, ASCE, Evanston, IL, June, 1989, ASCE Geotechnical Special Publication No. 23, pp. 96–106.

3

Shallow Foundations: Ultimate Bearing Capacity

3.1 Introduction

To perform satisfactorily, shallow foundations must have two main characteristics:

1. They have to be safe against overall shear failure in the soil that supports them.
2. They cannot undergo excessive displacement, or settlement. (The term *excessive* is relative, because the degree of settlement allowed for a structure depends on several considerations.)

The load per unit area of the foundation at which shear failure in soil occurs is called the *ultimate bearing capacity*, which is the subject of this chapter.

3.2 General Concept

Consider a strip foundation with a width of B resting on the surface of a dense sand or stiff cohesive soil, as shown in Figure 3.1a. Now, if a load is gradually applied to the foundation, settlement will increase. The variation of the load per unit area, with the foundation settlement on the foundation q, is also shown in Figure 3.1a. At a certain point—when the load per unit area equals q_u—a sudden failure in the soil supporting the foundation will take place, and the failure surface in the soil will extend to the ground surface. This load per unit area, q_u, is usually referred to as the *ultimate bearing capacity of the foundation*. When such sudden failure in soil takes place, it is called *general shear failure*.

If the foundation under consideration rests on sand or clayey soil of medium compaction (Figure 3.1b), an increase in the load on the foundation will also be accompanied by an increase in settlement. However, in this case the failure surface in the soil will gradually extend outward from the foundation, as shown by the solid lines in Figure 3.1b. When the load per unit area on the foundation equals $q_{u(1)}$, movement of the foundation will be accompanied by sudden jerks. A considerable movement of the foundation is then required for the failure surface in soil to extend to the ground surface (as shown by the broken lines in the figure). The load per unit area at which this happens is the *ultimate bearing capacity, q_u*. Beyond that point, an

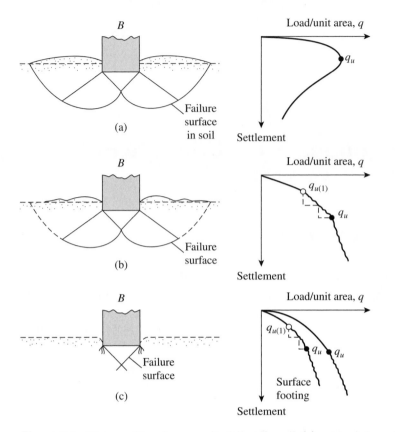

Figure 3.1   Nature of bearing capacity failure in soil: (a) general shear failure: (b) local shear failure; (c) punching shear failure (Redrawn after Vesic, 1973)

increase in load will be accompanied by a large increase in foundation settlement. The load per unit area of the foundation, $q_{u(1)}$, is referred to as the *first failure load* (Vesic, 1963). Note that a peak value of q is not realized in this type of failure, which is called the *local shear failure* in soil.

If the foundation is supported by a fairly loose soil, the load–settlement plot will be like the one in Figure 3.1c. In this case, the failure surface in soil will not extend to the ground surface. Beyond the ultimate failure load, q_u, the load–settlement plot will be steep and practically linear. This type of failure in soil is called the *punching shear failure.*

Vesic (1963) conducted several laboratory load-bearing tests on circular and rectangular plates supported by a sand at various relative densities of compaction, D_r. The variations of $q_{u(1)}/\frac{1}{2}\gamma B$ and $q_u/\frac{1}{2}\gamma B$ obtained from those tests, where B is the diameter of a circular plate or width of a rectangular plate and γ is a dry unit weight of sand, are shown in Figure 3.2. It is important to note from this figure that, for $D_r \geq$ about 70%, the general shear type of failure in soil occurs.

On the basis of experimental results, Vesic (1973) proposed a relationship for the mode of bearing capacity failure of foundations resting on sands. Figure 3.3

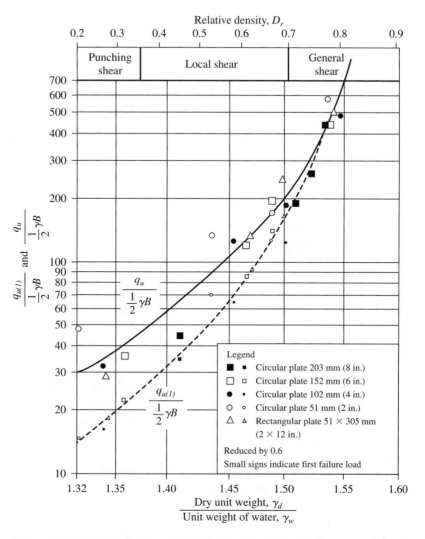

Figure 3.2 Variation of $q_{u(1)}/0.5\gamma B$ and $q_u/0.5\gamma B$ for circular and rectangular plates on the surface of a sand (Adapted from Vesic, 1963)

shows this relationship, which involves the notation

D_r = relative density of sand

D_f = depth of foundation measured from the ground surface

$$B^{\star} = \frac{2BL}{B + L} \qquad (3.1)$$

where

B = width of foundation
L = length of foundation

(*Note:* L is always greater than B.)

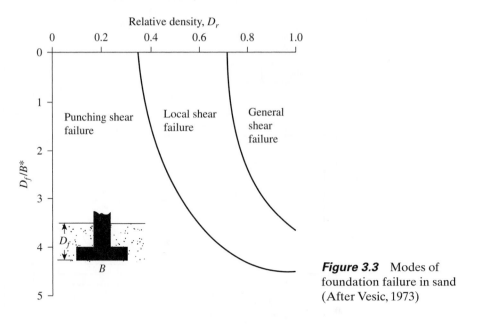

Figure 3.3   Modes of foundation failure in sand (After Vesic, 1973)

For square foundations, $B = L$; for circular foundations, $B = L =$ diameter, so

$$B^\star = B \tag{3.2}$$

Figure 3.4 shows the settlement S of the circular and rectangular plates on the surface of a sand at *ultimate load,* as described in Figure 3.2. The figure indicates a general range of S/B with the relative density of compaction of sand. So, in general, we can say that, for foundations at a shallow depth (i.e., small $D_f/B^\star$), the ultimate load may occur at a foundation settlement of 4 to 10% of B. This condition arises together with general shear failure in soil; however, in the case of local or punching shear failure, the ultimate load may occur at settlements of 15 to 25% of the width of the foundation (B).

3.3   Terzaghi's Bearing Capacity Theory

Terzaghi (1943) was the first to present a comprehensive theory for the evaluation of the ultimate bearing capacity of rough shallow foundations. According to this theory, a foundation is *shallow* if its depth, D_f (Figure 3.5), is less than or equal to its width. Later investigators, however, have suggested that foundations with D_f equal to 3 to 4 times their width may be defined as *shallow foundations.*

Terzaghi suggested that for a *continuous,* or *strip, foundation* (i.e., one whose width-to-length ratio approaches zero), the failure surface in soil at ultimate load may be assumed to be similar to that shown in Figure 3.5. (Note that this is the case of general shear failure, as defined in Figure 3.1a.) The effect of soil above the bottom of the foundation may also be assumed to be replaced by an equivalent surcharge, $q = \gamma D_f$ (where γ is a unit weight of soil). The failure zone under the foundation can be separated into three parts (see Figure 3.5):

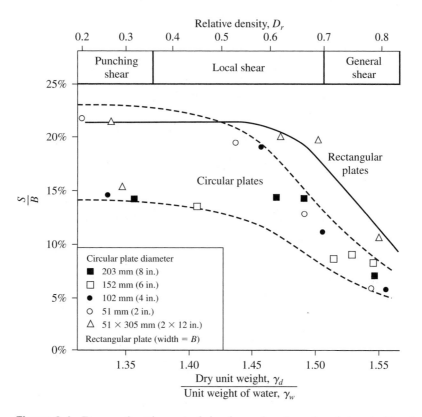

Figure 3.4   Range of settlement of circular and rectangular plates at ultimate load ($D_f/B = 0$) in sand (Modified from Vesic, 1963)

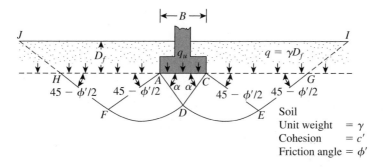

Figure 3.5   Bearing capacity failure in soil under a rough rigid continuous (strip) foundation

1. The *triangular zone ACD* immediately under the foundation
2. The *radial shear zones ADF and CDE*, with the curves *DE* and *DF* being arcs of a logarithmic spiral
3. Two triangular *Rankine passive zones AFH and CEG*

The angles *CAD* and *ACD* are assumed to be equal to the soil friction angle ϕ'. Note that, with the replacement of the soil above the bottom of the foundation by an equivalent surcharge q, the shear resistance of the soil along the failure surfaces *GI* and *HJ* was neglected.

Using equilibrium analysis, Terzaghi expressed the ultimate bearing capacity in the form

$$q_u = c'N_c + qN_q + \tfrac{1}{2}\gamma BN_\gamma \qquad \text{(continuous or strip foundation)} \qquad (3.3)$$

where

$$c' = \text{cohesion of soil}$$
$$\gamma = \text{unit weight of soil}$$
$$q = \gamma D_f$$
$$N_c, N_q, N_\gamma = \text{bearing capacity factors that are nondimensional and are functions only}$$
of the soil friction angle ϕ'

The bearing capacity factors N_c, N_q, and N_γ are defined by

$$N_c = \cot \phi' \left[\frac{e^{2(3\pi/4-\phi'/2)\tan \phi'}}{2 \cos^2\left(\dfrac{\pi}{4} + \dfrac{\phi'}{2}\right)} - 1 \right] = \cot \phi'(N_q - 1) \qquad (3.4)$$

$$N_q = \frac{e^{2(3\pi/4-\phi'/2)\tan \phi'}}{2 \cos^2\left(45 + \dfrac{\phi'}{2}\right)} \qquad (3.5)$$

and

$$N_\gamma = \frac{1}{2}\left(\frac{K_{p\gamma}}{\cos^2 \phi'} - 1\right)\tan \phi' \qquad (3.6)$$

where $K_{p\gamma}$ = passive pressure coefficient.

The variations of the bearing capacity factors defined by Eqs. (3.4), (3.5), and (3.6) are given in Table 3.1.

To estimate the ultimate bearing capacity of *square* and *circular foundations*, Eq. (3.1) may be respectively modified to

$$q_u = 1.3c'N_c + qN_q + 0.4\gamma BN_\gamma \qquad \text{(square foundation)} \qquad (3.7)$$

Table 3.1 Terzaghi's Bearing Capacity Factors—Eqs. (3.4), (3.5), and (3.6)

ϕ'	N_c	N_q	N_γ[a]	ϕ'	N_c	N_q	N_γ[a]
0	5.70	1.00	0.00	26	27.09	14.21	9.84
1	6.00	1.10	0.01	27	29.24	15.90	11.60
2	6.30	1.22	0.04	28	31.61	17.81	13.70
3	6.62	1.35	0.06	29	34.24	19.98	16.18
4	6.97	1.49	0.10	30	37.16	22.46	19.13
5	7.34	1.64	0.14	31	40.41	25.28	22.65
6	7.73	1.81	0.20	32	44.04	28.52	26.87
7	8.15	2.00	0.27	33	48.09	32.23	31.94
8	8.60	2.21	0.35	34	52.64	36.50	38.04
9	9.09	2.44	0.44	35	57.75	41.44	45.41
10	9.61	2.69	0.56	36	63.53	47.16	54.36
11	10.16	2.98	0.69	37	70.01	53.80	65.27
12	10.76	3.29	0.85	38	77.50	61.55	78.61
13	11.41	3.63	1.04	39	85.97	70.61	95.03
14	12.11	4.02	1.26	40	95.66	81.27	115.31
15	12.86	4.45	1.52	41	106.81	93.85	140.51
16	13.68	4.92	1.82	42	119.67	108.75	171.99
17	14.60	5.45	2.18	43	134.58	126.50	211.56
18	15.12	6.04	2.59	44	151.95	147.74	261.60
19	16.56	6.70	3.07	45	172.28	173.28	325.34
20	17.69	7.44	3.64	46	196.22	204.19	407.11
21	18.92	8.26	4.31	47	224.55	241.80	512.84
22	20.27	9.19	5.09	48	258.28	287.85	650.67
23	21.75	10.23	6.00	49	298.71	344.63	831.99
24	23.36	11.40	7.08	50	347.50	415.14	1072.80
25	25.13	12.72	8.34				

[a] From Kumbhojkar (1993)

and

$$q_u = 1.3c'N_c + qN_q + 0.3\gamma BN_\gamma \quad \text{(circular foundation)} \tag{3.8}$$

In Eq. (3.7), B equals the dimension of each side of the foundation; in Eq. (3.8), B equals the diameter of the foundation.

For foundations that exhibit the local shear failure mode in soils, Terzaghi suggested the following modifications to Eqs. (3.3), (3.7), and (3.8):

$$q_u = \frac{2}{3}c'N_c' + qN_q' + \tfrac{1}{2}\gamma BN_\gamma' \qquad \text{(strip foundation)} \tag{3.9}$$

$$q_u = 0.867c'N_c' + qN_q' + 0.4\gamma BN_\gamma' \quad \text{(square foundation)} \tag{3.10}$$

$$q_u = 0.867c'N_c' + qN_q' + 0.3\gamma BN_\gamma' \quad \text{(circular foundation)} \tag{3.11}$$

Table 3.2   Terzaghi's Modified Bearing Capacity Factors N'_c, N'_q, and N'_γ

ϕ'	N'_c	N'_q	N'_γ	ϕ'	N'_c	N'_q	N'_γ
0	5.70	1.00	0.00	26	15.53	6.05	2.59
1	5.90	1.07	0.005	27	16.30	6.54	2.88
2	6.10	1.14	0.02	28	17.13	7.07	3.29
3	6.30	1.22	0.04	29	18.03	7.66	3.76
4	6.51	1.30	0.055	30	18.99	8.31	4.39
5	6.74	1.39	0.074	31	20.03	9.03	4.83
6	6.97	1.49	0.10	32	21.16	9.82	5.51
7	7.22	1.59	0.128	33	22.39	10.69	6.32
8	7.47	1.70	0.16	34	23.72	11.67	7.22
9	7.74	1.82	0.20	35	25.18	12.75	8.35
10	8.02	1.94	0.24	36	26.77	13.97	9.41
11	8.32	2.08	0.30	37	28.51	15.32	10.90
12	8.63	2.22	0.35	38	30.43	16.85	12.75
13	8.96	2.38	0.42	39	32.53	18.56	14.71
14	9.31	2.55	0.48	40	34.87	20.50	17.22
15	9.67	2.73	0.57	41	37.45	22.70	19.75
16	10.06	2.92	0.67	42	40.33	25.21	22.50
17	10.47	3.13	0.76	43	43.54	28.06	26.25
18	10.90	3.36	0.88	44	47.13	31.34	30.40
19	11.36	3.61	1.03	45	51.17	35.11	36.00
20	11.85	3.88	1.12	46	55.73	39.48	41.70
21	12.37	4.17	1.35	47	60.91	44.45	49.30
22	12.92	4.48	1.55	48	66.80	50.46	59.25
23	13.51	4.82	1.74	49	73.55	57.41	71.45
24	14.14	5.20	1.97	50	81.31	65.60	85.75
25	14.80	5.60	2.25				

N'_c, N'_q, and N'_γ, the *modified bearing capacity factors,* can be calculated by using the bearing capacity factor equations (for N_c, N_q, and N_γ, respectively) by replacing ϕ' by $\overline{\phi}' = \tan^{-1}(\frac{2}{3} \tan \phi')$. The variation of N'_c, N'_q, and N'_γ with the soil friction angle ϕ' is given in Table 3.2.

Terzaghi's bearing capacity equations have now been modified to take into account the effects of the foundation shape (B/L), depth of embedment (D_f), and the load inclination. This is given in Section 3.6. Many design engineers, however, still use Terzaghi's equation, which provides fairly good results considering the uncertainty of the soil conditions at various sites.

3.4   *Factor of Safety*

Calculating the gross *allowable load-bearing capacity* of shallow foundations requires the application of a factor of safety (FS) to the gross ultimate bearing capacity, or

$$q_{\text{all}} = \frac{q_u}{\text{FS}} \tag{3.12}$$

However, some practicing engineers prefer to use a factor of safety such that

$$\text{Net stress increase on soil} = \frac{\text{net ultimate bearing capacity}}{\text{FS}} \quad (3.13)$$

The net ultimate bearing capacity is defined as the ultimate pressure per unit area of the foundation that can be supported by the soil in excess of the pressure caused by the surrounding soil at the foundation level. If the difference between the unit weight of concrete used in the foundation and the unit weight of soil surrounding is assumed to be negligible, then

$$q_{net(u)} = q_u - q \quad (3.14)$$

where

$q_{net(u)}$ = net ultimate bearing capacity
$q = \gamma D_f$

So

$$q_{all(net)} = \frac{q_u - q}{\text{FS}} \quad (3.15)$$

The factor of safety as defined by Eq. (3.15) should be at least 3 in all cases.

Another type of factor of safety for the bearing capacity of shallow foundations is often used. It is the factor of safety with respect to shear failure (FS_{shear}). In most cases, a value of $FS_{shear} = 1.4$ to 1.6 is desirable along with a *minimum* factor of safety of 3 to 4 against gross or net ultimate bearing capacity. The following procedure should be used to calculate the net allowable load for a given FS_{shear}.

1. Let c' and ϕ' be the cohesion and the angle of friction, respectively, of soil and let FS_{shear} be the required factor of safety with respect to shear failure. So the developed cohesion and the angle of friction are

$$c'_d = \frac{c'}{FS_{shear}} \quad (3.16)$$

$$\phi'_d = \tan^{-1}\left(\frac{\tan \phi'}{FS_{shear}}\right) \quad (3.17)$$

2. The gross allowable bearing capacity can now be calculated according to Eqs. (3.3), (3.7), (3.8), with c'_d and ϕ'_d as the shear strength parameters of the soil. For example, the gross allowable bearing capacity of a continuous foundation according to Terzaghi's equation is

$$q_{all} = c'_d N_c + q N_q + \tfrac{1}{2}\gamma B N_\gamma \quad (3.18)$$

where N_c, N_q, and N_γ = bearing capacity factors for the friction angle, ϕ'_d

3. The net allowable bearing capacity is thus

$$q_{all(net)} = q_{all} - q = c'_d N_c + q(N_q - 1) + \tfrac{1}{2}\gamma B N_\gamma \quad (3.19)$$

Irrespective of the procedure by which the factor of safety is applied, the magnitude of FS should depend on the uncertainties and risks involved for the conditions encountered.

Example 3.1

A square foundation is 5 ft × 5 ft in plan. The soil supporting the foundation has a friction angle of $\phi' = 20°$ and $c' = 320 \text{ lb/ft}^2$. The unit weight of soil, γ, is 115 lb/ft^3. Determine the allowable gross load on the foundation with a factor of safety (FS) of 4. Assume that the depth of the foundation (D_f) is 3 ft and that general shear failure occurs in the soil.

Solution

From Eq. (3.7)

$$q_u = 1.3c'N_c + qN_q + 0.4\gamma BN_\gamma$$

From Table 3.1, for $\phi' = 20°$,

$$N_c = 17.69$$
$$N_q = 7.44$$
$$N_\gamma = 3.64$$

Thus,

$$q_u = (1.3)(320)(17.69) + (3 \times 115)(7.44) + (0.4)(115)(5)(3.64)$$
$$= 7359 + 2566.8 + 837.2 = 10{,}736 \text{ lb/ft}^2$$

So, the allowable load per unit area of the foundation is

$$q_{all} = \frac{q_u}{FS} = \frac{10{,}736}{4} \approx 2691 \text{ lb/ft}^2$$

Thus, the total allowable gross load is

$$Q = (2691) B^2 = (2691)(5 \times 5) = 67{,}275 \text{ lb}$$ ∎

3.5   *Modification of Bearing Capacity Equations for Water Table*

Equations (3.3) and (3.7) through (3.11) give the ultimate bearing capacity, based on the assumption that the water table is located well below the foundation. However, if the water table is close to the foundation, some modifications of the bearing capacity equations will be necessary. (See Figure 3.6.)

Case I. If the water table is located so that $0 \leq D_1 \leq D_f$, the factor q in the bearing capacity equations takes the form

$$q = \text{effective surcharge} = D_1\gamma + D_2(\gamma_{sat} - \gamma_w) \tag{3.20}$$

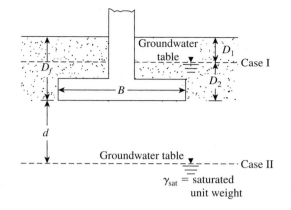

Figure 3.6 Modification of bearing capacity equations for water table

where

γ_{sat} = saturated unit weight of soil
γ_w = unit weight of water

Also, the value of γ in the last term of the equations has to be replaced by $\gamma' = \gamma_{sat} - \gamma_w$.

Case II. For a water table located so that $0 \leqslant d \leqslant B$,

$$q = \gamma D_f \qquad (3.21)$$

In this case, the factor γ in the last term of the bearing capacity equations must be replaced by the factor

$$\bar{\gamma} = \gamma' + \frac{d}{B}(\gamma - \gamma') \qquad (3.22)$$

The preceding modifications are based on the assumption that there is no seepage force in the soil.

Case III. When the water table is located so that $d \geqslant B$, the water will have no effect on the ultimate bearing capacity.

<table>
<tr><td>**3.6**</td><td></td></tr>
</table>

The General Bearing Capacity Equation

The ultimate bearing capacity equations (3.3), (3.7), and (3.8) are for continuous, square, and circular foundations only; they do not address the case of rectangular foundations $(0 < B/L < 1)$. Also, the equations do not take into account the shearing resistance along the failure surface in soil above the bottom of the foundation (the portion of the failure surface marked as *GI* and *HJ* in Figure 3.5). In addition, the load on the foundation may be inclined. To account for all these shortcomings, Meyerhof (1963) suggested the following form of the general bearing capacity equation:

$$q_u = c'N_cF_{cs}F_{cd}F_{ci} + qN_qF_{qs}F_{qd}F_{qi} + \frac{1}{2}\gamma BN_\gamma F_{\gamma s}F_{\gamma d}F_{\gamma i} \tag{3.23}$$

In this equation;

$$
\begin{aligned}
c' &= \text{cohesion} \\
q &= \text{effective stress at the level of the bottom of the foundation} \\
\gamma &= \text{unit weight of soil} \\
B &= \text{width of foundation } (= \text{diameter for a circular foundation}) \\
F_{cs}, F_{qs}, F_{\gamma s} &= \text{shape factors} \\
F_{cd}, F_{qd}, F_{\gamma d} &= \text{depth factors} \\
F_{ci}, F_{qi}, F_{\gamma i} &= \text{load inclination factors} \\
N_c, N_q, N_\gamma &= \text{bearing capacity factors}
\end{aligned}
$$

The equations for determining the various factors given in Eq. (3.23) are described briefly in the sections that follow. Note that the original equation for ultimate bearing capacity is derived only for the plane-strain case (i.e., for continuous foundations). The shape, depth, and load inclination factors are empirical factors based on experimental data.

Bearing Capacity Factors

The basic nature of the failure surface in soil suggested by Terzaghi now appears to have been borne out by laboratory and field studies of bearing capacity (Vesic, 1973). However, the angle α shown in Figure 3.5 is closer to $45 + \phi'/2$ than to ϕ'. If this change is accepted, the values of N_c, N_q, and N_γ for a given soil friction angle will also change from those given in Table 3.1. With $\alpha = 45 + \phi'/2$, it can be shown that

$$N_q = \tan^2\left(45 + \frac{\phi'}{2}\right)e^{\pi \tan \phi'} \tag{3.24}$$

and

$$N_c = (N_q - 1)\cot \phi' \tag{3.25}$$

Equation (3.25) for N_c was originally derived by Prandtl (1921), and Eq. (3.24) for N_q was presented by Reissner (1924). Caquot and Kerisel (1953) and Vesic (1973) gave the relation for N_γ as

$$N_\gamma = 2(N_q + 1)\tan \phi' \tag{3.26}$$

Table 3.3 Bearing Capacity Factors

ϕ'	N_c	N_q	N_γ	ϕ'	N_c	N_q	N_γ
0	5.14	1.00	0.00	26	22.25	11.85	12.54
1	5.38	1.09	0.07	27	23.94	13.20	14.47
2	5.63	1.20	0.15	28	25.80	14.72	16.72
3	5.90	1.31	0.24	29	27.86	16.44	19.34
4	6.19	1.43	0.34	30	30.14	18.40	22.40
5	6.49	1.57	0.45	31	32.67	20.63	25.99
6	6.81	1.72	0.57	32	35.49	23.18	30.22
7	7.16	1.88	0.71	33	38.64	26.09	35.19
8	7.53	2.06	0.86	34	42.16	29.44	41.06
9	7.92	2.25	1.03	35	46.12	33.30	48.03
10	8.35	2.47	1.22	36	50.59	37.75	56.31
11	8.80	2.71	1.44	37	55.63	42.92	66.19
12	9.28	2.97	1.69	38	61.35	48.93	78.03
13	9.81	3.26	1.97	39	67.87	55.96	92.25
14	10.37	3.59	2.29	40	75.31	64.20	109.41
15	10.98	3.94	2.65	41	83.86	73.90	130.22
16	11.63	4.34	3.06	42	93.71	85.38	155.55
17	12.34	4.77	3.53	43	105.11	99.02	186.54
18	13.10	5.26	4.07	44	118.37	115.31	224.64
19	13.93	5.80	4.68	45	133.88	134.88	271.76
20	14.83	6.40	5.39	46	152.10	158.51	330.35
21	15.82	7.07	6.20	47	173.64	187.21	403.67
22	16.88	7.82	7.13	48	199.26	222.31	496.01
23	18.05	8.66	8.20	49	229.93	265.51	613.16
24	19.32	9.60	9.44	50	266.89	319.07	762.89
25	20.72	10.66	10.88				

Table 3.3 shows the variation of the preceding bearing capacity factors with soil friction angles.

Shape Factors The equations for the shape factors F_{cs}, F_{qs}, and $F_{\gamma s}$ were recommended by De Beer (1970) and are

$$F_{cs} = 1 + \left(\frac{B}{L}\right)\left(\frac{N_q}{N_c}\right) \tag{3.27}$$

$$F_{qs} = 1 + \left(\frac{B}{L}\right)\tan\phi' \tag{3.28}$$

and

$$F_{\gamma s} = 1 - 0.4\left(\frac{B}{L}\right) \tag{3.29}$$

where L = length of the foundation ($L > B$).

The shape factors are empirical relations based on extensive laboratory tests.

Depth Factors   Hansen (1970) proposed the following equations for the depth factors:

$$F_{cd} = 1 + 0.4\left(\frac{D_f}{B}\right) \tag{3.30}$$

$$F_{qd} = 1 + 2 \tan \phi'(1 - \sin \phi')^2 \frac{D_f}{B} \tag{3.31}$$

$$F_{\gamma d} = 1 \tag{3.32}$$

Equations (3.30) and (3.31) are valid for $D_f/B \leq 1$. For a depth-of-embedment-to-foundation-width ratio greater than unity $(D_f/B > 1)$, the equations have to be modified to

$$F_{cd} = 1 + (0.4) \tan^{-1}\left(\frac{D_f}{B}\right) \tag{3.33}$$

$$F_{qd} = 1 + 2 \tan \phi'(1 - \sin\phi')^2 \tan^{-1}\left(\frac{D_f}{B}\right) \tag{3.34}$$

and

$$F_{\gamma d} = 1 \tag{3.35}$$

respectively. The factor $\tan^{-1}(D_f/B)$ is in radians in Eqs. (3.33) and (3.34).

Inclination Factors   Meyerhof (1963) and Hanna and Meyerhof (1981) suggested the following inclination factors for use in Eq. (3.23):

$$F_{ci} = F_{qi} = \left(1 - \frac{\beta^\circ}{90^\circ}\right)^2 \tag{3.36}$$

$$F_{\gamma i} = \left(1 - \frac{\beta}{\phi'}\right)^2 \tag{3.37}$$

Here, β = inclination of the load on the foundation with respect to the vertical.

Example 3.2

A square foundation ($B \times B$) has to be constructed as shown in Figure 3.7. Assume that $\gamma = 105$ lb/ft^3, $\gamma_{sat} = 118$ lb/ft^3, $D_f = 4$ ft, and $D_1 = 2$ ft. The gross allowable load, Q_{all}, with FS = 3 is 150,000 lb. The standard penetration resistance, N_{60} values are as follows:

Depth (ft)	N_{60} (blow/ft)
5	4
10	6
15	6
20	10
25	5

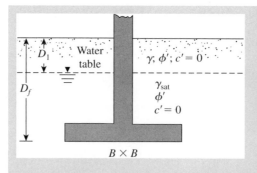

$B \times B$ **Figure 3.7** A square foundation

Determine the size of the footing. Use Eq. (3.23).

Solution
From Eqs. (2.13) and (2.14),

$$(N_1)_{60} = N_{60}\left(\frac{p_a}{\sigma'_o}\right)^{0.5} \tag{a}$$

Combining Eqs. (2.26) and (a) yields

$$\phi' = \left[20N_{60}\left(\frac{p_a}{\sigma'_o}\right)^{0.5}\right]^{0.5} + 20 \tag{b}$$

Thus, $p_a \approx 2000$ lb/ft². Now the following table can be prepared:

Depth (ft)	N_{60}	σ'_o (lb/ft²)	ϕ' (deg) [Eq. (b)]
5	4	$2 \times 105 + 3(118 - 62.4) = 376.8$	33.6
10	6	$376.8 + 5(118 - 62.4) = 654.8$	34.5
15	6	$654.8 + 5(118 - 62.4) = 932.8$	33.3
20	10	$932.8 + 5(118 - 62.4) = 1210.8$	36.0
25	5	$1210.8 + 5(118 - 62.4) = 1488.8$	30.8

Average $\phi' = 33.64° \approx 34°$

Next, we have

$$q_{all} = \frac{Q_{all}}{B^2} = \frac{150,000}{B^2} \text{ lb/ft}^2 \tag{c}$$

From Eq. (3.23) (with $c' = 0$), we obtain

$$q_{all} = \frac{q_u}{FS} = \frac{1}{3}\left(qN_qF_{qs}F_{qd} + \frac{1}{2}\gamma' BN_\gamma F_{\gamma s}F_{\gamma d}\right)$$

For $\phi' = 34°$, from Table 3.3, $N_q = 29.44$ and $N_\gamma = 41.06$. Hence,

$$F_{qs} = 1 + \frac{B}{L}\tan\phi' = 1 + \tan 34 = 1.67$$

$$F_{\gamma s} = 1 - 0.4\left(\frac{B}{L}\right) = 1 - 0.4 = 0.6$$

$$F_{qd} = 1 + 2\tan\phi'(1 - \sin\phi')^2\frac{D_f}{B} = 1 + 2\tan 34\,(1 - \sin 34)^2\frac{4}{B} = 1 + \frac{1.05}{B}$$

$$F_{\gamma d} = 1$$

and

$$q = (2)(105) + 2(118 - 62.4) = 321.2 \text{ lb/ft}^2$$

So

$$q_{all} = \frac{1}{3}\left[(321.2)(29.44)(1.67)\left(1 + \frac{1.05}{B}\right)\right.$$

$$\left. + \left(\frac{1}{2}\right)(118 - 62.4)(B)(41.06)(0.6)(1)\right] \tag{d}$$

$$= 5263.9 + \frac{5527.1}{B} + 228.3B$$

Combining Eqs. (c) and (d) results in

$$\frac{150,000}{B^2} = 5263.9 + \frac{5527.1}{B} + 228.3B$$

By trial and error, we find that $B \approx 4.5$ ft.   ∎

3.7   *Meyerhof's Bearing Capacity, Shape, Depth, and Inclination Factors*

In most solutions, presented in this text, the bearing capacity, shape, depth, and inclination factors presented in Section 3.6 will be used. However, many geotechnical engineers employ the various factors recommended by Meyerhof (1963) for use in Eq. (3.23). Table 3.4 is a summary of those factors.

Table 3.4 Meyerhof's Bearing Capacity, Shape, Depth, and Inclination Factors [Eq. (3.23)]

Factor	Relationship
Bearing capacity	
N_c	Equation (3.25)
N_q	Equation (3.24)
N_γ	$(N_q - 1) \tan (1.4 \phi')$; see Table 3.5
Shape	
For $\phi = 0$,	
$\quad F_{cs}$	$1 + 0.2 (B/L)$
$\quad F_{qs} = F_{\gamma s}$	1
For $\phi' \geqslant 10°$,	
$\quad F_{cs}$	$1 + 0.2 (B/L) \tan^2 (45 + \phi'/2)$
$\quad F_{qs} = F_{\gamma s}$	$1 + 0.1 (B/L) \tan^2 (45 + \phi'/2)$
Depth	
For $\phi = 0$,	
$\quad F_{cd}$	$1 + 0.2 (D_f/B)$
$\quad F_{qd} = F_{\gamma d}$	1
For $\phi' \geqslant 10°$	
$\quad F_{cd}$	$1 + 0.2 (D_f/B) \tan (45 + \phi'/2)$
$\quad F_{qd} = F_{\gamma d}$	$1 + 0.1 (D_f/B) \tan (45 + \phi'/2)$
Inclination	
$F_{ci} = F_{qi}$	Equation (3.36)
$F_{\gamma i}$	Equation (3.37)

Table 3.5 Meyerhof's Bearing Capacity Factor $N_\gamma = (N_q - 1) \tan (1.4 \phi')$

ϕ'	N_γ	ϕ'	N_γ	ϕ'	N_γ	ϕ'	N_γ
0	0.00	14	0.92	28	11.19	42	139.32
1	0.002	15	1.13	29	13.24	43	171.14
2	0.01	16	1.38	30	15.67	44	211.41
3	0.02	17	1.66	31	18.56	45	262.74
4	0.04	18	2.00	32	22.02	46	328.73
5	0.07	19	2.40	33	26.17	47	414.32
6	0.11	20	2.87	34	31.15	48	526.44
7	0.15	21	3.42	35	37.15	49	674.91
8	0.21	22	4.07	36	44.43	50	873.84
9	0.28	23	4.82	37	53.27	51	1143.93
10	0.37	24	5.72	38	64.07	52	1516.05
11	0.47	25	6.77	39	77.33	53	2037.26
12	0.60	26	8.00	40	93.69		
13	0.74	27	9.46	41	113.99		

3.8   Some Comments on Bearing Capacity Factor, N_γ, and Shape Factors

Bearing Capacity Factor, N_γ

At the present time, the bearing capacity factors N_q and N_c given by Eqs. (3.24) and (3.25) are well accepted for bearing capacity calculation. However, there are several relationships for N_γ that can be found in the literature. Some of these relationships are:

Vesic (1973):

$$N_\gamma = 2(N_q + 1)\tan \phi' \qquad \text{[Eq. (3.26)]}$$

Meyerhof (1953):

$$N_\gamma = (N_q - 1)\tan(1.4\,\phi') \quad \text{(see Table 3.5)}$$

Hansen (1970):

$$N_\gamma = 1.5(N_q - 1)\tan \phi'$$

Michalowski (1997):

$$N_\gamma = e^{(0.66 + 5.11 \tan \phi')}\tan \phi'$$

A comparison of these N_γ values for varying friction angles (ϕ') is shown in Figure 3.8. It can be seen from this figure that Vesic's and Michalowski's N_γ values are approximately the same, and they are the upper limits of all suggested relationships. These values are recommended for use in this text.

Shape Factors (F_{cs}, F_{qs}, and $F_{\gamma s}$)

The shape factors proposed by De Beer (1970) are given in Eqs. (3.27) to (3.29). Similarly, the shape factors suggested by Meyerhof are given in Table 3.4. These are conservative relationships acceptable for design purposes. However, it is interesting to compare the relationships for $F_{\gamma s}$. According to De Beer (1970) [see Eq. (3.29)]

$$F_{\gamma s} = 1 - 0.4\left(\frac{B}{L}\right)$$

Similarly, according to Meyerhof (1963) (see Table 3.4)

$$F_{\gamma s} = 1 + 0.1\left(\frac{B}{L}\right)\tan^2\left(45 + \frac{\phi'}{2}\right) \quad \text{(for } \phi' > 10°)$$

The two relationships stated above appear to contradict each other. For a given soil friction angle with an increase in B/L, De Beer's value of $F_{\gamma s}$ decreases, whereas the magnitude of $F_{\gamma s}$ suggested by Meyerhof increases. More recently, Zhu and

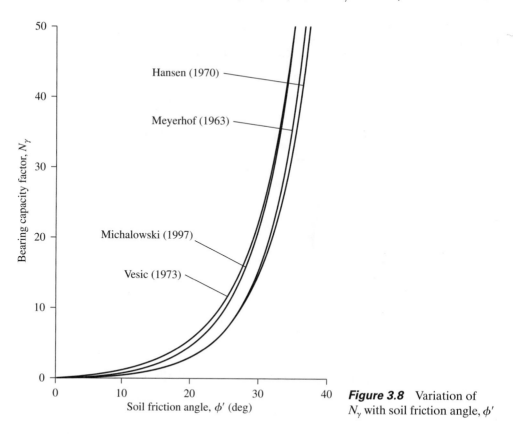

Figure 3.8 Variation of N_γ with soil friction angle, ϕ'

Michalowski (2005) evaluated the shape factors based on the elastoplastic model of soil and finite element analysis. They are as follows:

$$F_{cs} = 1 + (1.8 \tan^2 \phi' + 0.1)\left(\frac{B}{L}\right)^{0.5} \tag{3.38}$$

$$F_{qs} = 1 + 1.9 \tan^2 \phi'\left(\frac{B}{L}\right)^{0.5} \tag{3.39}$$

$$F_{\gamma s} = 1 + (0.6 \tan^2 \phi' - 0.25)\left(\frac{B}{L}\right) \quad (\text{for } \phi' \leq 30°) \tag{3.40}$$

and

$$F_{\gamma s} = 1 + (1.3 \tan^2 \phi' - 0.5)\left(\frac{L}{B}\right)^{1.5} e^{-(L/B)} (\text{for } \phi' > 30°) \tag{3.41}$$

Equations (3.38) through (3.41) have been derived based on sound theoretical background and may be used for bearing capacity calculation.

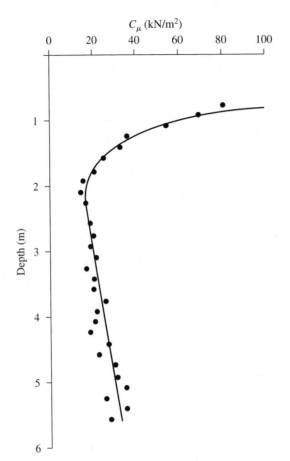

Figure 3.9   Variation of c_u with depth obtained from field vane shear test

A Case History for Bearing Capacity Failure

An excellent case of bearing capacity failure of a 6-m (20-ft) diameter concrete silo was provided by Bozozuk (1972). The concrete tower silo was 21 m (70 ft) high and was constructed over soft clay on a ring foundation. Figure 3.9 shows the variation of the undrained shear strength (c_u) obtained from field vane shear tests at the site. The groundwater table was located at about 0.6 m (2 ft) below the ground surface.

On September 30, 1970, just after it was filled to capacity for the first time with corn silage, the concrete tower silo suddenly overturned due to bearing capacity failure. Figure 3.10 shows the approximate profile of the failure surface in soil. The failure surface extended to about 7 m (23 ft) below the ground surface. Bozozuk (1972) provided the following average parameters for the soil in the failure zone and the foundation:

- Load per unit area on the foundation when failure occurred ≈ 160 kN/m^2
- Average plasticity index of clay (PI) ≈ 36

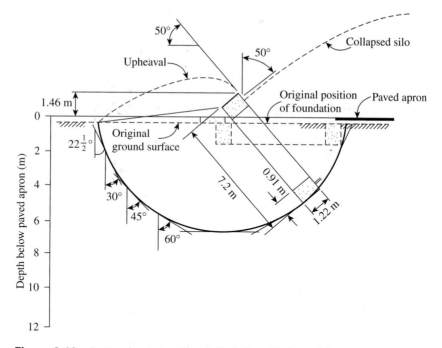

Figure 3.10 Approximate profile of silo failure (Adapted from Bozozuk, 1972)

- Average undrained shear strength (c_u) from 0.6 to 7 m depth obtained from field vane shear tests ≈ 27.1 kN/m^2
- From Figure 3.10, $B \approx 7.2$ m and $D_f \approx 1.52$ m.

We can now calculate the factor of safety against bearing capacity failure. From Eq. (3.23)

$$q_u = c' N_c F_{cs} F_{cd} F_{ci} + q N_c F_{qs} F_{qd} F_{qi} + \tfrac{1}{2} \gamma B \, N_\gamma F_{\gamma s} F_{\gamma d} F_{\gamma i}$$

For $\phi = 0$ condition and vertical loading, $c' = c_u$, $N_c = 5.14$, $N_q = 1$, $N_\gamma = 0$, and $F_{ci} = F_{qi} = F_{\gamma i} = 0$. Also, from Eqs. (3.27), (3.28), (3.30), and (3.31),

$$F_{cs} = 1 + \left(\frac{7.2}{7.2}\right)\left(\frac{1}{5.14}\right) = 1.195$$

$$F_{qs} = 1$$

$$F_{cd} = 1 + (0.4)\left(\frac{1.52}{7.2}\right) = 1.08$$

$$F_{qd} = 1$$

Thus,

$$q_u = (c_u)(5.14)(1.195)(1.08)(1) + (\gamma)(1.52)$$

Assuming $\gamma \approx 18$ kN/m^3

$$q_u = 6.63 c_u + 27.36 \tag{3.42}$$

According to Eqs. (2.34) and (2.35),

$$c_{u(\text{corrected})} = \lambda c_{u(\text{VST})}$$

$$\lambda = 1.7 - 0.54 \log [\text{PI}(\%)]$$

For this case, $\text{PI} \approx 36$ and $c_{u(\text{VST})} = 27.1 \text{ kN/m}^2$. So

$$c_{u(\text{corrected})} = \{1.7 - 0.54 \log [\text{PI}(\%)]\}c_{u(\text{VST})}$$

$$= (1.7 - 0.54 \log 36)(27.1) \approx 23.3 \text{ kN/m}^2$$

Substituting this value of c_u in Eq. (3.42)

$$q_u = (6.63)(23.3) + 27.36 = 181.8 \text{ kN/m}^2$$

The factor of safety against bearing capacity failure

$$\text{FS} = \frac{q_u}{\text{applied load per unit area}} = \frac{181.8}{160} = 1.14$$

This factor of safety is too low and approximately equals one, for which the failure occurred.

3.10   *Effect of Soil Compressibility*

In Section 3.3, Eqs. (3.3), (3.7), and (3.8), which apply to the case of general shear failure, were modified to Eqs. (3.9), (3.10), and (3.11) to take into account the change of failure mode in soil (i.e., local shear failure). The change of failure mode is due to soil compressibility, to account for which Vesic (1973) proposed the following modification of Eq. (3.23):

$$q_u = c'N_c F_{cs}F_{cd}F_{cc} + qN_q F_{qs}F_{qd}F_{qc} + \tfrac{1}{2}\gamma BN_\gamma F_{\gamma s}F_{\gamma d}F_{\gamma c} \qquad (3.43)$$

In this equation, F_{cc}, F_{qc}, and $F_{\gamma c}$ are soil compressibility factors.

The soil compressibility factors were derived by Vesic (1973) by analogy to the expansion of cavities. According to that theory, in order to calculate F_{cc}, F_{qc}, and $F_{\gamma c}$, the following steps should be taken:

Step 1.   Calculate the *rigidity index, I_r,* of the soil at a depth approximately $B/2$ below the bottom of the foundation, or

$$I_r = \frac{G_s}{c' + q' \tan \phi'} \qquad (3.44)$$

where

$G_s =$ shear modulus of the soil
$q =$ effective overburden pressure at a depth of $D_f + B/2$

Step 2.   The critical rigidity index, $I_{r(\text{cr})}$, can be expressed as

$$I_{r(\text{cr})} = \frac{1}{2}\left\{ \exp\left[\left(3.30 - 0.45\frac{B}{L}\right) \cot\left(45 - \frac{\phi'}{2}\right)\right]\right\} \qquad (3.45)$$

Table 3.6 Variation of $I_{r(cr)}$ with ϕ' and B/L[1]

ϕ' (deg)	$I_{r(cr)}$	
	$\dfrac{B}{L} = 0$	$\dfrac{B}{L} = 1$
0	13	8
5	18	11
10	25	15
15	37	20
20	55	30
25	89	44
30	152	70
35	283	120
40	592	225
45	1442	482
50	4330	1258

[1] After Vesic (1973)

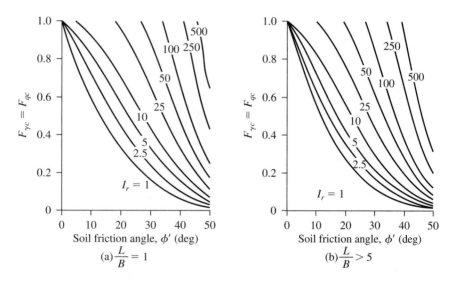

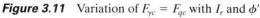

Figure 3.11 Variation of $F_{\gamma c} = F_{qc}$ with I_r and ϕ'

The variations of $I_{r(cr)}$ for $B/L = 0$ and $B/L = 1$ are given in Table 3.6.

Step 3. If $I_r \geqslant I_{r(cr)}$, then

$$F_{cc} = F_{qc} = F_{\gamma c} = 1$$

However, if $I_r < I_{r(cr)}$, then

$$F_{\gamma c} = F_{qc} = \exp\left\{\left(-4.4 + 0.6\,\frac{B}{L}\right)\tan\phi' + \left[\frac{(3.07\sin\phi')(\log 2I_r)}{1 + \sin\phi'}\right]\right\} \quad (3.46)$$

Figure 3.11 shows the variation of $F_{\gamma c} = F_{qc}$ [see Eq. (3.46)] with ϕ' and I_r. For $\phi = 0$,

$$F_{cc} = 0.32 + 0.12\frac{B}{L} + 0.60 \log I_r \qquad (3.47)$$

For $\phi' > 0$,

$$F_{cc} = F_{qc} - \frac{1 - F_{qc}}{N_q \tan \phi'} \qquad (3.48)$$

Example 3.3

For a shallow foundation, $B = 0.6$ m, $L = 1.2$ m, and $D_f = 0.6$ m. The known soil characteristics are as follows:

Soil:

$\phi' = 25°$
$c' = 48$ kN/m^2
$\gamma = 18$ kN/m^3
Modulus of elasticity, $E_s = 620$ kN/m^2
Poisson's ratio, $\mu_s = 0.3$

Calculate the ultimate bearing capacity.

Solution
From Eq. (3.44),

$$I_r = \frac{G_s}{c' + q' \tan \phi'}$$

However,

$$G_s = \frac{E_s}{2(1 + \mu_s)}$$

So

$$I_r = \frac{E_s}{2(1 + \mu_s)[c' + q' \tan \phi']}$$

Now,

$$q' = \gamma\left(D_f + \frac{B}{2}\right) = 18\left(0.6 + \frac{0.6}{2}\right) = 16.2 \text{ kN/m}^2$$

Thus,

$$I_r = \frac{620}{2(1 + 0.3)[48 + 16.2 \tan 25]} = 4.29$$

From Eq. (3.45),

$$I_{r(\text{cr})} = \frac{1}{2}\left\{\exp\left[\left(3.3 - 0.45\,\frac{B}{L}\right)\cot\left(45 - \frac{\phi'}{2}\right)\right]\right\}$$

$$= \frac{1}{2}\left\{\exp\left[\left(3.3 - 0.45\,\frac{0.6}{1.2}\right)\cot\left(45 - \frac{25}{2}\right)\right]\right\} = 62.41$$

Since $I_{r(\text{cr})} > I_r$, we use Eqs. (3.46) and (3.48) to obtain

$$F_{\gamma c} = F_{qc} = \exp\left\{\left(-4.4 + 0.6\,\frac{B}{L}\right)\tan\phi' + \left[\frac{(3.07\sin\phi')\log(2I_r)}{1 + \sin\phi'}\right]\right\}$$

$$= \exp\left\{\left(-4.4 + 0.6\,\frac{0.6}{1.2}\right)\tan 25\right.$$

$$\left. + \left[\frac{(3.07\sin 25)\log(2 \times 4.29)}{1 + \sin 25}\right]\right\} = 0.347$$

and

$$F_{cc} = F_{qc} - \frac{1 - F_{qc}}{N_q\tan\phi'}$$

For $\phi' = 25°$, $N_q = 10.66$ (see Table 3.3); therefore,

$$F_{cc} = 0.347 - \frac{1 - 0.347}{10.66\tan 25} = 0.216$$

Now, from Eq. (3.43),

$$q_u = c'N_cF_{cs}F_{cd}F_{cc} + qN_qF_{qs}F_{qd}F_{qc} + \tfrac{1}{2}\gamma BN_\gamma F_{\gamma s}F_{\gamma d}F_{\gamma c}$$

From Table 3.3, for $\phi' = 25°$, $N_c = 20.72$, $N_q = 10.66$, and $N_\gamma = 10.88$. Consequently,

$$F_{cs} = 1 + \left(\frac{N_q}{N_c}\right)\left(\frac{B}{L}\right) = 1 + \left(\frac{10.66}{20.72}\right)\left(\frac{0.6}{1.2}\right) = 1.257$$

$$F_{qs} = 1 + \frac{B}{L}\tan\phi' = 1 + \frac{0.6}{1.2}\tan 25 = 1.233$$

$$F_{\gamma s} = 1 - 0.4\,\frac{B}{L} = 1 - 0.4\,\frac{0.6}{1.2} = 0.8$$

$$F_{cd} = 1 + 0.4\left(\frac{D_f}{B}\right) = 1 + 0.4\left(\frac{0.6}{0.6}\right) = 1.4$$

$$F_{qd} = 1 + 2\tan\phi'(1 - \sin\phi')^2\left(\frac{D_f}{B}\right)$$

$$= 1 + 2\tan 25\,(1 - \sin 25)^2\left(\frac{0.6}{0.6}\right) = 1.311$$

and

$$F_{\gamma d} = 1$$

Thus,

$$q_u = (48)(20.72)(1.257)(1.4)(0.216) + (0.6 \times 18)(10.66)(1.233)(1.311)$$
$$(0.347) + (\tfrac{1}{2})(18)(0.6)(10.88)(0.8)(1)(0.347) = \mathbf{459 \ kN/m^2} \qquad \blacksquare$$

3.11   *Eccentrically Loaded Foundations*

In several instances, as with the base of a retaining wall, foundations are subjected to moments in addition to the vertical load, as shown in Figure 3.12a. In such cases, the distribution of pressure by the foundation on the soil is not uniform. The nominal distribution of pressure is

$$q_{max} = \frac{Q}{BL} + \frac{6M}{B^2 L} \qquad (3.49)$$

and

$$q_{min} = \frac{Q}{BL} - \frac{6M}{B^2 L} \qquad (3.50)$$

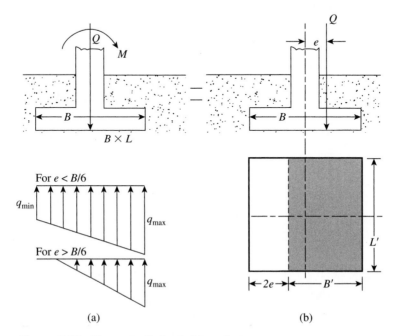

(a)   (b)

Figure 3.12   Eccentrically loaded foundations

where

Q = total vertical load
M = moment on the foundation

Figure 3.12b shows a force system equivalent to that shown in Figure 3.12a. The distance

$$e = \frac{M}{Q} \tag{3.51}$$

is the eccentricity. Substituting Eq. (3.51) into Eqs. (3.49) and (3.50) gives

$$q_{max} = \frac{Q}{BL}\left(1 + \frac{6e}{B}\right) \tag{3.52}$$

and

$$q_{min} = \frac{Q}{BL}\left(1 - \frac{6e}{B}\right) \tag{3.53}$$

Note that, in these equations, when the eccentricity e becomes $B/6$, q_{min} is zero. For $e > B/6$, q_{min} will be negative, which means that tension will develop. Because soil cannot take any tension, there will then be a separation between the foundation and the soil underlying it. The nature of the pressure distribution on the soil will be as shown in Figure 3.12a. The value of q_{max} is then

$$q_{max} = \frac{4Q}{3L(B - 2e)} \tag{3.54}$$

The exact distribution of pressure is difficult to estimate.

Figure 3.13 shows the nature of failure surface in soil for a surface strip foundation subjected to an eccentric load. The factor of safety for such type of loading against bearing capacity failure can be evaluated as

$$FS = \frac{Q_{ult}}{Q} \tag{3.55}$$

where Q_{ult} = ultimate load-carrying capacity.

The following section describes several theories for determining Q_{ult}.

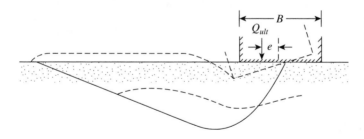

Figure 3.13 Nature of failure surface in soil supporting a strip foundation subjected to eccentric loading (*Note: $D_f = 0$; Q_{ult} is ultimate load per unit length of foundation)

3.12   ## Ultimate Bearing Capacity under Eccentric Loading—Meyerhof's Theory

In 1953, Meyerhof proposed a theory that is generally referred to as the *effective area method*.

The following is a step-by-step procedure for determining the ultimate load that the soil can support and the factor of safety against bearing capacity failure:

Step 1.   Determine the effective dimensions of the foundation (Figure 3.12b):

$$B' = \text{effective width} = B - 2e$$

$$L' = \text{effective length} = L$$

Note that if the eccentricity were in the direction of the length of the foundation, the value of L' would be equal to $L - 2e$. The value of B' would equal B. The smaller of the two dimensions (i.e., L' and B') is the effective width of the foundation.

Step 2.   Use Eq. (3.23) for the ultimate bearing capacity:

$$q'_u = c'N_cF_{cs}F_{cd}F_{ci} + qN_qF_{qs}F_{qd}F_{qi} + \tfrac{1}{2}\gamma B'N_\gamma F_{\gamma s}F_{\gamma d}F_{\gamma i} \tag{3.56}$$

To evaluate F_{cs}, F_{qs}, and $F_{\gamma s}$, use Eqs. (3.27) through (3.29) with *effective length* and *effective width* dimensions instead of L and B, respectively. To determine F_{cd}, F_{qd}, and $F_{\gamma d}$, use Eqs. (3.30) through (3.35). Do not replace B with B'.

Step 3.   The total ultimate load that the foundation can sustain is

$$Q_{\text{ult}} = q'_u \overbrace{(B')\,(L')}^{A'} \tag{3.57}$$

where A' = effective area.

Step 4.   The factor of safety against bearing capacity failure is

$$\text{FS} = \frac{Q_{\text{ult}}}{Q} \tag{}$$

Step 5.   Check the factor of safety against $q_{\max}$, or $\text{FS} = q'_u/q_{\max}$.

Foundations with Two-Way Eccentricity

Consider a situation in which a foundation is subjected to a vertical ultimate load Q_{ult} and a moment M, as shown in Figures 3.14a and b. For this case, the components of the moment M about the x- and y-axes can be determined as M_x and M_y, respectively. (See Figure 3.14.) This condition is equivalent to a load Q_{ult} placed eccentrically on the foundation with $x = e_B$ and $y = e_L$ (Figure 3.14d). Note that

$$e_B = \frac{M_y}{Q_{\text{ult}}} \tag{3.58}$$

and

$$e_L = \frac{M_x}{Q_{\text{ult}}} \tag{3.59}$$

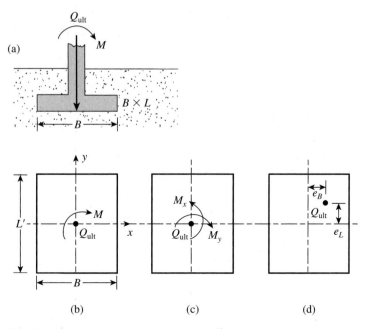

Figure 3.14   Analysis of foundation with two-way eccentricity

If Q_{ult} is needed, it can be obtained from Eq. (3.57); that is,

$$Q_{\text{ult}} = q'_u A'$$

where, from Eq. (3.56),

$$q'_u = c' N_c F_{cs} F_{cd} F_{ci} + q N_q F_{qs} F_{qd} F_{qi} + \tfrac{1}{2} \gamma B' N_\gamma F_{\gamma s} F_{\gamma d} F_{\gamma i}$$

and

$$A' = \text{effective area} = B'L'$$

As before, to evaluate F_{cs}, F_{qs}, and $F_{\gamma s}$ [Eqs. (3.27) through (3.29)], we use the effective length L' and effective width B' instead of L and B, respectively. To calculate F_{cd}, F_{qd}, and $F_{\gamma d}$, we use Eqs. (3.30) through (3.35); however, we do not replace B with B'. In determining the effective area A', effective width B', and effective length L', five possible cases may arise (Highter and Anders, 1985).

Case I. $e_L/L \geqslant \frac{1}{6}$ and $e_B/B \geqslant \frac{1}{6}$. The effective area for this condition is shown in Figure 3.15, or

$$A' = \tfrac{1}{2} B_1 L_1 \tag{3.60}$$

where

$$B_1 = B\left(1.5 - \frac{3e_B}{B} \right) \tag{3.61}$$

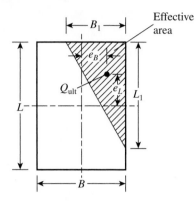

Figure 3.15 Effective area for the case of $e_L/L \geq \frac{1}{6}$ and $e_B/B \geq \frac{1}{6}$

and

$$L_1 = L\left(1.5 - \frac{3e_L}{L}\right) \tag{3.62}$$

The effective length L' is the larger of the two dimensions B_1 and L_1. So the effective width is

$$B' = \frac{A'}{L'} \tag{3.63}$$

Case II. $e_L/L < 0.5$ and $0 < e_B/B < \frac{1}{6}$. The effective area for this case, shown in Figure 3.16a, is

$$A' = \frac{1}{2}(L_1 + L_2)B \tag{3.64}$$

The magnitudes of L_1 and L_2 can be determined from Figure 3.16b. The effective width is

$$B' = \frac{A'}{L_1 \text{ or } L_2 \quad (\text{whichever is larger})} \tag{3.65}$$

The effective length is

$$L' = L_1 \text{ or } L_2 \quad (\text{whichever is larger}) \tag{3.66}$$

Case III. $e_L/L < \frac{1}{6}$ and $0 < e_B/B < 0.5$. The effective area, shown in Figure 3.17a, is

$$A' = \frac{1}{2}(B_1 + B_2)L \tag{3.67}$$

The effective width is

$$B' = \frac{A'}{L} \tag{3.68}$$

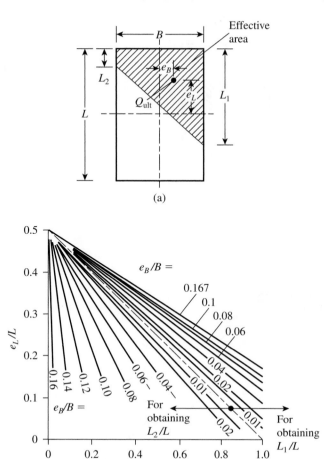

Figure 3.16 Effective area for the case of $e_L/L < 0.5$ and $0 < e_B/B < \frac{1}{6}$ (After Highter and Anders, 1985)

The effective length is

$$L' = L \tag{3.69}$$

The magnitudes of B_1 and B_2 can be determined from Figure 3.17b.

Case IV. $e_L/L < \frac{1}{6}$ and $e_B/B < \frac{1}{6}$. Figure 3.18a shows the effective area for this case. The ratio B_2/B, and thus B_2, can be determined by using the e_L/L curves that slope upward. Similarly, the ratio L_2/L, and thus L_2, can be determined by using the e_L/L curves that slope downward. The effective area is then

$$A' = L_2 B + \frac{1}{2}(B + B_2)(L - L_2) \tag{3.70}$$

The effective width is

$$B' = \frac{A'}{L} \tag{3.71}$$

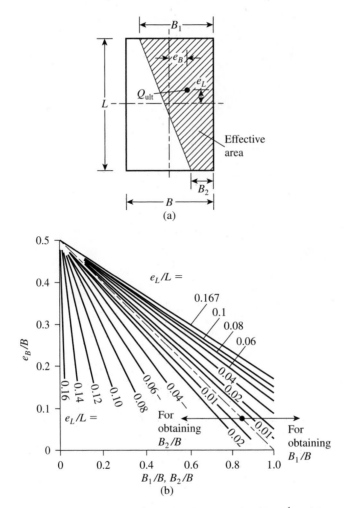

Figure 3.17 Effective area for the case of $e_L/L < \frac{1}{6}$ and $0 < e_B/B < 0.5$ (After Highter and Anders, 1985)

The effective length is

$$L' = L \qquad (3.72)$$

Case V. (Circular Foundation) In the case of circular foundations under eccentric loading (Figure 3.19a), the eccentricity is always one way. The effective area A' and the effective width B' for a circular foundation are given in a nondimensional form in Table 3.7. Once A' and B' are determined, the effective length can be obtained as

$$L' = \frac{A'}{B'}$$

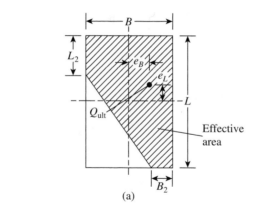

(a)

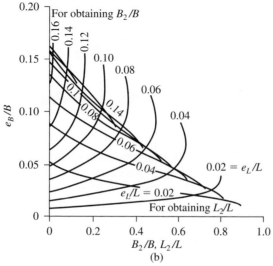

(b)

Figure 3.18 Effective area for the case of $e_L/L < \frac{1}{6}$ and $e_B/B < \frac{1}{6}$ (After Highter and Anders, 1985)

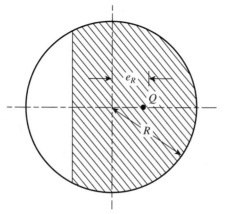

Figure 3.19 Effective area for circular foundation

Table 3.7   Variation of A'/R^2 and B'/R with e_R/R for Circular Foundations

e_R'/R	A'/R^2	B'/R
0.1	2.8	1.85
0.2	2.4	1.32
0.3	2.0	1.2
0.4	1.61	0.80
0.5	1.23	0.67
0.6	0.93	0.50
0.7	0.62	0.37
0.8	0.35	0.23
0.9	0.12	0.12
1.0	0	0

3.13   Eccentrically Loaded Foundation—Prakash and Saran's Theory

Prakash and Saran (1971) analyzed the problem of ultimate bearing capacity of eccentrically and vertically loaded continuous (strip) foundations by using the one-sided failure surface in soil, as shown in Figure 3.13. According to this theory, the ultimate load *per unit length of a continuous foundation* can be estimated as

$$Q_{ult} = B\left[c'N_{c(e)} + qN_{q(e)} + \frac{1}{2}\gamma B N_{\gamma(e)}\right] \qquad (3.73)$$

where $N_{c(e)}$, $N_{q(e)}$, $N_{\gamma(e)}$ = bearing capacity factors under eccentric loading.

The variations of $N_{c(e)}$, $N_{q(e)}$, and $N_{\gamma(e)}$ with soil friction angle ϕ' are given in Figures 3.20, 3.21, and 3.22. For rectangular foundations, the ultimate load can be given as

$$Q_{ult} = BL\left[c'N_{c(e)}F_{cs(e)} + qN_{q(e)}F_{qs(e)} + \frac{1}{2}\gamma B N_{\gamma(e)}F_{\gamma s(e)}\right] \qquad (3.74)$$

where $F_{cs(e)}$, $F_{qs(e)}$, and $F_{\gamma s(e)}$ = shape factors.

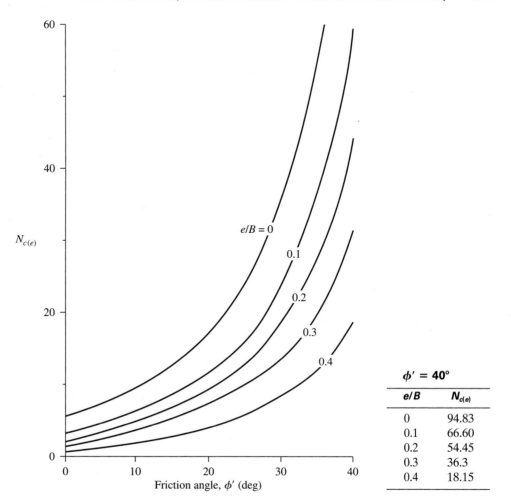

Figure 3.20   Variation of $N_{c(e)}$ with ϕ'

Prakash and Saran (1971) also recommended the following for the shape factors:

$$F_{cs(e)} = 1.2 - 0.025\frac{L}{B} \quad \text{(with a minimum of 1.0)} \tag{3.75}$$

$$F_{qs(e)} = 1 \tag{3.76}$$

and

$$F_{\gamma s(e)} = 1.0 + \left(\frac{2e}{B} - 0.68\right)\frac{B}{L} + \left[0.43 - \left(\frac{3}{2}\right)\left(\frac{e}{B}\right)\right]\left(\frac{B}{L}\right)^2 \tag{3.77}$$

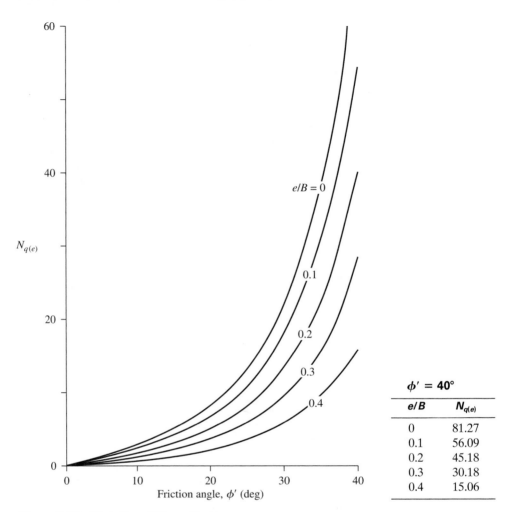

Figure 3.21   Variation of $N_{q(e)}$ with ϕ'

ϕ' = 40°	
e/B	$N_{q(e)}$
0	81.27
0.1	56.09
0.2	45.18
0.3	30.18
0.4	15.06

Example 3.4

A continuous foundation is shown in Figure 3.23. If the load eccentricity is 0.5 ft, determine the ultimate load, Q_{ult}, per unit length of the foundation. Use Meyerhof's effective area method.

Solution
For $c' = 0$, Eq. (3.56) gives

$$q'_u = qN_qF_{qs}F_{qd}F_{qi} + \frac{1}{2}\gamma B'N_\gamma F_{\gamma s}F_{\gamma d}F_{\gamma i}$$

where $q = (110)(4) = 440 \text{ lb/ft}^2$.

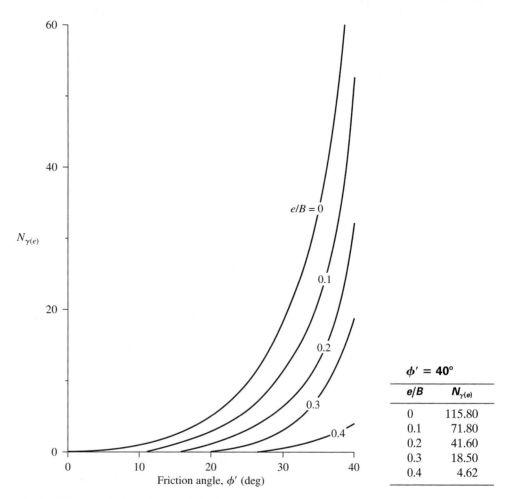

$\phi' = 40°$

e/B	$N_{\gamma(e)}$
0	115.80
0.1	71.80
0.2	41.60
0.3	18.50
0.4	4.62

Figure 3.22 Variation of $N_{\gamma(e)}$ with ϕ'

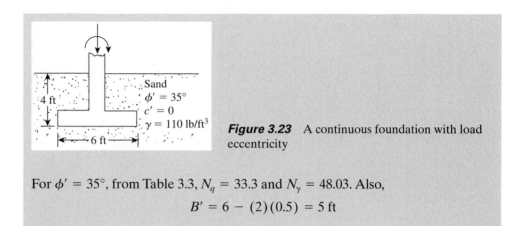

Figure 3.23 A continuous foundation with load eccentricity

For $\phi' = 35°$, from Table 3.3, $N_q = 33.3$ and $N_\gamma = 48.03$. Also,

$$B' = 6 - (2)(0.5) = 5 \text{ ft}$$

Because the foundation in question is a strip foundation, B'/L' is zero. Hence, $F_{qs} = 1, F_{\gamma s} = 1$,

$$F_{qi} = F_{\gamma i} = 1$$

$$F_{qd} = 1 + 2\tan\phi'(1 - \sin\phi')^2\,\frac{D_f}{B} = 1 + 0.255\left(\frac{4}{6}\right) = 1.17$$

$$F_{\gamma d} = 1$$

and

$$q_u' = (440)(33.3)(1)(1.17)(1) + \left(\frac{1}{2}\right)(110)(5)(48.03)(1)(1)(1) = 30{,}351\ \text{lb/ft}^2$$

Consequently,

$$Q_{\text{ult}} = (B')(1)(q_u') = (5)(1)(30{,}351) = 151{,}755\ \text{lb/ft} = \textbf{75.88 ton/ft} \qquad \blacksquare$$

Example 3.5

Solve Example 3.4 using Eq. (3.73).

Solution
Since $c' = 0$

$$Q_{\text{ult}} = B\left[qN_{q(e)} + \frac{1}{2}\gamma B N_{\gamma(e)}\right]$$

$$\frac{e}{B} = \frac{0.5}{6} = 0.083$$

For $\phi' = 35°$ and $e/B = 0.083$, Figures 3.21 and 3.22 give $N_{q(e)} = 33.4$ and $N_{\gamma(e)} \approx 26.8$. Hence,

$$Q_{\text{ult}} = 6[(440)(33.4) + (^1/_2)(110)(6)(26.8)] = 141{,}240\ \text{lb/ft} = \textbf{70.62 ton/ft}$$

Note: The ultimate load of 70.62 ton/ft is about 6% lower than that obtained using Meyerhof's method. This is due to the conservative assumption of depth factors to be unity. $\qquad \blacksquare$

Example 3.6

A square foundation is shown in Figure 3.24, with $e_L = 0.3$ m and $e_B = 0.15$ m. Assume two-way eccentricity, and determine the ultimate load, Q_{ult}.

Solution
We have

$$\frac{e_L}{L} = \frac{0.3}{1.5} = 0.2$$

and

$$\frac{e_B}{B} = \frac{0.15}{1.5} = 0.1$$

This case is similar to that shown in Figure 3.16a. From Figure 3.16b, for $e_L/L = 0.2$ and $e_B/B = 0.1$,

$$\frac{L_1}{L} \approx 0.85; \qquad L_1 = (0.85)(1.5) = 1.275 \text{ m}$$

and

$$\frac{L_2}{L} \approx 0.21; \qquad L_2 = (0.21)(1.5) = 0.315 \text{ m}$$

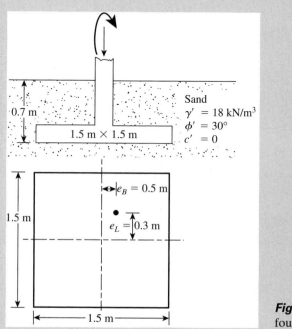

Sand
$\gamma' = 18$ kN/m³
$\phi' = 30°$
$c' = 0$

0.7 m

1.5 m × 1.5 m

$e_B = 0.5$ m

1.5 m

$e_L = 0.3$ m

1.5 m

Figure 3.24 An eccentrically loaded foundation

From Eq. (3.64),

$$A' = \tfrac{1}{2}(L_1 + L_2)B = \tfrac{1}{2}(1.275 + 0.315)(1.5) = 1.193 \text{ m}^2$$

From Eq. (3.66),

$$L' = L_1 = 1.275 \text{ m}$$

From Eq. (3.65),

$$B' = \frac{A'}{L'} = \frac{1.193}{1.275} = 0.936 \text{ m}$$

Note from Eq. (3.56) with $c' = 0$,

$$q'_u = qN_qF_{qs}F_{qd}F_{qi} + \tfrac{1}{2}\gamma B'N_\gamma F_{\gamma s}F_{\gamma d}F_{\gamma i}$$

where $q = (0.7)(18) = 12.6 \text{ kN/m}^2$.
For $\phi' = 30°$, from Table 3.3, $N_q = 18.4$ and $N_\gamma = 22.4$. Thus,

$$F_{qs} = 1 + \left(\frac{B'}{L'}\right)\tan\phi' = 1 + \left(\frac{0.936}{1.275}\right)\tan 30° = 1.424$$

$$F_{\gamma s} = 1 - 0.4\left(\frac{B'}{L'}\right) = 1 - 0.4\left(\frac{0.936}{1.275}\right) = 0.706$$

$$F_{qd} = 1 + 2\tan\phi'(1 - \sin\phi')^2\frac{D_f}{B} = 1 + \frac{(0.289)(0.7)}{1.5} = 1.135$$

and

$$F_{\gamma d} = 1$$

So

$$\begin{aligned}
Q_{\text{ult}} = A'q'_u &= A'(qN_qF_{qs}F_{qd} + \tfrac{1}{2}\gamma B'N_\gamma F_{\gamma s}F_{\gamma d}) \\
&= (1.193)[(12.6)(18.4)(1.424)(1.135) \\
&\quad + (0.5)(18)(0.936)(22.4)(0.706)(1)] \approx 606 \text{ kN} \quad\blacksquare
\end{aligned}$$

Example 3.7

A square foundation is shown in Figure 3.25. Assume that the one-way load eccentricity $e = 0.15$ m. Determine the ultimate load, Q_{ult}. Use Eq. (3.74).

Solution
Given $c' = 0$. So

$$Q_{\text{ult}} = BL\left[qN_{q(e)}F_{qs(e)} + \frac{1}{2}\gamma BN_{\gamma(e)}F_{\gamma s(e)}\right]$$

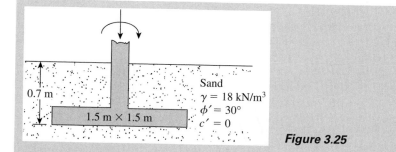

Figure 3.25

$B = L = 1.5$; $e/B = 0.15/1.5 = 0.1$; $\phi' = 30°$. From Figures 3.21 and 3.22, $N_{q(e)} = 18.4$ and $N_{\gamma(e)} = 11.58$. Also, $F_{qs(e)} = 1$. Thus,

$$F_{\gamma s(e)} = 1 + \left(\frac{2e}{B} - 0.68\right)\left(\frac{B}{L}\right) + \left[0.43 - \left(\frac{3}{2}\right)\left(\frac{e}{B}\right)\right]\left(\frac{B}{L}\right)^2$$

$$= 1 + [(2)(0.1) - (0.68)](1) + [(0.43) - (1.5)(0.1)](1)^2$$

$$= 0.8$$

Hence,

$$Q_{\text{ult}} = (1.5 \times 1.5)\left[(0.7 \times 18)(18.4)(1) + \frac{1}{2}(18)(1.5)(11.58)(0.8)\right]$$

$$= \textbf{803 kN}$$

Note: If this had been solved using Eq. (3.56), Q_{ult} would have been 988 kN. ∎

3.14   Bearing Capacity of a Continuous Foundation Subjected to Eccentric Inclined Loading

The problem of ultimate bearing capacity of a *continuous foundation* subjected to an eccentric inclined load was studied by Saran and Agarwal (1991). If a continuous foundation is located at a depth D_f below the ground surface and is subjected to an eccentric load (load eccentricity = e) inclined at an angle β to the vertical, the ultimate capacity can be expressed as

$$Q_{\text{ult}} = B\left[c'N_{c(ei)} + qN_{q(ei)} + \frac{1}{2}\gamma B N_{\gamma(ei)}\right] \tag{3.78}$$

where $N_{c(ei)}$, $N_{q(ei)}$, and $N_{\gamma(ei)}$ = bearing capacity factors

$$q = \gamma D_f$$

The variations of the bearing capacity factors with e/B, ϕ', and β derived by Saran and Agarwal are given in Figures 3.26, 3.27, and 3.28.

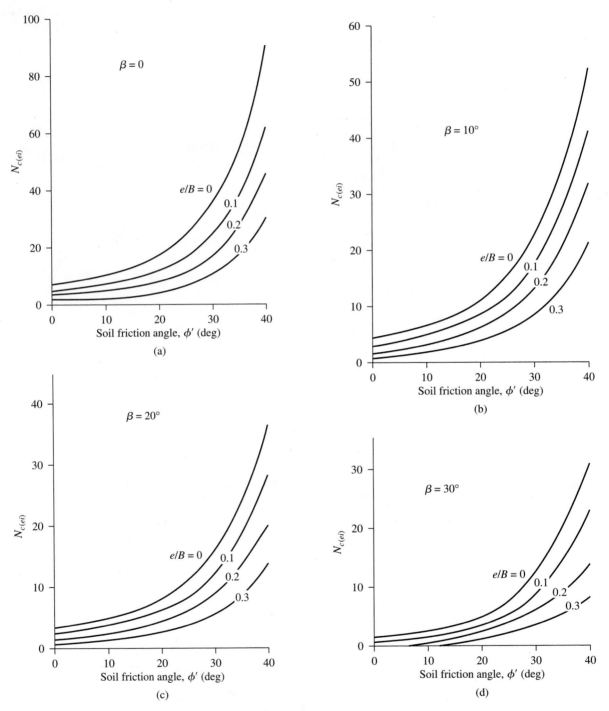

Figure 3.26   Variation of $N_{c(ei)}$ with ϕ', e/B, and β

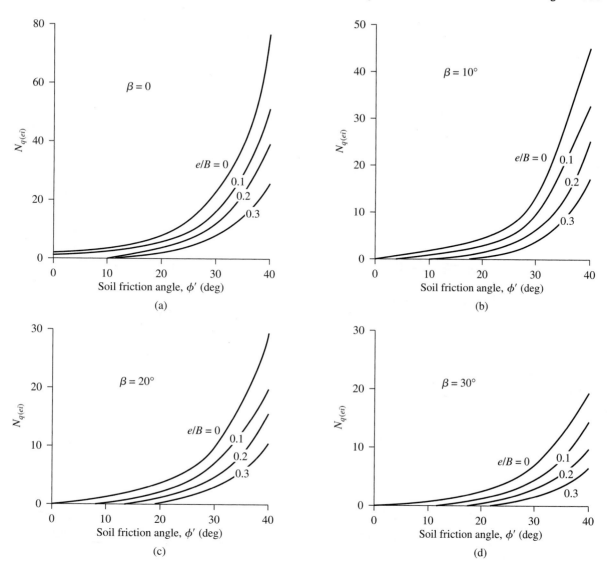

Figure 3.27 Variation of $N_{q(ei)}$ with ϕ', e/B, and β

Example 3.8

A continuous foundation is shown in Figure 3.29. Estimate the ultimate load, Q_{ult}.

Solution
With $c' = 0$, from Eq. (3.78),

$$Q_{ult} = \left[qN_{q(ei)} + \frac{1}{2}\gamma BN_{\gamma(ei)} \right]$$

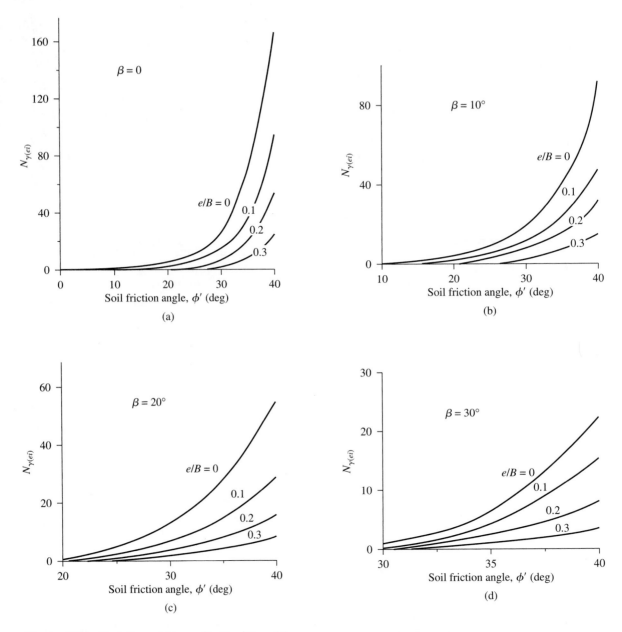

Figure 3.28 Variation of $N_{\gamma(ei)}$ with ϕ', e/B, and β

$B = 1.5$ m, $q = D_f\gamma = (1)(16) = 16$ kN/m², $e/B = 0.15/1.5 = 0.1$, and $\beta = 20°$. From Figures 3.27(c) and 3.28(c), $N_{q(ei)} = 14.2$ and $N_{\gamma(ei)} = 20$. Hence

$$Q_{ult} = (1.5)[(16)(14.2) + (\tfrac{1}{2})(16)(1.5)(20)] = \mathbf{700.8 \ kN/m}$$

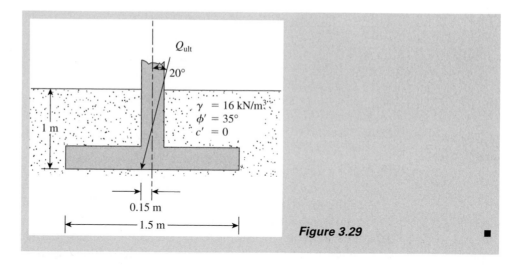

Figure 3.29

Problems

3.1 For the following cases, determine the allowable gross vertical load-bearing capacity of the foundation. Use Terzaghi's equation and assume general shear failure in soil. Use FS = 4.

Part	B	D_f	ϕ'	c'	γ	Foundation type
a.	3 ft	3 ft	28°	400 lb/ft²	110 lb/ft³	Continuous
b.	1.5 m	1.2 m	35°	0	17.8 kN/m³	Continuous
c.	3 m	2 m	30°	0	16.5 kN/m³	Square

3.2 A square column foundation has to carry a gross allowable load of 1805 kN (FS = 3). Given: D_f = 1.5 m, γ = 15.9 kN/m³, ϕ' = 34°, and c' = 0. Use Terzaghi's equation to determine the size of the foundation (B). Assume general shear failure.

3.3 Use the general bearing capacity equation [Eq. (3.23)] to solve the following:
   a. Problem 3.1a
   b. Problem 3.1b
   c. Problem 3.1c
   Use bearing capacity, shape, and depth factors given in Section 3.6.

3.4 The applied load on a shallow square foundation makes an angle of 15° with the vertical. Given: B = 5.5 ft, D_f = 4 ft, γ = 107 lb/ft³, ϕ' = 25°, and c' = 350 lb/ft². Use FS = 4 and determine the gross allowable load. Use Eq. (3.23). Use the bearing capacity, shape, depth, and inclination factors given in Section 3.6.

3.5 A column foundation (Figure P3.5) is 3 m × 2 m in plan. Given: D_f = 2 m, ϕ' = 25°, c' = 50 kN/m². Using Eq. (3.23) and FS = 4, determine the net allowable load [see Eq. (3.15)] the foundation could carry. Use bearing capacity, shape, and depth factors given in Section 3.6.

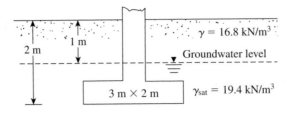

Figure P3.5

3.6   For a square foundation that is $B \times B$ in plan, $D_f = 3$ ft; vertical gross allowable load, $Q_{all} = 150{,}000$ lb; $\gamma = 115$ lb/ft³; $\phi' = 40°$; $c' = 0$; and FS $= 3$. Determine the size of the foundation. Use Eq. (3.23) and bearing capacity, shape, and depth factors given in Section 3.6.

3.7   A foundation measuring 8 ft $\times$ 8 ft has to be constructed in a granular soil deposit. Given: $D_f = 5$ ft and $\gamma = 110$ lb/ft³. Following are the results of a standard penetration test in that soil.

Depth (ft)	Field standard penetration number, N_{60}
5	11
10	14
15	16
20	21
25	24

   a. Use Eq. (2.25) to estimate an average friction angle, ϕ', for the soil. Use $p_a = 14.7$ lb/in².
   b. Using Eq. (3.23), estimate the gross ultimate load the foundation can carry. Use the bearing capacity, shape, and depth factors given in Section 3.6.

3.8   For the design of a shallow foundation, given the following:

   Soil: $\phi' = 20°$
                  $c' = 72$ kN/m²
                  Unit weight, $\gamma = 17$ kN/m³
                  Modulus of elasticity, $E_s = 1400$ kN/m²
                  Poisson's ratio, $\mu_s = 0.35$

   Foundation: $L = 2$ m
                  $B = 1$ m
                  $D_f = 1$ m

   Calculate the ultimate bearing capacity. Use Eq. (3.43).

3.9   An eccentrically loaded foundation is shown in Figure P3.9. Use FS of 4 and determine the maximum allowable load that the foundation can carry. Use Meyerhof's effective area method and the bearing capacity, shape, and depth factors given in Section 3.6.

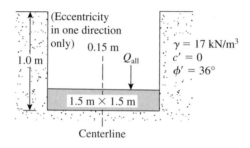

Figure P3.9

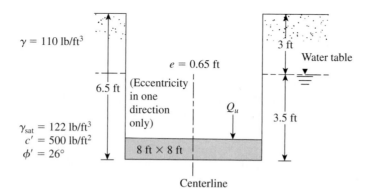

Figure P3.10

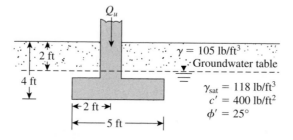

Figure P3.11

3.10 Repeat Problem 3.9 for the foundation shown in Figure P3.10.

3.11 An eccentrically loaded foundation is shown in Figure P3.11. Determine the ultimate load Q_u that the foundation can carry. Use the theory presented in Section 3.13.

3.12 A square footing is shown in Figure P3.12. Use FS = 6, and determine the size of the footing. Use the theory given in Section 3.13.

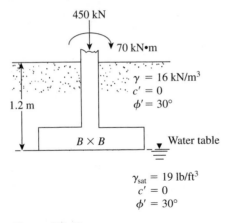

450 kN

70 kN•m

$\gamma = 16$ kN/m^3
$c' = 0$
$\phi' = 30°$

1.2 m

$B \times B$

Water table

$\gamma_{sat} = 19$ lb/ft^3
$c' = 0$
$\phi' = 30°$

Figure P3.12

3.13 The shallow foundation shown in Figure 3.14 measures 4 ft × 6 ft and is subjected to a centric load and a moment. If $e_B = 0.4$ ft, $e_L = 1.2$ ft, and the depth of the foundation is 3 ft, determine the allowable load the foundation can carry. Use a factor of safety of 4. For the soil, we are told that unit weight $\gamma = 115$ lb/ft^3, friction angle $\phi' = 35°$, and cohesion $c' = 0$. Use the bearing capacity, shape, and depth factors given in Section 3.6.

3.14 Redo Problem 3.13 with $e_L = 0.06$ ft and $e_B = 1.5$ ft.

References

Bjerrum, L. (1972). "Embankments on Soft Ground." *Proceedings of the Specialty Conference, American Society of Civil Engineers,* Vol. 2, pp. 1–54.

Bozozuk, M. (1972). "Foundation Failure of the Vankleek Hill Tower Site," *Proceedings,* Specialty Conference on Performance of Earth and Earth-Supported Structures, Vol. 1, Part 2, pp. 885–902.

Caquot, A., and Kerisel, J. (1953). "Sur le terme de surface dans le calcul des fondations en milieu pulverulent," *Proceedings, Third International Conference on Soil Mechanics and Foundation Engineering,* Zürich, Vol. I, pp. 336–337.

De Beer, E. E. (1970). "Experimental Determination of the Shape Factors and Bearing Capacity Factors of Sand," *Geotechnique,* Vol. 20, No. 4, pp. 387–411.

Hanna, A. M., and Meyerhof, G. G. (1981). "Experimental Evaluation of Bearing Capacity of Footings Subjected to Inclined Loads," *Canadian Geotechnical Journal,* Vol. 18, No. 4, pp. 599–603.

Hansen, J. B. (1970). *A Revised and Extended Formula for Bearing Capacity,* Bulletin 28, Danish Geotechnical Institute, Copenhagen.

Highter, W. H., and Anders, J. C. (1985). "Dimensioning Footings Subjected to Eccentric Loads," *Journal of Geotechnical Engineering,* American Society of Civil Engineers, Vol. 111, No. GT5, pp. 659–665.

Kumbhojkar, A. S. (1993). "Numerical Evaluation of Terzaghi's N_γ," *Journal of Geotechnical Engineering,* American Society of Civil Engineers, Vol. 119, No. 3, pp. 598–607.

Meyerhof, G. G. (1953). "The Bearing Capacity of Foundations Under Eccentric and Inclined Loads," *Proceedings, Third International Conference on Soil Mechanics and Foundation Engineering,* Zürich, Vol. 1, pp. 440–445.

Meyerhof, G. G. (1963). "Some Recent Research on the Bearing Capacity of Foundations," *Canadian Geotechnical Journal,* Vol. 1, No. 1, pp. 16–26.

Michalowski, R. L. (1977). "An Estimate of the Influence of Soil Weight on Bearing Capacity using Limit Analysis," *Soils and Foundations,* Vol. 37, No. 4, pp. 57–64.

Prakash, S., and Saran, S. (1971). "Bearing Capacity of Eccentrically Loaded Footings," *Journal of the Soil Mechanics and Foundations Division,* ASCE, Vol. 97, No. SM1, pp. 95–117.

Prandtl, L. (1921). "Über die Eindringungsfestigkeit (Härte) plastischer Baustoffe und die Festigkeit von Schneiden," *Zeitschrift für angewandte Mathematik und Mechanik,* Vol. 1, No. 1, pp. 15–20.

Reissner, H. (1924). "Zum Erddruckproblem," *Proceedings, First International Congress of Applied Mechanics,* Delft, pp. 295–311.

Saran, S., and Agarwal, R. B. (1991). "Bearing Capacity of Eccentrically Obliquely Loaded Footing," *Journal of Geotechnical Engineering,* ASCE, Vol. 117, No. 11, pp. 1669–1690.

Terzaghi, K. (1943). *Theoretical Soil Mechanics,* Wiley, New York.

Vesic, A. S. (1963). "Bearing Capacity of Deep Foundations in Sand," *Highway Research Record* No. 39, National Academy of Sciences, pp. 112–153.

Vesic, A. S. (1973). "Analysis of Ultimate Loads of Shallow Foundations," *Journal of the Soil Mechanics and Foundations Division,* American Society of Civil Engineers, Vol. 99, No. SM1, pp. 45–73.

Zhu, M., and Michalowski, R. L. (2005). "Shape Factors for Limit Loads on Square and Rectangular Footings," *Journal of Geotechnical and Geoenvironmental Engineering, ASCE,* Vol. 131, No. 2, pp. 223–231.

4

Ultimate Bearing Capacity of Shallow Foundations: Special Cases

4.1   Introduction

The ultimate bearing capacity problems described in Chapter 3 assume that the soil supporting the foundation is homogeneous and extends to a great depth below the bottom of the foundation. They also assume that the ground surface is horizontal. However, that is not true in all cases: It is possible to encounter a rigid layer at a shallow depth, or the soil may be layered and have different shear strength parameters. In some instances, it may be necessary to construct foundations on or near a slope, or it may be required to design a foundation subjected to uplifting load.

This chapter discusses bearing capacity problems relating to these special cases.

4.2   Foundation Supported by a Soil with a Rigid Base at Shallow Depth

Figure 4.1(a) shows a shallow, rough *continuous* foundation supported by a soil that extends to a great depth. Neglecting the depth factor, for vertical loading Eq. (3.23) will take the form

$$q_u = c'N_c + qN_q + \frac{1}{2}\gamma B N_\gamma \qquad (4.1)$$

The general approach for obtaining expressions for N_c, N_q, and N_γ was outlined in Chapter 3. The extent of the failure zone in soil, D, at ultimate load obtained in the derivation of N_c and N_q by Prandtl (1921) and Reissner (1924) is given in Figure 4.1(b). Similarly, the magnitude of D obtained by Lundgren and Mortensen (1953) in evaluating N_γ is given in the figure.

Now, if a rigid, rough base is located at a depth of $H < D$ below the bottom of the foundation, full development of the failure surface in soil will be restricted. In such a case, the soil failure zone and the development of slip lines at ultimate load will be as

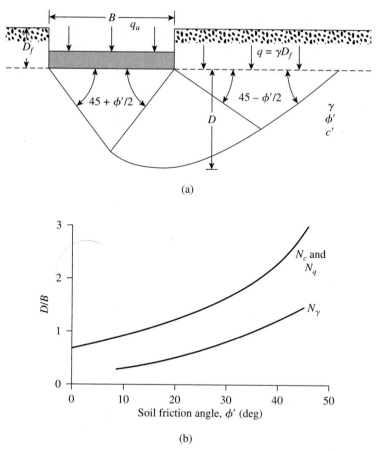

Figure 4.1
(a) Failure surface under a rough continuous foundation; (b) variation of D/B with soil friction angle ϕ'

shown in Figure 4.2. Mandel and Salencon (1972) determined the bearing capacity factors applicable to this case by numerical integration, using the theory of plasticity. According to their theory, the ultimate bearing capacity of a rough continuous foundation with a rigid, rough base located at a shallow depth can be given by the relation

$$q_u = c'N_c^* + qN_q^* + \frac{1}{2}\gamma B N_\gamma^* \qquad (4.2)$$

where

N_c^*, N_q^*, N_γ^* = modified bearing capacity factors

$\qquad B$ = width of foundation

$\qquad \gamma$ = unit weight of soil

Note that, for $H \geqslant D$, $N_c^* = N_c$, $N_q^* = N_q$, and $N_\gamma^* = N_\gamma$ (Lundgren and Mortensen, 1953). The variations of N_c^*, N_q^*, and N_γ^* with H/B and the soil friction angle ϕ' are given in Figures 4.3, 4.4, and 4.5, respectively.

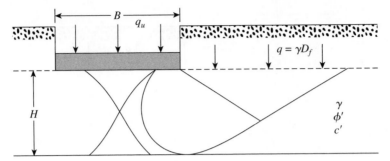

Figure 4.2   Failure surface under a rough, continuous foundation with a rigid, rough base located at a shallow depth

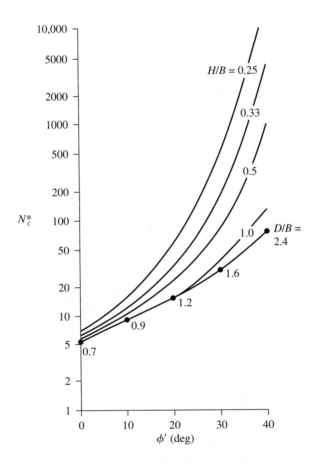

Figure 4.3   Mandel and Salencon's bearing capacity factor N_c^* [Eq. (4.2)]

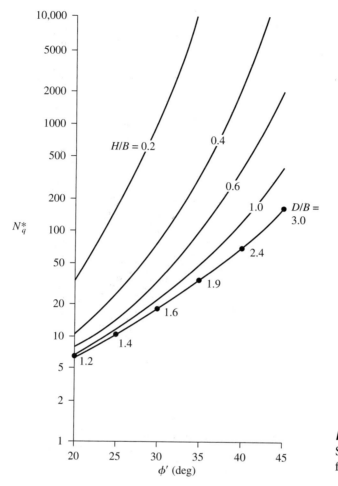

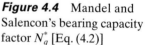

Figure 4.4 Mandel and Salencon's bearing capacity factor N_q^* [Eq. (4.2)]

Neglecting the depth factors, the ultimate bearing capacity of rough circular and rectangular foundations on a sand layer ($c' = 0$) with a rough, rigid base located at a shallow depth can be given as

$$q_u = qN_q^*F_{qs}^* + \frac{1}{2}\gamma B N_\gamma^* F_{\gamma s}^*$$ (4.3)

where $F_{qs}^*, F_{\gamma s}^* =$ modified shape factors.

The shape factors F_{qs}^* and $F_{\gamma s}^*$ are functions of H/B and ϕ'. On the basis of the work of Meyerhof and Chaplin (1953), and simplifying the assumption that, in radial planes, the stresses and shear zones are identical to those in transverse planes, Meyerhof (1974) proposed that

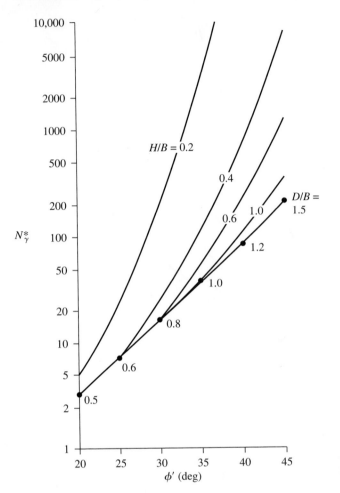

Figure 4.5 Mandel and Salencon's bearing capacity factor N_γ^* [Eq. (4.2)]

$$F_{qs}^* \approx 1 - m_1\left(\frac{B}{L}\right) \tag{4.4}$$

and

$$F_{\gamma s}^* \approx 1 - m_2\left(\frac{B}{L}\right) \tag{4.5}$$

where L = length of the foundation. The variations of m_1 and m_2 with H/B and ϕ' are shown in Figure 4.6.

For saturated clay (i.e., under the undrained condition, or $\phi = 0$), Eq. (4.2) will simplify to the form

$$q_u = c_u N_c^* + q \tag{4.6}$$

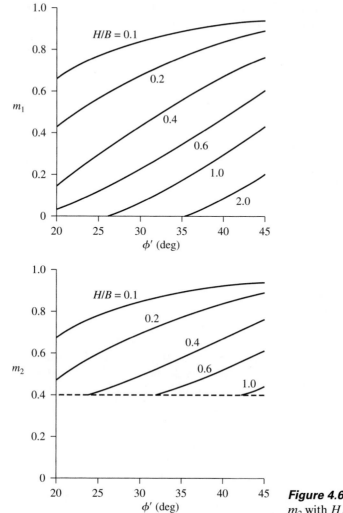

Figure 4.6 Variation of m_1 and m_2 with H/B and ϕ'

Mandel and Salencon (1972) performed calculations to evaluate N_c^* for *continuous foundations*. Similarly, Buisman (1940) gave the following relationship for obtaining the ultimate bearing capacity of square foundations:

$$q_{u(\text{square})} = \left(\pi + 2 + \frac{B}{2H} - \frac{\sqrt{2}}{2}\right)c_u + q \qquad \left(\text{for } \frac{B}{2H} - \frac{\sqrt{2}}{2} \geq 0\right) \qquad (4.7)$$

In this equation, c_u is the undrained shear strength.

Equation (4.7) can be rewritten as

$$q_{u(\text{square})} = 5.14\underbrace{\left(1 + \frac{0.5\dfrac{B}{H} - 0.707}{5.14}\right)}_{N_{c(\text{square})}^*}c_u + q$$

$$(4.8)$$

Table 4.1 Values of N_c^* for Continuous and Square Foundations ($\phi = 0$)

$\dfrac{B}{H}$	N_c^*	
	Square[a]	Continuous[b]
2	5.43	5.24
3	5.93	5.71
4	6.44	6.22
5	6.94	6.68
6	7.43	7.20
8	8.43	8.17
10	9.43	9.05

[a] Buisman's analysis (1940)
[b] Mandel and Salencon's analysis (1972)

Table 4.1 gives the values of N_c^* for continuous and square foundations.

Example 4.1

A square foundation measuring 2.5ft × 2.5 ft is constructed on a layer of sand. We are given that $D_f = 2$ ft, $\gamma = 110$ lb/ft³, $\phi' = 35°$, and $c' = 0$. A rock layer is located at a depth of 1.5 ft below the bottom of the foundation. Using a factor of safety of 4, determine the gross allowable load the foundation can carry.

Solution
From Eq. (4.3),

$$q_u = q N_q^* F_{qs}^* + \frac{1}{2}\gamma B N_\gamma^* F_{\gamma s}^*$$

and we also have

$$q = 110 \times 2 = 220 \text{ lb/ft}^2$$

For $\phi' = 35°$, $H/B = 1.5/2.5 = 0.6$, $N_q^* \approx 90$ (Figure 4.4), and $N_\gamma^* \approx 50$ (Figure 4.5), and we have

$$F_{qs}^* = 1 - m_1(B/L)$$

From Figure 4.6(a), for $\phi' = 35°$, $H/B = 0.6$, and the value of $m_1 = 0.34$, so

$$F_{qs}^* = 1 - (0.34)(1/1) = 0.66$$

Similarly,

$$F_{\gamma s}^* = 1 - m_2(B/L)$$

From Figure 4.6(b), $m_2 = 0.45$, so

$$F_{\gamma s}^* = 1 - (0.45)(1/1) = 0.55$$

Hence,

$$q_u = (220)(90)(0.66) + (1/2)(110)(2.5)(50)(0.55) = 16{,}849 \text{ lb/ft}^2$$

and

$$Q_{\text{all}} = \frac{q_u B^2}{\text{FS}} = \frac{(16{,}849)(2.5 \times 2.5)}{4} = \textbf{26{,}326 lb} \qquad \blacksquare$$

Example 4.2

Consider a square foundation 1 m × 1 m in plan located on a saturated clay layer underlain by a layer of rock. Given:

Clay: $c_u = 72 \text{ kN/m}^2$; Unit weight $\gamma = 18 \text{ kN/m}^3$
Distance between the bottom of foundation and the rock layer = 0.25 m
$D_f = 1$ m

Estimate the gross allowable bearing capacity of the foundation. Use FS = 3.

Solution
From Eq. (4.8),

$$q_u = 5.14\left(1 + \frac{0.5\dfrac{B}{H} - 0.707}{5.14}\right) c_u + q$$

For $B/H = 1/0.25 = 4$; $c_u = 72 \text{ kN/m}^2$; and $q = \gamma D_f = (18)(1) = 18 \text{ kN/m}^3$.

$$q_u = 5.14\left[1 + \frac{(0.5)(4) - 0.707}{5.14}\right]72 + 18 = 481.2 \text{ kN/m}^2$$

$$q_{\text{all}} = \frac{q_u}{\text{FS}} = \frac{481.2}{3} = \textbf{160.4 kN/m}^2 \qquad \blacksquare$$

4.3

Bearing Capacity of Layered Soils: Stronger Soil Underlain by Weaker Soil

The bearing capacity equations presented in Chapter 3 involve cases in which the soil supporting the foundation is homogeneous and extends to a considerable depth. The cohesion, angle of friction, and unit weight of soil were assumed to remain constant for the bearing capacity analysis. However, in practice, layered soil profiles are often encountered. In such instances, the failure surface at ultimate load may extend through two or more soil layers, and a determination of the ultimate bearing capacity in layered soils can be made in only a limited number of cases. This section features the procedure for estimating the bearing capacity for layered soils proposed by Meyerhof and Hanna (1978) and Meyerhof (1974).

Figure 4.7 shows a shallow continuous foundation supported by a *stronger soil layer*, underlain by a weaker soil that extends to a great depth. For the two soil layers, the physical parameters are as follows:

Layer	Unit weight	Soil friction angle	Cohesion
Top	γ_1	ϕ_1'	c_1'
Bottom	γ_2	ϕ_2'	c_2'

At ultimate load per unit area (q_u), the failure surface in soil will be as shown in the figure. If the depth H is relatively small compared with the foundation width B, a punching shear failure will occur in the top soil layer, followed by a general shear failure in the bottom soil layer. This is shown in Figure 4.7(a). However, if the depth H is relatively large, then the failure surface will be completely located in the top soil layer, which is the upper limit for the ultimate bearing capacity. This is shown in Figure 4.7b.

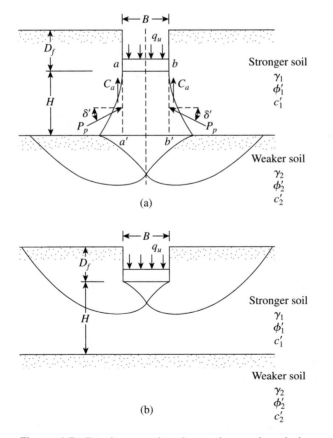

Figure 4.7   Bearing capacity of a continuous foundation on layered soil

The ultimate bearing capacity for this problem, as shown in Figure 4.7a, can be given as

$$q_u = q_b + \frac{2(C_a + P_p \sin \delta')}{B} - \gamma_1 H \tag{4.9}$$

where

B = width of the foundation
C_a = adhesive force
P_p = passive force per unit length of the faces aa' and bb'
q_b = bearing capacity of the bottom soil layer
δ' = inclination of the passive force P_p with the horizontal

Note that, in Eq. (4.9),

$$C_a = c_a' H$$

where c_a' = adhesion.
Equation (4.9) can be simplified to the form

$$q_u = q_b + \frac{2c_a' H}{B} + \gamma_1 H^2 \left(1 + \frac{2D_f}{H}\right) \frac{K_{pH} \tan \delta'}{B} - \gamma_1 H \tag{4.9a}$$

where K_{pH} = horizontal component of passive earth pressure coefficient. However, let

$$K_{pH} \tan \delta' = K_s \tan \phi_1' \tag{4.10}$$

where K_s = punching shear coefficient. Then,

$$q_u = q_b + \frac{2c_a' H}{B} + \gamma_1 H^2 \left(1 + \frac{2D_f}{H}\right) \frac{K_s \tan \phi_1'}{B} - \gamma_1 H \tag{4.11}$$

The punching shear coefficient, K_s, is a function of q_2/q_1 and ϕ_1', or, specifically,

$$K_s = f\left(\frac{q_2}{q_1}, \phi_1'\right)$$

Note that q_1 and q_2 are the ultimate bearing capacities of a continuous foundation of width B under vertical load on the surfaces of homogeneous thick beds of upper and lower soil, or

$$q_1 = c_1' N_{c(1)} + \tfrac{1}{2}\gamma_1 B N_{\gamma(1)} \tag{4.12}$$

and

$$q_2 = c_2' N_{c(2)} + \tfrac{1}{2}\gamma_2 B N_{\gamma(2)} \tag{4.13}$$

where

$N_{c(1)}, N_{\gamma(1)}$ = bearing capacity factors for friction angle ϕ_1' (Table 3.3)
$N_{c(2)}, N_{\gamma(2)}$ = bearing capacity factors for friction angle ϕ_2' (Table 3.3)

Observe that, for the top layer to be a stronger soil, q_2/q_1 should be less than unity. The variation of K_s with q_2/q_1 and ϕ_1' is shown in Figure 4.8. The variation of c_a'/c_1' with q_2/q_1 is shown in Figure 4.9. If the height H is relatively large, then the failure

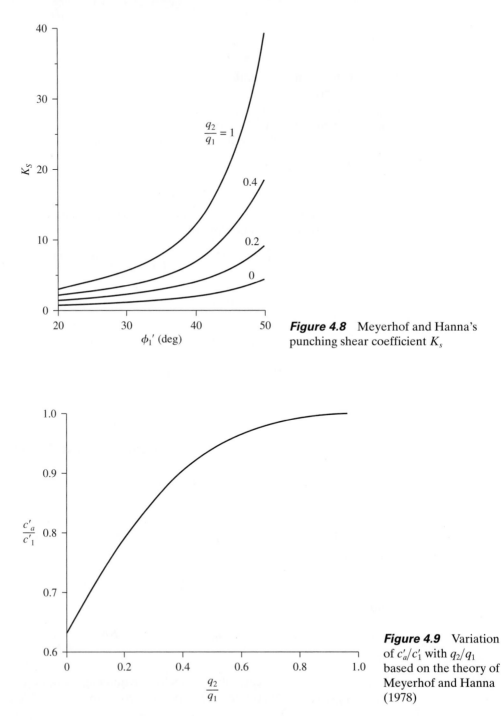

Figure 4.8 Meyerhof and Hanna's punching shear coefficient K_s

Figure 4.9 Variation of c_a'/c_1' with q_2/q_1 based on the theory of Meyerhof and Hanna (1978)

surface in soil will be completely located in the stronger upper-soil layer (Figure 4.7b). For this case,

$$q_u = q_t = c_1' N_{c(1)} + q N_{q(1)} + \tfrac{1}{2} \gamma_1 B N_{\gamma(1)}. \tag{4.14}$$

where $N_{q(1)}$ = bearing capacity factor for $\phi' = \phi_1'$ (Table 3.3) and $q = \gamma_1 D_f$.
    Combining Eqs. (4.11) and (4.14) yields

$$q_u = q_b + \frac{2c_a' H}{B} + \gamma_1 H^2 \left(1 + \frac{2D_f}{H} \right) \frac{K_s \tan \phi_1'}{B} - \gamma_1 H \leq q_t \tag{4.15}$$

For rectangular foundations, the preceding equation can be extended to the form

$$q_u = q_b + \left(1 + \frac{B}{L} \right) \left(\frac{2c_a' H}{B} \right)$$
$$+ \gamma_1 H^2 \left(1 + \frac{B}{L} \right) \left(1 + \frac{2D_f}{H} \right) \left(\frac{K_s \tan \phi_1'}{B} \right) - \gamma_1 H \leq q_t \tag{4.16}$$

where

$$q_b = c_2' N_{c(2)} F_{cs(2)} + \gamma_1 (D_f + H) N_{q(2)} F_{qs(2)} + \frac{1}{2} \gamma_2 B N_{\gamma(2)} F_{\gamma s(2)} \tag{4.17}$$

and

$$q_t = c_1' N_{c(1)} F_{cs(1)} + \gamma_1 D_f N_{q(1)} F_{qs(1)} + \frac{1}{2} \gamma_1 B N_{\gamma(1)} F_{\gamma s(1)} \tag{4.18}$$

in which

$F_{cs(1)}, F_{qs(1)}, F_{\gamma s(1)}$ = shape factors with respect to top soil layer (Eqs. 3.27 through 3.29)

$F_{cs(2)}, F_{qs(2)}, F_{\gamma s(2)}$ = shape factors with respect to bottom soil layer (Eqs. 3.27 through 3.29)

Special Cases

1. *Top layer is strong sand and bottom layer is saturated soft clay* ($\phi_2 = 0$). From Eqs. (4.16), (4.17), and (4.18),

$$q_b = \left(1 + 0.2 \frac{B}{L} \right) 5.14 c_2 + \gamma_1 (D_f + H) \tag{4.19}$$

and

$$q_t = \gamma_1 D_f N_{q(1)} F_{qs(1)} + \tfrac{1}{2}\gamma_1 B N_{\gamma(1)} F_{\gamma s(1)}$$

(4.20)

Hence,

$$q_u = \left(1 + 0.2\frac{B}{L}\right)5.14c_2 + \gamma_1 H^2\left(1 + \frac{B}{L}\right)\left(1 + \frac{2D_f}{H}\right)\frac{K_s \tan \phi_1'}{B}$$
$$+ \gamma_1 D_f \leqslant \gamma_1 D_f N_{q(1)} F_{qs(1)} + \frac{1}{2}\gamma_1 B N_{\gamma(1)} F_{\gamma s(1)}$$

(4.21)

where c_2 = undrained cohesion.

For a determination of K_s from Figure 4.8,

$$\frac{q_2}{q_1} = \frac{c_2 N_{c(2)}}{\tfrac{1}{2}\gamma_1 B N_{\gamma(1)}} = \frac{5.14c_2}{0.5\gamma_1 B N_{\gamma(1)}}$$

(4.22)

2. *Top layer is stronger sand and bottom layer is weaker sand* ($c_1' = 0, c_2' = 0$). *The ultimate bearing capacity can be given as*

$$q_u = \left[\gamma_1(D_f + H)N_{q(2)}F_{qs(2)} + \frac{1}{2}\gamma_2 B N_{\gamma(2)}F_{\gamma s(2)}\right]$$
$$+ \gamma_1 H^2\left(1 + \frac{B}{L}\right)\left(1 + \frac{2D_f}{H}\right)\frac{K_s \tan \phi_1'}{B} - \gamma_1 H \leqslant q_t$$

(4.23)

where

$$q_t = \gamma_1 D_f N_{q(1)} F_{qs(1)} + \frac{1}{2}\gamma_1 B N_{\gamma(1)} F_{\gamma s(1)}$$

(4.24)

Then

$$\frac{q_2}{q_1} = \frac{\tfrac{1}{2}\gamma_2 B N_{\gamma(2)}}{\tfrac{1}{2}\gamma_1 B N_{\gamma(1)}} = \frac{\gamma_2 N_{\gamma(2)}}{\gamma_1 N_{\gamma(1)}}$$

(4.25)

3. *Top layer is stronger saturated clay* ($\phi_1 = 0$) *and bottom layer is weaker saturated clay* ($\phi_2 = 0$). *The ultimate bearing capacity can be given as*

$$q_u = \left(1 + 0.2\frac{B}{L}\right)5.14c_2 + \left(1 + \frac{B}{L}\right)\left(\frac{2c_a H}{B}\right) + \gamma_1 D_f \leqslant q_t$$

(4.26)

where

$$q_t = \left(1 + 0.2\frac{B}{L}\right)5.14c_1 + \gamma_1 D_f \qquad (4.27)$$

and c_1 and c_2 are undrained cohesions. For this case,

$$\frac{q_2}{q_1} = \frac{5.14c_2}{5.14c_1} = \frac{c_2}{c_1} \qquad (4.28)$$

Example 4.3

Refer to Figure 4.7(a) and consider the case of a continuous foundation with $B = 2$ m, $D_f = 1.2$ m, and $H = 1.5$ m. The following are given for the two soil layers:

Top sand layer:

Unit weight $\gamma_1 = 17.5$ kN/m³
$\qquad \phi_1' = 40°$
$\qquad c_1' = 0$

Bottom clay layer:

Unit weight $\gamma_2 = 16.5$ kN/m³
$\qquad \phi_2' = 0$
$\qquad c_2 = 30$ kN/m²

Determine the gross ultimate load per unit length of the foundation.

Solution
For this case, Eqs. (4.21) and (4.22) apply. For $\phi_1' = 40°$, from Table 3.3, $N_\gamma = 109.41$ and

$$\frac{q_2}{q_1} = \frac{c_2 N_{c(2)}}{0.5\gamma_1 B N_{\gamma(1)}} = \frac{(30)(5.14)}{(0.5)(17.5)(2)(109.41)} = 0.081$$

From Figure 4.8, for $c_2 N_{c(2)}/0.5\gamma_1 B N_{\gamma(1)} = 0.081$ and $\phi_1' = 40°$, the value of $K_s \approx 2.5$. Equation (4.21) then gives

$$q_u = \left[1 + (0.2)\left(\frac{B}{L}\right)\right]5.14c_2 + \left(1 + \frac{B}{L}\right)\gamma_1 H^2\left(1 + \frac{2D_f}{H}\right)K_s\frac{\tan\phi_1'}{B} + \gamma_1 D_f$$

$$= [1 + (0.2)(0)](5.14)(30) + (1 + 0)(17.5)(1.5)^2$$

$$\times \left[1 + \frac{(2)(1.2)}{1.5}\right](2.5)\frac{\tan 40}{2.0} + (17.5)(1.2)$$

$$= 154.2 + 107.4 + 21 = 282.6 \text{ kN/m}^2$$

Again, from Eq. (4.21),

$$q_t = \gamma_1 D_f N_{q(1)} F_{qs(1)} + \tfrac{1}{2} \gamma_1 B N_{\gamma(1)} F_{\gamma s(1)}$$

From Table 3.3, for $\phi_1' = 40°$, $N_\gamma = 109.4$ and $N_q = 64.20$.
From Eqs. (3.28) and (3.29),

$$F_{qs(1)} = 1 + \left(\frac{B}{L}\right) \tan \phi_1' = 1 + (0) \tan 40 = 1$$

and

$$F_{\gamma s(1)} = 1 - 0.4 \frac{B}{L} = 1 - (0.4)(0) = 1$$

so that

$$q_t = (17.5)(1.2)(64.20)(1) + (\tfrac{1}{2})(17.5)(2)(109.4)(1) = 3262.7 \text{ kN/m}^2$$

Hence,

$$q_u = 282.6 \text{ kN/m}^2$$
$$Q_u = (282.6)(B) = (282.6)(2) = \mathbf{565.2 \text{ kN/m}}$$ ■

Example 4.4

A foundation 1.5 m $\times$ 1 m is located at a depth D_f of 1 m in a stronger clay. A softer clay layer is located at a depth H of 1 m, measured from the bottom of the foundation. For the top clay layer,

Undrained shear strength $= 120 \text{ kN/m}^2$
Unit weight $= 16.8 \text{ kN/m}^3$

and for the bottom clay layer,

Undrained shear strength $= 48 \text{ kN/m}^2$
Unit weight $= 16.2 \text{ kN/m}^3$

Determine the gross allowable load for the foundation with an FS of 4.

Solution
For this problem, Eqs. (4.26), (4.27), and (4.28) will apply, or

$$q_u = \left(1 + 0.2\frac{B}{L}\right)5.14c_2 + \left(1 + \frac{B}{L}\right)\left(\frac{2c_a H}{B}\right) + \gamma_1 D_f$$

$$\leq \left(1 + 0.2\frac{B}{L}\right)5.14c_1 + \gamma_1 D_f$$

We are given the following data:

$$B = 1 \text{ m} \qquad H = 1 \text{ m} \qquad D_f = 1 \text{ m}$$
$$L = 1.5 \text{ m} \qquad \gamma_1 = 16.8 \text{ kN/m}^3$$

From Figure 4.9 for $c_2/c_1 = 48/120 = 0.4$, the value of $c_a/c_1 \approx 0.9$, so

$$c_a = (0.9)(120) = 108 \text{ kN/m}^2$$

and

$$q_u = \left[1 + (0.2)\left(\frac{1}{1.5}\right) \right](5.14)(48) + \left(1 + \frac{1}{1.5} \right)\left[\frac{(2)(108)(1)}{1} \right] + (16.8)(1)$$

$$= 279.6 + 360 + 16.8 = 656.4 \text{ kN/m}^2$$

As a check, we have, from Eq. (4.27),

$$q_t = \left[1 + (0.2)\left(\frac{1}{1.5}\right) \right](5.14)(120) + (16.8)(1)$$

$$= 699 + 16.8 = 715.8 \text{ kN/m}^2$$

Thus, $q_u = 656.4 \text{ kN/m}^2$ (i.e., the smaller of the two values just calculated), and

$$q_{\text{all}} = \frac{q_u}{\text{FS}} = \frac{656.4}{4} = 164.1 \text{ kN/m}^2$$

The total allowable load is therefore

$$(q_{\text{all}})(1 \times 1.5) = \textbf{246.15 kN} \qquad \blacksquare$$

4.4 Closely Spaced Foundations—Effect on Ultimate Bearing Capacity

In Chapter 3, theories relating to the ultimate bearing capacity of single rough continuous foundations supported by a homogeneous soil extending to a great depth were discussed. However, if foundations are placed close to each other with similar soil conditions, the ultimate bearing capacity of each foundation may change due to the interference effect of the failure surface in the soil. This was theoretically investigated by Stuart (1962) for *granular soils*. It was assumed that the geometry of the rupture surface in the soil mass would be the same as that assumed by Terzaghi (Figure 3.5). According to Stuart, the following conditions may arise (Figure 4.10).

Case I. (Figure 4.10a) If the center-to-center spacing of the two foundations is $x \geqslant x_1$, the rupture surface in the soil under each foundation will not overlap. So the ultimate bearing capacity of each continuous foundation can be given by Terzaghi's equation [Eq. (3.3)]. For $(c' = 0)$

$$q_u = q N_q + \tfrac{1}{2}\gamma B N_\gamma \qquad (4.29)$$

where N_q, N_γ = Terzaghi's bearing capacity factors (Table 3.1).

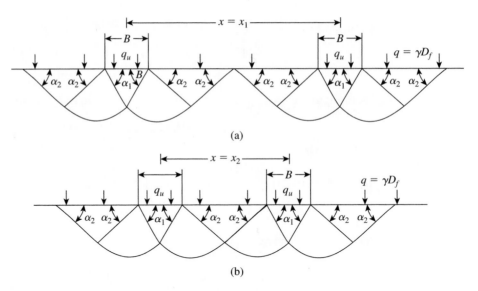

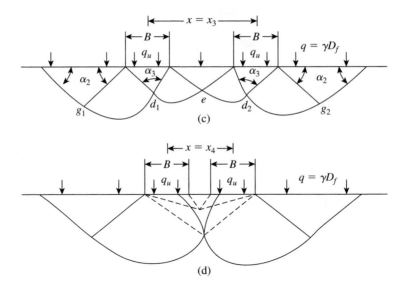

Figure 4.10 Assumptions for the failure surface in granular soil under two closely spaced rough continuous foundations
(*Note*: $\alpha_1 = \phi'$, $\alpha_2 = 45 - \phi'/2$, $\alpha_3 = 180 - 2\phi'$)

Case II. (Figure 4.10b) If the center-to-center spacing of the two foundations ($x = x_2 < x_1$) are such that the Rankine passive zones just overlap, then the magnitude of q_u will still be given by Eq. (4.29). However, the foundation settlement at ultimate load will change (compared to the case of an isolated foundation).

Case III. (Figure 4.10c) This is the case where the center-to-center spacing of the two continuous foundations is $x = x_3 < x_2$. Note that the triangular wedges in the soil under the foundations make angles of $180° - 2\phi'$ at points d_1 and d_2. The arcs of the logarithmic spirals d_1g_1 and d_1e are tangent to each other at d_1. Similarly, the arcs of the logarithmic spirals d_2g_2 and d_2e are tangent to each other at d_2. For this case, the ultimate bearing capacity of each foundation can be given as ($c' = 0$)

$$q_u = qN_q\zeta_q + \tfrac{1}{2}\gamma BN_\gamma\,\zeta_\gamma \tag{4.30}$$

where $\zeta_q,\ \zeta_\gamma$ = efficiency ratios.

The efficiency ratios are functions of x/B and soil friction angle ϕ'. The theoretical variations of ζ_q and ζ_γ are given in Figure 4.11.

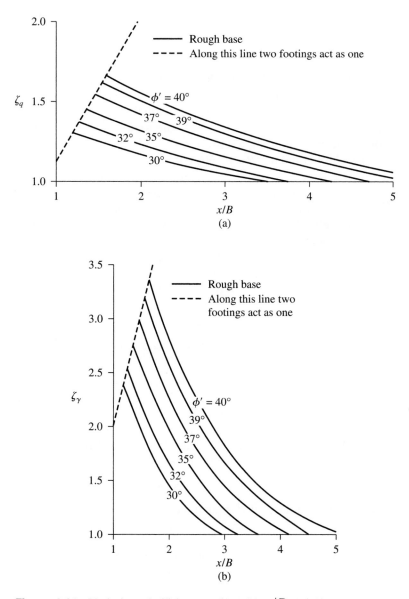

Figure 4.11 Variation of efficiency ratios with x/B and ϕ'

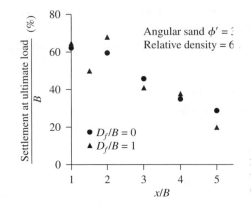

Figure 4.12   Settlement at ultimate load for two closely spaced continuous foundations (Redrawn from Das and Larbi-Cherif, 1982)

Case IV. (Figure 4.10d): If the spacing of the foundation is further reduced such that $x = x_4 < x_3$, blocking will occur and the pair of foundations will act as a single foundation. The soil between the individual units will form an inverted arch which travels down with the foundation as the load is applied. When the two foundations touch, the zone of arching disappears and the system behaves as a single foundation with a width equal to $2B$. The ultimate bearing capacity for this case can be given by Eq. (4.29), with B being replaced by $2B$ in the second term.

　　The ultimate bearing capacity of two continuous foundations spaced close to each other may increase since the efficiency ratios are greater than one. However, when the closely spaced foundations are subjected to a similar load per unit area, the settlement S_e will be larger when compared to that for an isolated foundation. Figure 4.12 shows the laboratory model test results of Das and Larbi-Cherif (1982) as they relate to the settlement at ultimate load for two closely spaced continuous foundations. The settlement decreases with the increase in x/B and becomes practically constant at $x/B \leq 4$.

4.5   *Bearing Capacity of Foundations on Top of a Slope*

In some instances, shallow foundations need to be constructed on top of a slope. In Figure 4.13, the height of the slope is H, and the slope makes an angle β with the horizontal. The edge of the foundation is located at a distance b from the top of the slope. At ultimate load, q_u, the failure surface will be as shown in the figure.

　　Meyerhof (1957) developed the following theoretical relation for the ultimate bearing capacity for *continuous foundations:*

$$q_u = c'N_{cq} + \tfrac{1}{2}\gamma B N_{\gamma q} \tag{4.31}$$

For purely granular soil, $c' = 0$, thus,

$$q_u = \tfrac{1}{2}\gamma B N_{\gamma q} \tag{4.32}$$

Again, for purely cohesive soil, $\phi = 0$ (the undrained condition); hence,

$$q_u = c N_{cq} \tag{4.33}$$

where c = undrained cohesion.

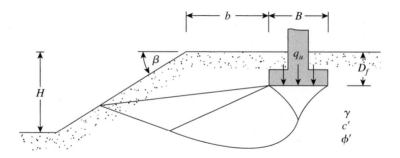

Figure 4.13 Shallow foundation on top of a slope

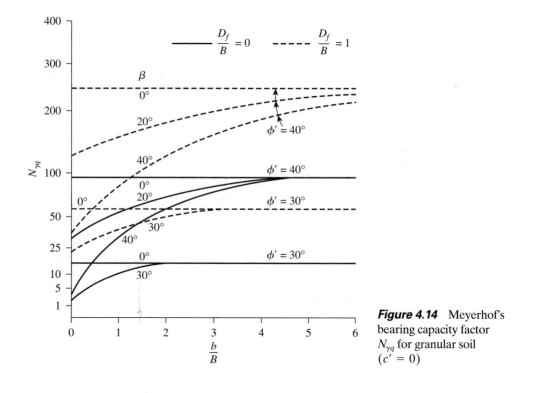

Figure 4.14 Meyerhof's bearing capacity factor $N_{\gamma q}$ for granular soil ($c' = 0$)

The variations of $N_{\gamma q}$ and N_{cq} defined by Eqs. (4.32) and (4.33) are shown in Figures 4.14 and 4.15, respectively. In using N_{cq} in Eq. (4.33) as given in Figure 4.15, the following points need to be kept in mind:

1. The term

$$N_s = \frac{\gamma H}{c} \tag{4.34}$$

is defined as the stability number.

2. If $B < H$, use the curves for $N_s = 0$.

3. If $B \geq H$, use the curves for the calculated stability number N_s.

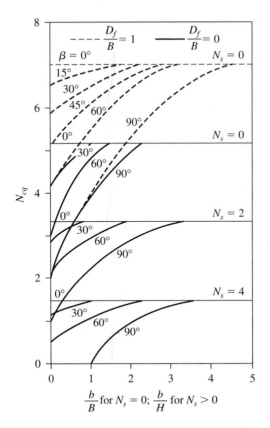

Figure 4.15 Meyerhof's bearing capacity factor N_{cq} for purely cohesive soil

Example 4.5

In Figure 4.13, for a shallow continuous foundation in a clay, the following data are given: $B = 1.2$ m; $D_f = 1.2$ m; $b = 0.8$ m; $H = 6.2$ m; $\beta = 30°$; unit weight of soil $= 17.5$ kN/m³; $\phi = 0$; and $c = 50$ kN/m². Determine the gross allowable bearing capacity with a factor of safety FS $= 4$.

Solution
Since $B < H$, we will assume the stability number $N_s = 0$. From Eq. (4.33),

$$q_u = cN_{cq}$$

We are given that

$$\frac{D_f}{B} = \frac{1.2}{1.2} = 1$$

and

$$\frac{b}{B} = \frac{0.8}{1.2} = 0.67$$

For $\beta = 30°$, $D_f/B = 1$ and $b/B = 0.75$, Figure 4.15 gives $N_{cq} = 6.3$. Hence,

$$q_u = (50)(6.3) = 315 \text{ kN/m}^2$$

and

$$q_{all} = \frac{q_u}{FS} = \frac{315}{4} = \textbf{78.8 kN/m}^2$$ ∎

Example 4.6

Figure 4.16 shows a continuous foundation on a slope of a granular soil. Estimate the ultimate bearing capacity.

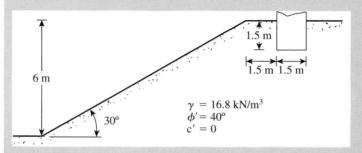

γ = 16.8 kN/m³
$\phi' = 40°$
c' = 0

Figure 4.16   Foundation on a granular slope

Solution

For granular soil ($c' = 0$), from Eq. (4.32),

$$q_u = \tfrac{1}{2}\gamma B N_{\gamma q}$$

We are given that $b/B = 1.5/1.5 = 1$, $D_f/B = 1.5/1.5 = 1$, $\phi' = 40°$, and $\beta = 30°$.
From Figure 4.14, $N_{\gamma q} \approx 120$. So,

$$q_u = \tfrac{1}{2}(16.8)(1.5)(120) = \textbf{1512 kN/m}^2$$ ∎

4.6   *Bearing Capacity of Foundations on a Slope*

A theoretical solution for the ultimate bearing capacity of a shallow foundation located on the face of a slope was developed by Meyerhof (1957). Figure 4.17 shows the nature of the plastic zone developed under a rough continuous foundation of width B. In Figure 4.17, *abc* is an elastic zone, *acd* is a radial shear zone, and *ade* is a

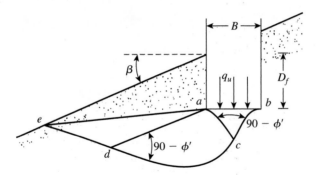

Figure 4.17 Nature of plastic zone under a rough continuous foundation on the face of a slope

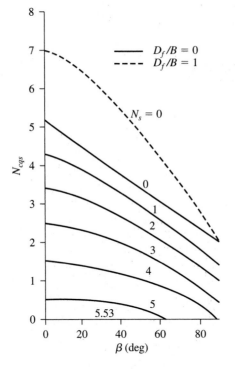

Figure 4.18 Variation of N_{cqs} with β. (*Note:* $N_s = \gamma H/c$)

mixed shear zone. Based on this solution, the ultimate bearing capacity can be expressed as

$$q_u = cN_{cqs} \text{ (for purely cohesive soil, that is, } \phi = 0) \tag{4.35}$$

and

$$q_u = \tfrac{1}{2}\gamma BN_{\gamma qs} \text{ (for granular soil, that is } c' = 0) \tag{4.36}$$

The variations of N_{cqs} and $N_{\gamma qs}$ with slope angle β are given in Figures 4.18 and 4.19.

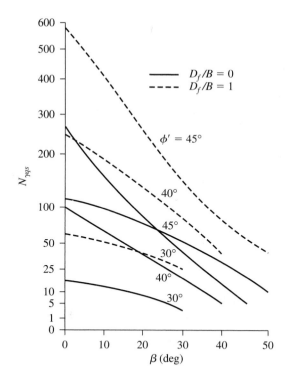

Figure 4.19 Variation of $N_{\gamma qs}$ with β

4.7 *Uplift Capacity of Foundations*

Foundations may be subjected to uplift forces under special circumstances. During the design process for those foundations, it is desirable to provide a sufficient factor of safety against failure by uplift. This section will provide the relationships for the uplift capacity of foundations in granular and cohesive soils.

Foundations in Granular Soil (c = 0)

Figure 4.20 shows a shallow continuous foundation that is being subjected to an uplift force. At ultimate load, Q_u, the failure surface in soil will be as shown in the figure. The ultimate load can be expressed in the form of a nondimensional breakout factor, F_q. Or

$$F_q = \frac{Q_u}{A\gamma D_f} \qquad (4.37)$$

where A = area of the foundation.

The breakout factor is a function of the soil friction angle ϕ' and D_f/B. For a given soil friction angle, F_q increases with D_f/B to a maximum at $D_f/B = (D_f/B)_{cr}$ and remains constant thereafter. For foundations subjected to uplift, $D_f/B \le (D_f/B)_{cr}$ is considered a shallow foundation condition. When a foundation has an embedment ratio of $D_f/B > (D_f/B)_{cr}$, it is referred to as a deep foundation.

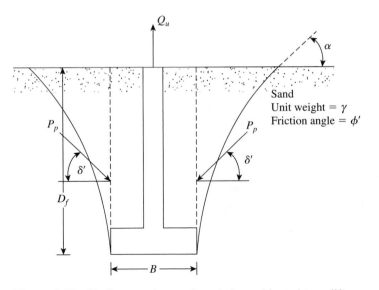

Figure 4.20 Shallow continuous foundation subjected to uplift

Meyerhof and Adams (1968) provided relationships to estimate the ultimate up-lifting load Q_u for shallow [that is, $D_f/B \leq (D_f/B)_{cr}$], circular, and rectangular foundations. Using these relationships and Eq. (4.37), Das and Seeley (1975) expressed the breakout factor F_q in the following form

$$F_q = 1 + 2\left[1 + m\left(\frac{D_f}{B}\right)\right]\left(\frac{D_f}{B}\right)K_u \tan \phi' \tag{4.38}$$

(for shallow circular foundations)

$$F_q = 1 + \left\{\left[1 + 2m\left(\frac{D_f}{B}\right)\right]\left(\frac{B}{L}\right) + 1\right\}\left(\frac{D_f}{B}\right)K_u \tan \phi' \tag{4.39}$$

(for shallow rectangular foundations)

where

m = a coefficient which is a function of ϕ'
K_u = nominal uplift coefficient

The variations of K_u, m, and $(D_f/B)_{cr}$ for square and circular foundations are given in Table 4.2 (Meyerhof and Adams, 1968).

For rectangular foundations, Das and Jones (1982) recommended that

$$\left(\frac{D_f}{B}\right)_{cr\text{-rectangular}} = \left(\frac{D_f}{B}\right)_{cr\text{-square}}\left[0.133\left(\frac{L}{B}\right) + 0.867\right] \leq 1.4\left(\frac{D_f}{B}\right)_{cr\text{-square}} \tag{4.40}$$

Using the values of K_u, m, and $(D_f/B)_{cr}$ in Eq. (4.38), the variations of F_q for square and circular foundations have been calculated and are shown in Figure 4.21. A step-by-step procedure to estimate the uplift capacity of foundations in granular soil follows.

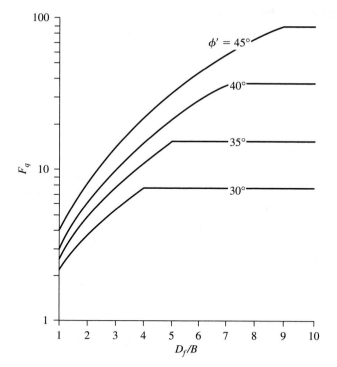

Figure 4.21 Variation of F_q with D_f/B and ϕ'

Table 4.2 Variation of K_u, m, and $(D_f/B)_{cr}$

Soil friction angle, ϕ' (deg)	K_u	m	(D_f/B_{cr}) for square and circular foundations
20	0.856	0.05	2.5
25	0.888	0.10	3
30	0.920	0.15	4
35	0.936	0.25	5
40	0.960	0.35	7
45	0.960	0.50	9

Step 1. Determine, D_f, B, L, and ϕ'.
Step 2. Calculate D_f/B.
Step 3. Using Table 4.2 and Eq. (4.40), calculate $(D_f/B)_{cr}$.
Step 4. If D_f/B is less than or equal to $(D_f/B)_{cr}$, it is a shallow foundation.
Step 5. If $D_f/B > (D_f/B)_{cr}$, it is a deep foundation.
Step 6. For shallow foundations, use D_f/B calculated in Step 2 in Eq. (4.38) or (4.39) to estimate F_q. Thus, $Q_u = F_q A \gamma D_f$.
Step 7. For deep foundations, substitute $(D_f/B)_{cr}$ for D_f/B in Eq. (4.38) or (4.39) to obtain F_q, from which the ultimate load Q_u may be obtained.

Foundations in Cohesive Soil ($\phi' = 0$)

The ultimate uplift capacity, Q_u, of a foundation in a purely cohesive soil can be expressed as

$$Q_u = A(\gamma D_f + c_u F_c) \tag{4.41}$$

where

A = area of the foundation
c_u = undrained shear strength of soil
F_c = breakout factor

As in the case of foundations in granular soil, the breakout factor F_c increases with embedment ratio and reaches a maximum value of $F_c = F_c^*$ at $D_f/B = (D_f/B)_{cr}$ and remains constant thereafter.

Das (1978) also reported some model test results with square and rectangular foundations. Based on these test results, it was proposed that

$$\left(\frac{D_f}{B}\right)_{\text{cr-square}} = 0.107 c_u + 2.5 \le 7 \tag{4.42}$$

where

$\left(\dfrac{D_f}{B}\right)_{\text{cr-square}}$ = critical embedment ratio of square (or circular) foundations

c_u = undrained cohesion, in kN/m^2

It was also observed by Das (1980) that

$$\left(\frac{D_f}{B}\right)_{\text{cr-rectangular}} = \left(\frac{D_f}{B}\right)_{\text{cr-square}}\left[0.73 + 0.27\left(\frac{L}{B}\right)\right] \le 1.55\left(\frac{D_f}{B}\right)_{\text{cr-square}} \tag{4.43}$$

where

$\left(\dfrac{D_f}{B}\right)_{\text{cr-rectangular}}$ = critical embedment ratio of rectangular foundations

L = length of foundation

Based on these findings, Das (1980) proposed an empirical procedure to obtain the breakout factors for shallow and deep foundations. According to this procedure, α' and β' are two nondimensional factors defined as

$$\alpha' = \frac{\dfrac{D_f}{B}}{\left(\dfrac{D_f}{B}\right)_{cr}} \tag{4.44}$$

and

$$\beta' = \frac{F_c}{F_c^*} \tag{4.45}$$

For a given foundation, the critical embedment ratio can be calculated using Eqs. (4.42) and (4.43). The magnitude of F_c^* can be given by the following empirical relationship

$$F_{c\text{-rectangular}}^* = 7.56 + 1.44\left(\frac{B}{L}\right) \tag{4.46}$$

where $F_{c\text{-rectangular}}^*$ = breakout factor for deep rectangular foundations

Figure 4.22 shows the experimentally derived plots (upper limit, lower limit, and average of β' and α'. Following is a step-by-step procedure to estimate the ultimate uplift capacity.

Step 1. Determine the representative value of the undrained cohesion, c_u.
Step 2. Determine the critical embedment ratio using Eqs. (4.42) and (4.43).
Step 3. Determine the D_f/B ratio for the foundation.
Step 4. If $D_f/B > (D_f/B)_{cr}$, as determined in Step 2, it is a deep foundation. However, if $D_f/B \le (D_f/B)_{cr}$, it is a shallow foundation.
Step 5. For $D_f/B > (D_f/B)_{cr}$

$$F_c = F_c^* = 7.56 + 1.44\left(\frac{B}{L}\right)$$

Thus,

$$Q_u = A\left\{\left[7.56 + 1.44\left(\frac{B}{L}\right)\right]c_u + \gamma D_f\right\} \tag{4.47}$$

where A = area of the foundation.

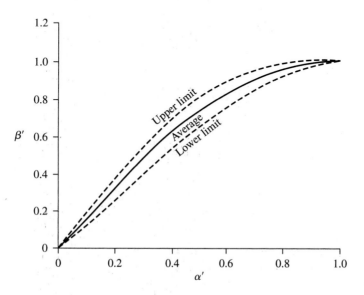

Figure 4.22 Plot of β' versus α'

Step 6. For $D_f/B \le (D_f/B)_{cr}$

$$Q_u = A(\beta' F_c^* c_u + \gamma D_f) = A\left\{\beta'\left[7.56 + 1.44\left(\frac{B}{L}\right)\right]c_u + \gamma D_f\right\} \quad (4.48)$$

The value of β' can be obtained from the average curve of Figure 4.22. The outlined procedure outlined above gives fairly good results for estimating the net ultimate uplift capacity of foundations and agrees reasonably well with the theoretical solution of Merifield et al. (2003).

Example 4.7

Consider a circular foundation in sand. Given for the foundation: diameter, $B = 1.5$ m and depth of embedment, $D_f = 1.5$ m. Given for the sand: unit weight, $\gamma = 17.4$ kN/m³, and friction angle, $\phi' = 35°$. Calculate the ultimate bearing capacity.

Solution
$D_f/B = 1.5/1.5 = 1$ and $\phi' = 35°$. For circular foundation, $(D_f/B)_{cr} = 5$. Hence, it is a shallow foundation. From Eq. (4.38)

$$F_q = 1 + 2\left[1 + m\left(\frac{D_f}{B}\right)\right]\left(\frac{D_f}{B}\right)K_u \tan\phi'$$

For $\phi' = 35°, m = 0.25$, and $K_u = 0.936$ (Table 4.2). So

$$F_q = 1 + 2[1 + (0.25)(1)](1)(0.936)(\tan 35) = 2.638$$

So

$$Q_u = F_q\gamma A D_f = (2.638)(17.4)\left[\left(\frac{\pi}{4}\right)(1.5)^2\right](1.5) = \mathbf{121.7 \ kN} \quad \blacksquare$$

Example 4.8

A rectangular foundation in a saturated clay measures 1.5 m × 3 m. Given: $D_f = 1.8$ m, $c_u = 52$ kN/m², and $\gamma = 18.9$ kN/m³. Estimate the ultimate uplift capacity.

Solution
From Eq. (4.42)

$$\left(\frac{D_f}{B}\right)_{\text{cr-square}} = 0.107c_u + 2.5 = (0.107)(52) + 2.5 = 8.06$$

So use $(D_f/B)_{\text{cr-square}} = 7$. Again from Eq. (4.43),

$$\left(\frac{D_f}{B}\right)_{\text{cr-rectangular}} = \left(\frac{D_f}{B}\right)_{\text{cr-square}}\left[0.73 + 0.27\left(\frac{L}{B}\right)\right]$$

$$= 7\left[0.73 + 0.27\left(\frac{3}{1.5}\right)\right] = 8.89$$

Check: $\qquad\qquad 1.55\left(\frac{D_f}{B}\right)_{\text{cr-square}} = (1.55)(7) = 10.85$

So use $(D_f/B)_{\text{cr-rectangular}} = 8.89$. The actual embedment ratio is $D_f/B = 1.8/1.5 = 1.2$.

Hence, this is a shallow foundation.

$$\alpha' = \frac{\dfrac{D_f}{B}}{\left(\dfrac{D_f}{B}\right)_{\text{cr}}} = \frac{1.2}{8.89} = 0.135$$

Referring to the average curve of Figure 4.22, for $\alpha' = 0.135$, the magnitude of $\beta' = 0.2$. From Eq. (4.48),

$$Q_u = A\left\{\beta'\left[7.56 + 1.44\left(\frac{B}{L}\right)\right]c_u + \gamma D_f\right\}$$

$$= (1.5)(3)\left\{(0.2)\left[7.56 + 1.44\left(\frac{1.5}{3}\right)\right](52) + (18.9)(1.8)\right\} = \textbf{540.6 kN} \quad\blacksquare$$

Problems

4.1 A rectangular foundation is shown in Figure P4.1. Determine the gross allowable load the foundation can carry, given that $B = 1.5$ m, $L = 2.5$ m, $D_f = 1.2$ m, $H = 1.5$ m, $\phi' = 40°$, $c' = 0$, and $\gamma = 17$ kN/m³. Use FS = 3.

4.2 Repeat Problem 4.1 with the following data: $B = 1.5$ m, $L = 1.5$ m, $D_f = 1$ m, $H = 0.6$ m, $\phi' = 35°$, $c' = 0$, and $\gamma = 15$ kN/m³. Use FS = 3.

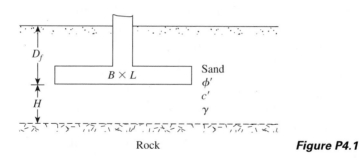

Rock   *Figure P4.1*

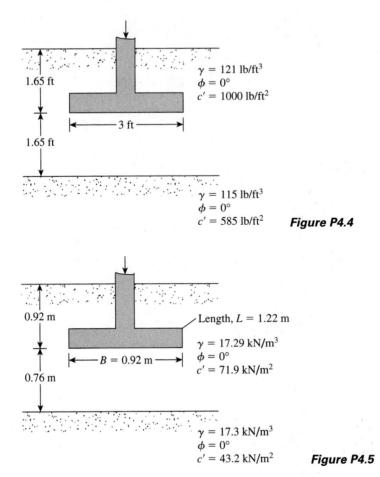

$\gamma = 121 \text{ lb/ft}^3$
$\phi = 0°$
$c' = 1000 \text{ lb/ft}^2$

1.65 ft

3 ft

1.65 ft

$\gamma = 115 \text{ lb/ft}^3$
$\phi = 0°$
$c' = 585 \text{ lb/ft}^2$ **Figure P4.4**

0.92 m

Length, $L = 1.22$ m

$B = 0.92$ m

$\gamma = 17.29 \text{ kN/m}^3$
$\phi = 0°$
$c' = 71.9 \text{ kN/m}^2$

0.76 m

$\gamma = 17.3 \text{ kN/m}^3$
$\phi = 0°$
$c' = 43.2 \text{ kN/m}^2$ **Figure P4.5**

4.3 A continuous foundation having a width of 1.4 m is supported by a saturated clay layer of limited depth underlain by a rock layer. Given that $D_f = 1$ m, $H = 0.7$ m, $c_u = 105 \text{ kN/m}^2$, and $\gamma = 18 \text{ kN/m}^3$, estimate the ultimate bearing capacity of the foundation.

4.4 A strip foundation in a two-layered clay is shown in Figure P4.4. Find the gross allowable bearing capacity. Use a factor of safety of 3.

4.5 Find the gross ultimate load that the footing shown in Figure P4.5 can carry.

4.6 Figure P4.6 shows a continuous foundation.
   a. If $H = 1.5$ m, determine the ultimate bearing capacity q_u.
   b. At what minimum value of H/B will the clay layer not have any effect on the ultimate bearing capacity of the foundation?

4.7 A square foundation on a layered sand is shown in Figure P4.7. Determine the net allowable load that the foundation can support. Use FS = 4.

4.8 Two continuous shallow foundations are constructed alongside each other in a granular soil. Given, for the foundation: $B = 3$ ft, $D_f = 3$ ft, center-to-center

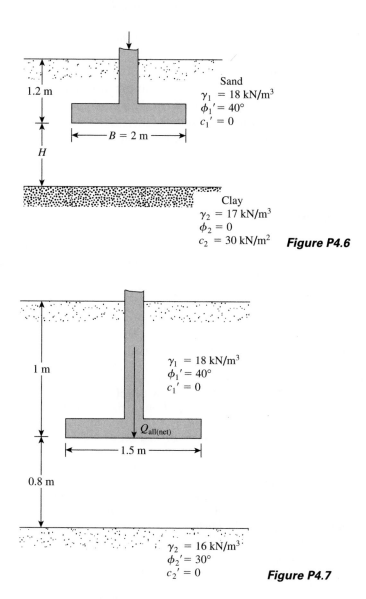

Sand
$\gamma_1 = 18$ kN/m³
$\phi_1' = 40°$
$c_1' = 0$

$B = 2$ m

H

Clay
$\gamma_2 = 17$ kN/m³
$\phi_2 = 0$
$c_2 = 30$ kN/m²   *Figure P4.6*

$\gamma_1 = 18$ kN/m³
$\phi_1' = 40°$
$c_1' = 0$

1 m

$Q_{all(net)}$

1.5 m

0.8 m

$\gamma_2 = 16$ kN/m³
$\phi_2' = 30°$
$c_2' = 0$ *Figure P4.7*

spacing = 6 ft. The soil friction angle, $\phi' = 35°$. Estimate the net allowable bearing capacity of the foundations. Use a factor of safety of FS = 4 and a unit weight of soil $\gamma = 110$ lb/ft³.

4.9 A continuous foundation with a width of 4 ft is located on a slope made of clay soil. Refer to Figure 4.13 and let $D_f = 4$ ft, $H = 15$ ft, $b = 6$ ft, $\gamma = 118$ lb/ft³, $c = 1500$ lb/ft², $\phi = 0$, and $\beta = 50°$.

a. Determine the allowable bearing capacity of the foundation. Let FS = 3.

b. Plot a graph of the ultimate bearing capacity q_u if b is changed from 0 to 20 ft.

4.10 A continuous foundation is to be constructed near a slope made of granular soil (see Figure 4.13). If $B = 4$ ft, $b = 6$ ft, $H = 15$ ft, $D_f = 4$ ft, $\beta = 30°$,

$\phi' = 40°$, and $\gamma = 110 \, \text{lb/ft}^3$, estimate the allowable bearing capacity of the foundation. Use FS = 4.

4.11 A square foundation in a sand deposit measures 4 ft × 4 ft in plan. Given: $D_f = 5$ ft, soil friction angle $= 35°$, and unit weight of soil $= 112 \, \text{lb/ft}^3$. Estimate the ultimate uplift capacity of the foundation.

4.12 A foundation measuring 1.2 m × 2.4 m in plan is constructed in a saturated clay. Given: depth of embedment of the foundation $= 2$ m, unit weight of soil $= 18 \, \text{kN/m}^3$, and undrained cohesion of clay $= 74 \, \text{kN/m}^2$. Estimate the ultimate uplift capacity of the foundation.

References

Buisman, A. S. K. (1940). *Grondmechanica*, Waltman, Delft, the Netherlands.

Das, B. M. (1978). "Model Tests for Uplift Capacity of Foundations in Clay," *Soils and Foundations*, Vol. 18, No. 2, pp. 17–24.

Das, B. M. (1980). "A Procedure for estimation of Ultimate Uplift Capacity of Foundations in Clay," *Soils and Foundations*, Vol. 20, No. 1, pp. 77–82.

Das, B. M., and Jones, A. D. (1982). "Uplift Capacity of Rectangular Foundations in Sand," *Transportation Research Record 884*, National Research Council, Washington, D.C., pp. 54–58.

Das, B. M., and Larbi-Cherif, S. (1982). "Bearing Capacity of Two Closely Spaced Shallow Foundations on Sand." *Soils and Foundations*, Vol. 23, No. 1, pp. 1–7.

Das, B. M., and Seeley, G. R. (1975). "Breakout Resistance of Horizontal Anchors," *Journal of Geotechnical Engineering Division*, ASCE, Vol. 101, No. 9, pp. 999–1003.

Lundgren, H., and Mortensen, K. (1953). "Determination by the Theory of Plasticity on the Bearing Capacity of Continuous Footings on Sand," *Proceedings, Third International Conference on Soil Mechanics and Foundation Engineering*, Zurich, Vol. 1, pp. 409–412.

Mandel, J., and Salencon, J. (1972). "Force portante d'un sol sur une assise rigide (étude théorique)," *Geotechnique*, Vol. 22, No. 1, pp. 79–93.

Merifield, R. S., Lyamin, A., and Sloan, S. W. (2003). "Three Dimensional Lower Bound Solutions for the Stability of Plate Anchors in Clay," *Journal of Geotechnical and Geoenvironmental Engineering*, ASCE, Vo. 129, No. 3, pp. 243–253.

Meyerhof, G. G. (1957). "The Ultimate Bearing Capacity of Foundations on Slopes," *Proceedings, Fourth International Conference on Soil Mechanics and Foundation Engineering*, London, Vol. 1, pp. 384–387.

Meyerhof, G. G. (1974). "Ultimate Bearing Capacity of Footings on Sand Layer Overlying Clay," *Canadian Geotechnical Journal*, Vol. 11, No. 2, pp. 224–229.

Meyerhof, G. G., and Adams, J. I. (1968). "The Ultimate Uplift Capacity of Foundations," *Canadian Geotechnical Journal*, Vol. 5, No. 4, pp. 225–244.

Meyerhof, G. G., and Chaplin, T. K. (1953). "The Compression and Bearing Capacity of Cohesive Soils," *British Journal of Applied Physics*, Vol. 4, pp. 20–26.

Meyerhof, G. G., and Hanna, A. M. (1978). "Ultimate Bearing Capacity of Foundations on Layered Soil under Inclined Load," *Canadian Geotechnical Journal*, Vol. 15, No. 4, pp. 565–572.

Prandtl, L. (1921). "Über die Eindringungsfestigkeit (Härte) plastischer Baustoffe und die Festigkeit von Schneiden," *Zeitschrift für angewandte Mathematik und Mechanik*, Vol. 1, No. 1, pp. 15–20.

Reissner, H. (1924). "Zum Erddruckproblem," *Proceedings, First International Congress of Applied Mechanics*, Delft, the Netherlands, pp. 295–311.

5

Shallow Foundations: Allowable Bearing Capacity and Settlement

5.1 Introduction

It was mentioned in Chapter 3 that, in many cases, the allowable settlement of a shallow foundation may control the allowable bearing capacity. The allowable settlement itself may be controlled by local building codes. Thus, the allowable bearing capacity will be the smaller of the following two conditions:

$$q_{all} = \begin{cases} \dfrac{q_u}{\text{FS}} \\ \text{or} \\ q_{\text{allowable settlement}} \end{cases}$$

The settlement of a foundation can be divided into two major categories: (a) elastic, or immediate, settlement and (b) consolidation settlement. Immediate, or elastic, settlement of a foundation takes place during or immediately after the construction of the structure. Consolidation settlement occurs over time. Pore water is extruded from the void spaces of saturated clayey soils submerged in water. The total settlement of a foundation is the sum of the elastic settlement and the consolidation settlement.

Consolidation settlement comprises two phases: *primary* and *secondary*. The fundamentals of primary consolidation settlement were explained in detail in Chapter 1. Secondary consolidation settlement occurs after the completion of primary consolidation caused by slippage and reorientation of soil particles under a sustained load. Primary consolidation settlement is more significant than secondary settlement in inorganic clays and silty soils. However, in organic soils, secondary consolidation settlement is more significant.

For the calculation of foundation settlement (both elastic and consolidation), it is required that we estimate the vertical stress increase in the soil mass due to the net load applied on the foundation. Hence, this chapter is divided into the following three parts:

1. Procedure for calculation of vertical stress increase
2. Elastic settlement calculation
3. Consolidation settlement calculation

Vertical Stress Increase in a Soil Mass Caused by Foundation Load

5.2   *Stress Due to a Concentrated Load*

In 1885, Boussinesq developed the mathematical relationships for determining the normal and shear stresses at any point inside *homogeneous, elastic,* and *isotropic* mediums due to a *concentrated point load* located at the surface, as shown in Figure 5.1. According to his analysis, the *vertical stress increase* at point A caused by a point load of magnitude P is given by

$$\Delta\sigma = \frac{3P}{2\pi z^2 \left[1 + \left(\dfrac{r}{z}\right)^2 \right]^{5/2}} \tag{5.1}$$

where

$$r = \sqrt{x^2 + y^2}$$
x, y, z = coordinates of the point A

Note that Eq. (5.1) is not a function of Poisson's ratio of the soil.

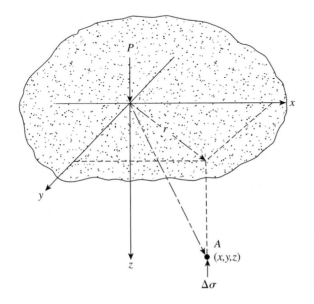

Figure 5.1   Vertical stress at a point A caused by a point load on the surface

Stress Due to a Circularly Loaded Area

The Boussinesq equation (5.1) can also be used to determine the vertical stress below the center of a flexible circularly loaded area, as shown in Figure 5.2. Let the radius of the loaded area be $B/2$, and let q_o be the uniformly distributed load per unit area. To determine the stress increase at a point A, located at a depth z below the center of the circular area, consider an elemental area on the circle. The load on this elemental area may be taken to be a point load and expressed as $q_o r\, d\theta\, dr$. The stress increase at A caused by this load can be determined from Eq. (5.1) as

$$d\sigma = \frac{3(q_o r\, d\theta\, dr)}{2\pi z^2 \left[1 + \left(\dfrac{r}{z}\right)^2\right]^{5/2}} \tag{5.2}$$

The total increase in stress caused by the entire loaded area may be obtained by integrating Eq. (5.2), or

$$\Delta\sigma = \int d\sigma = \int_{\theta=0}^{\theta=2\pi} \int_{r=0}^{r=B/2} \frac{3(q_o r\, d\theta\, dr)}{2\pi z^2 \left[1 + \left(\dfrac{r}{z}\right)^2\right]^{5/2}}$$

$$= q_o \left\{ 1 - \frac{1}{\left[1 + \left(\dfrac{B}{2z}\right)^2\right]^{3/2}} \right\} \tag{5.3}$$

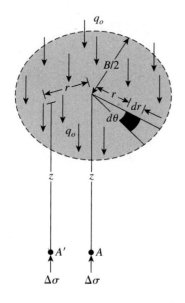

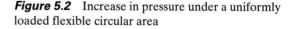

Figure 5.2 Increase in pressure under a uniformly loaded flexible circular area

Table 5.1 Variation of $\Delta\sigma/q_o$ for a Uniformly Loaded Flexible Circular Area

z/(B/2)	r/(B/2)					
	0	0.2	0.4	0.6	0.8	1.0
0	1.000	1.000	1.000	1.000	1.000	1.000
0.1	0.999	0.999	0.998	0.996	0.976	0.484
0.2	0.992	0.991	0.987	0.970	0.890	0.468
0.3	0.976	0.973	0.963	0.922	0.793	0.451
0.4	0.949	0.943	0.920	0.860	0.712	0.435
0.5	0.911	0.902	0.869	0.796	0.646	0.417
0.6	0.864	0.852	0.814	0.732	0.591	0.400
0.7	0.811	0.798	0.756	0.674	0.545	0.367
0.8	0.756	0.743	0.699	0.619	0.504	0.366
0.9	0.701	0.688	0.644	0.570	0.467	0.348
1.0	0.646	0.633	0.591	0.525	0.434	0.332
1.2	0.546	0.535	0.501	0.447	0.377	0.300
1.5	0.424	0.416	0.392	0.355	0.308	0.256
2.0	0.286	0.286	0.268	0.248	0.224	0.196
2.5	0.200	0.197	0.191	0.180	0.167	0.151
3.0	0.146	0.145	0.141	0.135	0.127	0.118
4.0	0.087	0.086	0.085	0.082	0.080	0.075

Similar integrations could be performed to obtain the vertical stress increase at A', located a distance r from the center of the loaded area at a depth z (Ahlvin and Ulery, 1962). Table 5.1 gives the variation of $\Delta\sigma/q_o$ with $r/(B/2)$ and $z/(B/2)$ [for $0 \leq r/(B/2) \leq 1$]. Note that the variation of $\Delta\sigma/q_o$ with depth at $r/(B/2) = 0$ can be obtained from Eq. (5.3).

5.4 Stress below a Rectangular Area

The integration technique of Boussinesq's equation also allows the vertical stress at any point A below the corner of a flexible rectangular loaded area to be evaluated. (See Figure 5.3.) To do so, consider an elementary area $dA = dx\,dy$ on the flexible loaded area. If the load per unit area is q_o, the total load on the elemental area is

$$dP = q_o\,dx\,dy \tag{5.4}$$

This elemental load, dP, may be treated as a point load. The increase in vertical stress at point A caused by dP may be evaluated by using Eq. (5.1). Note, however, the need to substitute $dP = q_o\,dx\,dy$ for P and $x^2 + y^2$ for r^2 in that equation. Thus,

The stress increase at A caused by $dP = \dfrac{3q_o\,(dx\,dy)z^3}{2\pi(x^2 + y^2 + z^2)^{5/2}}$

The total stress increase $\Delta\sigma$ caused by the entire loaded area at point A may now be obtained by integrating the preceding equation:

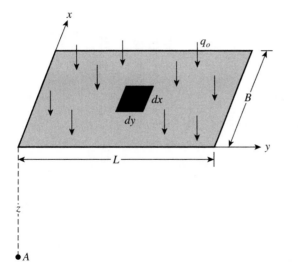

Figure 5.3 Determination of stress below the corner of a flexible rectangular loaded area

$$\Delta\sigma = \int_{y=0}^{L} \int_{x=0}^{B} \frac{3q_o\,(dx\,dy)z^3}{2\pi(x^2 + y^2 + z^2)^{5/2}} = q_oI \tag{5.5}$$

Here,

$$I = \text{influence factor} = \frac{1}{4\pi}\left(\frac{2mn\sqrt{m^2 + n^2 + 1}}{m^2 + n^2 + m^2n^2 + 1} \cdot \frac{m^2 + n^2 + 2}{m^2 + n^2 + 1}\right.$$

$$\left. + \tan^{-1}\frac{2mn\sqrt{m^2 + n^2 + 1}}{m^2 + n^2 + 1 - m^2n^2}\right) \tag{5.6}$$

When $m^2 + n^2 + 1 < m^2n^2$, the argument of $\tan^{-1}$ becomes negative. In that case,

$$I = \text{influence factor} = \frac{1}{4\pi}\left[\frac{2mn\sqrt{m^2 + n^2 + 1}}{m^2 + n^2 + m^2n^2 + 1} \cdot \frac{m^2 + n^2 + 2}{m^2 + n^2 + 1}\right.$$

$$\left. + \tan^{-1}\left(\pi - \frac{2mn\sqrt{m^2 + n^2 + 1}}{m^2 + n^2 + 1 - m^2n^2}\right)\right] \tag{5.6a}$$

where

$$m = \frac{B}{z} \tag{5.7}$$

and

$$n = \frac{L}{z} \tag{5.8}$$

The variations of the influence values with m and n are given in Table 5.2.

Table 5.2 Variation of Influence Value I [Eq. (5.6)][a]

						n						
m	0.1	0.2	0.3	0.4	0.5	0.6	0.7	0.8	0.9	1.0	1.2	1.4
0.1	0.00470	0.00917	0.01323	0.01678	0.01978	0.02223	0.02420	0.02576	0.02698	0.02794	0.02926	0.03007
0.2	0.00917	0.01790	0.02585	0.03280	0.03866	0.04348	0.04735	0.05042	0.05283	0.05471	0.05733	0.05894
0.3	0.01323	0.02585	0.03735	0.04742	0.05593	0.06294	0.06858	0.07308	0.07661	0.07938	0.08323	0.08561
0.4	0.01678	0.03280	0.04742	0.06024	0.07111	0.08009	0.08734	0.09314	0.09770	0.10129	0.10631	0.10941
0.5	0.01978	0.03866	0.05593	0.07111	0.08403	0.09473	0.10340	0.11035	0.11584	0.12018	0.12626	0.13003
0.6	0.02223	0.04348	0.06294	0.08009	0.09473	0.10688	0.11679	0.12474	0.13105	0.13605	0.14309	0.14749
0.7	0.02420	0.04735	0.06858	0.08734	0.10340	0.11679	0.12772	0.13653	0.14356	0.14914	0.15703	0.16199
0.8	0.02576	0.05042	0.07308	0.09314	0.11035	0.12474	0.13653	0.14607	0.15371	0.15978	0.16843	0.17389
0.9	0.02698	0.05283	0.07661	0.09770	0.11584	0.13105	0.14356	0.15371	0.16185	0.16835	0.17766	0.18357
1.0	0.02794	0.05471	0.07938	0.10129	0.12018	0.13605	0.14914	0.15978	0.16835	0.17522	0.18508	0.19139
1.2	0.02926	0.05733	0.08323	0.10631	0.12626	0.14309	0.15703	0.16843	0.17766	0.18508	0.19584	0.20278
1.4	0.03007	0.05894	0.08561	0.10941	0.13003	0.14749	0.16199	0.17389	0.18357	0.19139	0.20278	0.21020
1.6	0.03058	0.05994	0.08709	0.11135	0.13241	0.15028	0.16515	0.17739	0.18737	0.19546	0.20731	0.21510
1.8	0.03090	0.06058	0.08804	0.11260	0.13395	0.15207	0.16720	0.17967	0.18986	0.19814	0.21032	0.21836
2.0	0.03111	0.06100	0.08867	0.11342	0.13496	0.15326	0.16856	0.18119	0.19152	0.19994	0.21235	0.22058
2.5	0.03138	0.06155	0.08948	0.11450	0.13628	0.15483	0.17036	0.18321	0.19375	0.20236	0.21512	0.22364
3.0	0.03150	0.06178	0.08982	0.11495	0.13684	0.15550	0.17113	0.18407	0.19470	0.20341	0.21633	0.22499
4.0	0.03158	0.06194	0.09007	0.11527	0.13724	0.15598	0.17168	0.18469	0.19540	0.20417	0.21722	0.22600
5.0	0.03160	0.06199	0.09014	0.11537	0.13737	0.15612	0.17185	0.18488	0.19561	0.20440	0.21749	0.22632
6.0	0.03161	0.06201	0.09017	0.11541	0.13741	0.15617	0.17191	0.18496	0.19569	0.20449	0.21760	0.22644
8.0	0.03162	0.06202	0.09018	0.11543	0.13744	0.15621	0.17195	0.18500	0.19574	0.20455	0.21767	0.22652
10.0	0.03162	0.06202	0.09019	0.11544	0.13745	0.15622	0.17196	0.18502	0.19576	0.20457	0.21769	0.22654
∞	0.03162	0.06202	0.09019	0.11544	0.13745	0.15623	0.17197	0.18502	0.19577	0.20458	0.21770	0.22656

Table 5.2 (Continued)

m						n					
	1.6	1.8	2.0	2.5	3.0	4.0	5.0	6.0	8.0	10.0	∞
0.1	0.03058	0.03090	0.03111	0.03138	0.03150	0.03158	0.03160	0.03161	0.03162	0.03162	0.03162
0.2	0.05994	0.06058	0.06100	0.06155	0.06178	0.06194	0.06199	0.06201	0.06202	0.06202	0.06202
0.3	0.08709	0.08804	0.08867	0.08948	0.08982	0.09007	0.09014	0.09017	0.09018	0.09019	0.09019
0.4	0.11135	0.11260	0.11342	0.11450	0.11495	0.11527	0.11537	0.11541	0.11543	0.11544	0.11544
0.5	0.13241	0.13395	0.13496	0.13628	0.13684	0.13724	0.13737	0.13741	0.13744	0.13745	0.13745
0.6	0.15028	0.15207	0.15326	0.15483	0.15550	0.15598	0.15612	0.15617	0.15621	0.15622	0.15623
0.7	0.16515	0.16720	0.16856	0.17036	0.17113	0.17168	0.17185	0.17191	0.17195	0.17196	0.17197
0.8	0.17739	0.17967	0.18119	0.18321	0.18407	0.18469	0.18488	0.18496	0.18500	0.18502	0.18502
0.9	0.18737	0.18986	0.19152	0.19375	0.19470	0.19540	0.19561	0.19569	0.19574	0.19576	0.19577
1.0	0.19546	0.19814	0.19994	0.20236	0.20341	0.20417	0.20440	0.20449	0.20455	0.20457	0.20458
1.2	0.20731	0.21032	0.21235	0.21512	0.21633	0.21722	0.21749	0.21760	0.21767	0.21769	0.21770
1.4	0.21510	0.21836	0.22058	0.22364	0.22499	0.22600	0.22632	0.22644	0.22652	0.22654	0.22656
1.6	0.22025	0.22372	0.22610	0.22940	0.23088	0.23200	0.23236	0.23249	0.23258	0.23261	0.23263
1.8	0.22372	0.22736	0.22986	0.23334	0.23495	0.23617	0.23656	0.23671	0.23681	0.23684	0.23686
2.0	0.22610	0.22986	0.23247	0.23614	0.23782	0.23912	0.23954	0.23970	0.23981	0.23985	0.23987
2.5	0.22940	0.23334	0.23614	0.24010	0.24196	0.24344	0.24392	0.24412	0.24425	0.24429	0.24432
3.0	0.23088	0.23495	0.23782	0.24196	0.24394	0.24554	0.24608	0.24630	0.24646	0.24650	0.24654
4.0	0.23200	0.23617	0.23912	0.24344	0.24554	0.24729	0.24791	0.24817	0.24836	0.24842	0.24846
5.0	0.23236	0.23656	0.23954	0.24392	0.24608	0.24791	0.24857	0.24885	0.24907	0.24914	0.24919
6.0	0.23249	0.23671	0.23970	0.24412	0.24630	0.24817	0.24885	0.24916	0.24939	0.24946	0.24952
8.0	0.23258	0.23681	0.23981	0.24425	0.24646	0.24836	0.24907	0.24939	0.24964	0.24973	0.24980
10.0	0.23261	0.23684	0.23985	0.24429	0.24650	0.24842	0.24914	0.24946	0.24973	0.24981	0.24989
∞	0.23263	0.23686	0.23987	0.24432	0.24654	0.24846	0.24919	0.24952	0.24980	0.24989	0.25000

[a] After Newmark, 1935.

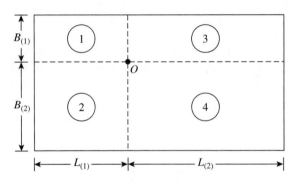

Figure 5.4 Stress below any point of a loaded flexible rectangular area

The stress increase at any point below a rectangular loaded area can also be found by using Eq. (5.5) in conjunction with Figure 5.4. To determine the stress at a depth z below point O, divide the loaded area into four rectangles, with O the corner common to each. Then use Eq. (5.5) to calculate the increase in stress at a depth z below O caused by each rectangular area. The total stress increase caused by the entire loaded area may now be expressed as

$$\Delta\sigma = q_o \left(I_1 + I_2 + I_3 + I_4 \right) \tag{5.9}$$

where I_1, I_2, I_3, and I_4 = the influence values of rectangles 1, 2, 3, and 4, respectively.

In most cases, the vertical stress below the center of a rectangular area is of importance. This can be given by the relationship

$$\Delta\sigma = q_o I_c \tag{5.10}$$

where

$$I_c = \frac{2}{\pi} \left[\frac{m_1 n_1}{\sqrt{1 + m_1^2 + n_1^2}} \frac{1 + m_1^2 + 2n_1^2}{(1 + n_1^2)(m_1^2 + n_1^2)} \right]$$

$$+ \sin^{-1} \frac{m_1}{\sqrt{m_1^2 + n_1^2}\sqrt{1 + n_1^2}} \tag{5.11}$$

$$m_1 = \frac{L}{B} \tag{5.12}$$

$$n_1 = \frac{z}{\left(\dfrac{B}{2}\right)} \tag{5.13}$$

The variation of I_c with m_1 and n_1 is given in Table 5.3.

Foundation engineers often use an approximate method to determine the increase in stress with depth caused by the construction of a foundation. The method

Table 5.3 Variation of I_c with m_1 and n_1

n_i	m_1									
	1	2	3	4	5	6	7	8	9	10
0.20	0.994	0.997	0.997	0.997	0.997	0.997	0.997	0.997	0.997	0.997
0.40	0.960	0.976	0.977	0.977	0.977	0.977	0.977	0.977	0.977	0.977
0.60	0.892	0.932	0.936	0.936	0.937	0.937	0.937	0.937	0.937	0.937
0.80	0.800	0.870	0.878	0.880	0.881	0.881	0.881	0.881	0.881	0.881
1.00	0.701	0.800	0.814	0.817	0.818	0.818	0.818	0.818	0.818	0.818
1.20	0.606	0.727	0.748	0.753	0.754	0.755	0.755	0.755	0.755	0.755
1.40	0.522	0.658	0.685	0.692	0.694	0.695	0.695	0.696	0.696	0.696
1.60	0.449	0.593	0.627	0.636	0.639	0.640	0.641	0.641	0.641	0.642
1.80	0.388	0.534	0.573	0.585	0.590	0.591	0.592	0.592	0.593	0.593
2.00	0.336	0.481	0.525	0.540	0.545	0.547	0.548	0.549	0.549	0.549
3.00	0.179	0.293	0.348	0.373	0.384	0.389	0.392	0.393	0.394	0.395
4.00	0.108	0.190	0.241	0.269	0.285	0.293	0.298	0.301	0.302	0.303
5.00	0.072	0.131	0.174	0.202	0.219	0.229	0.236	0.240	0.242	0.244
6.00	0.051	0.095	0.130	0.155	0.172	0.184	0.192	0.197	0.200	0.202
7.00	0.038	0.072	0.100	0.122	0.139	0.150	0.158	0.164	0.168	0.171
8.00	0.029	0.056	0.079	0.098	0.113	0.125	0.133	0.139	0.144	0.147
9.00	0.023	0.045	0.064	0.081	0.094	0.105	0.113	0.119	0.124	0.128
10.00	0.019	0.037	0.053	0.067	0.079	0.089	0.097	0.103	0.108	0.112

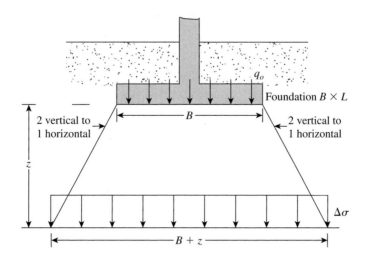

Figure 5.5 2:1 method of finding stress increase under a foundation

is referred to as the *2:1 method*. (See Figure 5.5.) According to this method, the increase in stress at depth z is

$$\Delta\sigma = \frac{q_0 \times B \times L}{(B + z)(L + z)} \qquad (5.14)$$

Note that Eq. (5.14) is based on the assumption that the stress from the foundation spreads out along lines with a *vertical-to-horizontal slope of 2:1.*

Example 5.1

A flexible rectangular area measures 5 ft × 10 ft in plan. It supports a load of 2000 lb/ft^2.

Determine the vertical stress increase due to the load at a depth of 12.5 ft below the center of the rectangular area.

Solution
Refer to Figure 5.4. For this case,

$$B_1 = B_2 = \frac{5}{2} = 2.5 \text{ ft}$$

$$L_1 = L_2 = \frac{10}{2} = 5 \text{ ft}$$

From Eqs. (5.7) and (5.8),

$$m = \frac{B_1}{z} = \frac{B_2}{z} = \frac{2.5}{12.5} = 0.2$$

$$n = \frac{L_1}{z} = \frac{L_2}{z} = \frac{5}{12.5} = 0.4$$

From Table 5.2, from $m = 0.2$ and $n = 0.4$, the value of $I = 0.0328$. Thus,

$$\Delta\sigma = q_0(4I) = (2000)(4)(0.0328) = \textbf{262.4 lb/ft}^2$$

Alternate Solution
From Eq. (5.10),

$$\Delta\sigma = q_o I_c$$

$$m_1 = \frac{L}{B} = \frac{10}{5} = 2$$

$$n_1 = \frac{z}{\left(\dfrac{B}{2}\right)} = \frac{12.5}{\left(\dfrac{5}{2}\right)} = 5$$

From Table 5.3, for $m_1 = 2$ and $n_1 = 5$, the value of $I_c = 0.131$. Thus,

$$\Delta\sigma = (2000)(0.131) = \textbf{262 lb/ft}^2 \qquad \blacksquare$$

5.5   *Average Vertical Stress Increase Due to a Rectangularly Loaded Area*

In Section 5.4, the vertical stress increase below the corner of a uniformly loaded rectangular area was given as

$$\Delta\sigma = q_o I$$

In many cases, one must find the average stress increase, $\Delta\sigma_{av}$, below the corner of a uniformly loaded rectangular area with limits of $z = 0$ to $z = H$, as shown in Figure 5.6. This can be evaluated as

$$\Delta\sigma_{av} = \frac{1}{H} \int_0^H (q_o I) \, dz = q_o I_a \qquad (5.15)$$

where

$$I_a = f(m_2, n_2) \qquad (5.16)$$

$$m_2 = \frac{B}{H} \qquad (5.17)$$

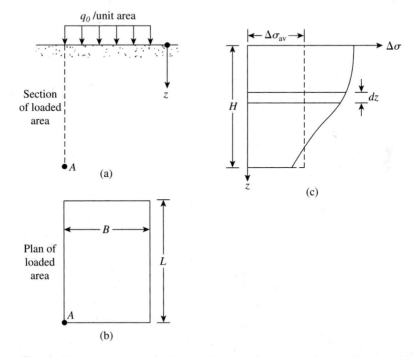

Figure 5.6   Average vertical stress increase due to a rectangularly loaded flexible area

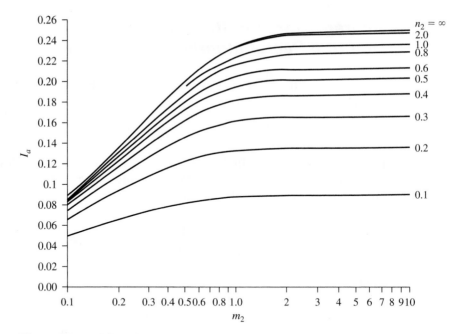

Figure 5.7   Griffiths' influence factor I_a

and

$$n_2 = \frac{L}{H} \qquad (5.18)$$

The variation of I_a with m_2 and n_2 is shown in Figure 5.7, as proposed by Griffiths (1984).

In estimating the consolidation settlement under a foundation, it may be required to determine the average vertical stress increase in only a given layer—that is, between $z = H_1$ and $z = H_2$, as shown in Figure 5.8. This can be done as (Griffiths, 1984)

$$\Delta\sigma_{av(H_2/H_1)} = q_o \left[\frac{H_2 I_{a(H_2)} - H_1 I_{a(H_1)}}{H_2 - H_1} \right] \qquad (5.19)$$

where

$\Delta\sigma_{av(H_2/H_1)}$ = average stress increase immediately below the corner of a uniformly loaded rectangular area between depths $z = H_1$ and $z = H_2$

$$I_{a(H_2)} = I_a \text{ for } z = 0 \text{ to } z = H_2 = f\left(m_2 = \frac{B}{H_2}, n_2 = \frac{L}{H_2} \right)$$

$$I_{a(H_1)} = I_a \text{ for } z = 0 \text{ to } z = H_1 = f\left(m_2 = \frac{B}{H_1}, n_2 = \frac{L}{H_1} \right)$$

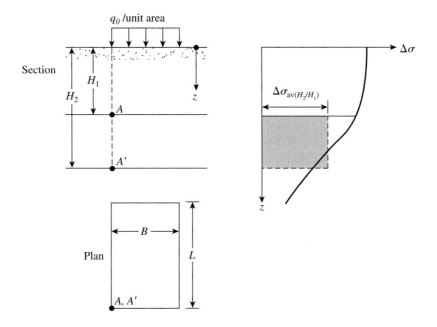

Figure 5.8 Average pressure increase between $z = H_1$ and $z = H_2$ below the corner of a uniformly loaded rectangular area

Example 5.2

Refer to Figure 5.9. Determine the *average* stress increase below the center of the loaded area between $z = 3$ m to $z = 5$ m (that is, between points A and A').

Solution
Refer to Figure 5.9. The loaded area can be divided into four rectangular areas, each measuring 1.5 m × 1.5 m ($L \times B$). Using Eq. (5.19), the average stress increase (between the required depths) below the corner of each rectangular area can be given as

$$\Delta\sigma_{\text{av} (H_2/H_1)} = q_o \left[\frac{H_2 I_{a(H_2)} - H_1 I_{a(H_1)}}{H_2 - H_1} \right] = 100 \left[\frac{(5) I_{a(H_2)} - (3) I_{a(H_1)}}{5 - 3} \right]$$

For $I_{a(H_2)}$:

$$m_2 = \frac{B}{H_2} = \frac{1.5}{5} = 0.3$$

$$n_2 = \frac{L}{H_2} = \frac{1.5}{5} = 0.3$$

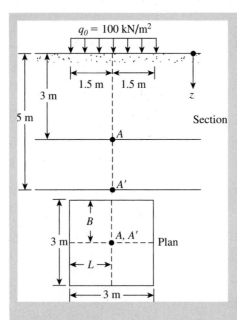

Figure 5.9   Determination of average increase in stress below a rectangular area

Referring to Figure 5.7, for $m_2 = 0.3$ and $n_2 = 0.3$, $I_{a(H_2)} = 0.126$. For $I_{a(H_1)}$:

$$m_2 = \frac{B}{H_1} = \frac{1.5}{3} = 0.5$$

$$n_2 = \frac{L}{H_1} = \frac{1.5}{3} = 0.5$$

Referring to Figure 5.7, $I_{a(H_1)} = 0.175$, so

$$\Delta\sigma_{av\,(H_2/H_1)} = 100\left[\frac{(5)(0.126) - (3)(0.175)}{5 - 3}\right] = \textbf{5.25 kN/m}^2$$

The stress increase between $z = 3$ m to $z = 5$ m below the center of the loaded area is equal to

$$4\Delta\sigma_{av\,(H_2/H_1)} = (4)(5.25) = \textbf{21 kN/m}^2 \qquad \blacksquare$$

5.6   *Stress Increase under an Embankment*

Figure 5.10 shows the cross section of an embankment of height H. For this two-dimensional loading condition, the vertical stress increase may be expressed as

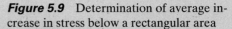

$$\Delta\sigma = \frac{q_o}{\pi}\left[\left(\frac{B_1 + B_2}{B_2}\right)(\alpha_1 + \alpha_2) - \frac{B_1}{B_2}(\alpha_2)\right] \qquad (5.20)$$

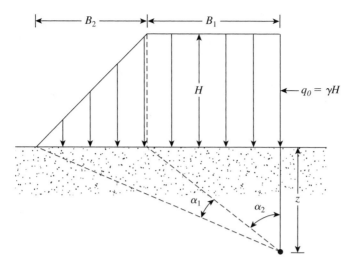

Figure 5.10 Embankment loading

where

$q_o = \gamma H$
γ = unit weight of the embankment soil
H = height of the embankment

$$\alpha_1 = \tan^{-1}\left(\frac{B_1 + B_2}{z}\right) - \tan^{-1}\left(\frac{B_1}{z}\right) \tag{5.21}$$

$$\alpha_2 = \tan^{-1}\left(\frac{B_1}{z}\right) \tag{5.22}$$

(Note that α_1 and α_2 are in radians.)

For a detailed derivation of Eq. (5.20), see Das (1997). A simplified form of the equation is

$$\Delta\sigma = q_o I' \tag{5.23}$$

where I' = a function of B_1/z and B_2/z.

The variation of I' with B_1/z and B_2/z is shown in Figure 5.11. An application of this diagram is given in Example 5.3.

Example 5.3

An embankment is shown in Figure 5.12a. Determine the stress increase under the embankment at points A_1 and A_2.

Solution
We have

$$\gamma H = (17.5)(7) = 122.5 \text{ kN/m}^2$$

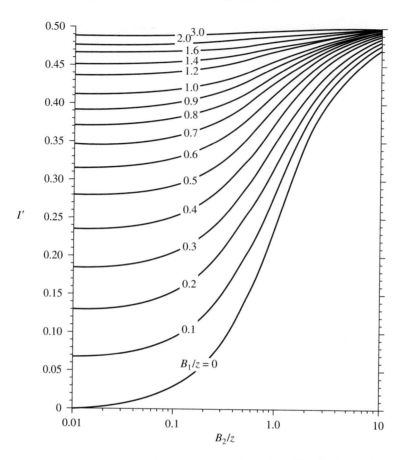

Figure 5.11   Influence value I' for embankment loading (After Osterberg, 1957)

Stress Increase at A_1

The left side of Figure 5.12b indicates that $B_1 = 2.5$ m and $B_2 = 14$ m, so

$$\frac{B_1}{z} = \frac{2.5}{5} = 0.5$$

and

$$\frac{B_2}{z} = \frac{14}{5} = 2.8$$

According to Figure 5.11, in this case $I' = 0.445$. Because the two sides in Figure 5.12b are symmetrical, the value of I' for the right side will also be 0.445, so

$$\Delta\sigma = \Delta\sigma_{(1)} + \Delta\sigma_{(2)} = q_o[I'_{(\text{left side})} + I'_{(\text{right side})}]$$

$$= 122.5[0.445 + 0.445] = \mathbf{109.03 \ kN/m^2}$$

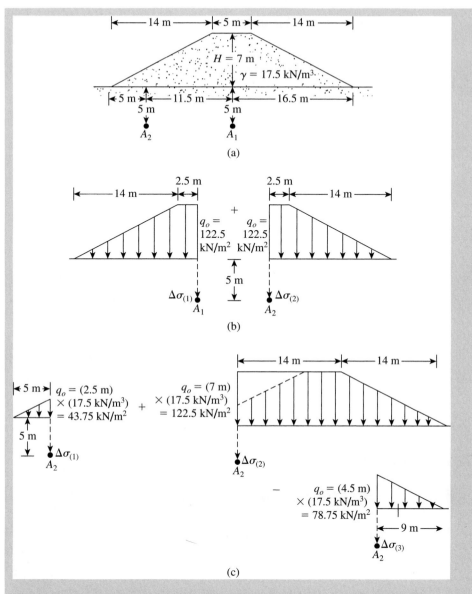

Figure 5.12   Stress increase due to embankment loading

Stress Increase at A_2

In Figure 5.12c, for the left side, $B_2 = 5$ m and $B_1 = 0$, so

$$\frac{B_2}{z} = \frac{5}{5} = 1$$

and

$$\frac{B_1}{z} = \frac{0}{5} = 0$$

According to Figure 5.11, for these values of B_2/z and B_1/z, $I' = 0.24$; hence,

$$\Delta\sigma_{(1)} = 43.75(0.24) = 10.5 \text{ kN/m}^2$$

For the middle section,

$$\frac{B_2}{z} = \frac{14}{5} = 2.8$$

and

$$\frac{B_1}{z} = \frac{14}{5} = 2.8$$

Thus, $I' = 0.495$, so

$$\Delta\sigma_{(2)} = 0.495(122.5) = 60.64 \text{ kN/m}^2$$

For the right side,

$$\frac{B_2}{z} = \frac{9}{5} = 1.8$$

$$\frac{B_1}{z} = \frac{0}{5} = 0$$

and $I' = 0.335$, so

$$\Delta\sigma_{(3)} = (78.75)(0.335) = 26.38 \text{ kN/m}^2$$

The total stress increase at point A_2 is

$$\Delta\sigma = \Delta\sigma_{(1)} + \Delta\sigma_{(2)} - \Delta\sigma_{(3)} = 10.5 + 60.64 - 26.38 = \textbf{44.76 kN/m}^2 \quad \blacksquare$$

Elastic Settlement

5.7   *Elastic Settlement Based on the Theory of Elasticity*

The elastic settlement of a shallow foundation can be estimated by using the theory of elasticity. From Hooke's law, as applied to Figure 5.13, we obtain

$$S_e = \int_0^H \varepsilon_z dz = \frac{1}{E_s} \int_0^H (\Delta\sigma_z - \mu_s\Delta\sigma_x - \mu_s\Delta\sigma_y)dz \qquad (5.24)$$

where

S_e = elastic settlement
E_s = modulus of elasticity of soil
H = thickness of the soil layer
μ_s = Poisson's ratio of the soil
$\Delta\sigma_x, \Delta\sigma_y, \Delta\sigma_z$ = stress increase due to the net applied foundation load in the x, y, and z directions, respectively

Theoretically, if the foundation is perfectly flexible (see Figure 5.14 and Bowles, 1987), the settlement may be expressed as

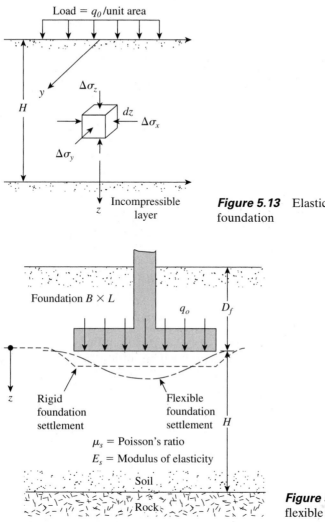

Figure 5.13 Elastic settlement of shallow foundation

Figure 5.14 Elastic settlement of flexible and rigid foundations

$$S_e = q_o(\alpha B')\frac{1 - \mu_s^2}{E_s}I_sI_f \qquad (5.25)$$

where

q_o = net applied pressure on the foundation
μ_s = Poisson's ratio of soil
E_s = average modulus of elasticity of the soil under the foundation, measured from
$\qquad z = 0$ to about $z = 4B$
B' = $B/2$ for center of foundation
$\quad$ = B for corner of foundation

$$I_s = \text{shape factor (Steinbrenner, 1934)}$$

$$= F_1 + \frac{1 - 2\mu_s}{1 - \mu_s}F_2 \tag{5.26}$$

$$F_1 = \frac{1}{\pi}(A_0 + A_1) \tag{5.27}$$

$$F_2 = \frac{n'}{2\pi}\tan^{-1}A_2 \tag{5.28}$$

$$A_0 = m'\ln\frac{\left(1 + \sqrt{m'^2 + 1}\right)\sqrt{m'^2 + n'^2}}{m'\left(1 + \sqrt{m'^2 + n'^2 + 1}\right)} \tag{5.29}$$

$$A_1 = \ln\frac{\left(m' + \sqrt{m'^2 + 1}\right)\sqrt{1 + n'^2}}{m' + \sqrt{m'^2 + n'^2 + 1}} \tag{5.30}$$

$$A_2 = \frac{m'}{n'\sqrt{m'^2 + n'^2 + 1}} \tag{5.31}$$

$$I_f = \text{depth factor (Fox, 1948)} = f\left(\frac{D_f}{B}, \mu_s, \text{ and } \frac{L}{B}\right) \tag{5.32}$$

$\alpha = $ a factor that depends on the location on the foundation where settlement is being calculated

To calculate settlement at the *center* of the foundation, we use

$$\alpha = 4$$
$$m' = \frac{L}{B}$$

and

$$n' = \frac{H}{\left(\dfrac{B}{2}\right)}$$

To calculate settlement at a *corner* of the foundation,

$$\alpha = 1$$
$$m' = \frac{L}{B}$$

and

$$n' = \frac{H}{B}$$

The variations of F_1 and F_2 [see Eqs. (5.27) and (5.28)] with m' and n' are given in Tables 5.4 and 5.5. Also, the variation of I_f with D_f/B (for $\mu_s = 0, 0.3, 0.4,$ and 0.5) is shown in Figure 5.15. These values are also given in tabular form by Bowles (1987).

Table 5.4   Variation of F_1 with m' and n'

					m'					
n'	1.0	1.2	1.4	1.6	1.8	2.0	2.5	3.0	3.5	4.0
0.25	0.014	0.013	0.012	0.011	0.011	0.011	0.010	0.010	0.010	0.010
0.50	0.049	0.046	0.044	0.042	0.041	0.040	0.038	0.038	0.037	0.037
0.75	0.095	0.090	0.087	0.084	0.082	0.080	0.077	0.076	0.074	0.074
1.00	0.142	0.138	0.134	0.130	0.127	0.125	0.121	0.118	0.116	0.115
1.25	0.186	0.183	0.179	0.176	0.173	0.170	0.165	0.161	0.158	0.157
1.50	0.224	0.224	0.222	0.219	0.216	0.213	0.207	0.203	0.199	0.197
1.75	0.257	0.259	0.259	0.258	0.255	0.253	0.247	0.242	0.238	0.235
2.00	0.285	0.290	0.292	0.292	0.291	0.289	0.284	0.279	0.275	0.271
2.25	0.309	0.317	0.321	0.323	0.323	0.322	0.317	0.313	0.308	0.305
2.50	0.330	0.341	0.347	0.350	0.351	0.351	0.348	0.344	0.340	0.336
2.75	0.348	0.361	0.369	0.374	0.377	0.378	0.377	0.373	0.369	0.365
3.00	0.363	0.379	0.389	0.396	0.400	0.402	0.402	0.400	0.396	0.392
3.25	0.376	0.394	0.406	0.415	0.420	0.423	0.426	0.424	0.421	0.418
3.50	0.388	0.408	0.422	0.431	0.438	0.442	0.447	0.447	0.444	0.441
3.75	0.399	0.420	0.436	0.447	0.454	0.460	0.467	0.458	0.466	0.464
4.00	0.408	0.431	0.448	0.460	0.469	0.476	0.484	0.487	0.486	0.484
4.25	0.417	0.440	0.458	0.472	0.481	0.484	0.495	0.514	0.515	0.515
4.50	0.424	0.450	0.469	0.484	0.495	0.503	0.516	0.521	0.522	0.522
4.75	0.431	0.458	0.478	0.494	0.506	0.515	0.530	0.536	0.539	0.539
5.00	0.437	0.465	0.487	0.503	0.516	0.526	0.543	0.551	0.554	0.554
5.25	0.443	0.472	0.494	0.512	0.526	0.537	0.555	0.564	0.568	0.569
5.50	0.448	0.478	0.501	0.520	0.534	0.546	0.566	0.576	0.581	0.584
5.75	0.453	0.483	0.508	0.527	0.542	0.555	0.576	0.588	0.594	0.597
6.00	0.457	0.489	0.514	0.534	0.550	0.563	0.585	0.598	0.606	0.609
6.25	0.461	0.493	0.519	0.540	0.557	0.570	0.594	0.609	0.617	0.621
6.50	0.465	0.498	0.524	0.546	0.563	0.577	0.603	0.618	0.627	0.632
6.75	0.468	0.502	0.529	0.551	0.569	0.584	0.610	0.627	0.637	0.643
7.00	0.471	0.506	0.533	0.556	0.575	0.590	0.618	0.635	0.646	0.653
7.25	0.474	0.509	0.538	0.561	0.580	0.596	0.625	0.643	0.655	0.662
7.50	0.477	0.513	0.541	0.565	0.585	0.601	0.631	0.650	0.663	0.671
7.75	0.480	0.516	0.545	0.569	0.589	0.606	0.637	0.658	0.671	0.680
8.00	0.482	0.519	0.549	0.573	0.594	0.611	0.643	0.664	0.678	0.688
8.25	0.485	0.522	0.552	0.577	0.598	0.615	0.648	0.670	0.685	0.695
8.50	0.487	0.524	0.555	0.580	0.601	0.619	0.653	0.676	0.692	0.703
8.75	0.489	0.527	0.558	0.583	0.605	0.623	0.658	0.682	0.698	0.710
9.00	0.491	0.529	0.560	0.587	0.609	0.627	0.663	0.687	0.705	0.716
9.25	0.493	0.531	0.563	0.589	0.612	0.631	0.667	0.693	0.710	0.723
9.50	0.495	0.533	0.565	0.592	0.615	0.634	0.671	0.697	0.716	0.719
9.75	0.496	0.536	0.568	0.595	0.618	0.638	0.675	0.702	0.721	0.735
10.00	0.498	0.537	0.570	0.597	0.621	0.641	0.679	0.707	0.726	0.740
20.00	0.529	0.575	0.614	0.647	0.677	0.702	0.756	0.797	0.830	0.858
50.00	0.548	0.598	0.640	0.678	0.711	0.740	0.803	0.853	0.895	0.931
100.00	0.555	0.605	0.649	0.688	0.722	0.753	0.819	0.872	0.918	0.956

Table 5.4 (Continued)

					m'					
n'	4.5	5.0	6.0	7.0	8.0	9.0	10.0	25.0	50.0	100.0
0.25	0.010	0.010	0.010	0.010	0.010	0.010	0.010	0.010	0.010	0.010
0.50	0.036	0.036	0.036	0.036	0.036	0.036	0.036	0.036	0.036	0.036
0.75	0.073	0.073	0.072	0.072	0.072	0.072	0.071	0.071	0.071	0.071
1.00	0.114	0.113	0.112	0.112	0.112	0.111	0.111	0.110	0.110	0.110
1.25	0.155	0.154	0.153	0.152	0.152	0.151	0.151	0.150	0.150	0.150
1.50	0.195	0.194	0.192	0.191	0.190	0.190	0.189	0.188	0.188	0.188
1.75	0.233	0.232	0.229	0.228	0.227	0.226	0.225	0.223	0.223	0.223
2.00	0.269	0.267	0.264	0.262	0.261	0.260	0.259	0.257	0.256	0.256
2.25	0.302	0.300	0.296	0.294	0.293	0.291	0.291	0.287	0.287	0.287
2.50	0.333	0.331	0.327	0.324	0.322	0.321	0.320	0.316	0.315	0.315
2.75	0.362	0.359	0.355	0.352	0.350	0.348	0.347	0.343	0.342	0.342
3.00	0.389	0.386	0.382	0.378	0.376	0.374	0.373	0.368	0.367	0.367
3.25	0.415	0.412	0.407	0.403	0.401	0.399	0.397	0.391	0.390	0.390
3.50	0.438	0.435	0.430	0.427	0.424	0.421	0.420	0.413	0.412	0.411
3.75	0.461	0.458	0.453	0.449	0.446	0.443	0.441	0.433	0.432	0.432
4.00	0.482	0.479	0.474	0.470	0.466	0.464	0.462	0.453	0.451	0.451
4.25	0.516	0.496	0.484	0.473	0.471	0.471	0.470	0.468	0.462	0.460
4.50	0.520	0.517	0.513	0.508	0.505	0.502	0.499	0.489	0.487	0.487
4.75	0.537	0.535	0.530	0.526	0.523	0.519	0.517	0.506	0.504	0.503
5.00	0.554	0.552	0.548	0.543	0.540	0.536	0.534	0.522	0.519	0.519
5.25	0.569	0.568	0.564	0.560	0.556	0.553	0.550	0.537	0.534	0.534
5.50	0.584	0.583	0.579	0.575	0.571	0.568	0.585	0.551	0.549	0.548
5.75	0.597	0.597	0.594	0.590	0.586	0.583	0.580	0.565	0.583	0.562
6.00	0.611	0.610	0.608	0.604	0.601	0.598	0.595	0.579	0.576	0.575
6.25	0.623	0.623	0.621	0.618	0.615	0.611	0.608	0.592	0.589	0.588
6.50	0.635	0.635	0.634	0.631	0.628	0.625	0.622	0.605	0.601	0.600
6.75	0.646	0.647	0.646	0.644	0.641	0.637	0.634	0.617	0.613	0.612
7.00	0.656	0.658	0.658	0.656	0.653	0.650	0.647	0.628	0.624	0.623
7.25	0.666	0.669	0.669	0.668	0.665	0.662	0.659	0.640	0.635	0.634
7.50	0.676	0.679	0.680	0.679	0.676	0.673	0.670	0.651	0.646	0.645
7.75	0.685	0.688	0.690	0.689	0.687	0.684	0.681	0.661	0.656	0.655
8.00	0.694	0.697	0.700	0.700	0.698	0.695	0.692	0.672	0.666	0.665
8.25	0.702	0.706	0.710	0.710	0.708	0.705	0.703	0.682	0.676	0.675
8.50	0.710	0.714	0.719	0.719	0.718	0.715	0.713	0.692	0.686	0.684
8.75	0.717	0.722	0.727	0.728	0.727	0.725	0.723	0.701	0.695	0.693
9.00	0.725	0.730	0.736	0.737	0.736	0.735	0.732	0.710	0.704	0.702
9.25	0.731	0.737	0.744	0.746	0.745	0.744	0.742	0.719	0.713	0.711
9.50	0.738	0.744	0.752	0.754	0.754	0.753	0.751	0.728	0.721	0.719
9.75	0.744	0.751	0.759	0.762	0.762	0.761	0.759	0.737	0.729	0.727
10.00	0.750	0.758	0.766	0.770	0.770	0.770	0.768	0.745	0.738	0.735
20.00	0.878	0.896	0.925	0.945	0.959	0.969	0.977	0.982	0.965	0.957
50.00	0.962	0.989	1.034	1.070	1.100	1.125	1.146	1.265	1.279	1.261
100.00	0.990	1.020	1.072	1.114	1.150	1.182	1.209	1.408	1.489	1.499

Table 5.5 Variation of F_2 with m' and n'

					m'					
n'	1.0	1.2	1.4	1.6	1.8	2.0	2.5	3.0	3.5	4.0
0.25	0.049	0.050	0.051	0.051	0.051	0.052	0.052	0.052	0.052	0.052
0.50	0.074	0.077	0.080	0.081	0.083	0.084	0.086	0.086	0.0878	0.087
0.75	0.083	0.089	0.093	0.097	0.099	0.101	0.104	0.106	0.107	0.108
1.00	0.083	0.091	0.098	0.102	0.106	0.109	0.114	0.117	0.119	0.120
1.25	0.080	0.089	0.096	0.102	0.107	0.111	0.118	0.122	0.125	0.127
1.50	0.075	0.084	0.093	0.099	0.105	0.110	0.118	0.124	0.128	0.130
1.75	0.069	0.079	0.088	0.095	0.101	0.107	0.117	0.123	0.128	0.131
2.00	0.064	0.074	0.083	0.090	0.097	0.102	0.114	0.121	0.127	0.131
2.25	0.059	0.069	0.077	0.085	0.092	0.098	0.110	0.119	0.125	0.130
2.50	0.055	0.064	0.073	0.080	0.087	0.093	0.106	0.115	0.122	0.127
2.75	0.051	0.060	0.068	0.076	0.082	0.089	0.102	0.111	0.119	0.125
3.00	0.048	0.056	0.064	0.071	0.078	0.084	0.097	0.108	0.116	0.122
3.25	0.045	0.053	0.060	0.067	0.074	0.080	0.093	0.104	0.112	0.119
3.50	0.042	0.050	0.057	0.064	0.070	0.076	0.089	0.100	0.109	0.116
3.75	0.040	0.047	0.054	0.060	0.067	0.073	0.086	0.096	0.105	0.113
4.00	0.037	0.044	0.051	0.057	0.063	0.069	0.082	0.093	0.102	0.110
4.25	0.036	0.042	0.049	0.055	0.061	0.066	0.079	0.090	0.099	0.107
4.50	0.034	0.040	0.046	0.052	0.058	0.063	0.076	0.086	0.096	0.104
4.75	0.032	0.038	0.044	0.050	0.055	0.061	0.073	0.083	0.093	0.101
5.00	0.031	0.036	0.042	0.048	0.053	0.058	0.070	0.080	0.090	0.098
5.25	0.029	0.035	0.040	0.046	0.051	0.056	0.067	0.078	0.087	0.095
5.50	0.028	0.033	0.039	0.044	0.049	0.054	0.065	0.075	0.084	0.092
5.75	0.027	0.032	0.037	0.042	0.047	0.052	0.063	0.073	0.082	0.090
6.00	0.026	0.031	0.036	0.040	0.045	0.050	0.060	0.070	0.079	0.087
6.25	0.025	0.030	0.034	0.039	0.044	0.048	0.058	0.068	0.077	0.085
6.50	0.024	0.029	0.033	0.038	0.042	0.046	0.056	0.066	0.075	0.083
6.75	0.023	0.028	0.032	0.036	0.041	0.045	0.055	0.064	0.073	0.080
7.00	0.022	0.027	0.031	0.035	0.039	0.043	0.053	0.062	0.071	0.078
7.25	0.022	0.026	0.030	0.034	0.038	0.042	0.051	0.060	0.069	0.076
7.50	0.021	0.025	0.029	0.033	0.037	0.041	0.050	0.059	0.067	0.074
7.75	0.020	0.024	0.028	0.032	0.036	0.039	0.048	0.057	0.065	0.072
8.00	0.020	0.023	0.027	0.031	0.035	0.038	0.047	0.055	0.063	0.071
8.25	0.019	0.023	0.026	0.030	0.034	0.037	0.046	0.054	0.062	0.069
8.50	0.018	0.022	0.026	0.029	0.033	0.036	0.045	0.053	0.060	0.067
8.75	0.018	0.021	0.025	0.028	0.032	0.035	0.043	0.051	0.059	0.066
9.00	0.017	0.021	0.024	0.028	0.031	0.034	0.042	0.050	0.057	0.064
9.25	0.017	0.020	0.024	0.027	0.030	0.033	0.041	0.049	0.056	0.063
9.50	0.017	0.020	0.023	0.026	0.029	0.033	0.040	0.048	0.055	0.061
9.75	0.016	0.019	0.023	0.026	0.029	0.032	0.039	0.047	0.054	0.060
10.00	0.016	0.019	0.022	0.025	0.028	0.031	0.038	0.046	0.052	0.059
20.00	0.008	0.010	0.011	0.013	0.014	0.016	0.020	0.024	0.027	0.031
50.00	0.003	0.004	0.004	0.005	0.006	0.006	0.008	0.010	0.011	0.013
100.00	0.002	0.002	0.002	0.003	0.003	0.003	0.004	0.005	0.006	0.006

Table 5.5 (Continued)

					m'					
n'	4.5	5.0	6.0	7.0	8.0	9.0	10.0	25.0	50.0	100.0
0.25	0.053	0.053	0.053	0.053	0.053	0.053	0.053	0.053	0.053	0.053
0.50	0.087	0.087	0.088	0.088	0.088	0.088	0.088	0.088	0.088	0.088
0.75	0.109	0.109	0.109	0.110	0.110	0.110	0.110	0.111	0.111	0.111
1.00	0.121	0.122	0.123	0.123	0.124	0.124	0.124	0.125	0.125	0.125
1.25	0.128	0.130	0.131	0.132	0.132	0.133	0.133	0.134	0.134	0.134
1.50	0.132	0.134	0.136	0.137	0.138	0.138	0.139	0.140	0.140	0.140
1.75	0.134	0.136	0.138	0.140	0.141	0.142	0.142	0.144	0.144	0.145
2.00	0.134	0.136	0.139	0.141	0.143	0.144	0.145	0.147	0.147	0.148
2.25	0.133	0.136	0.140	0.142	0.144	0.145	0.146	0.149	0.150	0.150
2.50	0.132	0.135	0.139	0.142	0.144	0.146	0.147	0.151	0.151	0.151
2.75	0.130	0.133	0.138	0.142	0.144	0.146	0.147	0.152	0.152	0.153
3.00	0.127	0.131	0.137	0.141	0.144	0.145	0.147	0.152	0.153	0.154
3.25	0.125	0.129	0.135	0.140	0.143	0.145	0.147	0.153	0.154	0.154
3.50	0.122	0.126	0.133	0.138	0.142	0.144	0.146	0.153	0.155	0.155
3.75	0.119	0.124	0.131	0.137	0.141	0.143	0.145	0.154	0.155	0.155
4.00	0.116	0.121	0.129	0.135	0.139	0.142	0.145	0.154	0.155	0.156
4.25	0.113	0.119	0.127	0.133	0.138	0.141	0.144	0.154	0.156	0.156
4.50	0.110	0.116	0.125	0.131	0.136	0.140	0.143	0.154	0.156	0.156
4.75	0.107	0.113	0.123	0.130	0.135	0.139	0.142	0.154	0.156	0.157
5.00	0.105	0.111	0.120	0.128	0.133	0.137	0.140	0.154	0.156	0.157
5.25	0.102	0.108	0.118	0.126	0.131	0.136	0.139	0.154	0.156	0.157
5.50	0.099	0.106	0.116	0.124	0.130	0.134	0.138	0.154	0.156	0.157
5.75	0.097	0.103	0.113	0.122	0.128	0.133	0.136	0.154	0.157	0.157
6.00	0.094	0.101	0.111	0.120	0.126	0.131	0.135	0.153	0.157	0.157
6.25	0.092	0.098	0.109	0.118	0.124	0.129	0.134	0.153	0.157	0.158
6.50	0.090	0.096	0.107	0.116	0.122	0.128	0.132	0.153	0.157	0.158
6.75	0.087	0.094	0.105	0.114	0.121	0.126	0.131	0.153	0.157	0.158
7.00	0.085	0.092	0.103	0.112	0.119	0.125	0.129	0.152	0.157	0.158
7.25	0.083	0.090	0.101	0.110	0.117	0.123	0.128	0.152	0.157	0.158
7.50	0.081	0.088	0.099	0.108	0.115	0.121	0.126	0.152	0.156	0.158
7.75	0.079	0.086	0.097	0.106	0.114	0.120	0.125	0.151	0.156	0.158
8.00	0.077	0.084	0.095	0.104	0.112	0.118	0.124	0.151	0.156	0.158
8.25	0.076	0.082	0.093	0.102	0.110	0.117	0.122	0.150	0.156	0.158
8.50	0.074	0.080	0.091	0.101	0.108	0.115	0.121	0.150	0.156	0.158
8.75	0.072	0.078	0.089	0.099	0.107	0.114	0.119	0.150	0.156	0.158
9.00	0.071	0.077	0.088	0.097	0.105	0.112	0.118	0.149	0.156	0.158
9.25	0.069	0.075	0.086	0.096	0.104	0.110	0.116	0.149	0.156	0.158
9.50	0.068	0.074	0.085	0.094	0.102	0.109	0.115	0.148	0.156	0.158
9.75	0.066	0.072	0.083	0.092	0.100	0.107	0.113	0.148	0.156	0.158
10.00	0.065	0.071	0.082	0.091	0.099	0.106	0.112	0.147	0.156	0.158
20.00	0.035	0.039	0.046	0.053	0.059	0.065	0.071	0.124	0.148	0.156
50.00	0.014	0.016	0.019	0.022	0.025	0.028	0.031	0.071	0.113	0.142
100.00	0.007	0.008	0.010	0.011	0.013	0.014	0.016	0.039	0.071	0.113

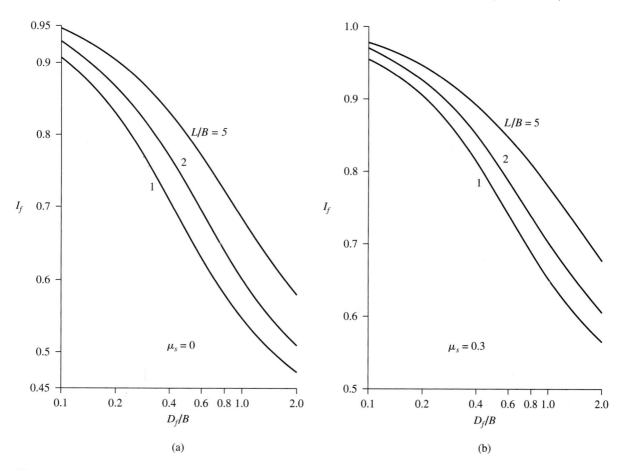

Figure 5.15 Variation of I_f with D_f/B, L/B, and μ_s

The elastic settlement of a *rigid foundation* can be estimated as

$$S_{e(\text{rigid})} \approx 0.93 S_{e(\text{flexible, center})} \tag{5.33}$$

Due to the nonhomogeneous nature of soil deposits, the magnitude of E_s may vary with depth. For that reason, Bowles (1987) recommended using a weighted average of E_s in Eq. (5.25), or

$$E_s = \frac{\sum E_{s(i)} \Delta z}{\bar{z}} \tag{5.34}$$

where

$E_{s(i)}$ = soil modulus of elasticity within a depth Δz
$\bar{z}$ = H or $5B$, whichever is smaller

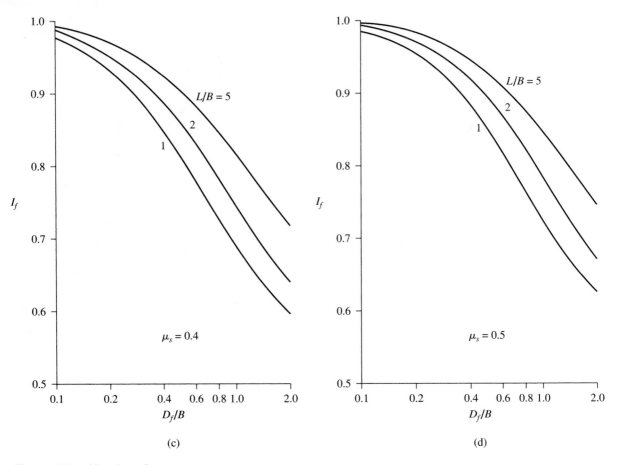

$$I_f$$

$$\mu_s = 0.4$$

$$L/B = 5$$

(c)

$$I_f$$

$$\mu_s = 0.5$$

$$L/B = 5$$

(d)

Figure 5.15 (Continued)

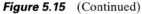

Example 5.4

A rigid shallow foundation 1 m × 2 m is shown in Figure 5.16. Calculate the elastic settlement at the center of the foundation.

Solution
We are given that $B = 1$ m and $L = 2$ m. Note that $\bar{z} = 5$ m $= 5B$. From Eq. (5.34),

$$E_s = \frac{\Sigma E_{s(i)} \Delta z}{\bar{z}}$$

$$= \frac{(10,000)(2) + (8,000)(1) + (12,000)(2)}{5} = 10,400 \text{ kN/m}^2$$

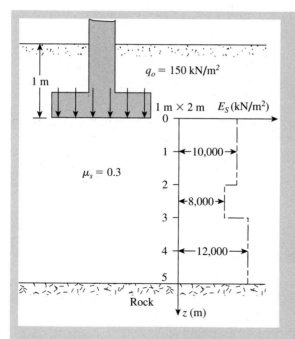

Figure 5.16 Elastic settlement below the center of a foundation

For the *center of the foundation,*

$$\alpha = 4$$

$$m' = \frac{L}{B} = \frac{2}{1} = 2$$

and

$$n' = \frac{H}{\left(\dfrac{B}{2}\right)} = \frac{5}{\left(\dfrac{1}{2}\right)} = 10$$

From Tables 5.4 and 5.5, $F_1 = 0.641$ and $F_2 = 0.031$. From Eq. (5.26),

$$I_s = F_1 + \frac{2 - \mu_s}{1 - \mu_s} F_2$$

$$= 0.641 + \frac{2 - 0.3}{1 - 0.3}(0.031) = 0.716$$

Again, $D_f/B = 1/1 = 1$, $L/B = 2$, and $\mu_s = 0.3$. From Figure 5.15b, $I_f = 0.709$. Hence,

$$S_{e(\text{flexible})} = q_o(\alpha B')\frac{1 - \mu_s^2}{E_s}I_s I_f$$

$$= (150)\left(4 \times \frac{1}{2}\right)\left(\frac{1 - 0.3^2}{10,400}\right)(0.716)(0.709) = 0.0133 \text{ m} = 13.3 \text{ mm}$$

Since the foundation is rigid, from Eq. (5.33) we obtain

$$S_{e(\text{rigid})} = (0.93)(13.3) = \textbf{12.4 mm}$$ ∎

5.8   *Elastic Settlement of Foundations on Saturated Clay*

Janbu et al. (1956) proposed an equation for evaluating the average settlement of flexible foundations on saturated clay soils (Poisson's ratio, $\mu_s = 0.5$). For the notation used in Figure 5.17, this equation is

$$S_e = A_1 A_2 \frac{q_o B}{E_s} \tag{5.35}$$

where A_1 is a function of H/B and L/B and A_2 is a function of D_f/B

Christian and Carrier (1978) modified the values of A_1 and A_2 to some extent as presented in Figure 5.17.

5.9   *Improved Equation for Elastic Settlement*

In 1999, Mayne and Poulos presented an improved formula for calculating the elastic settlement of foundations. The formula takes into account the rigidity of the foundation, the depth of embedment of the foundation, the increase in the modulus of elasticity of the soil with depth, and the location of rigid layers at a limited depth. To use Mayne and Poulos's equation, one needs to determine the equivalent diameter B_e of a rectangular foundation, or

$$B_e = \sqrt{\frac{4BL}{\pi}} \tag{5.36}$$

where

B = width of foundation
L = length of foundation

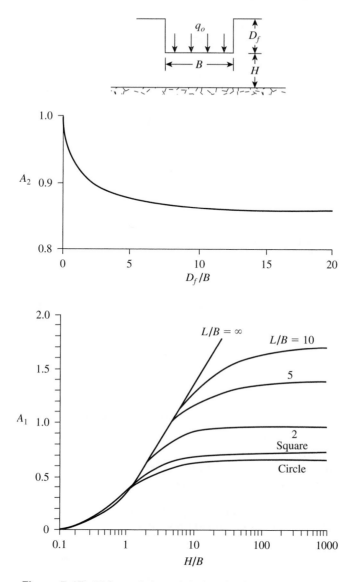

Figure 5.17 Values of A_1 and A_2 for elastic settlement calculation—Eq. (5.35) (After Christian and Carrier, 1978)

For circular foundations,

$$B_e = B \tag{5.37}$$

where B = diameter of foundation.

Figure 5.18 shows a foundation with an equivalent diameter B_e located at a depth D_f below the ground surface. Let the thickness of the foundation be t and the modulus of elasticity of the foundation material be E_f. A rigid layer is located at a

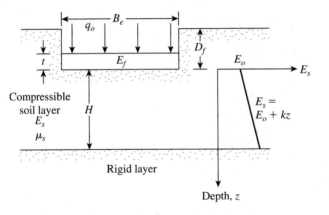

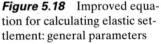

Figure 5.18 Improved equation for calculating elastic settlement: general parameters

depth H below the bottom of the foundation. The modulus of elasticity of the compressible soil layer can be given as

$$E_s = E_o + kz \tag{5.38}$$

With the preceding parameters defined, the elastic settlement below the center of the foundation is

$$S_e = \frac{q_o B_e I_G I_F I_E}{E_o}\left(1 - \mu_s^2\right) \tag{5.39}$$

where

I_G = influence factor for the variation of E_s with depth

$$= f\left(\beta = \frac{E_o}{kB_e}, \frac{H}{B_e}\right)$$

I_F = foundation rigidity correction factor
I_E = foundation embedment correction factor

Figure 5.19 shows the variation of I_G with $\beta = E_o/kB_e$ and H/B_e. The foundation rigidity correction factor can be expressed as

$$I_F = \frac{\pi}{4} + \cfrac{1}{4.6 + 10\left(\cfrac{E_f}{E_o + \cfrac{B_e}{2}k}\right)\left(\cfrac{2t}{B_e}\right)^3} \tag{5.40}$$

Similarly, the embedment correction factor is

$$I_E = 1 - \cfrac{1}{3.5\exp(1.22\mu_s - 0.4)\left(\cfrac{B_e}{D_f} + 1.6\right)} \tag{5.41}$$

Figures 5.20 and 5.21 show the variation of I_F and I_E with terms expressed in Eqs. (5.40) and (5.41).

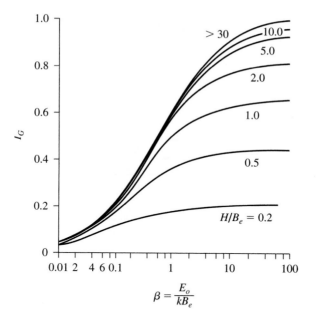

$$\beta = \frac{E_o}{kB_e}$$

Figure 5.19 Variation of I_G with β

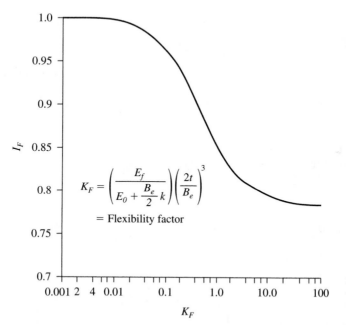

$$K_F = \left(\frac{E_f}{E_0 + \frac{B_e}{2}k}\right)\left(\frac{2t}{B_e}\right)^3$$

= Flexibility factor

Figure 5.20 Variation of rigidity correction factor I_F with flexibility factor K_F. [Eq. (5.40)]

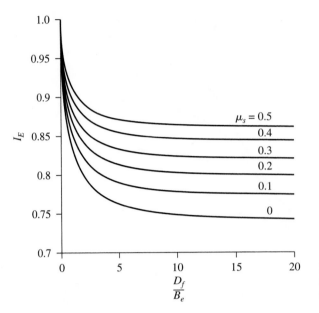

Figure 5.21   Variation of embedment correction factor I_E with D_f/B_e [Eq (5.41)]

Example 5.5

For a shallow foundation supported by a silty clay, as shown in Figure 5.18,

Length = L = 10 ft
Width = B = 5 ft
Depth of foundation = D_f = 5 ft
Thickness of foundation = t = 1 ft
Load per unit area = q_o = 5000 lb/ft²
$E_f = 2.3 \times 10^6$ lb/in²

The silty clay soil has the following properties:

H = 12 ft
μ_s = 0.3
E_o = 1400 lb/in²
k = 25 lb/in²/ft

Estimate the elastic settlement of the foundation.

Solution
From Eq. (5.36), the equivalent diameter is

$$B_e = \sqrt{\frac{4BL}{\pi}} = \sqrt{\frac{(4)(5)(10)}{\pi}} = 7.98 \text{ ft}$$

so

$$\beta = \frac{E_o}{kB_e} = \frac{1400}{(25)(7.98)} = 7.02$$

and

$$\frac{H}{B_e} = \frac{12}{7.98} = 1.5$$

From Figure 5.19, for $\beta = 7.02$ and $H/B_e = 1.5$, the value of $I_G \approx 0.69$. From Eq. (5.40),

$$I_F = \frac{\pi}{4} + \frac{1}{4.6 + 10\left(\dfrac{E_f}{E_o + \dfrac{B_e}{2}k}\right)\left(\dfrac{2t}{B_e}\right)^3}$$

$$= \frac{\pi}{4} + \frac{1}{4.6 + 10\left[\dfrac{2.3 \times 10^6}{1400 + \left(\dfrac{7.98}{2}\right)(25)}\right]\left[\dfrac{(2)(1.0)}{7.98}\right]^3} = 0.789$$

From Eq. (5.41),

$$I_E = 1 - \frac{1}{3.5 \exp(1.22\mu_s - 0.4)\left(\dfrac{B_e}{D_f} + 1.6\right)}$$

$$= 1 - \frac{1}{3.5 \exp[(1.22)(0.3) - 0.4]\left(\dfrac{7.98}{5} + 1.6\right)} = 0.908$$

From Eq. (5.39),

$$S_e = \frac{q_o B_e I_G I_F I_E}{E_o}(1 - \mu_s^2)$$

so, with $q_o = 5000$ lb/ft^2, it follows that

$$S_e = \frac{(5000)(7.98)(0.69)(0.789)(0.908)}{(1400)(144)}(1 - 0.3^2) = 0.089 \text{ ft} = \textbf{1.07 in.} \quad \blacksquare$$

5.10   *Settlement of Sandy Soil: Use of Strain Influence Factor*

The settlement of granular soils can also be evaluated by the use of a semiempirical *strain influence factor* proposed by Schmertmann et al. (1978). According to this method, the settlement is

$$S_e = C_1 C_2 (\bar{q} - q) \sum_0^{z_2} \frac{I_z}{E_s} \Delta z \tag{5.42}$$

where

I_z = strain influence factor
C_1 = a correction factor for the depth of foundation embedment = $1 - 0.5$
     $[q/(\bar{q} - q)]$
C_2 = a correction factor to account for creep in soil
     = $1 + 0.2 \log$ (time in years/0.1)
$\bar{q}$ = stress at the level of the foundation
$q = \gamma D_f$

The recommended variation of the strain influence factor I_z for square $(L/B = 1)$ or circular foundations and for foundations with $L/B \geq 10$ is shown in Figure 5.22. The I_z diagrams for $1 < L/B < 10$ can be interpolated. The procedure to calculate elastic settlement using Eq. (5.42) is given here (Figure 5.23).

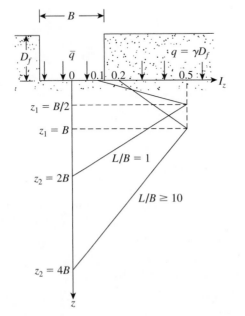

Figure 5.22   Variation of the strain influence factor, I_z

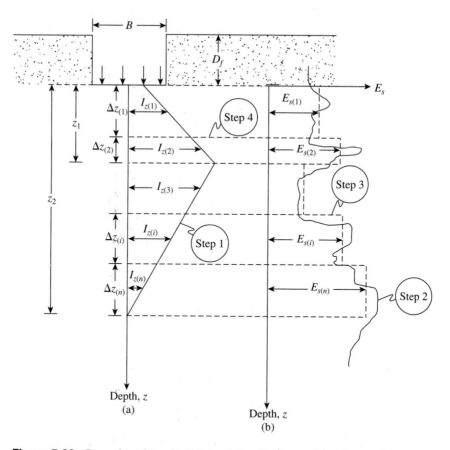

Figure 5.23 Procedure for calculation of S_e using the strain influence factor

Step 1. Plot the foundation and the variation of I_z with depth to scale (Figure 5.23a).

Step 2. Using the correlation from standard penetration resistance (N_{60}) or cone penetration resistance (q_c), plot the actual variation of E_s with depth (Figure 5.23b). Schmertmann et al. (1978) suggested $E_s \approx 3.5q_c$.

Step 3. Approximate the actual variation of E_s into a number of layers of soil having a constant E_s, such as $E_{s(1)}, E_{s(2)}, \ldots, E_{s(i)}, \ldots E_{s(n)}$ (Figure 5.23b).

Step 4. Divide the soil layer from $z = 0$ to $z = z_2$ into a number of layers by drawing horizontal lines. The number of layers will depend on the break in continuity in the I_z and E_s diagrams.

Step 5. Prepare a table (such as Table 5.6) to obtain $\Sigma \dfrac{I_z}{E_s} \Delta z$.

Step 6. Calculate C_1 and C_2.

Step 7. Calculate S_e from Eq. (5.42).

Table 5.6 Calculation of $\Sigma \dfrac{I_z}{E_s} \Delta z$

Layer No.	Δz	E_s	I_z at the middle of the layer	$\dfrac{I_z}{E_s} \Delta z$
1	$\Delta z_{(1)}$	$E_{s(1)}$	$I_{z(1)}$	$\dfrac{I_{z(1)}}{E_{s(1)}} \Delta z_1$
2	$\Delta z_{(2)}$	$E_{s(2)}$	$I_{z(2)}$	
$\vdots$	$\vdots$	$\vdots$	$\vdots$	
i	$\Delta z_{(i)}$	$E_{s(i)}$	$I_{z(i)}$	$\dfrac{I_{z(i)}}{E_{s(i)}} \Delta z_i$
$\vdots$	$\vdots$	$\vdots$	$\vdots$	$\vdots$
n	$\Delta z_{(n)}$	$E_{s(n)}$	$I_{z(n)}$	$\dfrac{I_{z(n)}}{E_{s(n)}} \Delta z_n$
				$\Sigma \dfrac{I_z}{E_s} \Delta z$

Example 5.6

Figure 5.24 shows a shallow continuous foundation along with the variation of E_s with depth obtained from cone penetration tests (broken line in Figure 5.24b). Given: $B = 8$ ft, $D_f = 4$ ft, $\gamma = 110$ lb/ft^3, and $\bar{q} = 25$ lb/in^2. Estimate the elastic settlement using the strain influence factor method described in Section 5.10.

Solution

Given $L/B > 10$. Based on this, the I_z diagram is plotted in Figure 5.24a. The approximate variation of E_s is shown in Figure 5.24b. The soil below the foundation has been divided into 5 layers. Now Table 5.7 can be prepared.

Since $\gamma = 110$ lb/ft^3, $q = \gamma D_f = (4)(110) = 440$ lb/ft$^2 = 3.06$ lb/in^2. Given $\bar{q} = 25$ lb/in^2. Thus, $\bar{q} - q = 25 - 3.06 = 21.94$ lb/in^2. Also,

$$C_1 = 1 - 0.5 \left(\frac{q}{\bar{q} - q} \right) = 1 - 0.5 \left(\frac{3.06}{21.94} \right) = 0.93$$

Assume the time for creep is 10 years. Hence,

$$C_2 = 1 + 0.2 \log\left(\frac{10}{0.1} \right) = 1.4$$

Table 5.7   Elastic settlement Calculation

Layer No.	Δz (in.)	E_s (lb/in.^{2})	z to the middle of the layer (in.)	I_z at the middle of the layer	$\dfrac{I_z}{E_s}\Delta z$ (in^3/lb)
1	48	750	24	0.275	0.0176
2	48	1250	72	0.425	0.016
3	96	1250	144	0.417	0.032
4	48	1000	216	0.292	0.014
5	144	2000	312	0.125	0.009
	$\Sigma 384$ in. $= 4B$				$\Sigma 0.0886$ in^3/lb

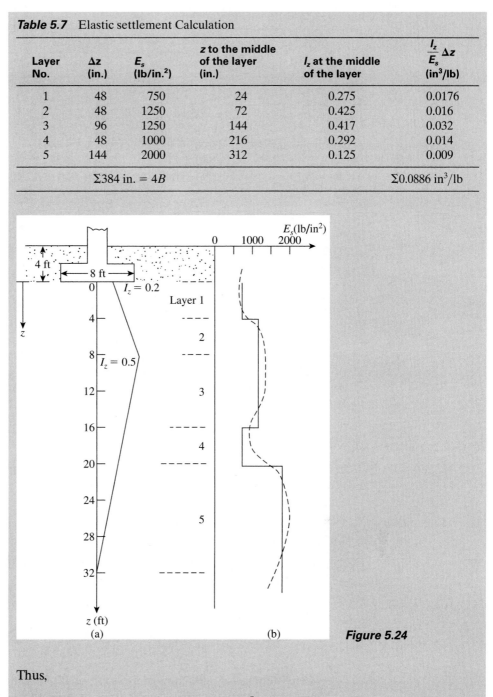

Figure 5.24

Thus,

$$S_e = C_1 C_2 (\overline{q} - q) \sum \frac{I_z}{E_s} \Delta z$$

$$= (0.93)(1.4)(21.94)(0.0886) = \mathbf{2.53\ in.}$$ ∎

Table 5.8   Elastic Parameters of Various Soils

Type of soil	Modulus of elasticity, E_s		Poisson's ratio, μ_s
	MN/m²	**lb/in²**	
Loose sand	10.5–24.0	1500–3500	0.20–0.40
Medium dense sand	17.25–27.60	2500–4000	0.25–0.40
Dense sand	34.50–55.20	5000–8000	0.30–0.45
Silty sand	10.35–17.25	1500–2500	0.20–0.40
Sand and gravel	69.00–172.50	10,000–25,000	0.15–0.35
Soft clay	4.1–20.7	600–3000	
Medium clay	20.7–41.4	3000–6000	0.20–0.50
Stiff clay	41.4–96.6	6000–14,000	

5.11   *Range of Material Parameters for Computing Elastic Settlement*

Relationships for calculating elastic settlement presented in Sections 5.7 through 5.10 involve elastic parameters E_s and μ_s for soils. Table 5.8 shows some approximate values of modulus of elasticity and Poisson's ratio for various soils. However, it must be realized that actual estimation of E_s is difficult and challenging. The reliability of the elastic settlement calculation primarily depends on that.

As a first approximation, the magnitude of E_s in sandy soil can be approximated according to Eq. (2.27), or

$$\frac{E_s}{p_a} = \alpha N_{60} \tag{5.43}$$

where

p_a = atmospheric pressure $\approx 100 \text{ kN/m}^2$ ($\approx 2000 \text{ lb/ft}^2$)

$$\alpha = \begin{cases} 5 \text{ for sands with fines} \\ 10 \text{ for clean normally consolidated sand} \\ 15 \text{ for clean overconsolidated sand} \end{cases}$$

Schmertmann et al. (1978) further suggested that the following correlations for sand may be used with the strain influence factors described in Section 5.10.

$$E_s = 2.5q_u \text{ (for square and circular foundations)} \tag{5.44a}$$

and

$$E_s = 3.5q_u \text{ (for strip foundations)} \tag{5.44b}$$

[*Note:* Any consistent set of units may be used in Eqs. (5.44a) and (5.44b).]

The modulus of elasticity (E_s) for clays can, in general, be given as

$$E_s = \beta c_u \tag{5.45}$$

where c_u = undrained shear strength.

Table 5.9 Range of β for Clay [Eq. (5.45)][a]

Plasticity index	β				
	OCR = 1	OCR = 2	OCR = 3	OCR = 4	OCR = 5
< 30	1500–600	1380–500	1200–580	950–380	730–300
30 to 50	600–300	550–270	580–220	380–180	300–150
> 50	300–150	270–120	220–100	180–90	150–75

[a]Interpolated from Duncan and Buchignani (1976)

The parameter β is primarily a function of the plasticity index and overconsolidation ratio. Table 5.9 provides a general range for β based on that proposed by Duncan and Buchignani (1976). In any case, proper judgment should be used in selecting the magnitude of β.

5.12 Settlement of Foundation on Sand Based on Standard Penetration Resistance

Meyerhof's Method

Meyerhof (1956) proposed a correlation for the *net bearing pressure* for foundations with the standard penetration resistance, N_{60}. The net pressure has been defined as

$$q_{net} = \bar{q} - \gamma D_f$$

where $\bar{q}$ = stress at the level of the foundation.

According to Meyerhof's theory, for 25 mm (1 in.) of estimated maximum settlement,

$$q_{net}(\text{kip/ft}^2) = \frac{N_{60}}{4} \quad (\text{for } B \leq 4 \text{ ft}) \tag{5.46a}$$

and

$$q_{net}(\text{kip/ft}^2) = \frac{N_{60}}{6}\left(\frac{B+1}{B}\right)^2 \quad (\text{for } B > 4 \text{ ft}) \tag{5.46b}$$

Since the time that Meyerhof proposed his original correlations, researchers have observed that its results are rather conservative. Later, Meyerhof (1965) suggested that the net allowable bearing pressure should be increased by about 50%. Bowles (1977) proposed that the modified form of the bearing equations be expressed as

$$q_{net}(\text{kip/ft}^2) = \frac{N_{60}}{2.5} F_d S_e \quad (\text{for } B \leq 4 \text{ ft}) \tag{5.47a}$$

and

$$q_{net}(\text{kip/ft}^2) = \frac{N_{60}}{4}\left(\frac{B+1}{B}\right)^2 F_d S_e \quad (\text{for } B > 4 \text{ ft}) \tag{5.47b}$$

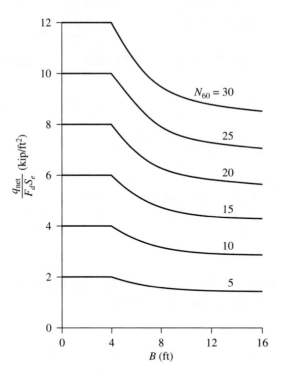

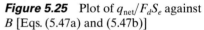

Figure 5.25 Plot of q_{net}/F_dS_e against B [Eqs. (5.47a) and (5.47b)]

where

F_d = depth factor = $1 + 0.33(D_f/B)$
B = foundation width, in feet
S_e = settlement, in inches

Figure 5.25 shows a plot of q_{net}/F_dS_e for various values of B and N_{60}. Thus,

$$S_e(\text{in.}) = \frac{2.5q_{net}(\text{kip/ft}^2)}{N_{60}F_d} \quad (\text{for } B \leqslant 4 \text{ ft}) \tag{5.48}$$

and

$$S_e(\text{in.}) = \frac{4q_{net}(\text{kip/ft}^2)}{N_{60}F_d}\left(\frac{B}{B+1}\right)^2 \quad (\text{for } B > 4 \text{ ft}) \tag{5.49}$$

In SI, units, Eqs. (5.47a) and (5.47b) can be written as

$$q_{net}(\text{kN/m}^2) = \frac{N_{60}}{0.05} F_d\left(\frac{S_e}{25}\right) \quad (\text{for } B \leqslant 1.22 \text{ m}) \tag{5.50}$$

and

$$q_{net}(\text{kN/m}^2) = \frac{N_{60}}{0.08}\left(\frac{B+0.3}{B}\right)^2 F_d\left(\frac{S_e}{25}\right) \quad (\text{for } B > 1.22 \text{ m}) \tag{5.51}$$

where B is in meters and S_e is in mm. Hence,

$$S_e(\text{mm}) = \frac{1.25q_{net}(\text{kN/m}^2)}{N_{60}F_d} \quad (\text{for } B \leqslant 1.22 \text{ m}) \tag{5.52}$$

and

$$S_e(\text{mm}) = \frac{2q_{\text{net}}(\text{kN/m}^2)}{N_{60}F_d}\left(\frac{B}{B + 0.3}\right)^2 \quad (\text{for } B > 1.22 \text{ m}) \tag{5.53}$$

The N_{60} referred to in Eqs. (5.46) through (5.53) is the standard penetration resistance between the bottom of the foundation and $2B$ below the bottom.

Burland and Burbidge's Method

Burland and Burbidge (1985) proposed a method of calculating the elastic settlement of sandy soil using the *field standard penetration number, N_{60}.* (See Chapter 2.) The method can be summarized as follows:

1. **Variation of Standard Penetration Number with Depth**
   Obtain the field penetration numbers (N_{60}) with depth at the location of the foundation. The following adjustments of N_{60} may be necessary, depending on the field conditions:
   For gravel or sandy gravel,

   $$N_{60(a)} \approx 1.25 N_{60} \tag{5.54}$$

   For fine sand or silty sand below the groundwater table and $N_{60} > 15$,

   $$N_{60(a)} \approx 15 + 0.5(N_{60} - 15) \tag{5.55}$$

   where $N_{60(a)}$ = adjusted N_{60} value.

2. **Determination of Depth of Stress Influence (z')**
   In determining the depth of stress influence, the following three cases may arise:

   Case I. If N_{60} [or $N_{60(a)}$] is approximately constant with depth, calculate z' from

   $$\frac{z'}{B_R} = 1.4\left(\frac{B}{B_R}\right)^{0.75} \tag{5.56}$$

   where

   $$B_R = \text{reference width}\begin{cases} = 1 \text{ ft (if } B \text{ is in ft)} \\ = 0.3 \text{ m (if } B \text{ is in m)} \end{cases}$$
   $$B = \text{width of the actual foundation}$$

   Case II. If N_{60} [or $N_{60(a)}$] is increasing with depth, use Eq. (5.56) to calculate z'.

   Case III. If N_{60} [or $N_{60(a)}$] is decreasing with depth, calculate $z' = 2B$ and z' = distance from the bottom of the foundation to the bottom of the soft soil layer (z''). Use $z' = 2B$ or $z' = z''$ (whichever is smaller).

3. **Calculation of Elastic Settlement S_e**
   The elastic settlement of the foundation, S_e, can be calculated from

   $$\frac{S_e}{B_R} = \alpha_1\alpha_2\alpha_3 \left[\frac{1.25\left(\frac{L}{B}\right)}{0.25 + \left(\frac{L}{B}\right)}\right]^2 \left(\frac{B}{B_R}\right)^{0.7}\left(\frac{q'}{p_a}\right) \tag{5.57}$$

where

α_1 = a constant
α_2 = compressibility index
α_3 = correction for the depth of influence
p_a = atmospheric pressure = $100 \text{ kN/m}^2 (\approx 2000 \text{ lb/ft}^2)$
L = length of the foundation

For *normally consolidated* sand,

$$\alpha_1 = 0.14 \tag{5.58}$$

and

$$\alpha_2 = \frac{1.71}{[\overline{N}_{60} \text{ or } \overline{N}_{60(a)}]^{1.4}} \tag{5.59}$$

where $\overline{N}_{60}$ or $\overline{N}_{60(a)}$ = average value of N_{60} or $N_{60(a)}$ in the depth of stress influence,

$$\alpha_3 = \frac{z''}{z'}\left(2 - \frac{z''}{z'}\right) \leqslant 1 \tag{5.60}$$

and

$$q' = q_{\text{net}} \tag{5.61}$$

in which q_{net} = net applied stress at the level of the foundation (i.e., the stress at the level of the foundation minus the overburden pressure).

For *overconsolidated sand with $q_{\text{net}} \leqslant \sigma_c'$*, the preconsolidation pressure,

$$\alpha_1 = 0.047 \tag{5.62}$$

and

$$\alpha_2 = \frac{0.57}{[\overline{N}_{60} \text{ or } \overline{N}_{60(a)}]^{1.4}} \tag{5.63}$$

For α_3, use Eq. (5.57):

$$q' = q_{\text{net}} \tag{5.64}$$

For *overconsolidated sand with $q_{\text{net}} > \sigma_c'$*,

$$\alpha_1 = 0.14 \tag{5.65}$$

For α_2, use Eq. (5.63), and for α_3, use Eq. (5.60). Finally, use

$$q' = q_{\text{net}} - 0.67\sigma_c' \tag{5.66}$$

Example 5.7

A shallow foundation measuring $1.75 \text{ m} \times 1.75 \text{ m}$ is to be constructed over a layer of sand. Given $D_f = 1 \text{ m}$; N_{60} is generally increasing with depth; $\overline{N}_{60}$ in the depth of stress influence = 10, $q_{\text{net}} = 120 \text{ kN/m}^2$. The sand is normally consolidated. Estimate the elastic settlement of the foundation. Use the Burland and Burbidge method.

Solution
From Eq. (5.56),

$$\frac{z'}{B_R} = 1.4\left(\frac{B}{B_R}\right)^{0.75}$$

Depth of stress influence,

$$z' = 1.4\left(\frac{B}{B_R}\right)^{0.75} B_R = (1.4)(0.3)\left(\frac{1.75}{0.3}\right)^{0.75} \approx 1.58 \text{ m}$$

From Eq. (5.57),

$$\frac{S_e}{B_R} = \alpha_1\alpha_2\alpha_3 \left[\frac{1.25\left(\dfrac{L}{B}\right)}{0.25 + \left(\dfrac{L}{B}\right)}\right]^2 \left(\frac{B}{B_R}\right)^{0.7}\left(\frac{q'}{p_a}\right)$$

For normally consolidated sand,

$$\alpha_1 = 0.14$$

$$\alpha_2 = \frac{1.71}{(N_{60})^{1.4}} = \frac{1.71}{(10)^{1.4}} = 0.068$$

$$\alpha_3 = 1$$

$$q' = q_{\text{net}} = 120 \text{ kN/m}^2$$

So,

$$\frac{S_e}{0.3} = (0.14)(0.068)(1)\left[\frac{(1.25)\left(\dfrac{1.75}{1.75}\right)}{0.25 + \left(\dfrac{1.75}{1.75}\right)}\right]^2 \left(\frac{1.75}{0.3}\right)^{0.7}\left(\frac{120}{100}\right)$$

$$S_e \approx 0.0118 \text{ m} = \textbf{11.8 mm} \quad\blacksquare$$

Example 5.8

Solve Example 5.7 using Meyerhof's method.

Solution
From Eq. (5.53),

$$S_e = \frac{2q_{\text{net}}}{(N_{60})(F_d)}\left(\frac{B}{B + 0.3}\right)^2$$

$$F_d = 1 + 0.33(D_f/B) = 1 + 0.33(1/1.75) = 1.19$$

$$S_e = \frac{(2)(120)}{(10)(1.19)}\left(\frac{1.75}{1.75 + 0.3}\right)^2 = \textbf{14.7 mm}$$ ∎

5.13   *General Comments on Elastic Settlement Prediction*

Idealized versions of elastic settlement prediction have been discussed in the preceding sections of this chapter. Predicting settlements of shallow foundations in sand is prone to uncertainties due to highly erratic density and compressibility variations and the difficulties associated with sampling and assessing the *in situ* characteristics of the granular soil deposits (for example, see Briaud and Gibbens, 1994). To illustrate this point, Sivakugan et al. (1998) compiled the settlement records for 79 foundations summarized by Jeyapalan and Boehm (1986) and Papadopoulos (1992) and compared them to those predicted by Schmertmann's strain influence factor method. The comparison between the predicted and observed settlements in shown in Figure 5.26. From this figure, it can be seen that predicted settlements

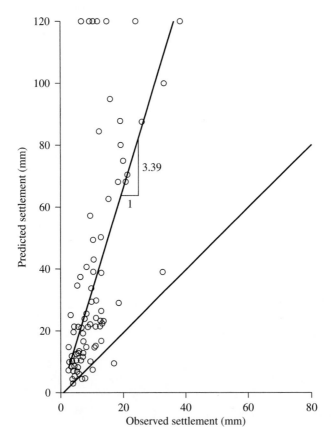

Figure 5.26   Predicted versus observed settlement of a shallow foundation or sand. Prediction by strain influence factor method [Adapted from Sivakugan and Johnson (2004)]

Table 5.10 Statistical Analysis of Settlement Prediction in Sand—Based on the Study of Sivakugan and Johnson (2004)

Predicted settlement (mm)	Probability of exceeding 25 mm settlement in the field	
	Schmertmann et al. (1978)	Burland and Burbidge (1985)
1	0	0
5	0	3%
10	2%	15%
15	13%	25%
20	20%	34%
25	27%	42%
30	32%	49%
35	37%	55%
40	42%	61%

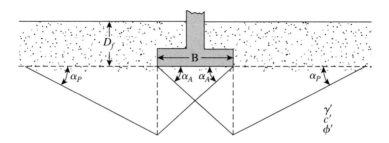

Figure 5.27 Failure surface in soil for static bearing capacity analysis. (*Note:* $\alpha_A = 45 + \phi'/2$ and $\alpha_p = 45 - \phi'/2$)

grossly overestimate the observed settlements. Sivakugan and Johnson (2004) conducted a statistical analysis of settlement predictions in granular soil. Table 5.10 gives a summary of this analysis. Hence, it is necessary to use proper judgment in predicting the actual settlement a foundation will undergo in the field.

5.14 Seismic Bearing Capacity and Settlement in Granular Soil

In some instances, shallow foundations may fail during seismic events. Published studies relating to the bearing capacity of shallow foundations in such instances are rare. In 1993, however, Richards et al. developed a seismic bearing capacity theory that we shall examine in this section. The theory is not yet supported by field data.

Figure 5.27 shows a failure surface in soil assumed for the subsequent analysis, under static conditions. Similarly, Figure 5.28 shows the failure surface under earthquake conditions. Note that, in the two figures,

$$\alpha_A, \alpha_{AE} = \text{inclination angles for active pressure conditions}$$

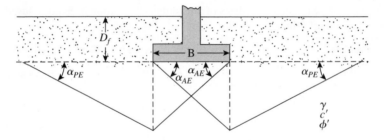

Figure 5.28 Failure surface in soil for seismic bearing capacity analysis

and

$$\alpha_P, \alpha_{PE} = \text{inclination angles for passive pressure conditions}$$

According to this theory, the ultimate bearing capacities for *continuous foundations* in granular soil are

$$q_u = qN_q + \tfrac{1}{2}\gamma B N_\gamma \ (static\ conditions) \tag{5.67}$$

and

$$q_{uE} = qN_{qE} + \tfrac{1}{2}\gamma B N_{\gamma E} \ (earthquake\ conditions) \tag{5.68}$$

where

$N_q, N_\gamma, N_{qE}, N_{\gamma E}$ = bearing capacity factors
$q = \gamma D_f$

Note that

$$N_q \text{ and } N_\gamma = f(\phi')$$

and

$$N_{qE} \text{ and } N_{\gamma E} = f(\phi', \tan\theta)$$

where

$$\tan\theta = \frac{k_h}{1 - k_v}$$

k_h = horizontal coefficient of acceleration due to an earthquake
k_v = vertical coefficient of acceleration due to an earthquake

The variations of N_q and N_γ with ϕ' are shown in Figure 5.29. Figure 5.30 shows the variations of $N_{\gamma E}/N_\gamma$ and N_{qE}/N_q with $\tan\theta$ and the soil friction angle ϕ'.

Under static conditions, bearing capacity failure can lead to a substantial sudden downward movement of the foundation. However, bearing capacity–related settlement in an earthquake takes place when the ratio $k_h/(1 - k_v)$ reaches the critical value $(k_h/1 - k_v)^*$. If $k_v = 0$, then $(k_h/1 - k_v)^*$ becomes equal to k_h^*. Figure 5.31

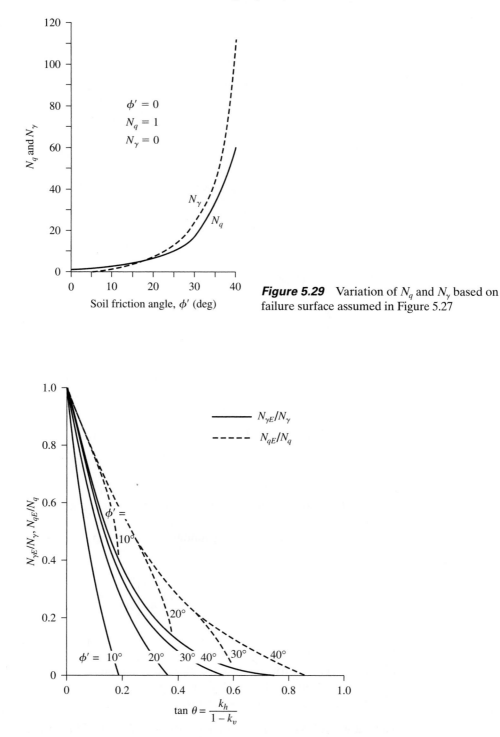

Figure 5.29 Variation of N_q and N_γ based on failure surface assumed in Figure 5.27

Figure 5.30 Variation of $N_{\gamma E}/N_\gamma$ and N_{qE}/N_q with $\tan \theta$

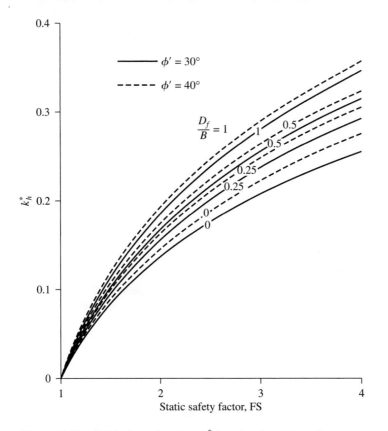

Figure 5.31 Critical acceleration k_h^* for $c' = 0$ and $k_v = 0$

shows the variation of k_h^* (for $k_v = 0$ and $c' = 0$ and in granular soil) with the factor of safety (FS) applied to the ultimate static bearing capacity [Eq. 5.67], with ϕ', and with D_f/B. (for $\phi' = 30°$ and $40°$).

The settlement of a strip foundation due to an earthquake can be estimated (Richards et al., 1993) as

$$S_{\text{Eq}}(\text{m}) = 0.174 \frac{V^2}{Ag} \left| \frac{k_h^*}{A} \right|^{-4} \tan \alpha_{AE} \tag{5.69}$$

where

V = peak velocity for the design earthquake (m/sec)
A = acceleration coefficient for the design earthquake
g = acceleration due to gravity (9.18 m/sec^2)

The values of k_h^* and α_{AE} can be obtained from Figure 5.31 and Table 5.11, respectively.

Table 5.11 Variation of $\tan \alpha_{AE}$ with k_h^* and soil friction angle ϕ' (Compiled from Richards et al., 1993)

	tan α_{AE}				
k_h^*	$\phi' = 20°$	$\phi' = 25°$	$\phi' = 30°$	$\phi' = 35°$	$\phi' = 40°$
0.05	1.10	1.24	1.39	1.57	1.75
0.10	0.97	1.13	1.26	1.44	1.63
0.15	0.82	1.00	1.15	1.32	1.48
0.20	0.71	0.87	1.02	1.18	1.35
0.25	0.56	0.74	0.92	1.06	1.23
0.30		0.61	0.77	0.94	1.10
0.35		0.47	0.66	0.84	0.98
0.40		0.32	0.55	0.73	0.88
0.45			0.42	0.63	0.79
0.50			0.27	0.50	0.68
0.55				0.44	0.60
0.60				0.32	0.50

Example 5.9

A strip foundation is to be constructed on a sandy soil with $B = 4$ ft, $D_f = 3$ ft, $\gamma = 110$ lb/ft^3, and $\phi' = 30°$.

a. Determine the gross ultimate bearing capacity q_{uE}. Assume that $k_v = 0$ and $k_h = 0.176$.
b. If the design earthquake parameters are $V = 1.3$ ft/sec and $A = 0.32$, determine the seismic settlement of the foundation. Use FS = 3 to obtain the static allowable bearing capacity.

Solution
Part a
From Figure 5.29, for $\phi' = 30°$, $N_q = 16.51$ and $N_\gamma = 23.76$. Also,

$$\tan \theta = \frac{k_h}{1 - k_v} = 0.176$$

For $\tan \theta = 0.176$, Figure 5.30 gives

$$\frac{N_{\gamma E}}{N_\gamma} = 0.4 \quad \text{and} \quad \frac{N_{qE}}{N_q} = 0.63$$

Thus,

$$N_{\gamma E} = (0.4)(23.76) = 9.5$$
$$N_{qE} = (0.63)(16.51) = 10.4$$

and

$$q_{uE} = qN_{qE} + \tfrac{1}{2}\gamma BN_{\gamma E}$$

$$= (3 \times 110)(10.4) + (\tfrac{1}{2})(110)(4)(9.5) \approx 5522 \text{ lb/ft}^2$$

Part b
For the foundation,

$$\frac{D_f}{B} = \frac{3}{4} = 0.75$$

From Figure 5.31, for $\phi' = 30°$, FS = 3, and $D_f/B = 0.75$, the value of $k_h^* \approx 0.26$. Also, from Table 5.11, for $k_h^* = 0.26$ and $\phi' = 30°$, the value of $\tan \alpha_{AE} \approx 0.92$. From Eq. (5.69), we have

$$S_{Eq} = 0.174 \left| \frac{k_h^*}{A} \right|^{-4} \tan \alpha_{AE} \left(\frac{V^2}{Ag} \right) \text{(meters)}$$

with

$$V = 1.3 \text{ ft} \approx 0.4 \text{ m}$$

it follows that

$$S_{Eq} = 0.174 \frac{(0.4)^2}{(0.32)(9.81)} \left| \frac{0.26}{0.32} \right|^{-4} (0.92) = 0.0187 \text{ m} = \textbf{0.74 in.} \quad \blacksquare$$

Consolidation Settlement

5.15   *Primary Consolidation Settlement Relationships*

As mentioned before, consolidation settlement occurs over time in saturated clayey soils subjected to an increased load caused by construction of the foundation. (See Figure 5.32.) On the basis of the one-dimensional consolidation settlement equations given in Chapter 1, we write

$$S_{c(p)} = \int \varepsilon_z dz$$

where

ε_z = vertical strain

$\quad = \dfrac{\Delta e}{1 + e_o}$

Δe = change of void ratio

$\quad = f(\sigma_o', \sigma_c', \text{and } \Delta\sigma')$

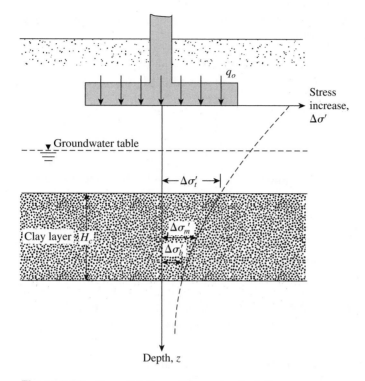

Figure 5.32 Consolidation settlement calculation

So,

$$S_{c(p)} = \frac{C_c H_c}{1 + e_o} \log \frac{\sigma'_o + \Delta\sigma'_{av}}{\sigma'_o} \qquad \text{(for normally consolidated clays)} \qquad [\text{Eq. (1.47)}]$$

$$S_{c(p)} = \frac{C_s H_c}{1 + e_o} \log \frac{\sigma'_o + \Delta\sigma'_{av}}{\sigma'_o} \qquad \text{(for overconsolidated clays with } \sigma'_o + \Delta\sigma'_{av} < \sigma'_c) \qquad [\text{Eq. (1.49)}]$$

$$S_{c(p)} = \frac{C_s H_c}{1 + e_o} \log \frac{\sigma'_c}{\sigma'_o} + \frac{C_c H_c}{1 + e_o} \log \frac{\sigma'_o + \Delta\sigma'_{av}}{\sigma'_c} \qquad \text{(for overconsolidated clays with } \sigma'_o < \sigma'_c < \sigma'_o + \Delta\sigma'_{av}) \qquad [\text{Eq. (1.51)}]$$

where

σ'_o = average effective pressure on the clay layer before the construction of the foundation

$\Delta\sigma'_{av}$ = average increase in effective pressure on the clay layer caused by the construction of the foundation

σ'_c = preconsolidation pressure

e_o = initial void ratio of the clay layer

C_c = compression index

C_s = swelling index

H_c = thickness of the clay layer

The procedures for determining the compression and swelling indexes were discussed in Chapter 1.

Note that the increase in effective pressure, $\Delta\sigma'$, on the clay layer is not constant with depth: The magnitude of $\Delta\sigma'$ will decrease with the increase in depth measured from the bottom of the foundation. However, the average increase in pressure may be approximated by

$$\Delta\sigma'_{av} = \tfrac{1}{6}(\Delta\sigma'_t + 4\Delta\sigma'_m + \Delta\sigma'_b) \tag{5.70}$$

where $\Delta\sigma'_t$, $\Delta\sigma'_m$, and $\Delta\sigma'_b$ are, respectively, the effective pressure increases at the *top, middle,* and *bottom* of the clay layer that are caused by the construction of the foundation.

The method of determining the pressure increase caused by various types of foundation load is discussed in Sections 5.2 through 5.6. $\Delta\sigma'_{av}$ can also be directly obtained from the method presented in Section 5.5.

5.16   *Three-Dimensional Effect on Primary Consolidation Settlement*

The consolidation settlement calculation presented in the preceding section is based on Eqs. (1.47), (1.49), and (1.51). These equations, as shown in Chapter 1, are in turn based on one-dimensional laboratory consolidation tests. The underlying assumption is that the increase in pore water pressure, Δu, immediately after application of the load equals the increase in stress, $\Delta\sigma$, at any depth. In this case,

$$S_{c(p)-oed} = \int \frac{\Delta e}{1 + e_o}\, dz = \int m_v \Delta\sigma'_{(1)} dz$$

where

$S_{c(p)-oed}$ = consolidated settlement calculated by using Eqs. (1.47), (1.49), and (1.51)
$\Delta\sigma'_{(1)}$ = effective vertical stress increase
m_v = volume coefficient of compressibility (see Chapter 1)

In the field, however, when a load is applied over a limited area on the ground surface, such an assumption will not be correct. Consider the case of a circular foundation on a clay layer, as shown in Figure 5.33. The vertical and the horizontal stress increases at a point in the layer immediately below the center of the foundation are $\Delta\sigma_{(1)}$ and $\Delta\sigma_{(3)}$, respectively. For a saturated clay, the pore water pressure increase at that depth (see Chapter 1) is

$$\Delta u = \Delta\sigma_{(3)} + A[\Delta\sigma_{(1)} - \Delta\sigma_{(3)}] \tag{5.71}$$

where A = pore water pressure parameter. For this case,

$$S_{c(p)} = \int m_v \Delta u\, dz = \int (m_v)\{\Delta\sigma_{(3)} + A[\Delta\sigma_{(1)} - \Delta\sigma_{(3)}]\}\, dz \tag{5.72}$$

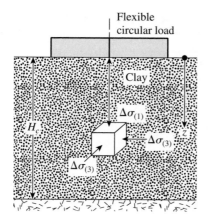

Figure 5.33 Circular foundation on a clay layer

Thus, we can write

$$K_{cir} = \frac{S_{c(p)}}{S_{c(p)-oed}} = \frac{\int_0^{H_c} m_v \Delta u \, dz}{\int_0^{H_c} m_v \Delta \sigma'_{(1)} dz} = A + (1 - A)\left[\frac{\int_0^{H_c} \Delta \sigma'_{(3)} dz}{\int_0^{H_c} \Delta \sigma'_{(1)} dz}\right] \quad (5.73)$$

where K_{cir} = settlement ratio for circular foundations.

The settlement ratio for a continuous foundation, K_{str}, can be determined in a manner similar to that for a circular foundation. The variation of K_{cir} and K_{str} with A and H_c/B is given in Figure 5.34. (*Note:* B = diameter of a circular foundation, and B = width of a continuous foundation.)

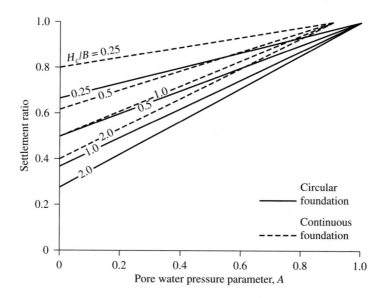

Figure 5.34 Settlement ratios for circular (K_{cir}) and continuous (K_{str}) foundations

Table 5.12   Variation of $K_{cr(OC)}$ with OCR and B/H_c

	$K_{cr(OC)}$		
OCR	**$B/H_c = 4.0$**	**$B/H_c = 1.0$**	**$B/H_c = 0.2$**
1	1	1	1
2	0.986	0.957	0.929
3	0.972	0.914	0.842
4	0.964	0.871	0.771
5	0.950	0.829	0.707
6	0.943	0.800	0.643
7	0.929	0.757	0.586
8	0.914	0.729	0.529
9	0.900	0.700	0.493
10	0.886	0.671	0.457
11	0.871	0.643	0.429
12	0.864	0.629	0.414
13	0.857	0.614	0.400
14	0.850	0.607	0.386
15	0.843	0.600	0.371
16	0.843	0.600	0.357

The preceding technique is generally referred to as the *Skempton–Bjerrum modification* (1957) for a consolidation settlement calculation.

Leonards (1976) examined the correction factor K_{cr} for a three-dimensional consolidation effect in the field for a circular foundation located over *overconsolidated clay.* Referring to Figure 5.33, we have

$$S_{c(p)} = K_{cr(OC)} \, S_{c(p)-oed} \tag{5.74}$$

where

$$K_{cr(OC)} = f\left(OCR, \frac{B}{H_c} \right) \tag{5.75}$$

in which

$$OCR = \text{overconsolidation ratio} = \frac{\sigma'_c}{\sigma'_o} \tag{5.76}$$

where

σ'_c = preconsolidation pressure
σ'_o = present average effective pressure

The interpolated values of $K_{cr(OC)}$ from Leonard's 1976 work are given in Table 5.12.

Example 5.10

A plan of a foundation 1 m $\times$ 2 m is shown in Figure 5.35. Estimate the consolidation settlement of the foundation, taking into account the three-dimensional effect. Given: $A = 0.6$.

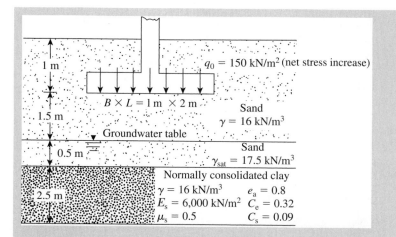

Figure 5.35   Calculation of primary consolidation settlement for a foundation

Solution

The clay is normally consolidated. Thus,

$$S_{c(p)-oed} = \frac{C_c H_c}{1 + e_o} \log \frac{\sigma_o' + \Delta\sigma_{av}'}{\sigma_o'}$$

so

$$\sigma_o' = (2.5)(16.5) + (0.5)(17.5 - 9.81) + (1.25)(16 - 9.81)$$

$$= 41.25 + 3.85 + 7.74 = 52.84 \text{ kN/m}^2$$

From Eq. (5.70),

$$\Delta\sigma_{av}' = \tfrac{1}{6}(\Delta\sigma_t' + 4\Delta\sigma_m' + \Delta\sigma_b')$$

Now the following table can be prepared (*Note:* $L = 2$ m; $B = 1$ m):

$m_1 = L/B$	z(m)	$z/(B/2) = n_1$	I_c^a	$\Delta\sigma' = q_o I_c^b$
2	2	4	0.190	$28.5 = \Delta\sigma_t'$
2	$2 + 2.5/2 = 3.25$	6.5	≈ 0.085	$12.75 = \Delta\sigma_m'$
2	$2 + 2.5 = 4.5$	9	0.045	$6.75 = \Delta\sigma_b'$

[a]Table 5.3
[b]Eq. (5.10)

Now,

$$\Delta\sigma_{av}' = \tfrac{1}{6}(28.5 + 4 \times 12.75 + 6.75) = 14.38 \text{ kN/m}^2$$

so

$$S_{c(p)-oed} = \frac{(0.32)(2.5)}{1 + 0.8} \log\left(\frac{52.84 + 14.38}{52.84}\right) = 0.0465 \text{ m}$$

$$= \mathbf{46.5 \text{ mm}}$$

Now assuming that the 2:1 method of stress increase (see Figure 5.5) holds good, the area of distribution of stress at the top of the clay layer will have dimensions

$$B' = \text{width} = B + z = 1 + (1.5 + 0.5) = 3 \text{ m}$$

and

$$L' = \text{width} = L + z = 2 + (1.5 + 0.5) = 4 \text{ m}$$

The diameter of an equivalent circular area, B_{eq}, can be given as

$$\frac{\pi}{4} B_{eq}^2 = B'L'$$

so that

$$B_{eq} = \sqrt{\frac{4B'L'}{\pi}} = \sqrt{\frac{(4)(3)(4)}{\pi}} = 3.91$$

Also,

$$\frac{H_c}{B_{eq}} = \frac{2.5}{3.91} = 0.64$$

From Figure 5.34, for $A = 0.6$ and $H_c/B_{eq} = 0.64$, the magnitude of $K_{cr} \approx 0.78$. Hence,

$$S_{e(p)} = K_{cr} S_{e(p)-\text{oed}} = (0.78)(46.5) \approx \mathbf{36.3 \text{ mm}} \qquad \blacksquare$$

5.17   *Settlement Due to Secondary Consolidation*

At the end of primary consolidation (i.e., after the complete dissipation of excess pore water pressure) some settlement is observed that is due to the plastic adjustment of soil fabrics. This stage of consolidation is called *secondary consolidation*. A plot of deformation against the logarithm of time during secondary consolidation is practically linear as shown in Figure 5.36. From the figure, the secondary compression index can be defined as

$$C_\alpha = \frac{\Delta e}{\log t_2 - \log t_1} = \frac{\Delta e}{\log (t_2/t_1)} \qquad (5.77)$$

where

C_α = secondary compression index
Δe = change of void ratio
t_1, t_2 = time

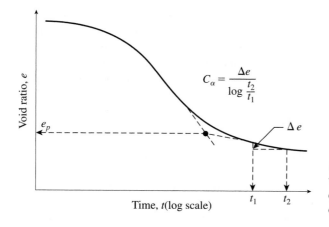

Figure 5.36 Variation of e with log t under a given load increment, and definition of secondary compression index

The magnitude of the secondary consolidation can be calculated as

$$S_{c(s)} = C'_\alpha H_c \log(t_2/t_1) \qquad (5.78)$$

where

$$C'_\alpha = C_\alpha/(1 + e_p) \qquad (5.79)$$
e_p = void ratio at the end of primary consolidation
H_c = thickness of clay layer

Mesri (1973) correlated C_α with the natural moisture content (w) of several soils, from which it appears that

$$C'_\alpha \approx 0.0001w \qquad (5.80)$$

where w = natural moisture content, in percent. For most overconsolidated soils, C'_a varies between 0.0005 to 0.001.

Mesri and Godlewski (1977) compiled the magnitude of C_α/C_c (C_c = compression index) for a number of soils. Based on their compilation, it can be summarized that

- For inorganic clays and silts:

$$C_\alpha/C_c \approx 0.04 \pm 0.01$$

- For organic clays and silts:

$$C_\alpha/C_c \approx 0.05 \pm 0.01$$

- For peats:

$$C_\alpha/C_c \approx 0.075 \pm 0.01$$

Secondary consolidation settlement is more important in the case of all organic and highly compressible inorganic soils. In overconsolidated inorganic clays, the secondary compression index is very small and of less practical significance.

There are several factors that might affect the magnitude of secondary consolidation, some of which are not yet very clearly understood (Mesri, 1973). The ratio of secondary to primary compression for a given thickness of soil layer is dependent on the ratio of the stress increment, $\Delta\sigma'$, to the initial effective overburden stress, σ'_o. For small $\Delta\sigma'/\sigma'_o$ ratios, the secondary-to-primary compression ratio is larger.

5.18   Field Load Test

The ultimate load-bearing capacity of a foundation, as well as the allowable bearing capacity based on tolerable settlement considerations, can be effectively determined from the field load test, generally referred to as the *plate load test* (ASTM, 2000; Test Designation D-1194-94). The plates that are used for tests in the field are usually made of steel and are 25 mm (1 in.) thick and 150 mm to 762 mm (6 in. to 30 in.) in diameter. Occasionally, square plates that are 305 mm × 305 mm (12 in. × 12 in.) are also used.

To conduct a plate load test, a hole is excavated with a minimum diameter of $4B$ (B is the diameter of the test plate) to a depth of D_f, the depth of the proposed foundation. The plate is placed at the center of the hole, and a load that is about one-fourth to one-fifth of the estimated ultimate load is applied to the plate in steps by means of a jack. A schematic diagram of the test arrangement is shown in Figure 5.37a. During each step of the application of the load, the settlement of the plate is observed on dial gauges. At least one hour is allowed to elapse between each

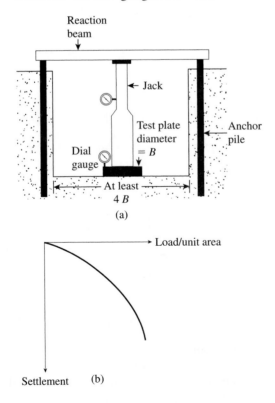

(a)

(b)

Figure 5.37 Plate load test: (a) test arrangement; (b) nature of load–settlement curve

Figure 5.38 Plate load test in the field

application. The test should be conducted until failure, or at least until the plate has gone through 25 mm (1 in.) of settlement. Figure 5.37b shows the nature of the load–settlement curve obtained from such tests, from which the ultimate load per unit area can be determined. Figure 5.38 shows a plate load test conducted in the field.

For tests in clay,

$$q_{u(F)} = q_{u(P)} \qquad (5.81)$$

where

$q_{u(F)}$ = ultimate bearing capacity of the proposed foundation
$q_{u(P)}$ = ultimate bearing capacity of the test plate

Equation (5.81) implies that the ultimate bearing capacity in clay is virtually independent of the size of the plate.

For tests in sandy soils,

$$q_{u(F)} = q_{u(P)} \frac{B_F}{B_P} \qquad (5.82)$$

where

B_F = width of the foundation
B_P = width of the test plate

The allowable bearing capacity of a foundation, based on settlement consider-ations and for a given intensity of load, q_o, is

$$S_F = S_P \frac{B_F}{B_P} \quad \text{(for clayey soil)} \tag{5.83}$$

and

$$S_F = S_P \left(\frac{2B_F}{B_F + B_P} \right)^2 \quad \text{(for sandy soil)} \tag{5.84}$$

The preceding relationship is based on the work of Terzaghi and Peck (1967).

Example 5.11

A shallow square foundation for a column is to be constructed on sand. The foun-dation must carry a net vertical mass of 102,000 kg. The standard penetration numbers (N_{60}) obtained from exploration are given in Figure 5.39. Assume that the depth of the foundation will be 1.5 m and the tolerable settlement is 25 mm. Determine the size of the foundation.

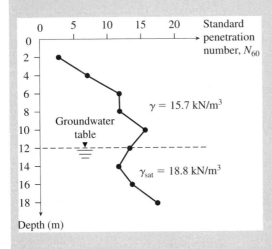

Figure 5.39   Variation of N_{60} with depth

Solution
Following is the variation of N_{60} with depth. Within a depth of 10 m, the average value of $N_{60} = (3 + 7 + 12 + 12 + 16)/5 = 10$.

Depth (m)	N_{60}
2	3
4	7
6	12
8	12
10	16
12	13
14	12
16	14
18	18

Now consider Eq. (5.51),

$$q_{net} = \frac{N_{60}}{0.08}\left(\frac{B + 0.3}{B}\right)^2 F_d\left(\frac{S_e}{25}\right)$$

With $S_e = 2.5$ mm and $N_{60} \approx 10$,

$$q_{net} = \frac{10}{0.08}\left(\frac{B + 0.3}{B}\right)^2 F_d\left(\frac{25}{25}\right) = 125\left(\frac{B + 0.3}{B}\right)^2 F_d$$

$$F_d = 1 + 0.33\frac{D_f}{B}$$

Also, given $Q_o = \dfrac{102,000 \text{ kg} \times 9.81}{1000} \approx 1000 \text{ kN}$, the following table can be prepared for trial calculations:

B (m)	F_d[a]	q_{net} (kN/m²)	$Q_o = q_{net} \times B^2$
2	1.248	206.3	825.2
2.25	1.22	195.9	991.7
2.3	1.215	194.1	1026.8

[a] $D_f = 1.5$ m

Because the required Q_o is 1000 kN, B will be approximately equal to **2.3 m**. ∎

5.19 Presumptive Bearing Capacity

Several building codes (e.g., the Uniform Building Code, Chicago Building Code, and New York City Building Code) specify the allowable bearing capacity of foundations on various types of soil. For minor construction, they often provide fairly acceptable guidelines. However, these bearing capacity values are based primarily on the *visual* classification of near-surface soils and generally do not take into consideration factors such as the stress history of the soil, the location of the water table, the depth of the foundation, and the tolerable settlement. So, for large construction projects, the codes' presumptive values should be used only as guides.

5.20   *Tolerable Settlement of Buildings*

In most instances of construction, the subsoil is not homogeneous and the load carried by various shallow foundations of a given structure can vary widely. As a result, it is reasonable to expect varying degrees of settlement in different parts of a given building. The *differential settlement* of the parts of a building can lead to damage of the superstructure. Hence, it is important to define certain parameters that quantify differential settlement and to develop limiting values for those parameters in order that the resulting structures be safe. Burland and Worth (1970) summarized the important parameters relating to differential settlement.

Figure 5.40 shows a structure in which various foundations, at *A*, *B*, *C*, *D*, and *E*, have gone through some settlement. The settlement at *A* is *AA'*, at *B* is *BB'*, etc. Based on this figure, the definitions of the various parameters are as follows:

S_T = total settlement of a given point

ΔS_T = difference in total settlement between any two points

α = gradient between two successive points

β = angular distortion = $\dfrac{\Delta S_{T(ij)}}{l_{ij}}$

(*Note*: l_{ij} = distance between points i and j)

ω = tilt

Δ = relative deflection (i.e. movement from a straight line joining two reference points)

$\dfrac{\Delta}{L}$ = deflection ratio

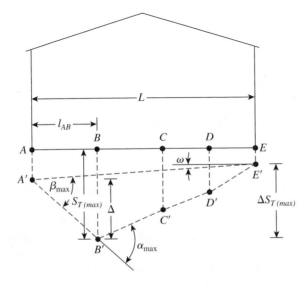

Figure 5.40 Definition of parameters for differential settlement

Since the 1950s, various researchers and building codes have recommended allowable values for the preceding parameters. A summary of several of these recommendations is presented next.

In 1956, Skempton and McDonald proposed the following limiting values for maximum settlement and maximum angular distortion, to be used for building purposes:

Maximum settlement, $S_{T(max)}$	
In sand	32 mm
In clay	45 mm
Maximum differential settlement, $\Delta S_{T(max)}$	
Isolated foundations in sand	51 mm
Isolated foundations in clay	76 mm
Raft in sand	51–76 mm
Raft in clay	76–127 mm
Maximum angular distortion, β_{max}	1/300

On the basis of experience, Polshin and Tokar (1957) suggested the following allowable deflection ratios for buildings as a function of L/H, the ratio of the length to the height of a building:

$$\Delta/L = 0.0003 \text{ for } L/H \leqslant 2$$

$$\Delta/L = 0.001 \text{ for } L/H = 8$$

The 1955 Soviet Code of Practice gives the following allowable values:

Type of building	L/H	Δ/L
Multistory buildings and civil dwellings	≤3	0.0003 (for sand) 0.0004 (for clay)
	≥5	0.0005 (for sand) 0.0007 (for clay)
One-story mills		0.001 (for sand and clay)

Bjerrum (1963) recommended the following limiting angular distortion, β_{max} for various structures:

Category of potential damage	β_{max}
Safe limit for flexible brick wall ($L/H > 4$)	1/150
Danger of structural damage to most buildings	1/150
Cracking of panel and brick walls	1/150
Visible tilting of high rigid buildings	1/250
First cracking of panel walls	1/300
Safe limit for no cracking of building	1/500
Danger to frames with diagonals	1/600

If the maximum allowable values of β_{max} are known, the magnitude of the allowable $S_{T(max)}$ can be calculated with the use of the foregoing correlations.

The European Committee for Standardization has also provided limiting values for serviceability and the maximum accepted foundation movements. (See Table 5.13.)

Table 5.13 Recommendations of European Committee for Standardization on Differential Settlement Parameters

Item	Parameter	Magnitude	Comments
Limiting values for	S_T	25 mm	Isolated shallow foundation
serviceability		50 mm	Raft foundation
(European Committee	ΔS_T	5 mm	Frames with rigid cladding
for Standardization,		10 mm	Frames with flexible cladding
1994a)		20 mm	Open frames
	β	1/500	—
Maximum acceptable	S_T	50	Isolated shallow foundation
foundation movement	ΔS_T	20	Isolated shallow foundation
(European Committee	β	$\approx 1/500$	—
for Standardization, 1994b)			

Problems

5.1 A flexible circular area is subjected to a uniformly distributed load of 3000 lb/ft². The diameter of the loaded area is 9.5 ft. Determine the stress increase in a soil mass at a point located 7.5 ft below the center of the loaded area.

5.2 Refer to Figure 5.4, which shows a flexible rectangular area. Given: $B_1 = 1.2$ m, $B_2 = 3$ m, $L_1 = 3$ m, and $L_2 = 6$ m. If the area is subjected to a uniform load of 110 kN/m², determine the stress increase at a depth of 8 m located immediately below point O.

5.3 Repeat Problem 5.2 with the following: $B_1 = 5$ ft, $B_2 = 10$ ft, $L_1 = 7$ ft, $L_2 = 12$ ft, and the uniform load on the flexible area = 2500 lb/ft². Determine the stress increase below point O at a depth of 20 ft.

5.4 Using Eq. (5.10), determine the stress increase ($\Delta \sigma$) from $z = 0$ to $z = 5$ m below the center of the area described in Problem 5.2.

5.5 Using Eq. (5.10), determine the stress increase ($\Delta \sigma$) from $z = 0$ to $z = 20$ ft below the center of the area described in Problem 5.3.

5.6 Refer to Figure P5.6. Using the procedure outlined in Section 5.5, determine the average stress increase in the clay layer below the center of the foundation due to the net foundation load of 50 ton.

5.7 Solve Problem 5.6 using the 2:1 method [Eqs. (5.14) and (5.70)].

5.8 Figure P5.8 shows an embankment load on a silty clay soil layer. Determine the stress increase at points A, B, and C, which are located at a depth of 15 ft below the ground surface.

5.9 A planned flexible load area (see Figure P5.9) is to be 3 m × 4.6 m and carries a uniformly distributed load of 180 kN/m². Estimate the elastic settlement below the center of the loaded area. Assume that $D_f = 2$ m and $H = \infty$. Use Eq. (5.25).

5.10 Redo Problem 5.9 assuming that $D_f = 5$ m and $H = 3$ m.

5.11 Figure 5.14 shows a foundation of 10 ft × 6.25 ft resting on a sand deposit. The net load per unit area at the level of the foundation, q_o, is 3000 lb/ft². For the sand, $\mu_s = 0.3$, $E_s = 3200$ lb/in.², $D_f = 2.5$ ft and $H = 32$ ft. Assume that the foundation is rigid, and determine the elastic settlement the foundation would undergo. Use Eqs. (5.25) and (5.33).

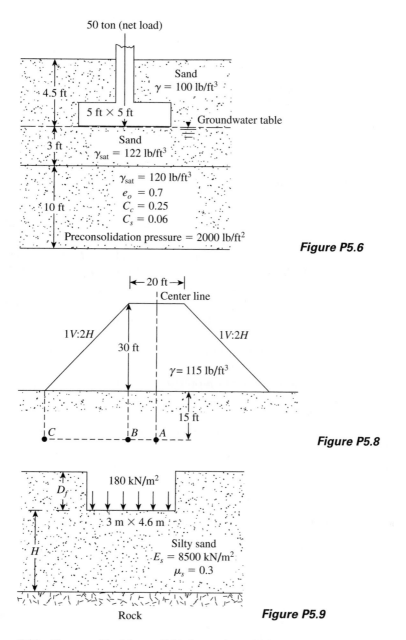

Figure P5.6

Figure P5.8

Figure P5.9

5.12 Repeat Problems 5.11 for a foundation of size = 2.1 m × 2.1 m and with $q_o = 230 \text{ kN/m}^2$, $D_f = 1.5$ m, $H = 12$ m; and soil conditions of $\mu_s = 0.4$, $E_s = 16,000 \text{ kN/m}^2$, and $\gamma = 18.1 \text{ kN/m}^3$.

5.13 Refer to Figure 5.17. A foundation measuring 1.5 m × 3 m is supported by a saturated clay. Given: $D_f = 1.2$ m, $H = 3$ m, E_s (clay) = 600 kN/m², and $q_o = 150 \text{ kN/m}^2$. Determine the elastic settlement of the foundation.

5.14 For a shallow foundation supported by a silty clay, as shown in Figure 5.18, the following are given:

Length, $L = 8$ ft
Width, $B = 2.5$ ft
Depth of foundation, $D_f = 2.5$ ft
Thickness of foundation, $t = 1$ ft
Load per unit area, $q_o = 3000$ lb/ft^2
$E_f = 2 \times 10^6$ lb/in^2

The silty clay soil has the following properties:

$H = 8$ ft
$\mu_s = 0.4$
$E_o = 1250$ lb/in^2
$k = 30$ lb/in^2/ft

Using Eq. (5.39), estimate the elastic settlement of the foundation.

5.15 A plan calls for a square foundation measuring 3 m × 3 m, supported by a layer of sand. (See figure 5.18.) Led $D_f = 1.5$ m, $t = 0.25$ m, $E_o = 16,000$ kN/m^2, $k = 400$ kN/m^2/m, $\mu_s = 0.3$, $H = 20$ m, $E_f = 15 \times 10^6$ kN/m^2, and $q_o = 150$ kN/m^2. Calculate the elastic settlement. Use Eq. (5.39).

5.16 Solve Problems 5.11 with Eq. (5.42). For the correction factor C_2, use a time of 5 years for creep and, for the unit weight of soil, use $\gamma = 115$ lb/ft^3. Assuming an I_z, plot the same as that for a square foundation.

5.17 Solve Problem 5.12 with Eq. (5.42). For the correction factor C_2, use a time of 5 years for creep.

5.18 A continuous foundation on a deposit of sand layer is shown in Figure P5.18 along with the variation of the modulus of elasticity of the soil (E_s). Assuming $\gamma = 115$ lb/ft^3 and $C_2 = 10$ years, calculate the elastic settlement of the foundation using the strain influence factor.

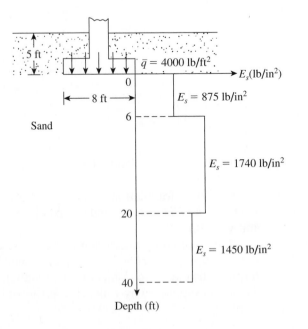

Depth (ft)

Figure P5.18

5.19 Following are the results of standard penetration tests in a granular soil deposit.

Depth (ft)	Standard penetration number, N_{60}
5	10
10	12
15	9
20	14
25	16

What will be the net allowable bearing capacity of a foundation planned to be 5 ft × 5 ft? Let $D_f = 3$ ft and the allowable settlement = 1 in., and use the relationships presented in Section 5.12.

5.20 A shallow foundation measuring 1 m × 2 m in plan is to be constructed over a normally consolidated sand layer. Given: $D_f = 1$ m, N_{60} increases with depth, $\overline{N}_{60}$ (in the depth of stress influence) = 8, and $q_{net} = 153$ kN/m². Estimate the elastic settlement using Burland and Burbidge's method.

5.21 Following are the average values of cone penetration resistance in a granular soil deposit:

Depth (m)	Cone penetration resistance, q_c (MN/m²)
2	2.1
4	4.2
6	5.2
8	7.3
10	8.7
15	14

Assume that $\gamma = 16.5$ kN/m³ and estimate the seismic ultimate bearing capacity (q_{uE}) for a continuous foundation with $B = 1.5$ m, $D_f = 1.0$ m, $k_h = 0.2$, and $k_v = 0$. Use Eqs. (2.47), (5.67), and (5.68).

5.22 In problem 5.21, if the design earthquake parameters are $V = 0.35$ m/sec and $A = 0.3$, determine the seismic settlement of the foundation. Assume that FS = 4 for use in obtaining the static allowable bearing capacity.

5.23 Estimate the consolidation settlement of the clay layer shown in Figure P5.6 using the results of Problem 5.6.

5.24 Estimate the consolidation settlement of the clay layer shown in Figure P5.6 using the results of Problem 5.7.

References

Ahlvin, R. G., and Ulery, H. H. (1962). *Tabulated Values of Determining the Composite Pattern of Stresses, Strains, and Deflections beneath a Uniform Load on a Homogeneous Half Space.* Highway Research Board Bulletin 342, pp. 1–13.

American Society for Testing and Materials (2000). *Annual Book of ASTM Standards,* Vol. 04.08, West Conshohocken, PA.

Bjerrum, L. (1963). "Allowable Settlement of Structures," *Proceedings, European Conference on Soil Mechanics and Foundation Engineering,* Wiesbaden, Germany, Vol. III, pp. 135–137.

Boussinesq, J. (1883). *Application des Potentials á L'Étude de L'Équilibre et du Mouvement des Solides Élastiques,* Gauthier-Villars, Paris.

Bowles, J. E. (1987). "Elastic Foundation Settlement on Sand Deposits," *Journal of Geotechnical Engineering,* ASCE, Vol. 113, No. 8, pp. 846–860.

Bowles, J. E. (1977). *Foundation Analysis and Design,* 2d ed., McGraw-Hill, New York.

Briaud, J. L., and Gibbens, R. M. (1994). "Predicted and Measured Behavior of Five Spread Footings on Sand," *Geotechnical Special Publications No. 41,* American Society of Civil Engineers.

Burland, J. B., and Burbidge, M. C. (1985). "Settlement of Foundations on Sand and Gravel," *Proceedings, Institute of Civil Engineers,* Part I, Vol. 7, pp. 1325–1381.

Christian, J. T., and Carrier, W. D. (1978). "Janbu, Bjerrum, and Kjaernsli's Chart Reinterpreted," *Canadian Geotechnical Journal,* Vol. 15, pp. 124–128.

Das, B. (1997). *Advanced Soil Mechanics,* 2d ed., Taylor and Francis, Washington, DC.

Duncan, J. M., and Buchignani, A. N. (1976). *An Engineering Manual for Settlement Studies,* Department of Civil Engineering, University of California, Berkeley.

European Committee for Standardization (1994a). *Basis of Design and Actions on Structures,* Eurocode 1, Brussels, Belgium.

European Committee for Standardization (1994b). *Geotechnical Design, General Rules—Part 1,* Eurocode 7, Brussels, Belgium.

Fox, E. N. (1948). "The Mean Elastic Settlement of a Uniformly Loaded Area at a Depth below the Ground Surface," *Proceedings, 2nd International Conference on Soil Mechanics and Foundation Engineering,* Rotterdam, Vol. 1, pp. 129–132.

Griffiths, D. V. (1984). "A Chart for Estimating the Average Vertical Stress Increase in an Elastic Foundation below a Uniformly Loaded Rectangular Area," *Canadian Geotechnical Journal,* Vol. 21, No. 4, 710–713.

Janbu, N., Bjerrum, L., and Kjaernsli, B. (1956). "Veiledning vedlosning av fundamentering—soppgaver," *Publication No. 18,* Norwegian Geotechnical Institute, pp. 30–32.

Jeyapalan, J. K., and Boehm, R. (1986). "Procedure for Predicting Settlements in Sand," in *Settlement of Shallow Foundations on Cohesionless Soils: Design and Performance, Geotechnical Special Technical Publication No. 5,* American Society of Civil Engineers, pp. 1–12.

Leonards, G. A. (1976). *Estimating Consolidation Settlement of Shallow Foundations on Overconsolidated Clay,* Special Report No. 163, Transportation Research Board, Washington, DC., pp. 13–16.

Mayne, P. W., and Poulos, H. G. (1999). "Approximate Displacement Influence Factors for Elastic Shallow Foundations," *Journal of Geotechnical and Geoenvironmental Engineering,* ASCE, Vol. 125, No. 6, 453–460.

Mesri, G. (1973). "Coefficient of Secondary Compression," *Journal of the Soil Mechanics and Foundations Division, American Society of Civil Engineers,* Vol. 99, No. SM1, 122–137.

Mesri, G., and Godlewski, P. M. (1977). "Time and Stress—Compressibility Interrelationship," *Journal of Geotechnical Engineering Division,* American Society of Civil Engineers, Vol. 103, No. GT5, pp. 417–430.

Meyerhof, G. G. (1956). "Penetration Tests and Bearing Capacity of Cohesionless Soils," *Journal of the Soil Mechanics and Foundations Division,* American Society of Civil Engineers, Vol. 82, No. SM1, pp. 1–19.

Meyerhof, G. G. (1965). "Shallow Foundations," *Journal of the Soil Mechanics and Foundations Division,* American Society of Civil Engineers, Vol. 91, No. SM2, pp. 21–31.

Mitchell, J. K., and Gardner, W. S. (1975). "*In Situ* Measurement of Volume Change Characteristics," *Proceedings, Specialty Conference,* American Society of Civil Engineers, Vol. 2, pp. 279–345.

Newmark, N. M. (1935). *Simplified Computation of Vertical Pressure in Elastic Foundation,* Circular 24, University of Illinois Engineering Experiment Station, Urbana, IL.

Osterberg, J. O. (1957). "Influence Values for Vertical Stresses in Semi-Infinite Mass Due to Embankment Loading," *Proceedings, Fourth International Conference on Soil Mechanics and Foundation Engineering,* London, Vol. 1, pp. 393–396.

Papadopoulos, B. P. (1992). "Settlement of Shallow Foundations on Cohesionless Soils," *Journal of Geotechnical Engineering,* American Society of Civil Engineers, Vol. 118, No. 3, pp. 377–393.

Polshin, D. E., and Tokar, R. A. (1957). "Maximum Allowable Nonuniform Settlement of Structures," *Proceedings, Fourth International Conference on Soil Mechanics and Foundation Engineering,* London, Vol. 1, pp. 402–405.

Richards, R., Jr., Elms, D. G., and Budhu, M. (1993). "Seismic Bearing Capacity and Settlement of Foundations," *Journal of Geotechnical Engineering,* American Society of Civil Engineers, Vol. 119, No. 4, pp. 662–674.

Schmertmann, J. H., Hartman, J. P., and Brown, P. R. (1978). "Improved Strain Influence Factor Diagrams," *Journal of the Geotechnical Engineering Division,* American Society of Civil Engineers, Vol. 104, No. GT8, pp. 1131–1135.

Sivakugan, N., Eckersley, J. D., and Li, H. (1998). "Settlement Prediction Using Neural Networks," *Australian Civil Engineering Transactions,* Vol. CE40, pp. 49–52.

Sivakugan, N., and Johnson, K. (2004). "Settlement Predictions in Granular Soils: A Probabilistic Approach," *Geotechnique,* Vol. 54, No. 7, pp. 499–502.

Skempton, A. W., and Bjerrum, L. (1957). " A Contribution to Settlement Analysis of Foundations in Clay," *Geotechnique,* London, Vol 7, p. 178.

Skempton, A. W., and McDonald, D. M. (1956). "The Allowable Settlement of Buildings," *Proceedings of Institute of Civil Engineers,* Vol. 5, Part III, p. 727.

Steinbrenner, W. (1934). "Tafeln zur Setzungsberechnung," *Die Strasse,* Vol. 1, pp. 121–124.

Terzaghi, K., and Peck, R. B. (1967). *Soil Mechanics in Engineering Practice,* 2d ed., Wiley, New York.

6

Mat Foundations

6.1   Introduction

Under normal conditions, square and rectangular footings such as those described in Chapters 3 and 4 are economical for supporting columns and walls. However, under certain circumstances, it may be desirable to construct a footing that supports a line of two or more columns. These footings are referred to as *combined footings*. When more than one line of columns is supported by a concrete slab, it is called a *mat foundation*. Combined footings can be classified generally under the following categories:

a. Rectangular combined footing
b. Trapezoidal combined footing
c. Strap footing

Mat foundations are generally used with soil that has a low bearing capacity. A brief overview of the principles of combined footings is given in Section 6.2, followed by a more detailed discussion on mat foundations.

6.2   Combined Footings

Rectangular Combined Footing

In several instances, the load to be carried by a column and the soil bearing capacity are such that the standard spread footing design will require extension of the column foundation beyond the property line. In such a case, two or more columns can be supported on a single rectangular foundation, as shown in Figure 6.1. If the net allowable soil pressure is known, the size of the foundation ($B \times L$) can be determined in the following manner:

a. Determine the area of the foundation

$$A = \frac{Q_1 + Q_2}{q_{net(all)}} \tag{6.1}$$

where

Q_1, Q_2 = column loads
$q_{net(all)}$ = net allowable soil bearing capacity

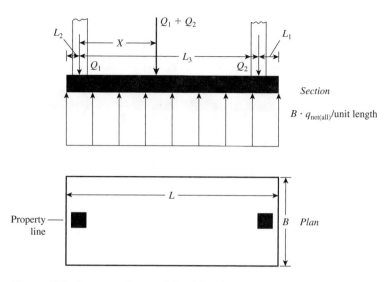

Figure 6.1 Rectangular combined footing

b. Determine the location of the resultant of the column loads. From Figure 6.1,

$$X = \frac{Q_2 L_3}{Q_1 + Q_2} \tag{6.2}$$

c. For a uniform distribution of soil pressure under the foundation, the resultant of the column loads should pass through the centroid of the foundation. Thus,

$$L = 2(L_2 + X) \tag{6.3}$$

where L = length of the foundation.

d. Once the length L is determined, the value of L_1 can be obtained as follows:

$$L_1 = L - L_2 - L_3 \tag{6.4}$$

Note that the magnitude of L_2 will be known and depends on the location of the property line.

e. The width of the foundation is then

$$B = \frac{A}{L} \tag{6.5}$$

Trapezoidal Combined Footing

Trapezoidal combined footing (see Figure 6.2) is sometimes used as an isolated spread foundation of columns carrying large loads where space is tight. The size of

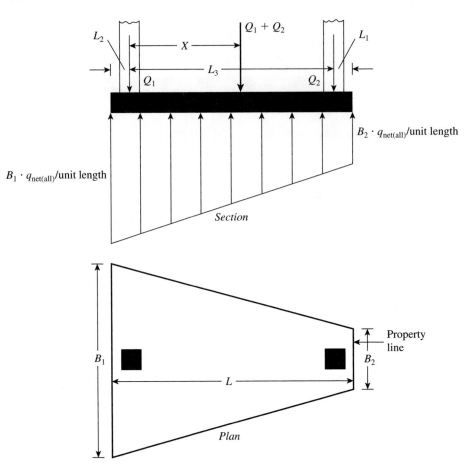

Figure 6.2   Trapezoidal combined footing

the foundation that will uniformly distribute pressure on the soil can be obtained in the following manner:

a. If the net allowable soil pressure is known, determine the area of the foundation:

$$A = \frac{Q_1 + Q_2}{q_{net(all)}}$$

From Figure 6.2,

$$A = \frac{B_1 + B_2}{2}L \tag{6.6}$$

b. Determine the location of the resultant for the column loads:

$$X = \frac{Q_2 L_3}{Q_1 + Q_2}$$

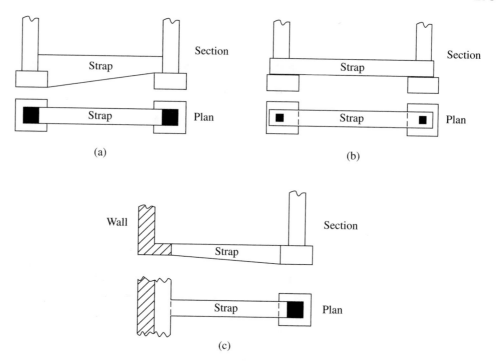

Figure 6.3   Cantilever footing—use of strap beam

c. From the property of a trapezoid,

$$X + L_2 = \left(\frac{B_1 + 2B_2}{B_1 + B_2}\right)\frac{L}{3} \qquad (6.7)$$

With known values of A, L, X, and L_2, solve Eqs. (6.6) and (6.7) to obtain B_1 and B_2. Note that, for a trapezoid,

$$\frac{L}{3} < X + L_2 < \frac{L}{2}$$

Cantilever Footing

Cantilever footing construction uses a *strap beam* to connect an eccentrically loaded column foundation to the foundation of an interior column. (See Figure 6.3). Cantilever footings may be used in place of trapezoidal or rectangular combined footings when the allowable soil bearing capacity is high and the distances between the columns are large.

6.3   Common Types of Mat Foundations

The mat foundation, which is sometimes referred to as a *raft foundation,* is a combined footing that may cover the entire area under a structure supporting several columns and walls. Mat foundations are sometimes preferred for soils that have low

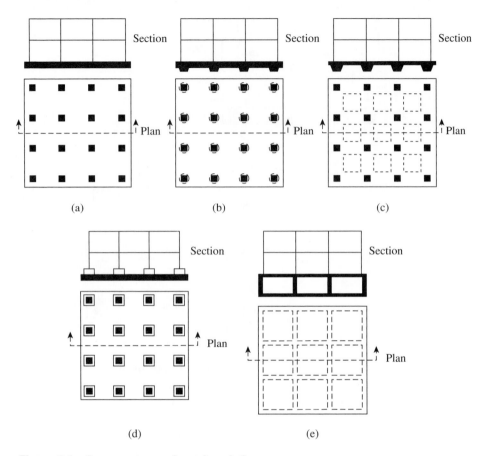

Figure 6.4   Common types of mat foundation

load-bearing capacities, but that will have to support high column or wall loads. Under some conditions, spread footings would have to cover more than half the building area, and mat foundations might be more economical. Several types of mat foundations are used currently. Some of the common ones are shown schematically in Figure 6.4 and include the following:

1. Flat plate (Figure 6.4a). The mat is of uniform thickness.
2. Flat plate thickened under columns (Figure 6.4b).
3. Beams and slab (Figure 6.4c). The beams run both ways, and the columns are located at the intersection of the beams.
4. Flat plates with pedestals (Figure 6.4d).
5. Slab with basement walls as a part of the mat (Figure 6.4e). The walls act as stiffeners for the mat.

Mats may be supported by piles, which help reduce the settlement of a structure built over highly compressible soil. Where the water table is high, mats are often placed over piles to control buoyancy. Figure 6.5 shows the difference between the depth D_f and the width B of isolated foundations and mat foundations.

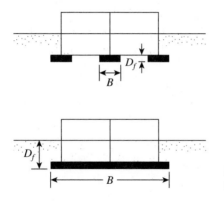

Figure 6.5 Comparison of isolated foundation and mat foundation (B = width, D_f = depth)

6.4 *Bearing Capacity of Mat Foundations*

The *gross ultimate bearing capacity* of a mat foundation can be determined by the same equation used for shallow foundations (see Section 3.6), or

$$q_u = c'N_cF_{cs}F_{cd}F_{ci} + qN_qF_{qs}F_{qd}F_{qi} + \tfrac{1}{2}\gamma BN_\gamma F_{\gamma s}F_{\gamma d}F_{\gamma i} \qquad \text{[Eq. (3.23)]}$$

(Chapter 3 gives the proper values of the bearing capacity factors, as well as the shape depth, and load inclination factors.) The term B in Eq. (3.23) is the smallest dimension of the mat. The *net ultimate capacity* of a mat foundation is

$$q_{\text{net}(u)} = q_u - q \qquad \text{[Eq. (3.14)]}$$

A suitable factor of safety should be used to calculate the net *allowable* bearing capacity. For mats on clay, the factor of safety should not be less than 3 under dead load or maximum live load. However, under the most extreme conditions, the factor of safety should be at least 1.75 to 2. For mats constructed over sand, a factor of safety of 3 should normally be used. Under most working conditions, the factor of safety against bearing capacity failure of mats on sand is very large.

For saturated clays with $\phi = 0$ and a vertical loading condition, Eq. (3.23) gives

$$q_u = c_u N_c F_{cs} F_{cd} + q \qquad (6.8)$$

where c_u = undrained cohesion. (*Note: N_c = 5.14, N_q = 1, and N_γ = 0.*)
From Eqs. (3.27) and (3.30), for $\phi = 0$,

$$F_{cs} = 1 + \frac{B}{L}\left(\frac{N_q}{N_c}\right) = 1 + \left(\frac{B}{L}\right)\left(\frac{1}{5.14}\right) = 1 + \frac{0.195B}{L}$$

and

$$F_{cd} = 1 + 0.4\left(\frac{D_f}{B}\right)$$

Substitution of the preceding shape and depth factors into Eq. (6.8) yields

$$q_u = 5.14c_u \left(1 + \frac{0.195B}{L}\right)\left(1 + 0.4\frac{D_f}{B}\right) + q \tag{6.9}$$

Hence, the net ultimate bearing capacity is

$$q_{net(u)} = q_u - q = 5.14c_u \left(1 + \frac{0.195B}{L}\right)\left(1 + 0.4\frac{D_f}{B}\right) \tag{6.10}$$

For FS = 3, the net allowable soil bearing capacity becomes

$$q_{net(all)} = \frac{q_{u(net)}}{FS} = 1.713c_u \left(1 + \frac{0.195B}{L}\right)\left(1 + 0.4\frac{D_f}{B}\right) \tag{6.11}$$

The net allowable bearing capacity for mats constructed over granular soil deposits can be adequately determined from the standard penetration resistance numbers. From Eq. (5.51), for shallow foundations,

$$q_{net}(kN/m^2) = \frac{N_{60}}{0.08}\left(\frac{B + 0.3}{B}\right)^2 F_d \left(\frac{S_e}{25}\right) \tag{Eq. (5.51)}$$

where

N_{60} = standard penetration resistance
B = width (m)
$F_d = 1 + 0.33(D_f/B) \leq 1.33$
S_e = settlement, (mm)

When the width B is large, the preceding equation can be approximated as

$$
\begin{aligned}
q_{net}(kN/m^2) &= \frac{N_{60}}{0.08} F_d \left(\frac{S_e}{25}\right) \\
&= \frac{N_{60}}{0.08}\left[1 + 0.33\left(\frac{D_f}{B}\right)\right]\left[\frac{S_e(mm)}{25}\right] \\
&\leq 16.63 N_{60}\left[\frac{S_e(mm)}{25}\right]
\end{aligned} \tag{6.12}
$$

In English units, Eq. (6.12) may be expressed as

$$
\begin{aligned}
q_{net(all)}(kip/ft^2) &= 0.25N_{60}\left[1 + 0.33\left(\frac{D_f}{B}\right)\right][S_e(in.)] \\
&\leq 0.33N_{60}[S_e(in.)]
\end{aligned} \tag{6.13}
$$

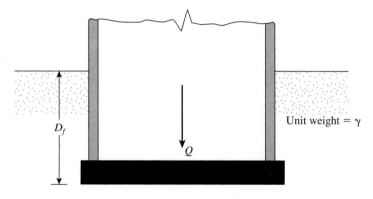

Figure 6.6 Definition of net pressure on soil caused by a mat foundation

Generally, shallow foundations are designed for a maximum settlement of 25 mm (1 in.) and a differential settlement of about 19 mm (0.75 in.).

However, the width of the raft foundations are larger than those of the isolated spread footings. As shown in Table 5.3, the depth of significant stress increase in the soil below a foundation depends on the width of the foundation. Hence, for a raft foundation, the depth of the zone of influence is likely to be much larger than that of a spread footing. Thus, the loose soil pockets under a raft may be more evenly distributed, resulting in a smaller differential settlement. Accordingly, the customary assumption is that, for a maximum raft settlement of 50 mm (2 in.), the differential settlement would be 19 mm (0.75 in.). Using this logic and conservatively assuming that $F_d = 1$, we can respectively approximate Eqs. (6.12) and (6.13) as

$$q_{net(all)} = q_{net}(\text{kN/m}^2) \approx 25N_{60} \tag{6.14}$$

and

$$q_{net(all)} = q_{net}(\text{kip/ft}^2) = 0.5N_{60} \tag{6.15}$$

The net allowable pressure applied on a foundation (see Figure 6.6) may be expressed as

$$q = \frac{Q}{A} - \gamma D_f \tag{6.16}$$

where

Q = dead weight of the structure and the live load
A = area of the raft

In all cases, q should be less than or equal to allowable q_{net}.

Example 6.1

Determine the net ultimate bearing capacity of a mat foundation measuring 45 ft × 30 ft on a saturated clay with $c_u = 1950$ lb/ft², $\phi = 0$, and $D_f = 6.5$ ft.

Solution

From Eq. (6.10),

$$q_{net(u)} = 5.14c_u \left[1 + \left(\frac{0.195B}{L} \right) \right]\left[1 + 0.4\frac{D_f}{B} \right]$$

$$= (5.14)(1950)\left[1 + \left(\frac{0.195 \times 30}{45} \right) \right]\left[1 + \left(\frac{0.4 \times 6.5}{30} \right) \right]$$

$$= \textbf{12,307 lb/ft}^2 \qquad\blacksquare$$

Example 6.2

What will be the net allowable bearing capacity of a mat foundation with dimensions of 15 m × 10 m constructed over a sand deposit? Here, $D_f = 2$ m, the allowable settlement is 25 mm, and the average penetration number $N_{60} = 10$.

Solution

From Eq. (6.12),

$$q_{net(all)} = \frac{N_{60}}{0.08}\left[1 + 0.33\left(\frac{D_f}{B} \right) \right]\left(\frac{S_e}{25} \right) \leq 16.63 N_{60}\left(\frac{S_e}{25} \right)$$

or

$$q_{net(all)} = \frac{10}{0.08}\left[1 + \frac{0.33 \times 2}{10} \right]\left(\frac{25}{25} \right) = \textbf{133.25 kN/m}^2 \qquad\blacksquare$$

6.5   *Differential Settlement of Mats*

In 1988, the American Concrete Institute Committee 336 suggested a method for calculating the differential settlement of mat foundations. According to this method, the rigidity factor K_r is calculated as

$$K_r = \frac{E'I_b}{E_s B^3} \tag{6.17}$$

where

E' = modulus of elasticity of the material used in the structure
E_s = modulus of elasticity of the soil
B = width of foundation
I_b = moment of inertia of the structure per unit length at right angles to B

The term $E'I_b$ can be expressed as

$$E'I_b = E'\left(I_F + \sum I_{b'} + \sum \frac{ah^3}{12} \right) \tag{6.18}$$

where

$E'I_b$ = flexural rigidity of the superstructure and foundation per unit length at right angles to B

$\Sigma E'I'_b$ = flexural rigidity of the framed members at right angles to B

$\Sigma(E'ah^3/12)$ = flexural rigidity of the shear walls

a = shear wall thickness

h = shear wall height

$E'I_F$ = flexibility of the foundation

Based on the value of K_r, the ratio (δ) of the differential settlement to the total settlement can be estimated in the following manner:

1. If $K_r > 0.5$, it can be treated as a rigid mat, and $\delta = 0$.
2. If $K_r = 0.5$, then $\delta \approx 0.1$.
3. If $K_r = 0$, then $\delta = 0.35$ for square mats $(B/L = 1)$ and $\delta = 0.5$ for long foundations $(B/L = 0)$.

6.6 *Field Settlement Observations for Mat Foundations*

Several field settlement observations for mat foundations are currently available in the literature. In this section, we compare the observed settlements for some mat foundations constructed over granular soil deposits with those obtained from Eqs. (6.12) and (6.13).

Meyerhof (1965) compiled the observed maximum settlements for mat foundations constructed on sand and gravel, as listed in Table 6.1. In Eq. (6.12), if the depth factor, $1 + 0.33(D_f/B)$, is assumed to be approximately unity, then

$$S_e(\text{mm}) \approx \frac{2q_{\text{net(all)}}}{N_{60}} \tag{6.19}$$

From the values of $q_{\text{net(all)}}$ and N_{60} given in Columns 6 and 5, respectively, of Table 6.1, the magnitudes of S_e were calculated and are given in Column 8.

Column 9 of Table 6.1 gives the ratios of calculated to measured values of S_e. These ratios vary from about 0.79 to 3.39. Thus, calculating the net allowable bearing capacity with the use of Eq. (6.12) or (6.13) will yield safe and conservative values.

6.7 *Compensated Foundation*

Figure 6.6 and Eq. (6.16) indicate that the net pressure increase in the soil under a mat foundation can be reduced by increasing the depth D_f of the mat. This approach is generally referred to as the *compensated foundation design* and is extremely useful when structures are to be built on very soft clays. In this design, a deeper basement is made below the higher portion of the superstructure, so that the net pressure increase in soil at any depth is relatively uniform. (See Figure 6.7.) From Eq. (6.16) and Figure 6.6, the net average applied pressure on soil is

$$q = \frac{Q}{A} - \gamma D_f$$

Table 6.1 Settlement of Mat Foundations on Sand and Gravel (Based on Meyerhof, 1965)

Case No. (1)	Structure (2)	Reference (3)	B m (ft) (4)	Average N_{60} (5)	$q_{net(all)}$ kN/m² (kip/ft²) (6)	Observed maximum settlement, S_e mm (in.) (7)	Calculated maximum settlement, S_e mm (in.) (8)	$\dfrac{\text{calculated } S_e}{\text{observed } S_e}$ (9)
1	T. Edison São Paulo, Brazil	Rios and Silva (1948)	18.29 (60)	15	229.8 (4.8)	15.24 (0.6)	30.64 (1.21)	2.01
2	Banco do Brazil São Paulo, Brazil	Rios and Silva (1948); Vargas (1961)	22.86 (75)	18	239.4 (5.0)	27.94 (1.1)	26.6 (1.05)	0.95
3	Iparanga São Paulo, Brazil	Vargas (1948)	9.14 (30)	9	304.4 (6.4)	35.56 (1.4)	67.64 (2.66)	1.9
4	C.B.I. Esplanda São Paulo, Brazil	Vargas (1961)	14.63 (48)	22	383.0 (8.0)	27.94 (1.1)	34.82 (1.37)	1.25
5	Riscala São Paulo, Brazil	Vargas (1948)	3.96 (13)	20	229.8 (4.8)	12.7 (0.5)	22.98 (0.9)	1.81
6	Thyssen Düsseldorf, Germany	Schultze (1962)	22.55 (74)	25	239.4 (5)	24.13 (0.95)	19.15 (0.75)	0.79
7	Ministry Düsseldorf, Germany	Schultze (1962)	15.85 (52)	20	220.2 (4.6)	20.32 (0.8)	22.02 (0.87)	1.08
8	Chimney Cologne, Germany	Schultze (1962)	20.42 (67)	10	172.4 (3.6)	10.16 (0.4)	34.48 (1.36)	3.39

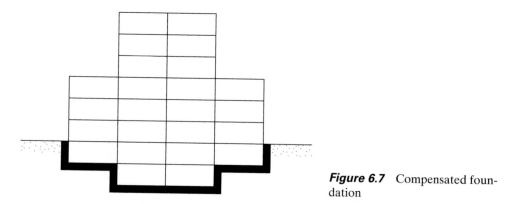

Figure 6.7 Compensated foundation

For no increase in the net pressure on soil below a mat foundation, q should be zero. Thus,

$$D_f = \frac{Q}{A\gamma} \tag{6.20}$$

This relation for D_f is usually referred to as the depth of a *fully compensated foundation*.

The factor of safety against bearing capacity failure for partially compensated foundations (i.e., $D_f < Q/A\gamma$) may be given as

$$\text{FS} = \frac{q_{\text{net}(u)}}{q} = \frac{q_{\text{net}(u)}}{\dfrac{Q}{A} - \gamma D_f} \tag{6.21}$$

where $q_{\text{net}(u)}$ = net ultimate bearing capacity.

For saturated clays, the factor of safety against bearing capacity failure can thus be obtained by substituting Eq. (6.10) into Eq. (6.21):

$$\text{FS} = \frac{5.14c_u\left(1 + \dfrac{0.195B}{L}\right)\left(1 + 0.4\dfrac{D_f}{B}\right)}{\dfrac{Q}{A} - \gamma D_f} \tag{6.22}$$

Example 6.3

The mat shown in Figure 6.6 has dimensions of 60 ft × 100 ft. The total dead and live load on the mat is 25×10^3 kip. The mat is placed over a saturated clay having a unit weight of 120 lb/ft^3 and c_u = 2800 lb/ft^2. Given that D_f = 5 ft, determine the factor of safety against bearing capacity failure.

Solution
From Eq. (6.22), the factor of safety

$$FS = \frac{5.14c_u\left(1 + \frac{0.195B}{L}\right)\left(1 + 0.4\frac{D_f}{B}\right)}{\frac{Q}{A} - \gamma D_f}$$

We are given that $c_u = 2800$ lb/ft², $D_f = 5$ ft, $B = 60$ ft, $L = 100$ ft, and $\gamma = 120$ lb/ft³. Hence,

$$FS = \frac{(5.14)(2800)\left[1 + \frac{(0.195)(60)}{100}\right]\left[1 + 0.4\left(\frac{5}{60}\right)\right]}{\left(\frac{25 \times 10^6\,\text{lb}}{60 \times 100}\right) - (120)(5)} = \textbf{4.66} \qquad \blacksquare$$

Example 6.4

Consider a mat foundation 90 ft × 120 ft in plan, as shown in Figure 6.8. The total dead load and live load on the raft is 45×10^3 kip. Estimate the consolidation settlement at the center of the foundation.

Solution
From Eq. (1.47)

$$S_{c(p)} = \frac{C_c H_c}{1 + e_o} \log\left(\frac{\sigma_o' + \Delta\sigma_{av}'}{\sigma_o'}\right)$$

$$\sigma_o' = (11)(100) + (40)(121.5 - 62.4) + \frac{18}{2}(118 - 62.4) \approx 3964\,\text{lb/ft}^2$$

$$H_c = 18 \times 12\,\text{in.}$$

$$C_c = 0.28$$

$$e_o = 0.9$$

For $Q = 45 \times 10^6$ lb, the net load per unit area is

$$q = \frac{Q}{A} - \gamma D_f = \frac{45 \times 10^6}{90 \times 120} - (100)(6) \approx 3567\,\text{lb/ft}^2$$

In order to calculate $\Delta\sigma_{av}'$ we refer to Section 5.5. The loaded area can be divided into four areas, each measuring 45 ft × 60 ft. Now using Eq. (5.19), we can calculate the average stress increase in the clay layer below the corner of each rectangular area, or

$$\Delta\sigma_{av(H_2/H_1)}' = q_o\left[\frac{H_2 I_{a(H_2)} - H_1 I_{a(H_1)}}{H_2 - H_1}\right]$$

$$= 3567\left[\frac{(5 + 40 + 18)I_{a(H_2)} - (5 + 40)I_{a(H_1)}}{18}\right]$$

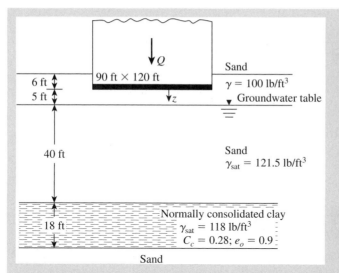

Figure 6.8 Consolidation settlement under a mat foundation

For $I_{a(H_2)}$,

$$m_2 = \frac{B}{H_2} = \frac{45}{5 + 40 + 18} = 0.71$$

$$n_2 = \frac{L}{H_2} = \frac{60}{63} = 0.95$$

From Fig. 5.7, for $m_2 = 0.71$ and $n_2 = 0.95$, the value of $I_{a(H_2)}$ is 0.21. Again, for $I_{a(H_1)}$,

$$m_2 = \frac{B}{H_1} = \frac{45}{45} = 1$$

$$n_2 = \frac{L}{H_1} = \frac{60}{45} = 1.33$$

From Figure 5.7, $I_{a(H_1)} = 0.225$, so

$$\Delta\sigma'_{av(H_2/H_1)} = 3567\left[\frac{(63)(0.21) - (45)(0.225)}{18}\right] = 615.3 \text{ lb/ft}^2$$

So, the stress increase below the center of the 90 ft × 120 ft area is $(4)(615.3) = 2461.2$ lb/ft². Thus

$$S_{c(p)} = \frac{(0.28)(18 \times 12)}{1 + 0.9}\log\left(\frac{3964 + 2461.2}{3964}\right) = \textbf{6.68 in.} \qquad \blacksquare$$

6.8   *Structural Design of Mat Foundations*

The structural design of mat foundations can be carried out by two conventional methods: the conventional rigid method and the approximate flexible method. Finite-difference and finite-element methods can also be used, but this section covers only the basic concepts of the first two design methods.

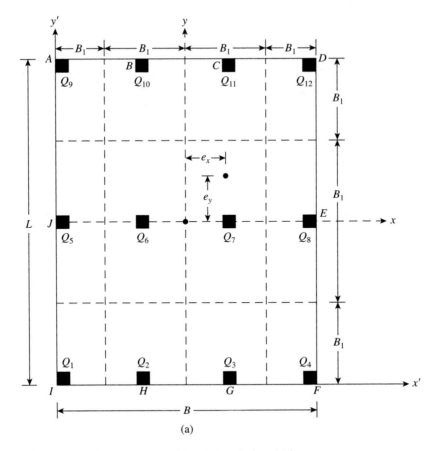

Figure 6.9 Conventional rigid mat foundation design

Conventional Rigid Method

The *conventional rigid method* of mat foundation design can be explained step by step with reference to Figure 6.9:

Step 1. Figure 6.9a shows mat dimensions of $L \times B$ and column loads of Q_1, $Q_2, Q_3, \ldots$. Calculate the total column load as

$$Q = Q_1 + Q_2 + Q_3 + \cdots \qquad (6.23)$$

Step 2. Determine the pressure on the soil, q, below the mat at points $A, B, C, D, \ldots$, by using the equation

$$q = \frac{Q}{A} \pm \frac{M_y x}{I_y} \pm \frac{M_x y}{I_x} \qquad (6.24)$$

where

$A = BL$
$I_x = (1/12)BL^3$ = moment of inertia about the x-axis
$I_y = (1/12)LB^3$ = moment of inertia about the y-axis
M_x = moment of the column loads about the x-axis = Qe_y
M_y = moment of the column loads about the y-axis = Qe_x

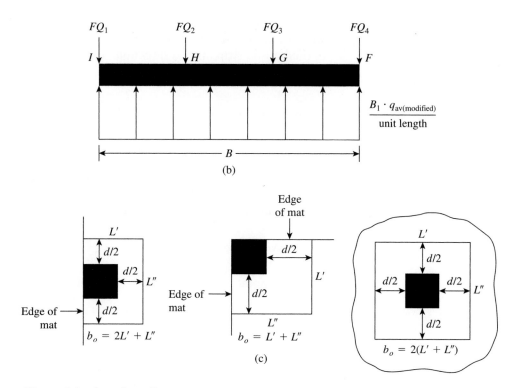

(b)

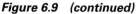

(c)

Figure 6.9 *(continued)*

The load eccentricities, e_x and e_y, in the x and y directions can be determined by using (x', y') coordinates:

$$x' = \frac{Q_1 x_1' + Q_2 x_2' + Q_3 x_3' + \cdots}{Q} \tag{6.25}$$

and

$$e_x = x' - \frac{B}{2} \tag{6.26}$$

Similarly,

$$y' = \frac{Q_1 y_1' + Q_2 y_2' + Q_3 y_3' + \cdots}{Q} \tag{6.27}$$

and

$$e_y = y' - \frac{L}{2} \tag{6.28}$$

Step 3. Compare the values of the soil pressures determined in Step 2 with the net allowable soil pressure to determine whether $q \leq q_{\text{all(net)}}$.

Step 4. Divide the mat into several strips in the x and y directions. (See Figure 6.9). Let the width of any strip be B_1.

Step 5. Draw the shear, *V*, and the moment, *M*, diagrams for each individual strip (in the *x* and *y* directions). For example, the average soil pressure of the bottom strip in the *x* direction of Figure 6.9a is

$$q_{av} \approx \frac{q_I + q_F}{2} \tag{6.29}$$

where q_I and q_F = soil pressures at points *I* and *F*, as determined from Step 2.

The total soil reaction is equal to $q_{av}B_1B$. Now obtain the total column load on the strip as $Q_1 + Q_2 + Q_3 + Q_4$. The sum of the column loads on the strip will not equal $q_{av}B_1B$, because the shear between the adjacent strips has not been taken into account. For this reason, the soil reaction and the column loads need to be adjusted, or

$$\text{Average load} = \frac{q_{av}B_1B + (Q_1 + Q_2 + Q_3 + Q_4)}{2} \tag{6.30}$$

Now, the modified average soil reaction becomes

$$q_{av(\text{modified})} = q_{av}\left(\frac{\text{average load}}{q_{av}B_1B}\right) \tag{6.31}$$

and the column load modification factor is

$$F = \frac{\text{average load}}{Q_1 + Q_2 + Q_3 + Q_4} \tag{6.32}$$

So the modified column loads are FQ_1, FQ_2, FQ_3, and FQ_4. This modified loading on the strip under consideration is shown in Figure 6.9b. The shear and the moment diagram for this strip can now be drawn, and the procedure is repeated in the *x* and *y* directions for all strips.

Step 6. Determine the effective depth *d* of the mat by checking for diagonal tension shear near various columns. According to ACI Code 318-95 (Section 11.12.2.1c, American Concrete Institute, 1995), for the critical section,

$$U = b_o d[\phi(0.34)\sqrt{f_c'}] \tag{6.33}$$

where

U = factored column load (MN), or (column load) × (load factor)
ϕ = reduction factor = 0.85
f_c' = compressive strength of concrete at 28 days (MN/m^2)

The units of b_o and *d* in Eq. (6.33) are in meters. In English units, Eq. (6.33) may be expressed as

$$U = b_o d(4\phi\sqrt{f_c'}) \tag{6.34}$$

where *U* is in lb, b_o and *d* are in in., and f_c' is in lb/in.^{2}.

The expression for b_o in terms of *d*, which depends on the location of the column with respect to the plan of the mat, can be obtained from Figure 6.9c.

Step 7. From the moment diagrams of all strips *in one direction* (*x* or *y*), obtain the *maximum* positive and negative moments per unit width (i.e., $M' = M/B_1$).

Step 8. Determine the areas of steel per unit width for positive and negative reinforcement in the *x* and *y* directions. We have

$$M_u = (M')(\text{load factor}) = \phi A_s f_y \left(d - \frac{a}{2} \right) \tag{6.35}$$

and

$$a = \frac{A_s f_y}{0.85 f'_c b} \tag{6.36}$$

where

A_s = area of steel per unit width
f_y = yield stress of reinforcement in tension
M_u = factored moment
$\phi = 0.9$ = reduction factor

Examples 6.5 and 6.6 illustrate the use of the conventional rigid method of mat foundation design.

Approximate Flexible Method

In the conventional rigid method of design, the mat is assumed to be infinitely rigid. Also, the soil pressure is distributed in a straight line, and the centroid of the soil pressure is coincident with the line of action of the resultant column loads. (See Figure 6.10a.) In the *approximate flexible method* of design, the soil is assumed to be equivalent to an infinite number of elastic springs, as shown in Figure 6.10b. This assumption is sometimes referred to as the *Winkler foundation*. The elastic constant of these assumed springs is referred to as the *coefficient of subgrade reaction, k*.

To understand the fundamental concepts behind flexible foundation design, consider a beam of width B_1 having infinite length, as shown in Figure 6.10c. The beam is subjected to a single concentrated load Q. From the fundamentals of mechanics of materials,

$$M = E_F I_F \frac{d^2 z}{dx^2} \tag{6.37}$$

where

M = moment at any section
E_F = modulus of elasticity of foundation material
I_F = moment of inertia of the cross section of the beam = $\left(\frac{1}{12}\right) B_1 h^3$ (see Figure 6.10c).

However,

$$\frac{dM}{dx} = \text{shear force} = V$$

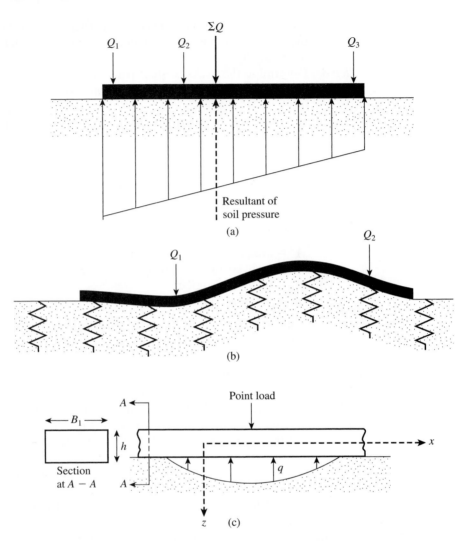

Figure 6.10 (a) Principles of design by conventional rigid method; (b) principles of approximate flexible method; (c) derivation of Eq. (6.41) for beams on elastic foundation

and

$$\frac{dV}{dx} = q = \text{soil reaction}$$

Hence,

$$\frac{d^2M}{dx^2} = q \qquad (6.38)$$

Combining Eqs. (6.37) and (6.38) yields

$$E_F I_F \frac{d^4z}{dx^4} = q \qquad (6.39)$$

However, the soil reaction is

$$q = -zk'$$

where

z = deflection
$k' = kB_1$
k = coefficient of subgrade reaction $(kN/m^3 \text{ or } lb/in^3)$

So,

$$E_F I_F = \frac{d^4z}{dx^4} = -zkB_1 \tag{6.40}$$

Solving Eq. (6.40) yields

$$z = e^{-\alpha x}(A' \cos \beta x + A'' \sin \beta x) \tag{6.41}$$

where A' and A'' are constants and

$$\beta = \sqrt[4]{\frac{B_1 k}{4E_F I_F}} \tag{6.42}$$

The unit of the term β, as defined by the preceding equation, is $(\text{length})^{-1}$. This parameter is very important in determining whether a mat foundation should be designed by the conventional rigid method or the approximate flexible method. According to the American Concrete Institute Committee 336 (1988), mats should be designed by the conventional rigid method if the spacing of columns in a strip is less than $1.75/\beta$. If the spacing of columns is larger than $1.75/\beta$, the approximate flexible method may be used.

To perform the analysis for the structural design of a flexible mat, one must know the principles involved in evaluating the coefficient of subgrade reaction, k. Before proceeding with the discussion of the approximate flexible design method, let us discuss this coefficient in more detail.

If a foundation of width B (see Figure 6.11) is subjected to a load per unit area of q, it will undergo a settlement Δ. The coefficient of subgrade modulus can be defined as

$$k = \frac{q}{\Delta} \tag{6.43}$$

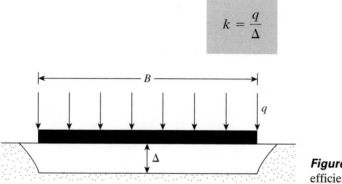

Figure 6.11 Definition of coefficient of subgrade reaction, k

The unit of k is kN/m^3 (or lb/in^3). The value of the coefficient of subgrade reaction is not a constant for a given soil, but rather depends on several factors, such as the length L and width B of the foundation and also the depth of embedment of the foundation. A comprehensive study by Terzaghi (1955) of the parameters affecting the coefficient of subgrade reaction indicated that the value of the coefficient decreases with the width of the foundation. In the field, load tests can be carried out by means of square plates measuring 0.3 m × 0.3 m (1 ft × 1 ft), and values of k can be calculated. The value of k can be related to large foundations measuring $B \times B$ in the following ways:

Foundations on Sandy Soils

For foundations on sandy soils,

$$k = k_{0.3}\left(\frac{B + 0.3}{2B}\right)^2 \tag{6.44}$$

where $k_{0.3}$ and k = coefficients of subgrade reaction of foundations measuring 0.3 m × 0.3 m and B (m) × B (m), respectively (unit is kN/m^3).
   In English units, Eq. (6.44) may be expressed as

$$k = k_1\left(\frac{B + 1}{2B}\right)^2 \tag{6.45}$$

where k_1 and k = coefficients of subgrade reaction of foundations measuring 1 ft × 1 ft and B (ft) × B (ft), respectively (unit is lb/in^3).

Foundations on Clays

For foundations on clays,

$$k\,(kN/m^3) = k_{0.3}\,(kN/m^3)\left[\frac{0.3\,(m)}{B\,(m)}\right] \tag{6.46}$$

The definition of k in Eq. (6.46) is the same as in Eq. (6.44).
   In English units,

$$k\,(lb/in^3) = k_1\,(lb/in^3)\left[\frac{1\,(ft)}{B\,(ft)}\right] \tag{6.47}$$

The definitions of k and k_1 are the same as in Eq. (6.45).

For rectangular foundations having dimensions of $B \times L$ (for similar soil and q),

$$k = \frac{k_{(B \times B)}\left(1 + 0.5\dfrac{B}{L}\right)}{1.5}$$ (6.48)

where

k = coefficient of subgrade modulus of the rectangular foundation ($L \times B$)

$k_{(B \times B)}$ = coefficient of subgrade modulus of a square foundation having dimension of $B \times B$

Equation (6.48) indicates that the value of k for a very long foundation with a width B is approximately $0.67k_{(B \times B)}$.

The modulus of elasticity of granular soils increases with depth. Because the settlement of a foundation depends on the modulus of elasticity, the value of k increases with the depth of the foundation.

Table 6.2 provides typical ranges of values for the coefficient of subgrade reaction, $k_{0.3}(k_1)$, for sandy and clayey soils.

For long beams, Vesic (1961) proposed an equation for estimating subgrade reaction, namely,

$$k' = Bk = 0.65 \sqrt[12]{\frac{E_s B^4}{E_F I_F}} \frac{E_s}{1 - \mu_s^2}$$

or

$$k = 0.65 \sqrt[12]{\frac{E_s B^4}{E_F I_F}} \frac{E_s}{B(1 - \mu_s^2)}$$ (6.49)

Table 6.2 Typical Subgrade Reaction Values, $k_{0.3}(k_1)$

Soil type	$k_{0.3}(k_1)$	
	MN/m³	lb/in.³
Dry or moist sand:		
Loose	8–25	30–90
Medium	25–125	90–450
Dense	125–375	450–1350
Saturated sand:		
Loose	10–15	35–55
Medium	35–40	125–145
Dense	130–150	475–550
Clay:		
Stiff	10–25	40–90
Very stiff	25–50	90–185
Hard	>50	>185

where

E_s = modulus of elasticity of soil
B = foundation width
E_F = modulus of elasticity of foundation material
I_F = moment of inertia of the cross section of the foundation
μ_s = Poisson's ratio of soil

For most practical purposes, Eq. (6.49) can be approximated as

$$k = \frac{E_s}{B(1 - \mu_s^2)} \tag{6.50}$$

Now that we have discussed the coefficient of subgrade reaction, we will proceed with the discussion of the approximate flexible method of designing mat foundations. This method, as proposed by the American Concrete Institute Committee 336 (1988), is described step by step. The use of the design procedure, which is based primarily on the theory of plates, allows the effects (i.e., moment, shear, and deflection) of a concentrated column load in the area surrounding it to be evaluated. If the zones of influence of two or more columns overlap, superposition can be employed to obtain the net moment, shear, and deflection at any point. The method is as follows:

Step 1. Assume a thickness h for the mat, according to Step 6 of the conventional rigid method. (*Note:* h is the *total* thickness of the mat.)

Step 2. Determine the flexural ridigity R of the mat as given by the formula

$$R = \frac{E_F h^3}{12(1 - \mu_F^2)} \tag{6.51}$$

where

E_F = modulus of elasticity of foundation material
μ_F = Poisson's ratio of foundation material

Step 3. Determine the radius of effective stiffness—that is,

$$L' = \sqrt[4]{\frac{R}{k}} \tag{6.52}$$

where k = coefficient of subgrade reaction. The zone of influence of any column load will be on the order of 3 to 4 L'.

Step 4. Determine the moment (in polar coordinates at a point) caused by a column load (see Figure 6.12a). The formulas to use are

$$M_r = \text{radial moment} = -\frac{Q}{4}\left[A_1 - \frac{(1 - \mu_F)A_2}{\dfrac{r}{L'}} \right] \tag{6.53}$$

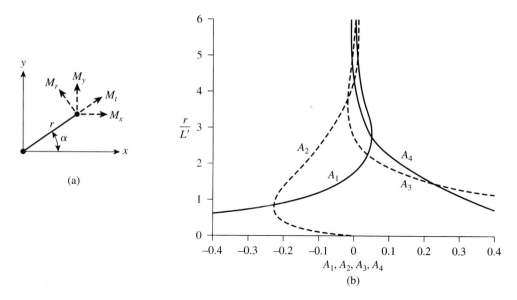

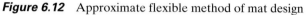

Figure 6.12   Approximate flexible method of mat design

and

$$M_t = \text{tangential moment} = -\frac{Q}{4}\left[\mu_F A_1 + \frac{(1 - \mu_F)A_2}{\dfrac{r}{L'}}\right] \quad (6.54)$$

where

r = radial distance from the column load
Q = column load
A_1, A_2 = functions of r/L'

The variations of A_1 and A_2 with r/L' are shown in Figure 6.12b. (For details see Hetenyi, 1946.)

In the Cartesian coordinate system (see Figure 6.12a),

$$M_x = M_t \sin^2\alpha + M_r \cos^2\alpha \quad (6.55)$$

and

$$M_y = M_t \cos^2\alpha + M_r \sin^2\alpha \quad (6.56)$$

Step 5.   For the unit width of the mat, determine the shear force V caused by a column load:

$$V = \frac{Q}{4L'}A_3 \quad (6.57)$$

The variation of A_3 with r/L' is shown in Figure 6.12b.

Step 6.   If the edge of the mat is located in the zone of influence of a column, determine the moment and shear along the edge. (Assume that the mat is continuous.) Moment and shear opposite in sign to those determined are applied at the edges to satisfy the known conditions.

Step 7. The deflection at any point is given by

$$\delta = \frac{QL'^2}{4R}A_4 \tag{6.58}$$

The variation of A_4 is presented in Figure 6.12b.

Example 6.5

The plan of a mat foundation with column loads is shown in Figure 6.13. Use Eq. (6.24) to calculate the soil pressures at points *A, B, C, D, E, F, G, H, I, J, K, L, M,* and *N.* The size of the mat is 76 ft × 96 ft, all columns are 24 in. × 24 in. in section,

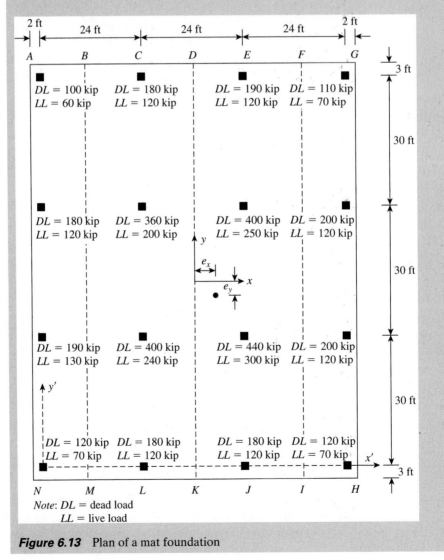

Figure 6.13   Plan of a mat foundation

and $q_{all(net)} = 1.5$ kip/ft^2. Verify that the soil pressures are less than the net allowable bearing capacity.

Solution

From Figure 6.13,

$$\text{Column dead load } (DL) = 100 + 180 + 190 + 110 + 180 + 360 + 400 + 200$$
$$+ 190 + 400 + 440 + 200 + 120 + 180 + 180 + 120$$
$$= 3550 \text{ kip}$$

and

$$\text{Column live load } (LL) = 60 + 120 + 120 + 70 + 120 + 200 + 250 + 120$$
$$+ 130 + 240 + 300 + 120 + 70 + 120 + 120 + 70$$
$$= 2230 \text{ kip}$$

So,

$$\text{Service load} = 3550 + 2230 = 5780 \text{ kip}$$

According to ACI 318-95 (Section 9.2), the factored load $U = (1.4)$ (Dead load) $+ (1.7)$ (Live load). So

$$\text{Factored load} = (1.4)(3550) + (1.7)(2230) = 8761 \text{ kip}$$

The moments of inertia of the foundation are

$$I_x = \tfrac{1}{12}(76)(96)^3 = 5603 \times 10^3 \text{ ft}^4$$

and

$$I_y = \tfrac{1}{12}(96)(76)^3 = 3512 \times 10^3 \text{ ft}^4$$

Also,

$$\Sigma M_{y'} = 0$$

so

$$5780x' = (24)(300 + 560 + 640 + 300) + (48)(310 + 650 + 740 + 300)$$
$$+ (72)(180 + 320 + 320 + 190)$$

or

$$x' = 36.664 \text{ ft}$$

and

$$e_x = 36.664 - 36.0 = 0.664 \text{ ft}$$

Similarly,

$$\Sigma M_{x'} = 0$$

so

$$5780y' = (30)(320 + 640 + 740 + 320) + (60)(300 + 560 + 650 + 320)$$
$$+ (90)(160 + 300 + 310 + 180)$$

or

$$y' = 44.273 \text{ ft}$$

and

$$e_y = 44.273 - \frac{90}{2} = -0.727 \text{ ft}$$

The moments caused by eccentricity are

$$M_x = Qe_y = (8761)(0.727) = 6369 \text{ kip-ft}$$

and

$$M_y = Qe_x = (8761)(0.664) = 5817 \text{ kip-ft}$$

From Eq. (6.24),

$$q = \frac{Q}{A} \pm \frac{M_y x}{I_y} \pm \frac{M_x y}{I_x}$$

$$= \frac{8761}{(76)(96)} \pm \frac{(5817)(x)}{3512 \times 10^3} \pm \frac{(6369)(y)}{5603 \times 10^3}$$

or

$$q = 1.20 \pm 0.0017x \pm 0.0011y \ (\text{kip/ft}^2)$$

Now the following table can be prepared:

Point	$\dfrac{Q}{A}$ (kip/ft^2)	x (ft)	$\pm 0.0017x$ (ft)	y (ft)	$\pm 0.0011y$ (ft)	q (kip/ft^2)
A	1.2	−38	−0.065	48	−0.053	**1.082**
B	1.2	−24	−0.041	48	−0.053	**1.106**
C	1.2	−12	−0.020	48	−0.053	**1.127**
D	1.2	0	0.0	48	−0.053	**1.147**
E	1.2	12	0.020	48	−0.053	**1.167**
F	1.2	24	0.041	48	−0.053	**1.188**
G	1.2	38	0.065	48	−0.053	**1.212**
H	1.2	38	0.065	−48	0.053	**1.318**
I	1.2	24	0.041	−48	0.053	**1.294**
J	1.2	12	0.020	−48	0.053	**1.273**
K	1.2	0	0.0	−48	0.053	**1.253**
L	1.2	−12	−0.020	−48	0.053	**1.233**
M	1.2	−24	−0.041	−48	0.053	**1.212**
N	1.2	−38	−0.065	−48	0.053	**1.188**

The soil pressures at all points are less than the given value of $q_{\text{all(net)}} = 1.5 \text{ kip/ft}^2$.

∎

Example 6.6

Use the results of Example 6.5 and the conventional rigid method.

a. Determine the thickness of the slab.
b. Divide the mat into four strips (that is, *ABMN, BCDKLM, DEFIJK,* and *FGHI*), and determine the average soil reactions at the ends of each strip.
c. Determine the reinforcement requirements in the *y* direction for $f'_c = 3000$ lb/in^2 and $f_y = 60,000$ lb/in^2.

Solution

Part a: Determination of Thickness of Mat
For the critical perimeter column, as shown in Figure 6.14 (ACI 318-95; Section 9.2.1),

$$U = 1.4(DL) + 1.7(LL) = (1.4)(190) + (1.7)(130) = 487 \text{ kip}$$

and

$$b_o = 2(36 + d/2) + (24 + d) = 96 + 2d \text{ (in.)}$$

From ACI 318-95,

$$\phi V_c \geq V_u$$

where

V_c = nominal shear strength of concrete
V_u = factored shear strength

we have

$$\phi V_c = \phi(4)\sqrt{f'_c}b_o d = (0.85)(4)(\sqrt{3000})(96 + 2d)d$$

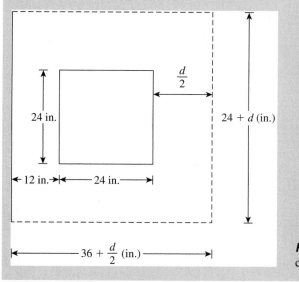

Figure 6.14   Critical perimeter column

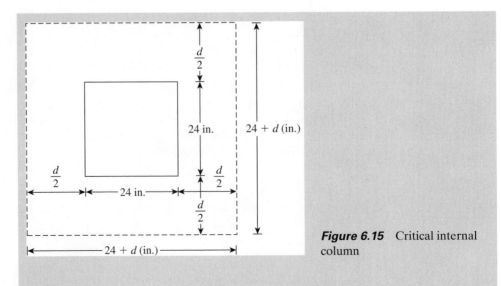

Figure 6.15   Critical internal column

so,

$$\frac{(0.85)(4)(\sqrt{3000})(96 + 2d)d}{1000} \geq 487$$

$$(96 + 2d)d \geq 2615.1$$

$$d \approx 19.4 \text{ in.}$$

For the critical internal column shown in Figure 6.15,

$$b_o = 4(24 + d) = 96 + 4d \text{ (in.)}$$

$$U = (1.4)(440) + (1.7)(300) = 1126 \text{ kip}$$

and

$$\frac{(0.85)(4)(\sqrt{3000})(96 + 4d)d}{1000} \geq 1126$$

$$(96 + 4d)d \geq 6046.4$$

$$d \approx 28.7 \text{ in.}$$

Accordingly, use $d = 29$ in.

With a minimum cover of 3 in. over the steel reinforcement and 1-in.-diameter steel bars, the total slab thickness is

$$h = 29 + 3 + 1 = \textbf{33 in.}$$

Part b: Average Soil Reaction
In Figure 6.13, for strip *ABMN* (width = 14 ft),

$$q_1 = \frac{q_{(\text{at}A)} + q_{(\text{at}B)}}{2} = \frac{1.082 + 1.106}{2} = \textbf{1.094 kip/ft}^2$$

and

$$q_2 = \frac{q_{(atM)} + q_{(atN)}}{2} = \frac{1.212 + 1.188}{2} = \textbf{1.20 kip/ft}^2$$

For strip $BCDKLM$ (width $= 24$ ft),

$$q_1 = \frac{1.106 + 1.127 + 1.147}{3} = \textbf{1.127 kip/ft}^2$$

and

$$q_2 = \frac{1.253 + 1.233 + 1.212}{3} = \textbf{1.233 kip/ft}^2$$

For strip $DEFIJK$ (width $= 24$ ft),

$$q_1 = \frac{1.147 + 1.167 + 1.188}{3} = \textbf{1.167 kip/ft}^2$$

and

$$q_2 = \frac{1.294 + 1.273 + 1.253}{3} = \textbf{1.273 kip/ft}^2$$

For strip $FGHI$ (width $= 14$ ft),

$$q_1 = \frac{1.188 + 1.212}{2} = \textbf{1.20 kip/ft}^2$$

and

$$q_2 = \frac{1.318 + 1.294}{2} = \textbf{1.306 kip/ft}^2$$

Check for $\Sigma F_V = 0$:

Soil reaction for strip $ABMN = \frac{1}{2}(1.094 + 1.20)(14)(96) = 1541.6$ kip
Soil reaction for strip $BCDKLM = \frac{1}{2}(1.127 + 1.233)(24)(96) = 2718.7$ kip
Soil reaction for strip $DEFIJK = \frac{1}{2}(1.167 + 1.273)(24)(96) = 2810.9$ kip
Soil reaction for strip $FGHJ = \frac{1}{2}(1.20 + 1.306)(14)(96) = 1684.0$ kip

$$\sum 8755.2 \text{ kip} \approx \sum \text{Column load} = 8761 \text{ kip—OK}$$

Part c: Reinforcement Requirements
Figure 6.16 gives the design of strip $BCDKLM$ and shows the load diagram, in which

$$Q_1 = (1.4)(180) + (1.7)(120) = 456 \text{ kip}$$
$$Q_2 = (1.4)(360) + (1.7)(200) = 844 \text{ kip}$$
$$Q_3 = (1.4)(400) + (1.7)(240) = 968 \text{ kip}$$

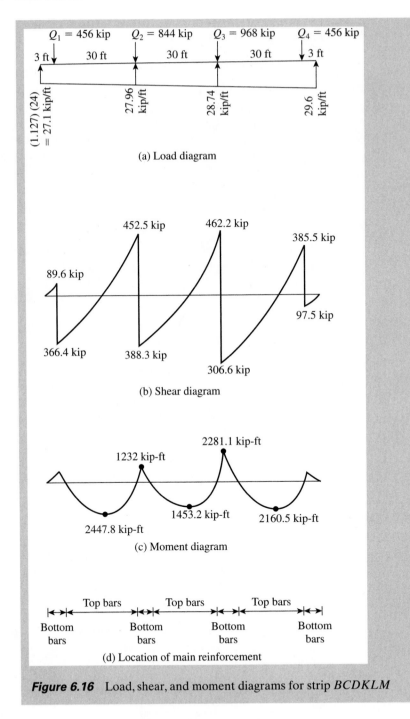

$Q_1 = 456$ kip $\quad Q_2 = 844$ kip $\quad Q_3 = 968$ kip $\quad Q_4 = 456$ kip

3 ft $\quad$ 30 ft $\quad$ 30 ft $\quad$ 30 ft $\quad$ 3 ft

(1.127) (24) = 27.1 kip/ft

27.96 kip/ft

28.74 kip/ft

29.6 kip/ft

(a) Load diagram

452.5 kip

462.2 kip

385.5 kip

89.6 kip

97.5 kip

366.4 kip

388.3 kip

306.6 kip

(b) Shear diagram

2281.1 kip-ft

1232 kip-ft

385.5

1453.2 kip-ft

2160.5 kip-ft

2447.8 kip-ft

(c) Moment diagram

Top bars $\quad$ Top bars $\quad$ Top bars

Bottom bars $\quad$ Bottom bars $\quad$ Bottom bars $\quad$ Bottom bars

(d) Location of main reinforcement

Figure 6.16   Load, shear, and moment diagrams for strip *BCDKLM*

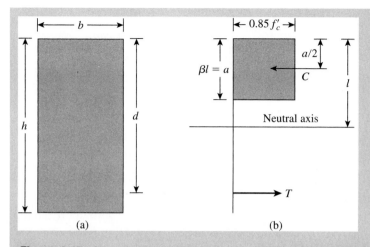

Figure 6.17   Rectangular section in bending; (a) section, (b) assumed stress distribution across the section

and

$$Q_4 = (1.4)(180) + (1.7)(120) = 456 \text{ kip}$$

The shear and moment diagrams are shown in Figures 6.16b and c, respectively. From Figure 6.16c, the maximum positive moment at the bottom of the foundation is 2281.1/24 = 95.05 kip-ft/ft.

   Note that Figure 6.17 shows the design concepts of a rectangular section in bending.

$$\sum \text{Compressive force}, C = 0.85 f'_c ab$$

$$\sum \text{Tensile force}, T = A_s f_y$$

and

$$C = T$$

For this case, $b = 1 \text{ ft} = 12 \text{ in.}$, so

$$(0.85)(3)(12)a = A_s(60)$$

and

$$A_s = 0.51a$$

From Eq. (6.35),

$$M_u = \phi A_s f_y \left(d - \frac{a}{2} \right)$$

and we have

$$(95.05)(12) = (0.9)(0.51a)(60)\left(29 - \frac{a}{2}\right)$$

or

$$a = 1.47 \text{ in.}$$

Thus,

$$A_s = (0.51)(1.47) = 0.75 \text{ in}^2$$

- Minimum reinforcement s_{min} (ACI 318-95, Section 10.5) = $200/f_y$ = $200/60,000 = 0.00333$
- Mainimum $A_s = (0.00333)(12)(29) = 1.16 \text{ in}^2/\text{ft}$. Hence, use minimum reinforcement with $A_s = 1.16 \text{ in}^2/\text{ft}$.
- **Use No. 9 bars at 10 in. center to center ($A_s = 1.2 \text{ in}^2/\text{ft}$) at the bottom of the foundation.**

From Figure 6.16c, the maximum negative moment is 2447.8 kip-ft/24 = 102 kip-ft/ft. By observation, $A_s \leqslant A_{s(min)}$.

- **Use No. 9 bars at 10 in. center to center at the top of the foundation.** ∎

Problems

6.1   Determine the net ultimate bearing capacity of mat foundations with the following characteristics:
  a. $c_u = 120 \text{ kN/m}^2$, $\phi = 0$, $B = 8$ m, $L = 18$ m, $D_f = 3$ m
  b. $c_u = 2500 \text{ lb/ft}^2$, $\phi = 0$, $B = 20$ ft, $L = 30$ ft, $D_f = 6.2$ ft.

6.2   Following are the results of a standard penetration test in the field (sandy soil):

Depth (m)	Field value of N_{60}
1.5	9
3.0	12
4.5	11
6.0	7
7.5	13
9.0	11
10.5	13

Estimate the net allowable bearing capacity of a mat foundation 6.5 m × 5 m in plan. Here, $D_f = 1.5$ m and allowable settlement = 50 mm. Assume that the unit weight of soil, $\gamma = 16.5 \text{ kN/m}^3$.

6.3   Repeat Problem 6.2 for an allowable settlement of 30 mm.

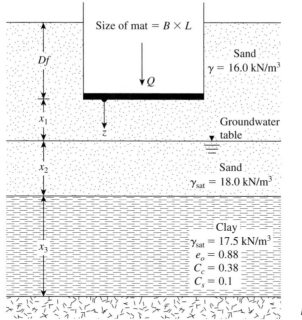

Size of mat = $B \times L$

Df

Sand
$\gamma = 16.0$ kN/m³

Q

x_1

z

Groundwater
table

x_2

Sand
$\gamma_{sat} = 18.0$ kN/m³

Clay
$\gamma_{sat} = 17.5$ kN/m³
$e_o = 0.88$
$C_c = 0.38$
$C_s = 0.1$

x_3

Figure P6.6

6.4   A mat foundation on a saturated clay soil has dimensions of 20 m × 20 m. Given: dead and live load = 48 MN, c_u = 30 kN/m², and γ_{clay} = 18.5 kN/m³.
   a. Find the depth, D_f, of the mat for a fully compensated foundation.
   b. What will be the depth of the mat (D_f) for a factor of safety of 2 against bearing capacity failure?

6.5   Repeat Problem 6.4 part b for c_u = 20 kN/m².

6.6   A mat foundation is shown in Figure P6.6. The design considerations are L = 12 m, B = 10 m, D_f = 2.2 m, Q = 30 MN, x_1 = 2 m, x_2 = 2 m, x_3 = 5.2 m, and preconsolidation pressure σ'_c = 105 kN/m². Calculate the consolidation settlement under the center of the mat.

6.7   For the mat foundation in Problem 6.6, estimate the consolidation settlement under the corner of the mat.

6.8   For the mat foundation shown in Figure P6.8, Q_1 = Q_3 = 40 tons, Q_4 = Q_5 = Q_6 = 60 tons, Q_2 = Q_9 = 45 tons, and Q_7 = Q_8 = 50 tons. All columns are 20 in. × 20 in. in cross section. Use the procedure outlined in section 6.8 to determine the pressure on the soil at points A, B, C, D, E, F, G, and H.

6.9   The plan of a mat foundation is shown in Figure P6.9. Calculate the soil pressure at points A, B, C, D, E, and F. (*Note:* All column sections are planned to be 0.5 m × 0.5 m.)

6.10 Divide the mat shown in Figure P6.9 into three strips, such as *AGHI* (B_1 = 4.25 m), *GIJH* (B_1 = 8 m), and *ICDJ* (B_1 = 4.25 m). Use the result of Problem 6.9, and determine the reinforcement requirements in the y direction. Here, f'_c = 20.7 MN/m², f_y = 413.7 MN/m², and the load factor is 1.7.

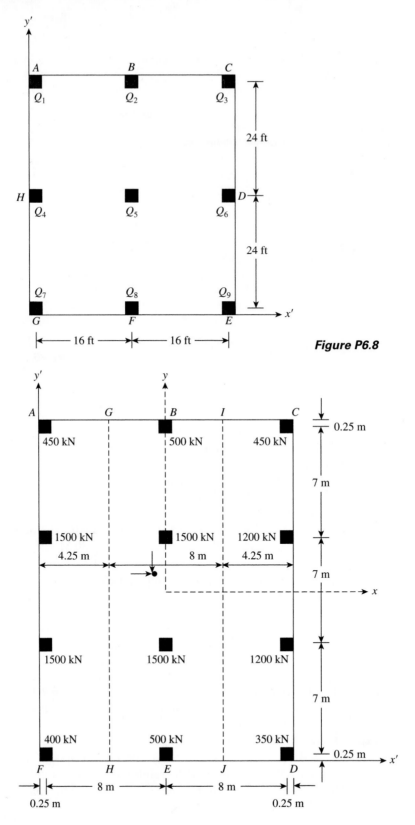

Figure P6.8

Figure P6.9

6.11 From the plate load test (plate dimensions 1 ft $\times$ 1 ft) in the field, the coefficient of subgrade reaction of a sandy soil is determined to be 55 lb/in^3. What will be the value of the coefficient of subgrade reaction on the same soil for a foundation with dimensions of 25 ft $\times$ 25 ft?

6.12 Refer to Problem 6.11. If the full-sized foundation had dimensions of 70 ft $\times$ 30 ft, what will be the value of the coefficient of subgrade reaction?

6.13 The subgrade reaction of a sandy soil obtained from the plate load test (plate dimensions 1 m $\times$ 0.7 m) is 18 kN/m^3. What will be the value of k on the same soil for a foundation measuring 5 m $\times$ 3.5 m?

References

American Concrete Institute (1995). *ACI Standard Building Code Requirements for Reinforced Concrete,* ACI 318–95, Farmington Hills, MI.

American Concrete Institute Committee 336 (1988). "Suggested Design Procedures for Combined Footings and Mats," *Journal of the American Concrete Institute,* Vol. 63, No. 10, pp. 1041–1077.

Hetenyi, M. (1946). *Beams of Elastic Foundations,* University of Michigan Press, Ann Arbor, MI.

Meyerhof, G. G. (1965). "Shallow Foundations," *Journal of the Soil Mechanics and Foundations Division,* American Society of Civil Engineers, Vol. 91, No. SM2, pp. 21–31.

Rios, L., and Silva, F. P. (1948). "Foundations in Downtown São Paulo (Brazil)," *Proceedings, Second International Conference on Soil Mechanics and Foundation Engineering,* Rotterdam, Vol. 4, p. 69.

Schultze, E. (1962). "Probleme bei der Auswertung von Setzungsmessungen," *Proceedings, Baugrundtagung,* Essen, Germany, p. 343.

Terzaghi, K. (1955). "Evaluation of the Coefficient of Subgrade Reactions," *Geotechnique,* Institute of Engineers, London, Vol. 5, No. 4, pp. 197–226.

Vargas, M. (1948). "Building Settlement Observations in São Paulo," *Proceedings Second International Conference on Soil Mechanics and Foundation Engineering,* Rotterdam, Vol. 4, p. 13.

Vargas, M. (1961). "Foundations of Tall Buildings on Sand in São Paulo (Brazil)," *Proceedings, Fifth International Conference on Soil Mechanics and Foundation Engineering,* Paris, Vol. 1, p. 841.

Vesic, A. S. (1961). "Bending of Beams Resting on Isotropic Solid," *Journal of the Engineering Mechanics Division,* American Society of Civil Engineers, Vol. 87, No. EM2, pp. 35–53.

7

Lateral Earth Pressure

Introduction

Vertical or near-vertical slopes of soil are supported by retaining walls, cantilever sheet-pile walls, sheet-pile bulkheads, braced cuts, and other, similar structures. The proper design of those structures requires an estimation of lateral earth pressure, which is a function of several factors, such as (a) the type and amount of wall movement, (b) the shear strength parameters of the soil, (c) the unit weight of the soil, and (d) the drainage conditions in the backfill. Figure 7.1 shows a retaining wall of height H. For similar types of backfill,

a. The wall may be restrained from moving (Figure 7.1a). The lateral earth pressure on the wall at any depth is called the *at-rest earth pressure*.
b. The wall may tilt away from the soil that is retained (Figure 7.1b). With sufficient wall tilt, a triangular soil wedge behind the wall will fail. The lateral pressure for this condition is referred to as *active earth pressure*.
c. The wall may be pushed into the soil that is retained (Figure 7.1c). With sufficient wall movement, a soil wedge will fail. The lateral pressure for this condition is referred to as *passive earth pressure*.

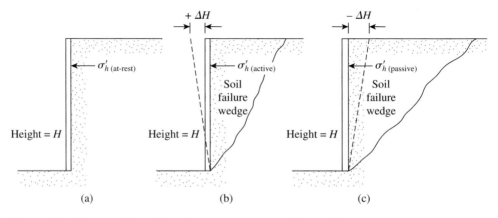

Figure 7.1 Nature of lateral earth pressure on a retaining wall

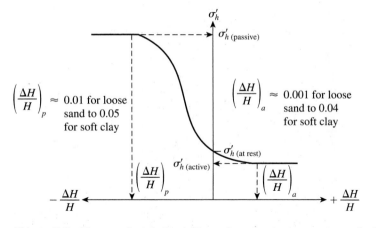

Figure 7.2 Nature of variation of lateral earth pressure at a certain depth

Figure 7.2 shows the nature of variation of the lateral pressure, σ'_h, at a certain depth of the wall with the magnitude of wall movement.

In the sections that follow, we will discuss various relationships to determine the at-rest, active, and passive pressures on a retaining wall. It is assumed that the reader has studied lateral earth pressure in the past, so this chapter will serve as a review.

7.2 *Lateral Earth Pressure at Rest*

Consider a vertical wall of height *H,* as shown in Figure 7.3, retaining a soil having a unit weight of γ. A uniformly distributed load, q/unit area, is also applied at the ground surface. The shear strength of the soil is

$$s = c' + \sigma' \tan \phi'$$

where

c' = cohesion
ϕ' = effective angle of friction
σ' = effective normal stress

At any depth z below the ground surface, the vertical subsurface stress is

$$\sigma'_o = q + \gamma z \tag{7.1}$$

If the *wall is at rest and is not allowed to move at all,* either away from the soil mass or into the soil mass (i.e., there is zero horizontal strain), the lateral pressure at a depth z is

$$\sigma_h = K_o \sigma'_o + u \tag{7.2}$$

where

u = pore water pressure
K_o = coefficient of at-rest earth pressure

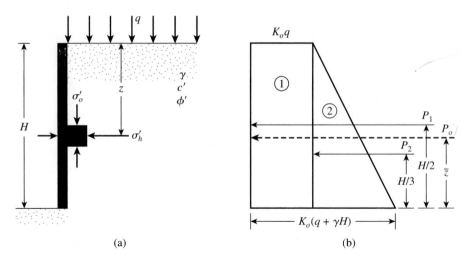

Figure 7.3 At-rest earth pressure

For normally consolidated soil, the relation for K_o (Jaky, 1944) is

$$K_o \approx 1 - \sin \phi' \tag{7.3}$$

Equation (7.3) is an empirical approximation.

For overconsolidated soil, the at-rest earth pressure coefficient may be expressed as (Mayne and Kulhawy, 1982)

$$K_o = (1 - \sin \phi')OCR^{\sin \phi'} \tag{7.4}$$

where OCR = overconsolidation ratio.

With a properly selected value of the at-rest earth pressure coefficient, Eq. (7.2) can be used to determine the variation of lateral earth pressure with depth z. Figure 7.3b shows the variation of σ'_h with depth for the wall depicted in Figure 7.3a. Note that if the surcharge $q = 0$ and the pore water pressure $u = 0$, the pressure diagram will be a triangle. The total force, P_o, *per unit length* of the wall given in Figure 7.3a can now be obtained from the area of the pressure diagram given in Figure 7.3b and is

$$P_o = P_1 + P_2 = qK_oH + \tfrac{1}{2}\gamma H^2 K_o \tag{7.5}$$

where

P_1 = area of rectangle 1
P_2 = area of triangle 2

The location of the line of action of the resultant force, P_o, can be obtained by taking the moment about the bottom of the wall. Thus,

$$\bar{z} = \frac{P_1\left(\dfrac{H}{2}\right) + P_2\left(\dfrac{H}{3}\right)}{P_o} \tag{7.6}$$

If the water table is located at a depth $z < H$, the at-rest pressure diagram shown in Figure 7.3b will have to be somewhat modified, as shown in Figure 7.4. If the effective unit weight of soil below the water table equals γ' (i.e., $\gamma_{sat} - \gamma_w$), then

at $z = 0$, $\sigma_h' = K_o\sigma_o' = K_o q$
at $z = H_1$, $\sigma_h' = K_o\sigma_o' = K_o(q + \gamma H_1)$

and

at $z = H_2$, $\sigma_h' = K_o\sigma_o' = K_o(q + \gamma H_1 + \gamma'H_2)$

Note that in the preceding equations, σ_o' and σ_h' are effective vertical and horizontal pressures, respectively. Determining the total pressure distribution on the wall requires adding the hydrostatic pressure u, which is zero from $z = 0$ to $z = H_1$ and is $H_2\gamma_w$ at $z = H_2$. The variation of σ_h' and u with depth is shown in Figure 7.4b. Hence, the total force per unit length of the wall can be determined from the area of the pressure diagram. Specifically,

$$P_o = A_1 + A_2 + A_3 + A_4 + A_5$$

where A = area of the pressure diagram.
So,

$$P_o = K_o q H_1 + \tfrac{1}{2}K_o\gamma H_1^2 + K_o(q + \gamma H_1)H_2 + \tfrac{1}{2}K_o\gamma'H_2^2 + \tfrac{1}{2}\gamma_w H_2^2 \tag{7.7}$$

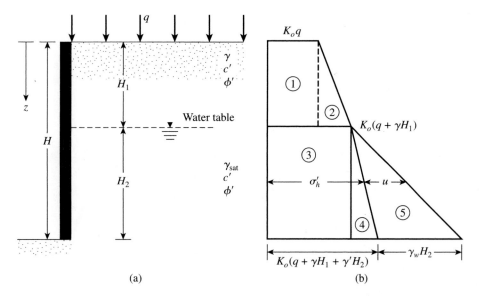

(a)   (b)

Figure 7.4   At-rest earth pressure with water table located at a depth $z < H$

Active Pressure

7.3 Rankine Active Earth Pressure

The lateral earth pressure described in Section 7.2 involves walls that do not yield at all. However, if a wall tends to move away from the soil a distance Δx, as shown in Figure 7.5a, the soil pressure on the wall at any depth will decrease. For a wall that is *frictionless,* the horizontal stress, σ'_h, at depth z will equal $K_o\sigma'_o(=K_o\gamma z)$ when Δx is zero. However, with $\Delta x > 0$, σ'_h will be less than $K_o\sigma'_o$.

The Mohr's circles corresponding to wall displacements of $\Delta x = 0$ and $\Delta x > 0$ are shown as circles a and b, respectively, in Figure 7.5b. If the displacement of the wall,

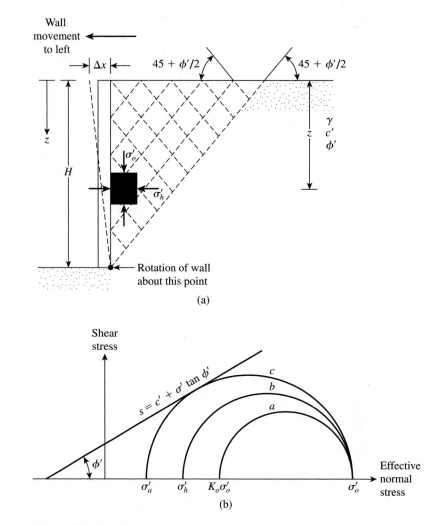

Figure 7.5 Rankine active pressure

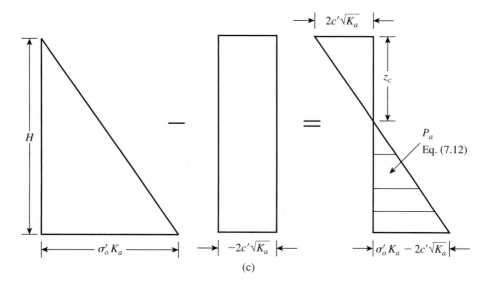

Figure 7.5 *(Continued)*

Δx, continues to increase, the corresponding Mohr's circle eventually will just touch the Mohr–Coulomb failure envelope defined by the equation

$$s = c' + \sigma' \tan \phi'$$

This circle, marked c in the figure, represents the failure condition in the soil mass; the horizontal stress then equals σ'_a, referred to as the *Rankine active pressure*. The *slip lines* (failure planes) in the soil mass will then make angles of $\pm(45 + \phi'/2)$ with the horizontal, as shown in Figure 7.5a.

Equation (1.69) relates the principal stresses for a Mohr's circle that touches the Mohr–Coulomb failure envelope:

$$\sigma'_1 = \sigma'_3 \tan^2\left(45 + \frac{\phi'}{2}\right) + 2c' \tan\left(45 + \frac{\phi'}{2}\right)$$

For the Mohr's circle c in Figure 7.5b,

Major principal stress, $\sigma'_1 = \sigma'_o$

and

Minor principal stress, $\sigma'_3 = \sigma'_a$

Thus,

$$\sigma'_o = \sigma'_a \tan^2\left(45 + \frac{\phi'}{2}\right) + 2c' \tan\left(45 + \frac{\phi'}{2}\right)$$

$$\sigma'_a = \frac{\sigma'_o}{\tan^2\left(45 + \dfrac{\phi'}{2}\right)} - \frac{2c'}{\tan\left(45 + \dfrac{\phi'}{2}\right)}$$

or

$$\sigma_a' = \sigma_o' \tan^2\left(45 - \frac{\phi'}{2}\right) - 2c' \tan\left(45 - \frac{\phi'}{2}\right)$$

$$= \sigma_o' K_a - 2c'\sqrt{K_a} \tag{7.8}$$

where $K_a = \tan^2(45 - \phi'/2)$ = Rankine active-pressure coefficient.

The variation of the active pressure with depth for the wall shown in Figure 7.5a is given in Figure 7.5c. Note that $\sigma_o' = 0$ at $z = 0$ and $\sigma_o' = \gamma H$ at $z = H$. The pressure distribution shows that at $z = 0$ the active pressure equals $-2c'\sqrt{K_a}$, indicating a tensile stress that decreases with depth and becomes zero at a depth $z = z_c$, or

$$\gamma z_c K_a - 2c'\sqrt{K_a} = 0$$

and

$$z_c = \frac{2c'}{\gamma\sqrt{K_a}} \tag{7.9}$$

The depth z_c is usually referred to as the *depth of tensile crack,* because the tensile stress in the soil will eventually cause a crack along the soil–wall interface. Thus, the total Rankine active force per unit length of the wall before the tensile crack occurs is

$$P_a = \int_0^H \sigma_a' dz = \int_0^H \gamma z K_a dz - \int_0^H 2c'\sqrt{K_a} dz$$

$$= \tfrac{1}{2}\gamma H^2 K_a - 2c'H\sqrt{K_a} \tag{7.10}$$

After the tensile crack appears, the force per unit length on the wall will be caused only by the pressure distribution between depths $z = z_c$ and $z = H$, as shown by the hatched area in Figure 7.5c. This force may be expressed as

$$P_a = \tfrac{1}{2}(H - z_c)(\gamma H K_a - 2c'\sqrt{K_a}) \tag{7.11}$$

or

$$P_a = \frac{1}{2}\left(H - \frac{2c'}{\gamma\sqrt{K_a}}\right)\left(\gamma H K_a - 2c'\sqrt{K_a}\right) \tag{7.12}$$

However, it is important to realize that the active earth pressure condition will be reached only if the wall is allowed to "yield" sufficiently. The necessary amount of outward displacement of the wall is about $0.001H$ to $0.004H$ for granular soil backfills and about $0.01H$ to $0.04H$ for cohesive soil backfills.

Note further that if the *total stress* shear strength parameters (c, ϕ) were used, an equation similar to Eq. (7.8) could have been derived, namely,

$$\sigma_a = \sigma_o \tan^2\left(45 - \frac{\phi}{2}\right) - 2c \tan\left(45 - \frac{\phi}{2}\right)$$

7.4   *A Generalized Case for Rankine Active Pressure*

In Section 7.3, the relationship was developed for Rankine active pressure for a retaining wall with a vertical back and a horizontal backfill. That can be extended to general cases of frictionless walls with inclined backs and inclined backfills. Some of these cases will be discussed in this section.

Granular Backfill

Figure 7.6 shows a retaining wall whose back is inclined at an angle θ with the vertical. The granular backfill is inclined at an angle α with the horizontal.

For a Rankine active case, the lateral earth pressure (σ'_a) at a depth z can be given as (Chu, 1991),

$$\sigma'_a = \frac{\gamma z \cos\alpha\sqrt{1 + \sin^2\phi' - 2\sin\phi'\cos\psi_a}}{\cos\alpha + \sqrt{\sin^2\phi' - \sin^2\alpha}} \tag{7.13}$$

where $\psi_a = \sin^{-1}\left(\dfrac{\sin\alpha}{\sin\phi'}\right) - \alpha + 2\theta.$ (7.14)

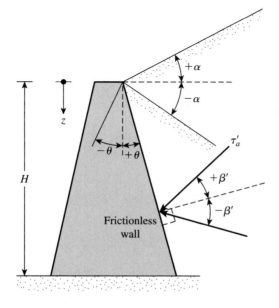

Figure 7.6   General case for a retaining wall with granular backfill

The pressure σ_a' will be inclined at an angle β' with the plane drawn at right angle to the backface of the wall, and

$$\beta' = \tan^{-1}\left(\frac{\sin\phi'\sin\psi_a}{1 - \sin\phi'\cos\psi_a}\right) \tag{7.15}$$

The active force P_a for unit length of the wall then can be calculated as

$$P_a = \frac{1}{2}\gamma H^2 K_a \tag{7.16}$$

where

$$K_a = \frac{\cos(\alpha - \theta)\sqrt{1 + \sin^2\phi' - 2\sin\phi'\cos\psi_a}}{\cos^2\theta\left(\cos\alpha + \sqrt{\sin^2\phi' - \sin^2\alpha}\right)}$$

$$= \text{Rankine active earth-pressure coefficient for generalized case} \quad (7.17)$$

The location and direction of the resultant force P_a is shown in Figure 7.7. Also shown in this figure is the failure wedge, ABC. Note that BC will be inclined at an angle η. Or

$$\eta = \frac{\pi}{4} + \frac{\phi'}{2} + \frac{\alpha}{2} - \frac{1}{2}\sin^{-1}\left(\frac{\sin\alpha}{\sin\phi'}\right) \tag{7.18}$$

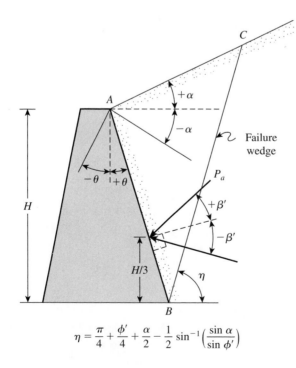

$$\eta = \frac{\pi}{4} + \frac{\phi'}{4} + \frac{\alpha}{2} - \frac{1}{2}\sin^{-1}\left(\frac{\sin\alpha}{\sin\phi'}\right)$$

Figure 7.7 Location and direction of Rankine active force

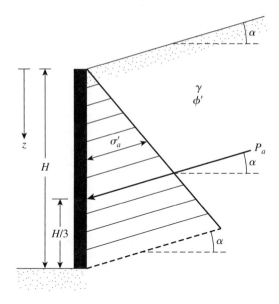

Figure 7.8 Notations for active
pressure—Eqs. (7.19), (7.20), (7.21)

Granular Backfill with Vertical Back Face

As a special case, for a vertical backface of a wall (that is, $\theta = 0$), as shown in Figure 7.8, Eqs. (7.13), (7.16) and (7.17) simplify to the following.

If the backfill of a frictionless retaining wall is a *granular soil* ($c' = 0$) and rises at an angle α with respect to the horizontal (see Figure 7.8), the *active earth-pressure coefficient* may be expressed in the form

$$K_a = \cos \alpha \frac{\cos \alpha - \sqrt{\cos^2 \alpha - \cos^2 \phi'}}{\cos \alpha + \sqrt{\cos^2 \alpha - \cos^2 \phi'}} \tag{7.19}$$

where ϕ' = angle of friction of soil.

At any depth z, the *Rankine active pressure* may be expressed as

$$\sigma_a' = \gamma z K_a \tag{7.20}$$

Also, the total force per unit length of the wall is

$$P_a = \tfrac{1}{2} \gamma H^2 K_a \tag{7.21}$$

Note that, in this case, the direction of the resultant force P_a is *inclined at an angle α with the horizontal* and intersects the wall at a distance $H/3$ from the base of the wall. Table 7.1 presents the values of K_a (active earth pressure) for varios values of α and ϕ'.

Backface with $c'-\phi'$ Soil Backfill

For a retaining wall with a *vertical back* ($\theta = 0$) and *inclined backfill* of $c'-\phi'$ soil (Mazindrani and Ganjali, 1997),

$$\sigma_a' = \gamma z K_a = \gamma z K_a' \cos \alpha \tag{7.22}$$

Table 7.1 Active Earth-Pressure Coefficient K_a [from Eq. (7.19)]

↓α(deg)	ϕ'(deg) → 28	30	32	34	36	38	40
0	0.361	0.333	0.307	0.283	0.260	0.238	0.217
5	0.366	0.337	0.311	0.286	0.262	0.240	0.219
10	0.380	0.350	0.321	0.294	0.270	0.246	0.225
15	0.409	0.373	0.341	0.311	0.283	0.258	0.235
20	0.461	0.414	0.374	0.338	0.306	0.277	0.250
25	0.573	0.494	0.434	0.385	0.343	0.307	0.275

where

$$K'_a = \frac{1}{\cos^2 \phi'} \left\{ \frac{2\cos^2 \alpha + 2\left(\dfrac{c'}{\gamma z}\right)\cos \phi' \sin \phi'}{-\sqrt{\left[4\cos^2\alpha(\cos^2 \alpha - \cos^2 \phi') + 4\left(\dfrac{c'}{\gamma z}\right)^2 \cos^2 \phi' + 8\left(\dfrac{c'}{\gamma z}\right)\cos^2 \alpha \sin \phi' \cos \phi'\right]}} - 1 \right\}$$

(7.23)

Some values of K'_a are given in Table 7.2. For a problem of this type, the depth of tensile crack is given as

$$z_c = \frac{2c'}{\gamma}\sqrt{\frac{1 + \sin \phi'}{1 - \sin \phi'}}$$

(7.24)

For this case, the active pressure is inclined at an angle α with the horizontal.

Table 7.2 Values of K'_a

ϕ' (deg)	α (deg)	$\dfrac{c'}{\gamma z}$ 0.025	0.05	0.1	0.5
15	0	0.550	0.512	0.435	−0.179
	5	0.566	0.525	0.445	−0.184
	10	0.621	0.571	0.477	−0.186
	15	0.776	0.683	0.546	−0.196
20	0	0.455	0.420	0.350	−0.210
	5	0.465	0.429	0.357	−0.212
	10	0.497	0.456	0.377	−0.218
	15	0.567	0.514	0.417	−0.229
25	0	0.374	0.342	0.278	−0.231
	5	0.381	0.348	0.283	−0.233
	10	0.402	0.366	0.296	−0.239
	15	0.443	0.401	0.321	−0.250
30	0	0.305	0.276	0.218	−0.244
	5	0.309	0.280	0.221	−0.246
	10	0.323	0.292	0.230	−0.252
	15	0.350	0.315	0.246	−0.263

Example 7.1

Refer to Figure 7.9a. Assume that the wall can yield sufficiently and determine the Rankine active force per unit length of the wall. Also determine the location of the resultant line of action.

Solution
If the cohesion, c', is equal to zero,

$$\sigma'_a = \sigma'_O K_a$$

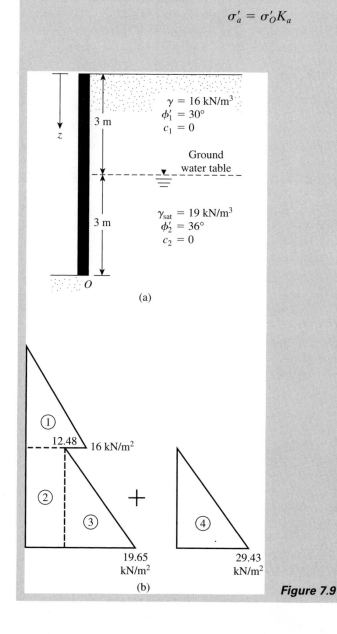

(a)

(b)

Figure 7.9

For the top soil layer, $\phi_1' = 30°$, so

$$K_{a(1)} = \tan^2\left(45 - \frac{\phi_1'}{2}\right) = \tan^2(45 - 15) = \frac{1}{3}$$

Similarly, for the bottom soil layer, $\phi_2' = 36°$, and

$$K_{a(2)} = \tan^2\left(45 - \frac{36}{2}\right) = 0.26$$

Because of the presence of the water table, the effective lateral pressure and the hydrostatic pressure have to be calculated separately.

At $z = 0$, $\sigma_O' = 0$, $\sigma_a' = 0$
At $z = 3$ m, $\sigma_O' = \gamma z = (16)(3) = 48$ kN/m²

At this depth, for the top soil layer,

$$\sigma_a' = K_{a(1)}\sigma_O' = \left(\tfrac{1}{3}\right)(48) = 16 \text{ kN/m}^2$$

Similarly, for the bottom soil layer,

$$\sigma_a' = K_{a(2)}\sigma_O' = (0.26)(48) = 12.48 \text{ kN/m}^2$$

At $z = 6$ m,

$$\sigma_O' = (\gamma)(3) + (\gamma_{\text{sat}} - \gamma_w)(3) = (16)(3) + (19 - 9.81)(3)$$

$$= 48 + 27.57 = 75.57 \text{ kN/m}^2$$

$$\sigma_a' = K_{a(2)}\sigma_O' = (0.26)(75.57) = 19.65 \text{ kN/m}^2$$

The hydrostatic pressure u is zero from $z = 0$ to $z = 3$ m. At $z = 6$ m, $u = 3\gamma_w = 3(9.81) = 29.43$ kN/m². The pressure distribution diagram is plotted in Figure 7.9b. The force per unit length

$$P_o = \text{Area 1} + \text{Area 2} + \text{Area 3} + \text{Area 4}$$

$$= \tfrac{1}{2}(3)(16) + (3)(12.48) + \tfrac{1}{2}(3)(19.65 - 12.48) + \tfrac{1}{2}(3)(29.43)$$

$$= 24 + 37.44 + 10.76 + 44.15 = 116.35 \text{ kN/m}$$

The distance of the line of action of the resultant from the bottom of the wall ($\bar{z}$) can be determined by taking the moments about the bottom of the wall (point O in Figure 7.9a), or

$$\bar{z} = \frac{(24)\left(3 + \dfrac{3}{3}\right) + (37.44)\left(\dfrac{3}{2}\right) + (10.76)\left(\dfrac{3}{3}\right) + (44.15)\left(\dfrac{3}{3}\right)}{116.35}$$

$$= \frac{96 + 56.16 + 10.76 + 44.15}{116.35} = \textbf{1.779 m} \qquad \blacksquare$$

Example 7.2

A 6-m-high retaining wall is to support a soil with unit weight $\gamma = 17.4 \text{ kN/m}^3$, soil friction angle $\phi' = 26°$, and cohesion $c' = 14.36 \text{ kN/m}^2$. Determine the Rankine active force per unit length of the wall both before and after the tensile crack occurs, and determine the line of action of the resultant in both cases.

Solution
For $\phi' = 26°$,

$$K_a = \tan^2\left(45 - \frac{\phi'}{2}\right) = \tan^2(45 - 13) = 0.39$$

$$\sqrt{K_a} = 0.625$$

$$\sigma_a' = \gamma H K_a - 2c'\sqrt{K_a}$$

From Figure 7.5c,

at $z = 0, \sigma_a' = -2c'\sqrt{K_a} = -2(14.36)(0.625) = -17.95 \text{ kN/m}^2$

and

at $z = 6 \text{ m}, \sigma_a' = (17.4)(6)(0.39) - 2(14.36)(0.625)$
    $= 40.72 - 17.95 = 22.77 \text{ kN/m}^2$

Active Force before the Tensile Crack Appeared: Eq. (7.10)

$$P_a = \tfrac{1}{2}\gamma H^2 K_a - 2c'H\sqrt{K_a}$$

$$= \tfrac{1}{2}(6)(40.72) - (6)(17.95) = 122.16 - 107.7 = 14.46 \text{ kN/m}$$

The line of action of the resultant can be determined by taking the moment of the area of the pressure diagrams about the bottom of the wall, or

$$P_a\bar{z} = (122.16)\left(\tfrac{6}{3}\right) - (107.7)\left(\tfrac{6}{2}\right)$$

Thus,

$$\bar{z} = \frac{244.32 - 323.1}{14.46} = -5.45 \text{ m}$$

Active Force after the Tensile Crack Appeared: Eq. (7.9)

$$z_c = \frac{2c'}{\gamma\sqrt{K_a}} = \frac{2(14.36)}{(17.4)(0.625)} = 2.64 \text{ m}$$

Using Eq. (7.11) gives

$$P_a = \tfrac{1}{2}(H - z_c)(\gamma H K_a - 2c'\sqrt{K_a}) = \tfrac{1}{2}(6 - 2.64)(22.77) = 38.25 \text{ kN/m}$$

Figure 7.5c indicates that the force $P_a = 38.25$ kN/m is the area of the hatched triangle. Hence, the line of action of the resultant will be located at a height $\bar{z} = (H - z_c)/3$ above the bottom of the wall, or

$$\bar{z} = \frac{6 - 2.64}{3} = \textbf{1.12 m}$$ ∎

Example 7.3

Refer to the retaining wall in Figure 7.7. The backfill is granular soil. Given:

$$\text{Wall:} \qquad H = 10 \text{ ft}$$
$$\theta = +10°$$
$$\text{Backfill:} \qquad \alpha = 15°$$
$$\phi' = 35°$$
$$c' = 0$$
$$\gamma = 110 \text{ lb/ft}^3$$

Determine the Rankine active force, P_a, and its location and direction.

Solution

From Eq. (7.14),

$$\psi_a = \sin^{-1}\left(\frac{\sin \alpha}{\sin \phi'}\right) - \alpha + 2\theta = \sin^{-1}\left(\frac{\sin 15}{\sin 35}\right) - 15 + (2)(10) = 31.82°$$

From Eq. (7.17),

$$K_a = \frac{\cos(\alpha - \theta)\sqrt{1 + \sin^2 \phi' - 2\sin\phi' \cos\psi_a}}{\cos^2\theta\left(\cos\alpha + \sqrt{\sin^2 \phi' - \sin^2 \alpha}\right)}$$

$$= \frac{\cos(15 - 10)\sqrt{1 + \sin^2 35 - (2)(\sin 35)(\sin 31.82)}}{\cos^2 10\left(\cos 15 + \sqrt{\sin^2 35 - \sin^2 15}\right)} = 0.59$$

$$P_a = \frac{1}{2} \gamma H^2 K_a = (\frac{1}{2})(110)(10)^2(0.59) = \textbf{3245 lb/ft}$$

From Eq. (7.15),

$$\beta' = \tan^{-1}\left(\frac{\sin \phi' \sin \psi_a}{1 - \sin \phi' \cos \psi_a}\right) = \tan^{-1}\left[\frac{(\sin 35)(\sin 31.82)}{1 - (\sin 35)(\cos 31.82)}\right] = \textbf{30.5°}$$

The force P_a will act at a distance of $10/3 = 3.33$ ft above the bottom of the wall and will be inclined at an angle of $+30.5°$ to the normal drawn to the back face of the wall. ∎

Coulomb's Active Earth Pressure

The Rankine active earth pressure calculations discussed in the preceding sections were based on the assumption that the wall is frictionless. In 1776, Coulomb proposed a theory for calculating the lateral earth pressure on a retaining wall with granular soil backfill. This theory takes wall friction into consideration.

To apply Coulomb's active earth pressure theory, let us consider a retaining wall with its back face inclined at an angle β with the horizontal, as shown in Figure 7.10a. The backfill is a granular soil that slopes at an angle α with the horizontal. Also, let δ' be the angle of friction between the soil and the wall (i.e., the angle of wall friction).

Under active pressure, the wall will move away from the soil mass (to the left in the figure). Coulomb assumed that, in such a case, the failure surface in the soil mass would be a plane (e.g., $BC_1, BC_2, \ldots$). So, to find the active force, consider a possible soil failure wedge ABC_1. The forces acting on this wedge (per unit length at right angles to the cross section shown) are as follows:

1. The weight of the wedge, W.
2. The resultant, R, of the normal and resisting shear forces along the surface, BC_1. The force R will be inclined at an angle ϕ' to the normal drawn to BC_1.
3. The active force per unit length of the wall, P_a, which will be inclined at an angle δ' to the normal drawn to the back face of the wall.

For equilibrium purposes, a force triangle can be drawn, as shown in Figure 7.10b. Note that θ_1 is the angle that BC_1 makes with the horizontal. Because the

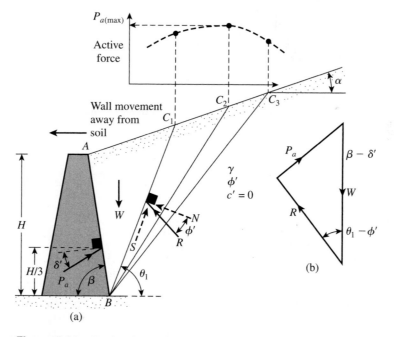

Figure 7.10   Coulomb's active pressure

magnitude of W, as well as the directions of all three forces, are known, the value of P_a can now be determined. Similarly, the active forces of other trial wedges, such as $ABC_2, ABC_3, \ldots$, can be determined. The maximum value of P_a thus determined is Coulomb's active force (see top part of Figure 7.10), which may be expressed as

$$P_a = \tfrac{1}{2}K_a\gamma H^2 \tag{7.25}$$

where

$$K_a = \text{Coulomb's active earth pressure coefficient}$$
$$= \frac{\sin^2 (\beta + \phi')}{\sin^2 \beta \sin (\beta - \delta') \left[1 + \sqrt{\dfrac{\sin (\phi' + \delta')\sin (\phi' - \alpha)}{\sin (\beta - \delta')\sin (\alpha + \beta)}} \right]^2} \tag{7.26}$$

and H = height of the wall.

The values of the active earth pressure coefficient, K_a, for a vertical retaining wall ($\beta = 90°$) with horizontal backfill ($\alpha = 0°$) are given in Table 7.3. Note that the line of action of the resultant force (P_a) will act at a distance $H/3$ above the base of the wall and will be inclined at an angle δ' to the normal drawn to the back of the wall.

In the actual design of retaining walls, the value of the wall friction angle δ' is assumed to be between $\phi'/2$ and $\tfrac{2}{3}\phi'$. The active earth pressure coefficients for various values of ϕ', α, and β with $\delta' = \tfrac{1}{2}\phi'$ and $\tfrac{2}{3}\phi'$ are respectively given in Tables 7.4 and 7.5. These coefficients are very useful design considerations.

Table 7.3 Values of K_a [Eq. (7.26)] for $\beta = 90°$ and $\alpha = 0°$

	δ' (deg)					
ϕ' (deg)	0	5	10	15	20	25
28	0.3610	0.3448	0.3330	0.3251	0.3203	0.3186
30	0.3333	0.3189	0.3085	0.3014	0.2973	0.2956
32	0.3073	0.2945	0.2853	0.2791	0.2755	0.2745
34	0.2827	0.2714	0.2633	0.2579	0.2549	0.2542
36	0.2596	0.2497	0.2426	0.2379	0.2354	0.2350
38	0.2379	0.2292	0.2230	0.2190	0.2169	0.2167
40	0.2174	0.2098	0.2045	0.2011	0.1994	0.1995
42	0.1982	0.1916	0.1870	0.1841	0.1828	0.1831

Table 7.4 Values of K_a [from Eq. (7.26)] for $\delta' = \frac{2}{3}\phi'$

α (deg)	φ' (deg)	β (deg)					
		90	85	80	75	70	65
0	28	0.3213	0.3588	0.4007	0.4481	0.5026	0.5662
	29	0.3091	0.3467	0.3886	0.4362	0.4908	0.5547
	30	0.2973	0.3349	0.3769	0.4245	0.4794	0.5435
	31	0.2860	0.3235	0.3655	0.4133	0.4682	0.5326
	32	0.2750	0.3125	0.3545	0.4023	0.4574	0.5220
	33	0.2645	0.3019	0.3439	0.3917	0.4469	0.5117
	34	0.2543	0.2916	0.3335	0.3813	0.4367	0.5017
	35	0.2444	0.2816	0.3235	0.3713	0.4267	0.4919
	36	0.2349	0.2719	0.3137	0.3615	0.4170	0.4824
	37	0.2257	0.2626	0.3042	0.3520	0.4075	0.4732
	38	0.2168	0.2535	0.2950	0.3427	0.3983	0.4641
	39	0.2082	0.2447	0.2861	0.3337	0.3894	0.4553
	40	0.1998	0.2361	0.2774	0.3249	0.3806	0.4468
	41	0.1918	0.2278	0.2689	0.3164	0.3721	0.4384
	42	0.1840	0.2197	0.2606	0.3080	0.3637	0.4302
5	28	0.3431	0.3845	0.4311	0.4843	0.5461	0.6190
	29	0.3295	0.3709	0.4175	0.4707	0.5325	0.6056
	30	0.3165	0.3578	0.4043	0.4575	0.5194	0.5926
	31	0.3039	0.3451	0.3916	0.4447	0.5067	0.5800
	32	0.2919	0.3329	0.3792	0.4324	0.4943	0.5677
	33	0.2803	0.3211	0.3673	0.4204	0.4823	0.5558
	34	0.2691	0.3097	0.3558	0.4088	0.4707	0.5443
	35	0.2583	0.2987	0.3446	0.3975	0.4594	0.5330
	36	0.2479	0.2881	0.3338	0.3866	0.4484	0.5221
	37	0.2379	0.2778	0.3233	0.3759	0.4377	0.5115
	38	0.2282	0.2679	0.3131	0.3656	0.4273	0.5012
	39	0.2188	0.2582	0.3033	0.3556	0.4172	0.4911
	40	0.2098	0.2489	0.2937	0.3458	0.4074	0.4813
	41	0.2011	0.2398	0.2844	0.3363	0.3978	0.4718
	42	0.1927	0.2311	0.2753	0.3271	0.3884	0.4625
10	28	0.3702	0.4164	0.4686	0.5287	0.5992	0.6834
	29	0.3548	0.4007	0.4528	0.5128	0.5831	0.6672
	30	0.3400	0.3857	0.4376	0.4974	0.5676	0.6516
	31	0.3259	0.3713	0.4230	0.4826	0.5526	0.6365
	32	0.3123	0.3575	0.4089	0.4683	0.5382	0.6219
	33	0.2993	0.3442	0.3953	0.4545	0.5242	0.6078
	34	0.2868	0.3314	0.3822	0.4412	0.5107	0.5942
	35	0.2748	0.3190	0.3696	0.4283	0.4976	0.5810
	36	0.2633	0.3072	0.3574	0.4158	0.4849	0.5682
	37	0.2522	0.2957	0.3456	0.4037	0.4726	0.5558
	38	0.2415	0.2846	0.3342	0.3920	0.4607	0.5437
	39	0.2313	0.2740	0.3231	0.3807	0.4491	0.5321
	40	0.2214	0.2636	0.3125	0.3697	0.4379	0.5207
	41	0.2119	0.2537	0.3021	0.3590	0.4270	0.5097
	42	0.2027	0.2441	0.2921	0.3487	0.4164	0.4990
15	28	0.4065	0.4585	0.5179	0.5868	0.6685	0.7670

(continued)

Table 7.4 (Continued)

α (deg)	ϕ' (deg)	β (deg) 90	85	80	75	70	65
	29	0.3881	0.4397	0.4987	0.5672	0.6483	0.7463
	30	0.3707	0.4219	0.4804	0.5484	0.6291	0.7265
	31	0.3541	0.4049	0.4629	0.5305	0.6106	0.7076
	32	0.3384	0.3887	0.4462	0.5133	0.5930	0.6895
	33	0.3234	0.3732	0.4303	0.4969	0.5761	0.6721
	34	0.3091	0.3583	0.4150	0.4811	0.5598	0.6554
	35	0.2954	0.3442	0.4003	0.4659	0.5442	0.6393
	36	0.2823	0.3306	0.3862	0.4513	0.5291	0.6238
	37	0.2698	0.3175	0.3726	0.4373	0.5146	0.6089
	38	0.2578	0.3050	0.3595	0.4237	0.5006	0.5945
	39	0.2463	0.2929	0.3470	0.4106	0.4871	0.5805
	40	0.2353	0.2813	0.3348	0.3980	0.4740	0.5671
	41	0.2247	0.2702	0.3231	0.3858	0.4613	0.5541
	42	0.2146	0.2594	0.3118	0.3740	0.4491	0.5415
20	28	0.4602	0.5205	0.5900	0.6714	0.7689	0.8880
	29	0.4364	0.4958	0.5642	0.6445	0.7406	0.8581
	30	0.4142	0.4728	0.5403	0.6195	0.7144	0.8303
	31	0.3935	0.4513	0.5179	0.5961	0.6898	0.8043
	32	0.3742	0.4311	0.4968	0.5741	0.6666	0.7799
	33	0.3559	0.4121	0.4769	0.5532	0.6448	0.7569
	34	0.3388	0.3941	0.4581	0.5335	0.6241	0.7351
	35	0.3225	0.3771	0.4402	0.5148	0.6044	0.7144
	36	0.3071	0.3609	0.4233	0.4969	0.5856	0.6947
	37	0.2925	0.3455	0.4071	0.4799	0.5677	0.6759
	38	0.2787	0.3308	0.3916	0.4636	0.5506	0.6579
	39	0.2654	0.3168	0.3768	0.4480	0.5342	0.6407
	40	0.2529	0.3034	0.3626	0.4331	0.5185	0.6242
	41	0.2408	0.2906	0.3490	0.4187	0.5033	0.6083
	42	0.2294	0.2784	0.3360	0.4049	0.4888	0.5930

Example 7.4

Consider the retaining wall shown in Figure 7.10a. Given: $H = 4.6$ m; unit weight of soil $= 16.5$ kN/m³; angle of friction of soil $= 30°$; wall friction-angle, $\delta' = \frac{2}{3}\phi'$, soil cohesion, $c' = 0$; $\alpha = 0$, and $\beta = 90°$. Calculate the Coulomb's active force per unit length of the wall.

Solution
From Eq. (7.25)

$$P_a = \tfrac{1}{2}\gamma H^2 K_a$$

From Table 7.4, for $\alpha = 0°$, $\beta = 90°$, $\phi' = 30°$, and $\delta' = \frac{2}{3}\phi' = 20°$, $K_a = 0.297$. Hence,

$$P_a = \tfrac{1}{2}(16.5)(4.6)^2(0.297) = \textbf{51.85 kN/m}$$ ■

Table 7.5 Values of K_a [from Eq. (7.26)] for $\delta' = \phi'/2$

α (deg)	φ' (deg)	β (deg)					
		90	85	80	75	70	65
0	28	0.3264	0.3629	0.4034	0.4490	0.5011	0.5616
	29	0.3137	0.3502	0.3907	0.4363	0.4886	0.5492
	30	0.3014	0.3379	0.3784	0.4241	0.4764	0.5371
	31	0.2896	0.3260	0.3665	0.4121	0.4645	0.5253
	32	0.2782	0.3145	0.3549	0.4005	0.4529	0.5137
	33	0.2671	0.3033	0.3436	0.3892	0.4415	0.5025
	34	0.2564	0.2925	0.3327	0.3782	0.4305	0.4915
	35	0.2461	0.2820	0.3221	0.3675	0.4197	0.4807
	36	0.2362	0.2718	0.3118	0.3571	0.4092	0.4702
	37	0.2265	0.2620	0.3017	0.3469	0.3990	0.4599
	38	0.2172	0.2524	0.2920	0.3370	0.3890	0.4498
	39	0.2081	0.2431	0.2825	0.3273	0.3792	0.4400
	40	0.1994	0.2341	0.2732	0.3179	0.3696	0.4304
	41	0.1909	0.2253	0.2642	0.3087	0.3602	0.4209
	42	0.1828	0.2168	0.2554	0.2997	0.3511	0.4177
5	28	0.3477	0.3879	0.4327	0.4837	0.5425	0.6115
	29	0.3337	0.3737	0.4185	0.4694	0.5282	0.5972
	30	0.3202	0.3601	0.4048	0.4556	0.5144	0.5833
	31	0.3072	0.3470	0.3915	0.4422	0.5009	0.5698
	32	0.2946	0.3342	0.3787	0.4292	0.4878	0.5566
	33	0.2825	0.3219	0.3662	0.4166	0.4750	0.5437
	34	0.2709	0.3101	0.3541	0.4043	0.4626	0.5312
	35	0.2596	0.2986	0.3424	0.3924	0.4505	0.5190
	36	0.2488	0.2874	0.3310	0.3808	0.4387	0.5070
	37	0.2383	0.2767	0.3199	0.3695	0.4272	0.4954
	38	0.2282	0.2662	0.3092	0.3585	0.4160	0.4840
	39	0.2185	0.2561	0.2988	0.3478	0.4050	0.4729
	40	0.2090	0.2463	0.2887	0.3374	0.3944	0.4620
	41	0.1999	0.2368	0.2788	0.3273	0.3840	0.4514
	42	0.1911	0.2276	0.2693	0.3174	0.3738	0.4410
10	28	0.3743	0.4187	0.4688	0.5261	0.5928	0.6719
	29	0.3584	0.4026	0.4525	0.5096	0.5761	0.6549
	30	0.3432	0.3872	0.4368	0.4936	0.5599	0.6385
	31	0.3286	0.3723	0.4217	0.4782	0.5442	0.6225
	32	0.3145	0.3580	0.4071	0.4633	0.5290	0.6071
	33	0.3011	0.3442	0.3930	0.4489	0.5143	0.5920
	34	0.2881	0.3309	0.3793	0.4350	0.5000	0.5775
	35	0.2757	0.3181	0.3662	0.4215	0.4862	0.5633
	36	0.2637	0.3058	0.3534	0.4084	0.4727	0.5495
	37	0.2522	0.2938	0.3411	0.3957	0.4597	0.5361
	38	0.2412	0.2823	0.3292	0.3833	0.4470	0.5230
	39	0.2305	0.2712	0.3176	0.3714	0.4346	0.5103
	40	0.2202	0.2604	0.3064	0.3597	0.4226	0.4979
	41	0.2103	0.2500	0.2956	0.3484	0.4109	0.4858
	42	0.2007	0.2400	0.2850	0.3375	0.3995	0.4740
15	28	0.4095	0.4594	0.5159	0.5812	0.6579	0.7498

(*continued*)

Table 7.5 (Continued)

α (deg)	φ' (deg)	90	85	80	75	70	65
					β (deg)		
	29	0.3908	0.4402	0.4964	0.5611	0.6373	0.7284
	30	0.3730	0.4220	0.4777	0.5419	0.6175	0.7080
	31	0.3560	0.4046	0.4598	0.5235	0.5985	0.6884
	32	0.3398	0.3880	0.4427	0.5059	0.5803	0.6695
	33	0.3244	0.3721	0.4262	0.4889	0.5627	0.6513
	34	0.3097	0.3568	0.4105	0.4726	0.5458	0.6338
	35	0.2956	0.3422	0.3953	0.4569	0.5295	0.6168
	36	0.2821	0.3282	0.3807	0.4417	0.5138	0.6004
	37	0.2692	0.3147	0.3667	0.4271	0.4985	0.5846
	38	0.2569	0.3017	0.3531	0.4130	0.4838	0.5692
	39	0.2450	0.2893	0.3401	0.3993	0.4695	0.5543
	40	0.2336	0.2773	0.3275	0.3861	0.4557	0.5399
	41	0.2227	0.2657	0.3153	0.3733	0.4423	0.5258
	42	0.2122	0.2546	0.3035	0.3609	0.4293	0.5122
20	28	0.4614	0.5188	0.5844	0.6608	0.7514	0.8613
	29	0.4374	0.4940	0.5586	0.6339	0.7232	0.8313
	30	0.4150	0.4708	0.5345	0.6087	0.6968	0.8034
	31	0.3941	0.4491	0.5119	0.5851	0.6720	0.7772
	32	0.3744	0.4286	0.4906	0.5628	0.6486	0.7524
	33	0.3559	0.4093	0.4704	0.5417	0.6264	0.7289
	34	0.3384	0.3910	0.4513	0.5216	0.6052	0.7066
	35	0.3218	0.3736	0.4331	0.5025	0.5851	0.6853
	36	0.3061	0.3571	0.4157	0.4842	0.5658	0.6649
	37	0.2911	0.3413	0.3991	0.4668	0.5474	0.6453
	38	0.2769	0.3263	0.3833	0.4500	0.5297	0.6266
	39	0.2633	0.3120	0.3681	0.4340	0.5127	0.6085
	40	0.2504	0.2982	0.3535	0.4185	0.4963	0.5912
	41	0.2381	0.2851	0.3395	0.4037	0.4805	0.5744
	42	0.2263	0.2725	0.3261	0.3894	0.4653	0.5582

7.6   *Active Earth Pressure for Earthquake Conditions*

Coulomb's active earth pressure theory (see Section 7.5) can be extended to take into account the forces caused by an earthquake. Figure 7.11 shows a condition of active pressure with a granular backfill ($c' = 0$). Note that the forces acting on the soil failure wedge in Figure 7.11 are essentially the same as those shown in Figure 7.10a with the addition of $k_h W$ and $k_v W$ in the horizontal and vertical direction respectively; k_h and k_v may be defined as

$$k_h = \frac{\text{horizontal earthquake acceleration component}}{\text{acceleration due to gravity, } g} \tag{7.27}$$

$$k_v = \frac{\text{vertical earthquake acceleration component}}{\text{acceleration due to gravity, } g} \tag{7.28}$$

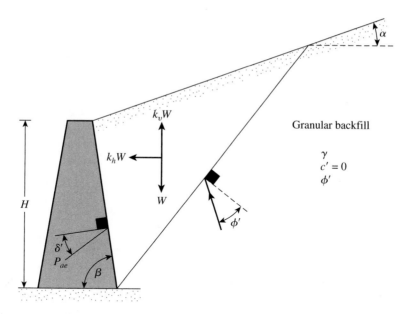

Figure 7.11 Derivation of Eq. (7.29)

As in Section 7.5, the relation for the active force per unit length of the wall (P_{ae}) can be determined as

$$P_{ae} = \tfrac{1}{2}\gamma H^2 (1 - k_v) K_{ae} \tag{7.29}$$

where

K_{ae} = active earth pressure coefficient

$$= \frac{\sin^2 (\phi' + \beta - \theta')}{\cos \theta' \sin^2 \beta \sin(\beta - \theta' - \delta') \left[1 + \sqrt{\dfrac{\sin (\phi' + \delta')\sin (\phi' - \theta' - \alpha)}{\sin (\beta - \delta' - \theta')\sin (\alpha + \beta)}} \right]^2}$$

$$\tag{7.30}$$

$$\theta' = \tan^{-1}\left[\frac{k_h}{(1 - k_v)} \right] \tag{7.31}$$

Note that for no earthquake condition

$$k_h = 0, \qquad k_v = 0, \qquad \text{and} \qquad \theta' = 0$$

Hence $K_{ae} = K_a$ [as given by Eq. (7.26)]. Some values of K_{ae} for $\beta = 90°$ and $k_v = 0$ are given in Table 7.6.

Equation (7.29) is usually referred to as the *Mononobe–Okabe* solution. Unlike the case shown in Figure 7.10a, the resultant earth pressure in this situation, as

Table 7.6 Values of K_{ae} [Eq. (7.30)] for $\beta = 90°$ and $k_v = 0$

k_h	δ' (deg)	α (deg)	ϕ' (deg) 28	30	35	40	45
0.1	0	0	0.427	0.397	0.328	0.268	0.217
0.2			0.508	0.473	0.396	0.382	0.270
0.3			0.611	0.569	0.478	0.400	0.334
0.4			0.753	0.697	0.581	0.488	0.409
0.5			1.005	0.890	0.716	0.596	0.500
0.1	0	5	0.457	0.423	0.347	0.282	0.227
0.2			0.554	0.514	0.424	0.349	0.285
0.3			0.690	0.635	0.522	0.431	0.356
0.4			0.942	0.825	0.653	0.535	0.442
0.5			—	—	0.855	0.673	0.551
0.1	0	10	0.497	0.457	0.371	0.299	0.238
0.2			0.623	0.570	0.461	0.375	0.303
0.3			0.856	0.748	0.585	0.472	0.383
0.4			—	—	0.780	0.604	0.486
0.5			—	—	—	0.809	0.624
0.1	$\phi'/2$	0	0.396	0.368	0.306	0.253	0.207
0.2			0.485	0.452	0.380	0.319	0.267
0.3			0.604	0.563	0.474	0.402	0.340
0.4			0.778	0.718	0.599	0.508	0.433
0.5			1.115	0.972	0.774	0.648	0.522
0.1	$\phi/2$	5	0.428	0.396	0.326	0.268	0.218
0.2			0.537	0.497	0.412	0.342	0.283
0.3			0.699	0.640	0.526	0.438	0.367
0.4			1.025	0.881	0.690	0.568	0.475
0.5			—	—	0.962	0.752	0.620
0.1	$\phi/2$	10	0.472	0.433	0.352	0.285	0.230
0.2			0.616	0.562	0.454	0.371	0.303
0.3			0.908	0.780	0.602	0.487	0.400
0.4			—	—	0.857	0.656	0.531
0.5			—	—	—	0.944	0.722
0.1	$\frac{2}{3}\phi$	0	0.393	0.366	0.306	0.256	0.212
0.2			0.486	0.454	0.384	0.326	0.276
0.3			0.612	0.572	0.486	0.416	0.357
0.4			0.801	0.740	0.622	0.533	0.462
0.5			1.177	1.023	0.819	0.693	0.600
0.1	$\frac{2}{3}\phi$	5	0.427	0.395	0.327	0.271	0.224
0.2			0.541	0.501	0.418	0.350	0.294
0.3			0.714	0.655	0.541	0.455	0.386
0.4			1.073	0.921	0.722	0.600	0.509
0.5			—	—	1.034	0.812	0.679
0.1	$\frac{2}{3}\phi$	10	0.472	0.434	0.354	0.290	0.237
0.2			0.625	0.570	0.463	0.381	0.317
0.3			0.942	0.807	0.624	0.509	0.423
0.4			—	—	0.909	0.699	0.573
0.5			—	—	—	1.037	0.800

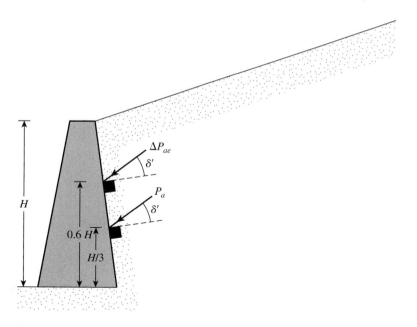

Figure 7.12 Determining the line of action of P_{ae}

calculated by Eq. (7.29) *does not act* at a distance of $H/3$ from the bottom of the wall. The following procedure may be used to obtain the location of the resultant force P_{ae}:

> *Step 1.* Calculate P_{ae} by using Eq. (7.29)
> *Step 2.* Calculate P_a by using Eq. (7.25)
> *Step 3.* Calculate

$$\Delta P_{ae} = P_{ae} - P_a \tag{7.32}$$

> *Step 4.* Assume that P_a acts at a distance of $H/3$ from the bottom of the wall (Figure 7.12)
> *Step 5.* Assume that ΔP_{ae} acts at a distance of $0.6H$ from the bottom of the wall (Figure 7.12)
> *Step 6.* Calculate the location of the resultant as

$$\bar{z} = \frac{(0.6H)(\Delta P_{ae}) + \left(\dfrac{H}{3}\right)(P_a)}{P_{ae}} \tag{7.33}$$

Example 7.5

Refer to Figure 7.13. For $k_v = 0$ and $k_h = 0.3$, determine:

a. P_{ae}
b. The location of the resultant, $\bar{z}$, from the bottom of the wall

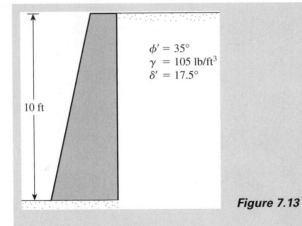

$\phi' = 35°$
$\gamma = 105 \text{ lb/ft}^3$
$\delta' = 17.5°$

10 ft

Figure 7.13

Solution
Part a
From Eq. (6.29),

$$P_{ae} = \tfrac{1}{2}\gamma H^2 (1 - k_v)\, K_{ae}$$

Here, $\gamma = 105$ lb/ft³, $H = 10$ ft, and $k_v = 0$. As $\delta' = \phi'/2$, we can use Table 7.6 to determine K_{ae}. For $k_h = 0.3$, $K_{ae} \approx 0.474$, so

$$P_{ae} = \tfrac{1}{2}(105)(10)^2(1 - 0)(0.474) = \textbf{2488.5 lb/ft}$$

Part b
From Eq. (6.25),

$$P_a = \tfrac{1}{2}\gamma H^2 K_a$$

From Eq. (7.26) with $\delta' = 17.5°$, $\beta = 90°$, and $\alpha = 0°$, $K_a \approx 0.246$ (Table 7.5).

$$P_a = \tfrac{1}{2}(105)(10)^2(0.246) = 1292 \text{ lb/ft}$$
$$\Delta P_{ae} = P_{ae} - P_a = 2488.5 - 1292 = 1196.5 \text{ lb/ft}$$

From Eq. (7.33),

$$\bar{z} = \frac{(0.6H)(\Delta P_{ae}) + (H/3)(P_a)}{P_{ae}}$$

$$= \frac{[(0.6)(10)](1196.5) + (10/3)(1292)}{2488.5} = \textbf{4.62 ft} \qquad \blacksquare$$

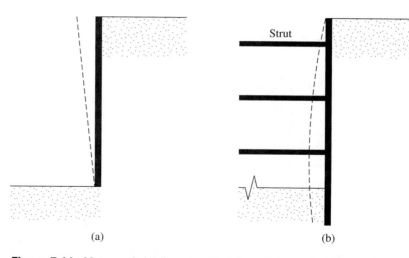

Figure 7.14   Nature of yielding of walls: (a) retaining wall; (b) braced cut

7.7   *Active Pressure for Wall Rotation about the Top: Braced Cut*

In the preceding sections, we have seen that a retaining wall rotates about its bottom. (See Figure 7.14a.) With sufficient yielding of the wall, the lateral earth pressure is approximately equal to that obtained by Rankine's theory or Coulomb's theory. In contrast to retaining walls, braced cuts show a different type of wall yielding. (See Figure 7.14b.) In this case, deformation of the wall gradually increases with the depth of excavation. The variation of the amount of deformation depends on several factors, such as the type of soil, the depth of excavation, and the workmanship involved. However, with very little wall yielding at the top of the cut, the lateral earth pressure will be close to the at-rest pressure. At the bottom of the wall, with a much larger degree of yielding, the lateral earth pressure will be substantially lower than the Rankine active earth pressure. As a result, the distribution of lateral earth pressure will vary substantially in comparison to the linear distribution assumed in the case of retaining walls.

The total lateral force per unit length of the wall, P_a, imposed on a wall may be evaluated theoretically by using Terzaghi's (1943) general wedge theory. (See Figure 7.15.) The failure surface is assumed to be the arc of a logarithmic spiral, defined as

$$r = r_o e^{\theta \tan \phi'} \tag{7.34}$$

where ϕ' = effective angle of friction of soil.

In the figure, H is the height of the cut, and the unit weight, angle of friction, and cohesion of the soil are equal to γ, ϕ', and c', respectively. Following are the

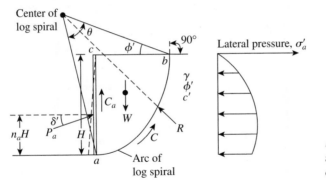

Figure 7.15   Braced cut analysis by general wedge theory: wall rotation about top

forces per unit length of the cut acting on the trial failure wedge:

1. Weight of the wedge, W
2. Resultant of the normal and shear forces along ab, R
3. Cohesive force along ab, C
4. Adhesive force along ac, C_a
5. P_a, which is the force acting a distance $n_a H$ from the bottom of the wall and is inclined at an angle δ' to the horizontal

The adhesive force is

$$C_a = c_a' H \qquad (7.35)$$

where c_a' = unit adhesion.

A detailed outline for the evaluation of P_a is beyond the scope of this text; those interested should check a soil mechanics text for more information. Kim and Preber (1969) provided tabulated values of $P_a/\frac{1}{2}\gamma H^2$ determined by using the principles of general wedge theory. Table 7.7 gives the variation of $P_a/0.5\gamma H^2$ for granular soil backfill obtained using the general wedge theory.

7.8   *Active Earth Pressure for Translation of Retaining Wall—Granular Backfill*

Under certain circumstances, retaining walls may undergo lateral translation, as shown in Figure 7.16. A solution to the distribution of active pressure for this case was provided by Dubrova (1963) and was also described by Harr (1966). The solution of Dubrova assumes the validity of Coulomb's solution [Eqs. (7.25) and (7.26)]. In order to understand this procedure, let us consider a vertical wall with a horizontal granular backfill (Figure 7.17). For rotation about the top of the wall, the resultant R of the normal and shear forces along the rupture line AC is inclined at an angle ϕ' to the normal drawn to AC. According to Dubrova there exists infinite number of

Table 7.7 Active Pressure for Wall Rotation—General Wedge Theory
(Granular Soil Backfill)

Soil friction angle, ϕ'(deg)	δ'/ϕ'	$P_a/0.5\,\gamma H^2$			
		$n_a = 0.3$	$n_a = 0.4$	$n_a = 0.5$	$n_a = 0.6$
25	0	0.371	0.405	0.447	0.499
	½	0.345	0.376	0.413	0.460
	⅔	0.342	0.373	0.410	0.457
	1	0.344	0.375	0.413	0.461
30	0	0.304	0.330	0.361	0.400
	½	0.282	0.306	0.334	0.386
	⅔	0.281	0.305	0.332	0.367
	1	0.289	0.313	0.341	0.377
35	0	0.247	0.267	0.290	0.318
	½	0.231	0.249	0.269	0.295
	⅔	0.232	0.249	0.270	0.296
	1	0.243	0.262	0.289	0.312
40	0	0.198	0.213	0.230	0.252
	½	0.187	0.200	0.216	0.235
	⅔	0.190	0.204	0.220	0.239
	1	0.197	0.211	0.228	0.248
45	0	0.205	0.220	0.237	0.259
	½	0.149	0.159	0.171	0.185
	⅔	0.153	0.164	0.176	0.196
	1	0.173	0.184	0.198	0.215

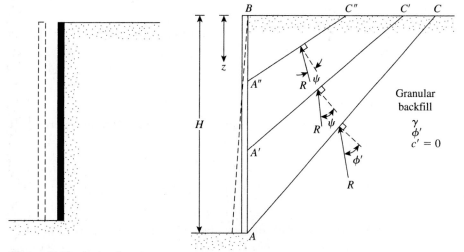

Figure 7.16 Lateral translation of retaining wall

Figure 7.17 Quasi-rupture lines behind a retaining wall

quasi-rupture lines such as $A'C', A''C'', \ldots$ for which the resultant force R is inclined at an angle ψ, where

$$\psi = \frac{\phi' z}{H} \qquad (7.36)$$

Now, refer to Eqs. (7.25) and (7.26) for Coulomb's active pressure. For $\beta = 90°$ and $\alpha = 0$, the relationship for Coulomb's active force can also be rewritten as

$$P_a = \frac{\gamma}{2 \cos \delta'} \left[\frac{H}{\frac{1}{\cos \phi'} + (\tan^2 \phi' + \tan \phi' \tan \delta')^{0.5}} \right]^2 \qquad (7.37)$$

The force against the wall at any z is then given as

$$P_a = \frac{\gamma}{2 \cos \delta'} \left[\frac{z}{\frac{1}{\cos \psi} + (\tan^2 \psi + \tan \psi \tan \delta')^{0.5}} \right]^2 \qquad (7.38)$$

The active pressure at any depth z for *wall rotation about the top* is

$$\sigma_a'(z) = \frac{dP_a}{dz} \approx \frac{\gamma}{\cos \delta'} \left[\frac{z \cos^2 \psi}{(1 + m \sin \psi)^2} - \frac{z^2 \phi' \cos^2 \psi}{H(1 + m \sin \psi)} (\sin \psi + m) \right] \qquad (7.39)$$

where $m = \left(1 + \frac{\tan \delta'}{\tan \psi} \right)^{0.5}$. $\qquad (7.40)$

For frictionless walls, $\delta' = 0$ and Eq. (7.39) simplifies to

$$\sigma_a'(z) = \gamma \tan^2 \left(45 - \frac{\psi}{2} \right) \left(z - \frac{\phi' z^2}{H \cos \psi} \right) \qquad (7.41)$$

For wall rotation about the bottom, a similar expression can be found in the form

$$\sigma_a'(z) = \frac{\gamma z}{\cos \delta'} \left(\frac{\cos \phi'}{1 + m \sin \phi'} \right)^2 \qquad (7.42)$$

For translation of the wall, the active pressure can then be taken as

$$\sigma_a'(z)_{\text{translation}} = \tfrac{1}{2} [\sigma_a'(z)_{\text{rotation about top}} + \sigma_a'(z)_{\text{rotation about bottom}}] \qquad (7.43)$$

Example 7.6

Consider a frictionless wall 16 ft high. For the granular backfill, $\gamma = 110$ lb/ft^3 and $\phi' = 36°$. Calculate and plot the variation of $\sigma_a(z)$ for a translation mode of the wall movement.

Solution

For a frictionless wall, $\delta' = 0$. Hence, m is equal to one [Eq. (7.40)]. So for rotation about the top, from Eq. (7.41),

$$\sigma_a'(z) = \sigma_{a(1)}' = \gamma \tan^2\left(45 - \frac{\phi' z}{2H}\right)\left[z - \frac{\phi' z^2}{H \cos\left(\dfrac{\phi' z}{H}\right)}\right]$$

For rotation about the bottom, from Eq. (7.42),

$$\sigma_a'(z) = \sigma_{a(2)}' = \gamma z \left(\frac{\cos \phi'}{1 + \sin \phi'}\right)^2$$

$$\sigma_a'(z)_{\text{translation}} = \frac{\sigma_{a(1)}' + \sigma_{a(2)}'}{2}$$

The following table can now be prepared with $\gamma = 110$ lb/ft^3, $\phi' = 36°$, and $H = 16$ ft.

z (ft)	$\sigma_{a(1)}'$ (lb/ft^2)	$\sigma_{a(1)}'$ (lb/ft^2)	$\sigma_{a(z)\ \text{translation}}'$ (lb/ft^2)
0	0	0	0
4	269.9	114.2	192.05
8	311.2	228.4	269.8
12	233.6	342.6	288.1
16	102.2	456.8	279.5

The plot of $\sigma_a(z)$ versus z is shown in Figure 7.18

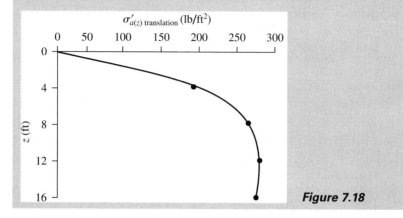

Figure 7.18

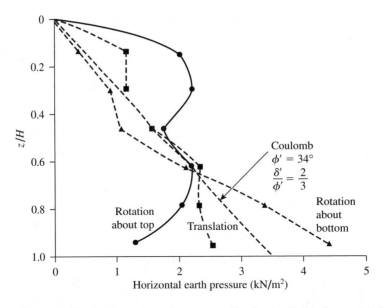

Figure 7.19 Horizontal earth-pressure distribution behind a model rigid retaining wall *(Note:* sand back fill, $\phi' \approx 34°$, $\delta' = \frac{2}{3}\phi'$, $\gamma = 15.4$ kN/m³) (based on Matsuzawa and Hazarika, 1996)

7.9 General Comments on Active Earth Pressure

Matsuzawa and Hazarika (1996) analyzed the active earth pressure against a rigid retaining wall subjected to different modes of movement. Figure 7.19 shows some laboratory test results of the horizontal pressure distribution ($\sigma'_a \cos \delta'$) behind retaining wall rotating about the bottom and top, and also undergoing translation. The backfill was sand for all cases with an average unit weight (γ) of 15.4 kN/m³, $\phi' \approx 34°$, and $\delta' = \frac{2}{3}\phi'$. It is interesting to note how the center of the pressure distribution diagram moves upward (measured from the bottom of the wall) from rotation about the bottom to translation and to the rotation about the top mode. However, the total horizontal active force in all three cases is approximately the same. Figure 7.20 shows the variation of $\bar{z}$ (distance from the resultant active force from the bottom of the wall) for all three modes of wall movement with the soil friction angle ϕ'.

Passive Pressure

7.10 Rankine Passive Earth Pressure

Figure 7.21a shows a vertical frictionless retaining wall with a horizontal backfill. At depth z, the effective vertical pressure on a soil element is $\sigma'_o = \gamma z$. Initially, if the wall does not yield at all, the lateral stress at that depth will be $\sigma'_h = K_o\sigma'_o$. This

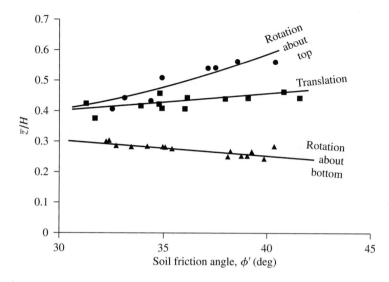

Figure 7.20 Variation of $\overline{z}/H$ for resultant active force (Based on the laboratory experimental results of Matsuzawa and Hazarika, 1996)

state of stress is illustrated by the Mohr's circle *a* in Figure 7.21b. Now, if the wall is pushed into the soil mass by an amount Δx, as shown in Figure 7.21a, the vertical stress at depth *z* will stay the same; however, the horizontal stress will increase. Thus, σ'_h will be greater than $K_o\sigma'_o$. The state of stress can now be represented by the Mohr's circle *b* in Figure 7.21b. If the wall moves farther inward (i.e., Δx is increased still more), the stresses at depth *z* will ultimately reach the state represented by Mohr's circle *c*. Note that this Mohr's circle touches the Mohr–Coulomb failure envelope, which implies that the soil behind the wall will fail by being pushed upward. The horizontal stress, σ'_h, at this point is referred to as the *Rankine passive pressure*, or $\sigma'_h = \sigma'_p$.

For Mohr's circle *c* in Figure 7.21b, the major principal stress is σ'_p, and the minor principal stress is σ'_o. Substituting these quantities into Eq. (1.69) yields

$$\sigma'_p = \sigma'_o \tan^2\left(45 + \frac{\phi'}{2}\right) + 2c'\tan\left(45 + \frac{\phi'}{2}\right) \tag{7.44}$$

Now, let

$$K_p = \text{Rankine passive earth pressure coefficient}$$
$$= \tan^2\left(45 + \frac{\phi'}{2}\right) \tag{7.45}$$

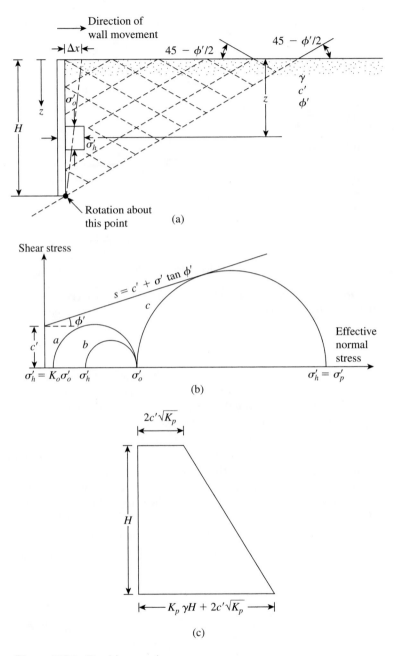

Figure 7.21 Rankine passive pressure

Then, from Eq. (7.44), we have

$$\sigma'_p = \sigma'_o K_p + 2c'\sqrt{K_p} \qquad (7.46)$$

Equation (7.46) produces (Figure 7.21c), the passive pressure diagram for the wall shown in Figure 7.21a. Note that at $z = 0$,

$$\sigma'_o = 0 \quad \text{and} \quad \sigma'_p = 2c'\sqrt{K_p}$$

and at $z = H$,

$$\sigma'_o = \gamma H \quad \text{and} \quad \sigma'_p = \gamma H K_p + 2c'\sqrt{K_p}$$

The passive force per unit length of the wall can be determined from the area of the pressure diagram, or

$$P_p = \tfrac{1}{2}\gamma H^2 K_p + 2c'H\sqrt{K_p} \tag{7.47}$$

The approximate magnitudes of the wall movements, Δx, required to develop failure under passive conditions are as follows:

Soil type	Wall movement for passive condition, Δx
Dense sand	$0.005H$
Loose sand	$0.01H$
Stiff clay	$0.01H$
Soft clay	$0.05H$

If the backfill behind the wall is a granular soil (i.e., $c' = 0$), then, from Eq. (7.47), the passive force per unit length of the wall will be

$$P_p = \frac{1}{2}\gamma H^2 K_p \tag{7.48}$$

The passive force on a *frictionless inclined* retaining wall (see Figure 7.22) with a horizontal granular backfill ($c' = 0$) can also be expressed by Eq. (7.48). The variation of K_p for this case, with wall inclination θ and effective soil friction angle ϕ', is given in Table 7.8 (Zhu and Qian, 2000).

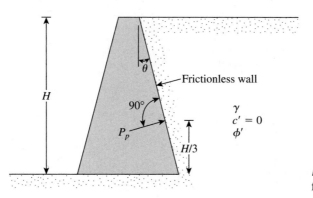

Figure 7.22 Passive force on a frictionless inclined retaining wall

Table 7.8 Variation of K_p [see Eq. (7.48) and Figure 7.22]*

ϕ' (deg)	θ (deg)						
	30	25	20	15	10	5	0
20	1.70	1.69	1.72	1.77	1.83	1.92	2.04
21	1.74	1.73	1.76	1.81	1.89	1.99	2.12
22	1.77	1.77	1.80	1.87	1.95	2.06	2.20
23	1.81	1.81	1.85	1.92	2.01	2.13	2.28
24	1.84	1.85	1.90	1.97	2.07	2.21	2.37
25	1.88	1.89	1.95	2.03	2.14	2.28	2.46
26	1.91	1.93	1.99	2.09	2.21	2.36	2.56
27	1.95	1.98	2.05	2.15	2.28	2.45	2.66
28	1.99	2.02	2.10	2.21	2.35	2.54	2.77
29	2.03	2.07	2.15	2.27	2.43	2.63	2.88
30	2.07	2.11	2.21	2.34	2.51	2.73	3.00
31	2.11	2.16	2.27	2.41	2.60	2.83	3.12
32	2.15	2.21	2.33	2.48	2.68	2.93	3.25
33	2.20	2.26	2.39	2.56	2.77	3.04	3.39
34	2.24	2.32	2.45	2.64	2.87	3.16	3.53
35	2.29	2.37	2.52	2.72	2.97	3.28	3.68
36	2.33	2.43	2.59	2.80	3.07	3.41	3.84
37	2.38	2.49	2.66	2.89	3.18	3.55	4.01
38	2.43	2.55	2.73	2.98	3.29	3.69	4.19
39	2.48	2.61	2.81	3.07	3.41	3.84	4.38
40	2.53	2.67	2.89	3.17	3.53	4.00	4.59
41	2.59	2.74	2.97	3.27	3.66	4.16	4.80
42	2.64	2.80	3.05	3.38	3.80	4.34	5.03
43	2.70	2.88	3.14	3.49	3.94	4.52	5.27
44	2.76	2.94	3.23	3.61	4.09	4.72	5.53
45	2.82	3.02	3.32	3.73	4.25	4.92	5.80

*Based on Zhu and Qian, 2000

Example 7.7

A 3-m high wall is shown in Figure 7.23a. Determine the Rankine passive force per unit length of the wall.

Solution
For the top layer

$$K_{p(1)} = \tan^2\left(45 + \frac{\phi_1'}{2}\right) = \tan^2(45 + 15) = 3$$

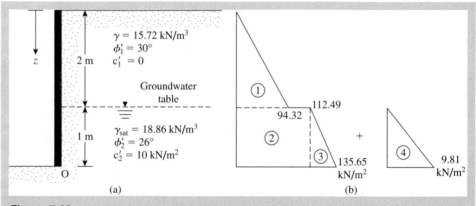

Figure 7.23

From the bottom soil layer

$$K_{p(2)} = \tan^2\left(45 + \frac{\phi_2'}{2}\right) = \tan^2(45 + 13) = 2.56$$

$$\sigma_p' = \sigma_O'K_p + 2c'\sqrt{K_p}$$

where

σ_O' = effective vertical stress
at $z = 0$, $\sigma_O' = 0$, $c_1' = 0$, $\sigma_p' = 0$
at $z = 2$ m, $\sigma_O' = (15.72)(2) = 31.44$ kN/m², $c_1' = 0$

So, for the top soil layer

$$\sigma_p' = 31.44 K_{p(1)} + 2(0)\sqrt{K_{p(1)}} = 31.44(3) = 94.32 \text{ kN/m}^2$$

At this depth, that is $z = 2$ m, for the bottom soil layer

$$\sigma_p' = \sigma_O'K_{p(2)} + 2c_2'\sqrt{K_{p(2)}} = 31.44(2.56) + 2(10)\sqrt{2.56}$$
$$= 80.49 + 32 = 112.49 \text{ kN/m}^2$$

Again, at $z = 3$ m,

$$\sigma_O' = (15.72)(2) + (\gamma_{\text{sat}} - \gamma_w)(1)$$
$$= 31.44 + (18.86 - 9.81)(1) = 40.49 \text{ kN/m}^2$$

Hence,

$$\sigma_p' = \sigma_O'K_{p(2)} + 2c_2'\sqrt{K_{p(2)}} = 40.49(2.56) + (2)(10)(1.6)$$
$$= \mathbf{135.65 \text{ kN/m}^2}$$

Note that, because a water table is present, the hydrostatic stress, u, also has to be taken into consideration. For $z = 0$ to 2 m, $u = 0$; $z = 3$ m, $u = (1)(\gamma_w) = 9.81$ kN/m^2.

The passive pressure diagram is plotted in Figure 6.26b. The passive force per unit length of the wall can be determined from the area of the pressure diagram as follows:

Area No.	Area	
1	$(\frac{1}{2})(2)(94.32)$	= 94.32
2	$(112.49)(1)$	= 112.49
3	$(\frac{1}{2})(1)(135.65 - 112.49)$	= 11.58
4	$(\frac{1}{2})(9.81)(1)$	= 4.905
		$P_P \approx 223.3$ **kN/m**

■

7.11 Rankine Passive Earth Pressure: Inclined Backfill

Granular Soil

For a frictionless vertical retaining wall (Figure 7.8) with a *granular backfill* ($c' = 0$), the Rankine passive pressure at any depth can be determined in a manner similar to that done in the case of active pressure in Section 7.4. The pressure is

$$\sigma_p' = \gamma z K_p \tag{7.49}$$

and the passive force is

$$P_p = \tfrac{1}{2}\gamma H^2 K_p \tag{7.50}$$

where

$$K_p = \cos\alpha \frac{\cos\alpha + \sqrt{\cos^2\alpha - \cos^2\phi'}}{\cos\alpha - \sqrt{\cos^2\alpha - \cos^2\phi'}} \tag{7.51}$$

As in the case of the active force, the resultant force, P_p, is inclined at an angle α with the horizontal and intersects the wall at a distance $H/3$ from the bottom of the wall. The values of K_p (the passive earth pressure coefficient) for various values of α and ϕ' are given in Table 7.9.

Table 7.9 Passive Earth Pressure Coefficient, K_p [from Eq. 7.51)]

	ϕ' (deg)→						
↓α (deg)	28	30	32	34	36	38	40
0	2.770	3.000	3.255	3.537	3.852	4.204	4.599
5	2.715	2.943	3.196	3.476	3.788	4.136	4.527
10	2.551	2.775	3.022	3.295	3.598	3.937	4.316
15	2.284	2.502	2.740	3.003	3.293	3.615	3.977
20	1.918	2.132	2.362	2.612	2.886	3.189	3.526
25	1.434	1.664	1.894	2.135	2.394	2.676	2.987

Table 7.10 Values of K'_p

ϕ' (deg)	α (deg)	$c'/\gamma z$			
		0.025	0.050	0.100	0.500
15	0	1.764	1.829	1.959	3.002
	5	1.716	1.783	1.917	2.971
	10	1.564	1.641	1.788	2.880
	15	1.251	1.370	1.561	2.732
20	0	2.111	2.182	2.325	3.468
	5	2.067	2.140	2.285	3.435
	10	1.932	2.010	2.162	3.339
	15	1.696	1.786	1.956	3.183
25	0	2.542	2.621	2.778	4.034
	5	2.499	2.578	2.737	3.999
	10	2.368	2.450	2.614	3.895
	15	2.147	2.236	2.409	3.726
30	0	3.087	3.173	3.346	4.732
	5	3.042	3.129	3.303	4.674
	10	2.907	2.996	3.174	4.579
	15	2.684	2.777	2.961	4.394

$c'-\phi'$ Soil

If the backfill of the frictionless vertical retaining wall is a $c'-\phi'$ soil (see Figure 7.8), then (Mazindrani and Ganjali, 1997)

$$\sigma'_a = \gamma z K_p = \gamma z K'_p \cos \alpha \tag{7.52}$$

where

$$K'_p = \frac{1}{\cos^2 \phi'} \left\{ \frac{2\cos^2 \alpha + 2\left(\dfrac{c'}{\gamma z}\right)\cos \phi' \sin \phi'}{+ \sqrt{4\cos^2 \alpha(\cos^2 \alpha - \cos^2 \phi') + 4\left(\dfrac{c'}{\gamma z}\right)^2 \cos^2 \phi' + 8\left(\dfrac{c'}{\gamma z}\right)\cos^2 \alpha \sin \phi' \cos \phi'}} - 1 \right\} \tag{7.53}$$

The variation of K'_p with ϕ', α, and $c'/\gamma z$ is given in Table 7.10 (Mazindrani and Ganjali, 1997).

7.12 Coulomb's Passive Earth Pressure

Coulomb (1776) also presented an analysis for determining the passive earth pressure (i.e., when the wall moves *into* the soil mass) for walls possessing friction (δ' = angle of wall friction) and retaining a granular backfill material similar to that discussed in Section 7.5.

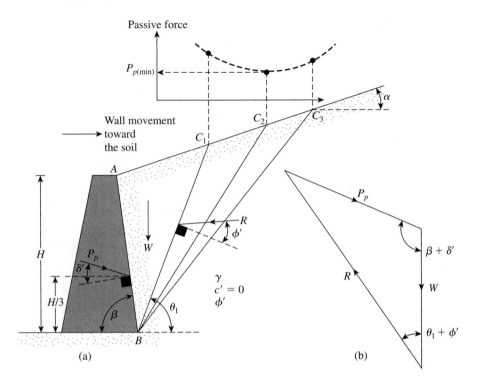

Figure 7.24 Coulomb's passive pressure

To understand the determination of Coulomb's passive force, P_p, consider the wall shown in Figure 7.24a. As in the case of active pressure, Coulomb assumed that the potential failure surface in soil is a plane. For a trial failure wedge of soil, such as ABC_1, the forces per unit length of the wall acting on the wedge are

1. The weight of the wedge, W
2. The resultant, R, of the normal and shear forces on the plane BC_1, and
3. The passive force, P_p

Figure 7.24 shows the force triangle at equilibrium for the trial wedge ABC_1. From this force triangle, the value of P_p can be determined, because the direction of all three forces and the magnitude of one force are known.

Similar force triangles for several trial wedges, such as ABC_1, ABC_2, ABC_3, ..., can be constructed, and the corresponding values of P_p can be determined. The top part of Figure 7.24 shows the nature of variation of the P_p values for different wedges. The *minimum value of P_p* in this diagram is *Coulombs passive force,* mathematically expressed as

$$P_p = \tfrac{1}{2}\gamma H^2 K_p \tag{7.54}$$

Table 7.11 Values of K_p [from Eq. (7.55)] for $\beta = 90°$ and $\alpha = 0°$

ϕ' (deg)	δ' (deg)				
	0	5	10	15	20
15	1.698	1.900	2.130	2.405	2.735
20	2.040	2.313	2.636	3.030	3.525
25	2.464	2.830	3.286	3.855	4.597
30	3.000	3.506	4.143	4.977	6.105
35	3.690	4.390	5.310	6.854	8.324
40	4.600	5.590	6.946	8.870	11.772

where

$$K_p = \text{Coulomb's passive pressure coefficient}$$

$$= \frac{\sin^2(\beta - \phi')}{\sin^2\beta \sin(\beta + \delta')\left[1 - \sqrt{\dfrac{\sin(\phi' + \delta')\sin(\phi' + \alpha)}{\sin(\beta + \delta')\sin(\beta + \alpha)}}\right]^2} \tag{7.55}$$

The values of the passive pressure coefficient, K_p, for various values of ϕ' and δ' are given in Table 7.11 ($\beta = 90°, \alpha = 0°$).

Note that the resultant passive force, P_p, will act at a distance $H/3$ from the bottom of the wall and will be inclined at an angle δ to the normal drawn to the back face of the wall.

7.13 *Comments on the Failure Surface Assumption for Coulomb's Pressure Calculations*

Coulomb's pressure calculation methods for active and passive pressure have been discussed in Sections 7.5 and 7.12. The fundamental assumption in these analyses is the acceptance of *plane failure surface*. However, for walls with friction, this assumption does not hold in practice. The nature of *actual* failure surface in the soil mass for active and passive pressure is shown in Figure 7.25a and b, respectively (for a vertical wall with a horizontal backfill). Note that the failure surface *BC* is curved and that the failure surface *CD* is a plane.

Although the actual failure surface in soil for the case of active pressure is somewhat different from that assumed in the calculation of the Coulomb pressure, the results are not greatly different. However, in the case of passive pressure, as the value of δ' increases, Coulomb's method of calculation gives increasingly erroneous values of P_p. This factor of error could lead to an unsafe condition because the values of P_p would become higher than the soil resistance.

Several studies have been conducted to determine the passive force P_p, assuming that the curved portion *BC* in Figure 7.25b is an arc of a circle, an ellipse, or a logarithmic spiral.

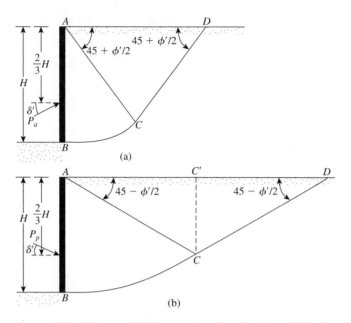

Figure 7.25 Nature of failure surface in soil with wall friction: (a) active pressure; (b) passive pressure

Shields and Tolunay (1973) analyzed the problem of passive pressure for a *vertical* wall with a *horizontal granular soil* backfill ($c' = 0$). This analysis was done by considering the stability of the wedge $ABCC'$ (see Figure 7.25b), using the *method of slices.* From Figure 7.25b, the passive force per unit length of the wall can be expressed as

$$P_p = \frac{1}{2}K_p\gamma H^2$$

The values of the passive earth-pressure coefficient, K_p, obtained by Shields and Tolunay are given in Figure 7.26. These are as good as any for design purposes.

7.14   *Passive Pressure under Earthquake Conditions*

The relationship for passive earth pressure on a retaining wall with a granular backfill and under earthquake conditions was evaluated by Subba Rao and Choudhury (2005) by the method of limit equilibrium using the pseudo-static approach. Figure 7.27 shows the nature of failure surface in soil considered in this analysis. The passive pressure, P_{pe}, can be expressed as

$$P_{pe} = [\tfrac{1}{2}\gamma H^2 K_{p\gamma(e)}]\frac{1}{\cos\delta'} \tag{7.56}$$

where $K_{p\gamma(e)}$ = passive earth-pressure coefficient in the normal direction to the wall.

$K_{p\gamma(e)}$ is a function of k_h and k_v that are, respectively, coefficient of horizontal and vertical acceleration due to earthquake. The variations of $K_{p\gamma(e)}$ for $\delta'/\phi' = 0.5$ and 1 are shown in Figures 7.28a and b. The passive pressure P_{pe} will be inclined at an angle δ' to the back face of the wall and will act at a distance of $H/3$ above the bottom of the wall.

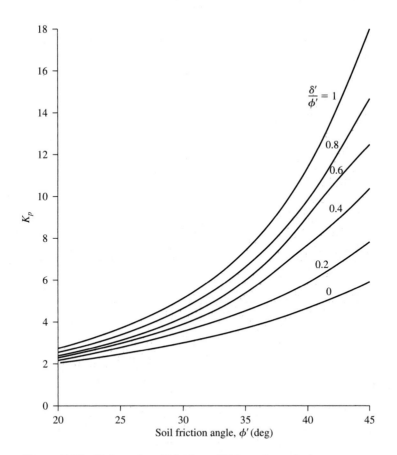

Figure 7.26 K_p based on Shields and Tolunay's analysis

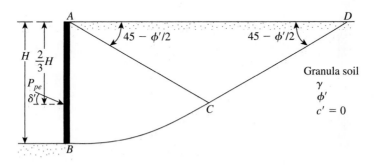

Figure 7.27   Nature of failure surface in soil considered in the analysis to determine P_{pe}

Problems

7.1   Refer to Figure 7.3a. Given: $H = 12$ ft, $q = 0$, $\gamma = 108 \, \text{lb/ft}^3$, $c' = 0$, and $\phi' = 30°$. Determine the at-rest lateral earth force per foot length of the wall. Also, find the location of the resultant. Use Eq. (7.4) and OCR = 2.

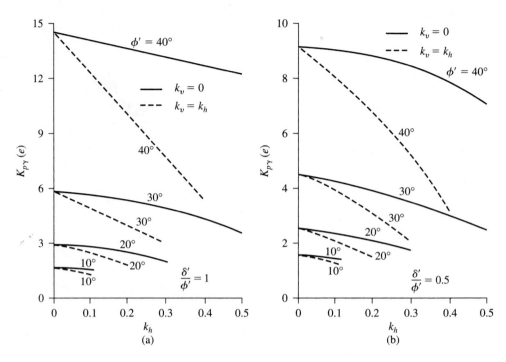

Figure 7.28   Variation of $K_{p\gamma(e)}$: (a) $\dfrac{\delta'}{\phi'} = 1$, (b) $\dfrac{\delta'}{\phi'} = 0.5$

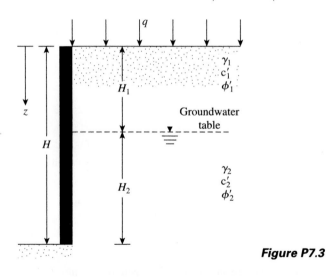

Figure P7.3

7.2   Repeat Problem 7.1 with the following: $H = 3.5$ m, $q = 20$ kN/m², $\gamma = 18.2$ kN/m³, $c' = 0$, $\phi' = 35°$, and OCR = 1.5.

7.3   Use Eq. (7.3), Figure P7.3, and the following values to determine the at-rest lateral earth force per unit length of the wall. Also find the location of the resultant. $H = 5$ m, $H_1 = 2$ m, $H_2 = 3$ m, $\gamma = 15.5$ kN/m³, $\gamma_{\text{sat}} = 18.5$ kN/m³, $\phi' = 34°$, $c' = 0$, $q = 20$ kN/m², and OCR = 1.

7.4   Refer to Figure 7.5a. Given the height of the retaining wall, H is 18 ft; the backfill is a saturated clay with $\phi' = 0°, c = 500$ lb/ft^2, $\gamma_{sat} = 120$ lb/ft^3,
   a. Determine the Rankine active pressure distribution diagram behind the wall.
   b. Determine the depth of the tensile crack, z_c.
   c. Estimate the Rankine active force per foot length of the wall before and after the occurrence of the tensile crack.

7.5   A vertical retaining wall (Figure 7.5a) is 6.3 m high with a horizontal backfill. For the backfill, assume that $\gamma = 17.9$ kN/m^3, $\phi' = 26°$, and $c' = 15$ kN/m^2, Determine the Rankine active force per unit length of the wall after the occurrence of the tensile crack.

7.6   Refer to Problem 7.3. For the retaining wall, determine the Rankine active force per unit length of the wall and the location of the line of action of the resultant.

7.7   In Figure P7.3., let $H_1 = 8.2$ ft, $H_2 = 14.8$ ft, $\gamma_1 = 107$ lb/ft^3, $q = 0$, $\phi_1' = 34°$, $c_1' = 0$, $\gamma_2 = 140$ lb/ft^3, $\phi' = 25°$, and $c' = 209$ lb/ft^2. Determine the Rankine active force per unit length of the wall.

7.8   Refer to Figure 7.8. For the retaining wall, $H = 7.5$ m, $\phi' = 32°$, $\alpha = 5°$, $\gamma = 18.2$ kN/m^3, and $c' = 0$.
   a. Determine the intensity of the Rankine active force at $z = 2$ m, 4 m, 6 m and 7.5 m.
   b. Determine the Rankine active force per meter length of the wall and also the location and direction of the resultant.

7.9   Refer to Figure 7.8. Given: $H = 22$ ft $\gamma = 115$ lb/ft^3, $\phi' = 25°, c' = 250$ lb/ft^2, and $\alpha = 10°$. Calculate the Rankine active force per unit length of the wall after the occurrence of the tensile crack.

7.10   Refer to Figure 7.10a. Given: $H = 12$ ft, $\gamma = 105$ lb/ft^3, $\phi' = 30°, c' = 0$, and $\beta = 85°$. Determine the Coulomb's active force per foot length of the wall and the location and direction of the resultant for the following cases:
   a. $\alpha = 10°$ and $\delta' = 20°$
   b. $\alpha = 20°$ and $\delta' = 15°$

7.11   Refer to Figure 7.11. Here, $H = 5$ m, $\gamma = 18.2$ kN/m^3, $\phi' = 30°, \delta' = 20°, c' = 0$, $\alpha = 10°$, and $\beta = 85°$. Determine the Coulomb's active force for earthquake conditions (P_{ae}) per meter length of the wall and the location and direction of the resultant. Given $k_h = 0.2$ and $k_v = 0$.

7.12   A retaining wall is shown in Figure P7.12. If the wall rotates about its top, determine the magnitude of the active force per unit length of the wall for $n_a = 0.3, 0.4$, and 0.5. Assume $\delta'/\phi' = 0.5$.

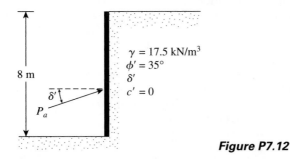

8 m

$\gamma = 17.5$ kN/m^3
$\phi' = 35°$
δ'
$c' = 0$

$\overline{\delta'}$
P_a

Figure P7.12

7.13 A vertical frictionless retaining wall is 6-m high with a horizontal granular backfill. Given: $\gamma = 16$ kN/m^3 and $\phi' = 30°$. For the translation mode of the wall, calculate the active pressure at depths $z = 1.5$ m, 3 m, 4.5 m, and 6 m.

7.14 Refer to Problem 7.4.
   a. Draw the Rankine passive pressure distribution diagram behind the wall.
   b. Estimate the Rankine passive force per foot length of the wall and also the location of the resultant.

7.15–7.16 Use Figure P7.3 and the following data to determine the Rankine passive force per unit length of the wall.

Prob.	H_1	H_2	γ_1	γ_2	ϕ_1'	ϕ_2'	c_1'	c_2'
7.15	8 ft	16 ft	110 lb/ft^3	140 lb/ft^3	38°	25°	0	209 lb/ft^2
7.16	8.2 ft	14.8 ft	107 lb/ft^3	125 lb/ft^3	28°	20°	350 lb/ft^2	100 lb/ft^2

7.17 In Figure 7.25b, which shows a vertical retaining wall with a horizontal backfill, let $H = 4$ m, $\gamma = 16.5$ kN/m^3, $\phi' = 35°$, and $\delta' = 10°$. Based on Shields and Tolunay's work (see Figure 7.26), what would be the passive force per meter length of the wall?

7.18 Consider a 4-m high retaining wall with a vertical back and horizontal granular backfill, as shown in Figure 7.27. Given: $\gamma = 18$ kN/m^3, $\phi' = 40°$, $c' = 0$, $\delta' = 20°$, $k_v = 0$ and $k_h = 0.2$. Determine the passive force P_{pe} per unit length of the wall taking the earthquake effect into consideration.

References

Chu, S. C. (1991). "Rankine Analysis of Active and Passive Pressures on Dry Sand," *Soils and Foundations,* Vol. 31, No. 4, pp. 115–120.

Coulomb, C. A. (1776). *Essai sur une Application des Règles de Maximis et Minimum à quelques Problemes de Statique Relatifs à l'Architecture,* Mem. Acad. Roy. des Sciences, Paris, Vol. 3, p. 38.

Dubrova, G. A. (1963). "Interaction of Soil and Structures," Izd. *Rechnoy Transport,* Moscow.

Harr, M. E. (1966). *Fundamentals of Theoretical Soil Mechanics,* McGraw-Hill, New York.

Kim, J. S., and Preber, T. (1969). "Earth Pressure against Braced Excavations," *Journal of the Soil Mechanics and Foundations Division,* ASCE, Vol. 96, No. 6, pp. 1581–1584.

Matsuzawa, H., and Hazarika, H. (1996). "Analysis of Active Earth Pressure Against Rigid Retaining Wall Subjected to Different Modes of Movement," *Soils and Foundation,* Tokyo, Japan, Vol. 36, No. 3, pp. 51–66.

Mayne, P. W., and Kulhawy, F. H. (1982). "K_o–OCR Relationships in Soil," *Journal of the Geotechnical Engineering Division,* ASCE, Vol. 108, No. GT6, pp. 851–872.

Mazindrani, Z. H., and Ganjali, M. H. (1997). "Lateral Earth Pressure Problem of Cohesive Backfill with Inclined Surface," *Journal of Geotechnical and Geoenvironmental Engineering,* ASCE, Vol. 123, No. 2, pp. 110–112.

Shields, D. H., and Tolunay, A. Z. (1973). "Passive Pressure Coefficients by Method of Slices," *Journal of the Soil Mechanics and Foundations Division,* ASCE, Vol. 99, No. SM12, 1043–1053.

Subba Rao, K. S., and Choudhury, D. (2005). "Seismic Passive Earth Pressures in Soil," *Journal of Geotechnical and Geoenvironmental Engineering,* American Society of Civil Engineers, Vo. 131, No. 1, pp. 131–135.

Terzaghi, K. (1943). *Theoretical Soil Mechanics,* Wiley, New York.

Zhu, D. Y., and Qian, Q. (2000). "Determination of Passive Earth Pressure Coefficient by the Method of Triangular Slices," *Canadian Geotechnical Journal,* Vol. 37, No. 2, pp. 485–491.

8

Retaining Walls

8.1 Introduction

In Chapter 7, you were introduced to various theories of lateral earth pressure. Those theories will be used in this chapter to design various types of retaining walls. In general, retaining walls can be divided into two major categories: (a) conventional retaining walls and (b) mechanically stabilized earth walls.

Conventional retaining walls can generally be classified into four varieties:

1. Gravity retaining walls
2. Semigravity retaining walls
3. Cantilever retaining walls
4. Counterfort retaining walls

Gravity retaining walls (Figure 8.1a) are constructed with plain concrete or stone masonry. They depend for stability on their own weight and any soil resting on the masonry. This type of construction is not economical for high walls.

In many cases, a small amount of steel may be used for the construction of gravity walls, thereby minimizing the size of wall sections. Such walls are generally referred to as *semigravity walls* (Figure 8.1b).

Cantilever retaining walls (Figure 8.1c) are made of reinforced concrete that consists of a thin stem and a base slab. This type of wall is economical to a height of about 8 m (25 ft).

Counterfort retaining walls (Figure 8.1d) are similar to cantilever walls. At regular intervals, however, they have thin vertical concrete slabs known as *counterforts* that tie the wall and the base slab together. The purpose of the counterforts is to reduce the shear and the bending moments.

To design retaining walls properly, an engineer must know the basic parameters—the *unit weight, angle of friction,* and *cohesion*—of the soil retained behind the wall and the soil below the base slab. Knowing the properties of the soil behind the wall enables the engineer to determine the lateral pressure distribution that has to be designed for.

There are two phases in the design of a conventional retaining wall. First, with the lateral earth pressure known, the structure as a whole is checked for *stability:* The

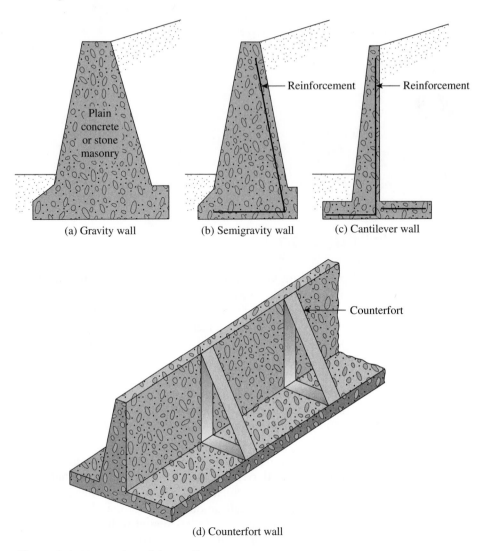

(a) Gravity wall   (b) Semigravity wall   (c) Cantilever wall

(d) Counterfort wall

Figure 8.1 Types of retaining wall

structure is examined for possible *overturning, sliding,* and *bearing capacity* failures. Second, each component of the structure is checked for *strength,* and the *steel reinforcement* of each component is determined.

This chapter presents the procedures for determining the stability of the retaining wall. Checks for strength can be found in any textbook on reinforced concrete.

Some retaining walls have their backfills stabilized mechanically by including reinforcing elements such as metal strips, bars, welded wire mats, geotextiles, and geogrids. These walls are relatively flexible and can sustain large horizontal and vertical displacements without much damage.

Gravity and Cantilever Walls

Proportioning Retaining Walls

In designing retaining walls, an engineer must assume some of their dimensions. Called *proportioning,* such assumptions allow the engineer to check trial sections of the walls for stability. If the stability checks yield undesirable results, the sections can be changed and rechecked. Figure 8.2 shows the general proportions of various retaining-wall components that can be used for initial checks.

Note that the top of the stem of any retaining wall should not be less than about 0.3 m. ($\approx$12 in.) for proper placement of concrete. The depth, D, to the bottom of the base slab should be a minimum of 0.6 m($\approx$2 ft). However, the bottom of the base slab should be positioned below the seasonal frost line.

For counterfort retaining walls, the general proportion of the stem and the base slab is the same as for cantilever walls. However, the counterfort slabs may be about 0.3 m ($\approx$12 in.) thick and spaced at center-to-center distances of $0.3H$ to $0.7H$.

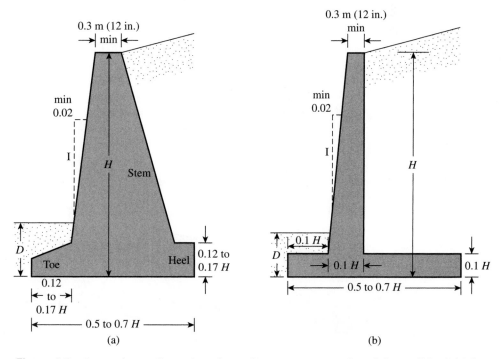

Figure 8.2 Approximate dimensions for various components of retaining wall for initial stability checks: (a) gravity wall; (b) cantilever wall

Application of Lateral Earth Pressure Theories to Design

The fundamental theories for calculating lateral earth pressure were presented in Chapter 7. To use these theories in design, an engineer must make several simple assumptions. In the case of cantilever walls, the use of the Rankine earth pressure theory for stability checks involves drawing a vertical line AB through point A, located at the edge of the heel of the base slab in Figure 8.3a. The Rankine active condition is assumed to exist along the vertical plane AB. Rankine active earth pressure equations may then be used to calculate the lateral pressure on the face AB of the wall. In the analysis of the wall's stability, the force $P_{a(\text{Rankine})}$, the weight of soil above the heel, and the weight W_c of the concrete all should be taken into consideration. The assumption for the development of Rankine active pressure along the soil face AB is theoretically correct if the shear zone bounded by the line AC is not obstructed by the stem of the wall. The angle, η, that the line AC makes with the vertical is

$$\eta = 45 + \frac{\alpha}{2} - \frac{\phi'}{2} - \sin^{-1}\left(\frac{\sin\alpha}{\sin\phi'}\right) \qquad (8.1)$$

A similar type of analysis may be used for gravity walls, as shown in Figure 8.3b. However, *Coulomb's active earth pressure theory* also may be used, as shown in Figure 8.3c. If it is used, the only forces to be considered are $P_{a(\text{Coulomb})}$ and the weight of the wall, W_c.

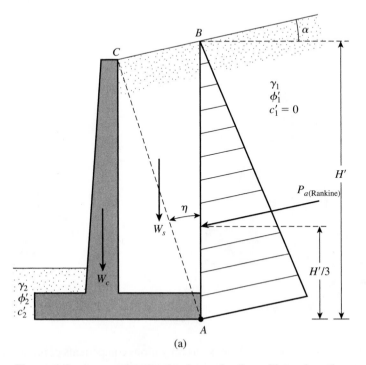

(a)

Figure 8.3 Assumption for the determination of lateral earth pressure: (a) cantilever wall; (b) and (c) gravity wall

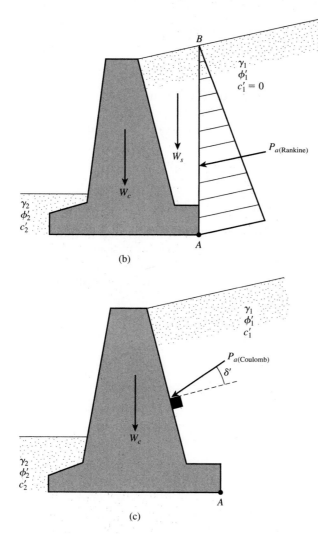

(b)

(c)

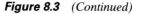

Figure 8.3   *(Continued)*

If Coulomb's theory is used, it will be necessary to know the range of the wall friction angle δ' with various types of backfill material. Following are some ranges of wall friction angle for masonry or mass concrete walls:

Backfill material	Range of δ' (deg)
Gravel	27–30
Coarse sand	20–28
Fine sand	15–25
Stiff clay	15–20
Silty clay	12–16

In the case of ordinary retaining walls, water table problems and hence hydrostatic pressure are not encountered. Facilities for drainage from the soils that are retained are always provided.

8.4   *Stability of Retaining Walls*

A retaining wall may fail in any of the following ways:

- It may *overturn* about its toe. (See Figure 8.4a.)
- It may *slide* along its base. (See Figure 8.4b.)
- It may fail due to the loss of *bearing capacity* of the soil supporting the base. (See Figure 8.4c.)
- It may undergo deep-seated shear failure. (See Figure 8.4d.)
- It may go through excessive settlement.

The checks for stability against overturning, sliding, and bearing capacity failure will be described in Sections 8.5, 8.6, and 8.7. The principles used to estimate settlement were covered in Chapter 5 and will not be discussed further. When a weak soil layer is located at a shallow depth—that is, within a depth of 1.5 times the width of the base slab of the retaining wall—the possibility of excessive settlement should be considered. In some cases, the use of lightweight backfill material behind the retaining wall may solve the problem.

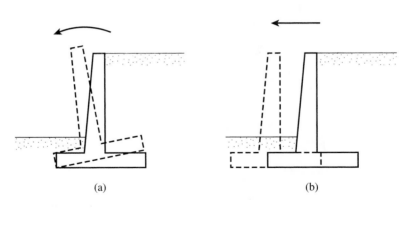

(a)   (b)

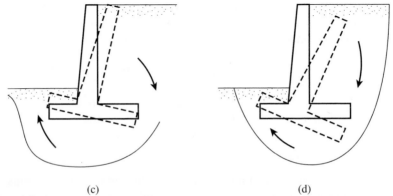

(c)   (d)

Figure 8.4   Failure of retaining wall: (a) by overturning; (b) by sliding; (c) by bearing capacity failure; (d) by deep-seated shear failure

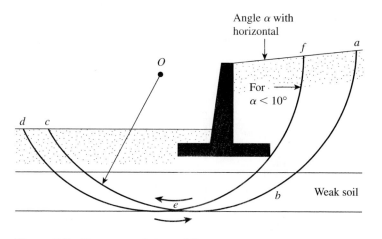

Figure 8.5   Deep-seated shear failure

Deep shear failure can occur along a cylindrical surface, such as *abc* shown in Figure 8.5, as a result of the existence of a weak layer of soil underneath the wall at a depth of about 1.5 times the width of the base slab of the retaining wall. In such cases, the critical cylindrical failure surface *abc* has to be determined by trial and error, using various centers such as *O*. The failure surface along which the minimum factor of safety is obtained is the *critical surface of sliding*. For the backfill slope with α less than about 10°, the critical failure circle apparently passes through the edge of the heel slab (such as *def* in the figure). In this situation, the minimum factor of safety also has to be determined by trial and error by changing the center of the trial circle.

8.5   *Check for Overturning*

Figure 8.6 shows the forces acting on a cantilever and a gravity retaining wall, based on the assumption that the Rankine active pressure is acting along a vertical plane *AB* drawn through the heel of the structure. P_p is the Rankine passive pressure; recall that its magnitude is [from Eq. (7.47)].

$$P_p = \tfrac{1}{2}K_p\gamma_2 D^2 + 2c_2'\sqrt{K_p}D$$

where

γ_2 = unit weight of soil in front of the heel and under the base slab
K_p = Rankine passive earth pressure coefficient = $\tan^2(45 + \phi_2'/2)$
c_2', ϕ_2' = cohesion and effective soil friction angle, respectively

The factor of safety against overturning about the toe—that is, about point *C* in Figure 8.6—may be expressed as

$$\text{FS}_{(\text{overturning})} = \frac{\Sigma M_R}{\Sigma M_O} \tag{8.2}$$

where

ΣM_O = sum of the moments of forces tending to overturn about point *C*
ΣM_R = sum of the moments of forces tending to resist overturning about point *C*

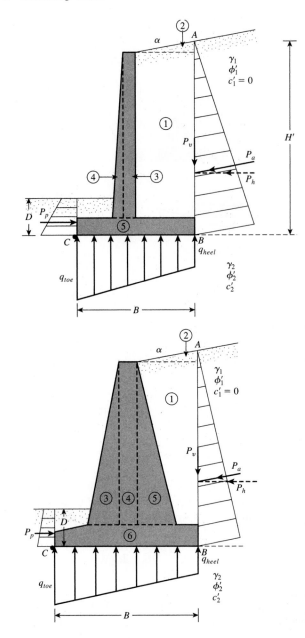

Figure 8.6 Check for overturning, assuming that the Rankine pressure is valid

The overturning moment is

$$\Sigma M_O = P_h\left(\frac{H'}{3}\right) \tag{8.3}$$

where $P_h = P_a \cos\alpha$.

Table 8.1   Procedure for Calculating $\Sigma\, M_R$

Section (1)	Area (2)	Weight/unit length of wall (3)	Moment arm measured from C (4)	Moment about C (5)
1	A_1	$W_1 = \gamma_1 \times A_1$	X_1	M_1
2	A_2	$W_2 = \gamma_2 \times A_2$	X_2	M_2
3	A_3	$W_3 = \gamma_c \times A_3$	X_3	M_3
4	A_4	$W_4 = \gamma_c \times A_4$	X_4	M_4
5	A_5	$W_5 = \gamma_c \times A_5$	X_5	M_5
6	A_6	$W_6 = \gamma_c \times A_6$	X_6	M_6
		P_v	B	M_v
		$\Sigma\, V$		$\Sigma\, M_R$

(*Note:* γ_l = unit weight of backfill
γ_c = unit weight of concrete)

To calculate the resisting moment, ΣM_R (neglecting P_p), a table such as Table 8.1 can be prepared. The weight of the soil above the heel and the weight of the concrete (or masonry) are both forces that contribute to the resisting moment. Note that the force P_v also contributes to the resisting moment. P_v is the vertical component of the active force P_a, or

$$P_v = P_a \sin\alpha$$

The moment of the force P_v about C is

$$M_v = P_v B = P_a \sin\alpha B \tag{8.4}$$

where B = width of the base slab.

Once $\Sigma\, M_R$ is known, the factor of safety can be calculated as

$$\text{FS}_{(overturning)} = \frac{M_1 + M_2 + M_3 + M_4 + M_5 + M_6 + M_v}{P_a \cos\alpha(H'/3)} \tag{8.5}$$

The usual minimum desirable value of the factor of safety with respect to overturning is 2 to 3.

Some designers prefer to determine the factor of safety against overturning with the formula

$$\text{FS}_{(overturning)} = \frac{M_1 + M_2 + M_3 + M_4 + M_5 + M_6}{P_a \cos\alpha(H'/3) - M_v} \tag{8.6}$$

8.6   *Check for Sliding along the Base*

The factor of safety against sliding may be expressed by the equation

$$\text{FS}_{(sliding)} = \frac{\Sigma\, F_{R'}}{\Sigma\, F_d} \tag{8.7}$$

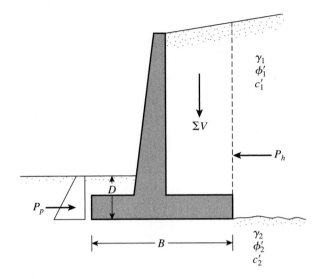

Figure 8.7 Check for sliding along the base

where

$\Sigma F_{R'}$ = sum of the horizontal resisting forces
ΣF_d = sum of the horizontal driving forces

Figure 8.7 indicates that the shear strength of the soil immediately below the base slab may be represented as

$$s = \sigma' \tan \delta' + c'_a$$

where

δ' = angle of friction between the soil and the base slab
c'_a = adhesion between the soil and the base slab

Thus, the maximum resisting force that can be derived from the soil per unit length of the wall along the bottom of the base slab is

$$R' = s(\text{area of cross section}) = s(B \times 1) = B\sigma' \tan \delta' + Bc'_a$$

However,

$$B\sigma' = \text{sum of the vertical force} = \Sigma V (\text{see Table 8.1})$$

so

$$R' = (\Sigma V) \tan \delta' + Bc'_a$$

Figure 8.7 shows that the passive force P_p is also a horizontal resisting force. Hence,

$$\Sigma F_{R'} = (\Sigma V) \tan \delta' + Bc'_a + P_p \tag{8.8}$$

The only horizontal force that will tend to cause the wall to slide (a *driving force*) is the horizontal component of the active force P_a, so

$$\Sigma F_d = P_a \cos \alpha \tag{8.9}$$

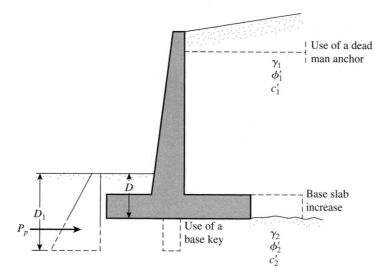

Figure 8.8   Alternatives for increasing the factor of safety with respect to sliding

Combining Eqs. (8.7), (8.8), and (8.9) yields

$$FS_{(sliding)} = \frac{(\Sigma V) \tan \delta' + Bc'_a + P_p}{P_a \cos \alpha} \tag{8.10}$$

A minimum factor of safety of 1.5 against sliding is generally required.

In many cases, the passive force P_p is ignored in calculating the factor of safety with respect to sliding. In general, we can write $\delta' = k_1 \phi'_2$ and $c'_a = k_2 c'_2$. In most cases, k_1 and k_2 are in the range from $\frac{1}{2}$ to $\frac{2}{3}$. Thus,

$$FS_{(sliding)} = \frac{(\Sigma V) \tan (k_1 \phi'_2) + Bk_2 c'_2 + P_p}{P_a \cos \alpha} \tag{8.11}$$

If the desired value of $FS_{(sliding)}$ is not achieved, several alternatives may be investigated (see Figure 8.8):

- Increase the width of the base slab (i.e., the heel of the footing).
- Use a key to the base slab. If a key is included, the passive force per unit length of the wall becomes

$$P_p = \frac{1}{2} \gamma_2 D_1^2 K_p + 2c'_2 D_1 \sqrt{K_p}$$

where $K_p = \tan^2 \left(45 + \frac{\phi'_2}{2} \right)$.

- Use a *deadman anchor* at the stem of the retaining wall.

8.7   *Check for Bearing Capacity Failure*

The vertical pressure transmitted to the soil by the base slab of the retaining wall should be checked against the ultimate bearing capacity of the soil. The nature of variation of the vertical pressure transmitted by the base slab into the soil is shown in Figure 8.9. Note that q_{toe} and q_{heel} are the *maximum* and the *minimum* pressures occurring at the ends of the toe and heel sections, respectively. The magnitudes of q_{toe} and q_{heel} can be determined in the following manner:

The sum of the vertical forces acting on the base slab is ΣV (see column 3 of Table 8.1), and the horizontal force $\mathbf{P}_h$ is $P_a \cos \alpha$. Let

$$\mathbf{R} = \Sigma \mathbf{V} + \mathbf{P}_h \tag{8.12}$$

be the resultant force. The net moment of these forces about point C in Figure 8.9 is

$$M_{net} = \Sigma M_R - \Sigma M_o \tag{8.13}$$

Note that the values of ΣM_R and ΣM_o were previously determined. [See column 5 of Table 8.1 and Eq. (8.3)]. Let the line of action of the resultant R intersect the base slab at E. Then the distance

$$\overline{CE} = \overline{X} = \frac{M_{net}}{\Sigma V} \tag{8.14}$$

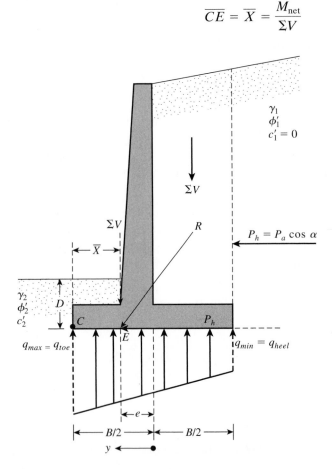

Figure 8.9   Check for bearing capacity failure

Hence, the eccentricity of the resultant R may be expressed as

$$e = \frac{B}{2} - \overline{CE} \tag{8.15}$$

The pressure distribution under the base slab may be determined by using simple principles from the mechanics of materials. First, we have

$$q = \frac{\Sigma V}{A} \pm \frac{M_{net} y}{I} \tag{8.16}$$

where

$M_{net} = $ moment $= (\Sigma V)e$

$\quad I = $ moment of inertia per unit length of the base section

$\quad\quad = \frac{1}{12}(1)(B^3)$

For maximum and minimum pressures, the value of y in Eq. (8.16) equals $B/2$. Substituting into Eq. (8.16) gives

$$q_{max} = q_{toe} = \frac{\Sigma V}{(B)(1)} + \frac{e(\Sigma V)\dfrac{B}{2}}{\left(\dfrac{1}{12}\right)(B^3)} = \frac{\Sigma V}{B}\left(1 + \frac{6e}{B}\right) \tag{8.17}$$

Similarly,

$$q_{min} = q_{heel} = \frac{\Sigma V}{B}\left(1 - \frac{6e}{B}\right) \tag{8.18}$$

Note that ΣV includes the weight of the soil, as shown in Table 8.1, and that when the value of the eccentricity e becomes greater than $B/6$, q_{min} [Eq. (8.18)] becomes negative. Thus, there will be some tensile stress at the end of the heel section. This stress is not desirable, because the tensile strength of soil is very small. If the analysis of a design shows that $e > B/6$, the design should be reproportioned and calculations redone.

The relationships pertaining to the ultimate bearing capacity of a shallow foundation were discussed in Chapter 3. Recall that [Eq. (3.56)].

$$q_u = c_2' N_c F_{cd} F_{ci} + q N_q F_{qd} F_{qi} + \tfrac{1}{2}\gamma_2 B' N_\gamma F_{\gamma d} F_{\gamma i} \tag{8.19}$$

where

$\quad q = \gamma_2 D$

$\quad B' = B - 2e$

$\quad F_{cd} = 1 + 0.4\,\dfrac{D}{B'}$

$\quad F_{qd} = 1 + 2\tan\phi_2'(1 - \sin\phi_2')^2\dfrac{D}{B'}$

$\quad F_{\gamma d} = 1$

$\quad F_{ci} = F_{qi} = \left(1 - \dfrac{\psi^\circ}{90^\circ}\right)^2$

$$F_{\gamma i} = \left(1 - \frac{\psi^\circ}{\phi_2'^\circ}\right)^2$$

$$\psi^\circ = \tan^{-1}\left(\frac{P_a \cos\alpha}{\Sigma V}\right)$$

Note that the shape factors F_{cs}, F_{qs}, and $F_{\gamma s}$ given in Chapter 3 are all equal to unity, because they can be treated as a continuous foundation. For this reason, the shape factors are not shown in Eq. (8.19).

Once the ultimate bearing capacity of the soil has been calculated by using Eq. (8.19), the factor of safety against bearing capacity failure can be determined:

$$FS_{(bearing\ capacity)} = \frac{q_u}{q_{max}} \tag{8.20}$$

Generally, a factor of safety of 3 is required. In Chapter 3, we noted that the ultimate bearing capacity of shallow foundations occurs at a settlement of about 10% of the foundation width. In the case of retaining walls, the width B is large. Hence, the ultimate load q_u will occur at a fairly large foundation settlement. A factor of safety of 3 against bearing capacity failure may not ensure that settlement of the structure will be within the tolerable limit in all cases. Thus, this situation needs further investigation.

An alternate relationship to Eq. (8.19) will be Eq. (3.78), or

$$q_u = c'N_{c(ei)}F_{cd} + qN_{q(ei)}F_{qd} + \tfrac{1}{2}\gamma_2 BN_{\gamma(ei)}F_{\gamma d}$$

Since $F_{\gamma d} = 1$,

$$q_u = c'N_{c(ei)}F_{cd} + qN_{q(ei)}F_{qd} + \tfrac{1}{2}\gamma_2 BN_{\gamma(ei)} \tag{8.21}$$

The bearing capacity factors, $N_{c(ei)}$, $N_{q(ei)}$, and $N_{\gamma(ei)}$ were given in Figures 3.26 through 3.28.

Example 8.1

The cross section of a cantilever retaining wall is shown in Figure 8.10. Calculate the factors of safety with respect to overturning, sliding, and bearing capacity.

Solution
From the figure,

$$H' = H_1 + H_2 + H_3 = 6\tan 10^\circ + 18 + 2.75 = 21.81 \text{ ft}$$

$$P_a = \frac{1}{2}\gamma_1 H'^2 K_a$$

For $\phi_1' = 34^\circ$ and $\alpha = 10^\circ$, the value of K_a is 0.294 (Table 7.1), so

$$P_a = \frac{\frac{1}{2}(117)(21.81)^2(0.294)}{1000} = 8.18 \text{ kip/ft}$$

$$P_v = P_a \sin 10^\circ = 1.42 \text{ kip/ft}$$

$$P_h = P_a \cos 10^\circ = 8.06 \text{ kip/ft}$$

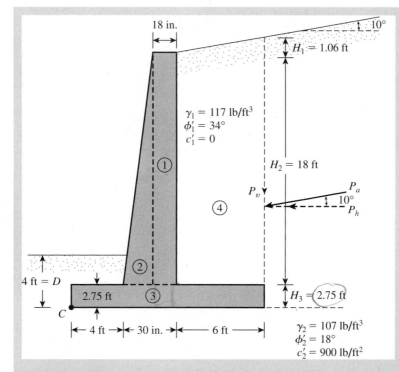

Figure 8.10 Calculation of stability of a retaining wall

Factor of Safety against Overturning

The following table can now be prepared for determination of the resisting moment.

Section	Weight (kip/ft)	Moment arm from C (ft)	Moment about C (kip-ft/ft)
1	$(1.5)(18)(0.15) = 4.05$	5.75	23.29
2	$\frac{1}{2}(1.0)(18)(0.15) = 1.35$	$4 + \frac{2}{3}(1) = 4.67$	6.3
3	$(12.5)(2.75)(0.15) = 5.156$	6.25	32.23
4	$\left(\dfrac{18 + 19.06}{2}\right)(6)(0.117) = 13.01$	$\approx 4 + 2.5 + \dfrac{6}{2} = 9.5$	123.6
	$P_v = 1.42$	12.5	17.75
	$\Sigma V = 24.986$		$\Sigma 203.17 = \Sigma M_R$

(*Note:* $\gamma_{\text{concrete}} = 150 \text{ lb/ft}^3$)

The overturning moment, M_O, is

$$M_O = P_h \frac{H'}{3} = (8.06)\left(\frac{21.81}{3}\right) = 58.6 \text{ kip/ft}$$

So

$$\text{FS}_{(\text{overturning})} = \frac{\Sigma M_R}{M_O} = \frac{203.17}{58.6} = \textbf{3.47} > \textbf{2—OK}$$

Factor of Safety against Sliding
From Eq. (8.11),

$$FS_{(sliding)} = \frac{(\Sigma V)\tan(k_1\phi'_2) + Bk_2c'_2 + P_p}{P_a\cos\alpha}$$

Let $k_1 = k_2 = \frac{2}{3}$ and $P_p = 0$. So

$$FS_{(sliding)} = \frac{(\Sigma V)\tan\left[\frac{2}{3}\phi'_2\right] + B\left(\frac{2}{3}c'_2\right)}{P_a\cos\alpha}$$

$$= \frac{(24.986)\tan\left[\frac{2}{3}(18)\right] + (12.5)\frac{2}{3}(0.9)}{8.06}$$

$$= \mathbf{1.59 > 1.5—OK}$$

Factor of Safety against Bearing Capacity Failure
Combining Eqs. (8.13), (8.14), and (8.15) yields

$$e = \frac{B}{2} - \frac{\Sigma M_R - \Sigma M_o}{\Sigma V} = 6.25 - \frac{203.17 - 58.6}{24.986}$$

$$= 0.464 \text{ ft} < \frac{B}{6} = \frac{12.5}{6} = 2.08 \text{ ft}$$

Again, from Eqs. (8.17) and (8.18),

$$q_{toe} = \frac{\Sigma V}{B}\left(1 + \frac{6e}{B}\right) = \frac{24.986}{12.5}\left[1 + \frac{(6)(0.464)}{12.5}\right] = 2.44 \text{ kip/ft}^2$$

The ultimate bearing capacity of the soil can be determined from Eq. (8.19):

$$q_u = c'_2N_cF_{cd}F_{ci} + qN_qF_{qd}F_{qi} + \frac{1}{2}\gamma_2B'N_\gamma F_{\gamma d}F_{\gamma i}$$

From Table 3.3 for $\phi'_2 = 18°$, $N_c = 13.1$, $N_q = 5.26$, and $N_\gamma = 4.07$,

$$q = \gamma_2 D = (4)(0.107) = 0.428 \text{ kip/ft}^2$$

$$B' = B - 2e = 12.5 - (2)(0.464) = 11.572 \text{ ft}$$

$$F_{cd} = 1 + 0.4\left(\frac{D}{B'}\right) = 1 + 0.4\left(\frac{4}{11.572}\right) = 1.138$$

$$F_{qd} = 1 + 2\tan\phi'_2(1 - \sin\phi'_2)^2\left(\frac{D}{B'}\right) = 1 + (0.31)\left(\frac{4}{11.572}\right) = 1.107$$

$$F_{\gamma d} = 1$$

$$F_{ci} = F_{qi} = \left(1 - \frac{\psi°}{90°}\right)^2$$

$$\psi = \tan^{-1}\left(\frac{P_a \cos\alpha}{\Sigma V}\right) = \left(\frac{8.06}{24.986}\right) = 17.88°.$$

So,

$$F_{ci} = F_{qi} = \left(1 - \frac{17.88}{90}\right)^2 = 0.642$$

$$F_{\gamma i} = \left(1 - \frac{\psi}{\phi_2'}\right)^2 = \left(1 - \frac{17.88}{18}\right)^2 \approx 0$$

Hence,

$$q_u = (0.9)(13.1)(1.138)(0.642) + (0.428)(5.26)(1.107)(0.642)$$
$$+ \tfrac{1}{2}(0.107)(11.572)(4.07)(1)(0)$$
$$= 10.21 \text{ kip/ft}^2$$

$$\text{FS}_{(\text{bearing capacity})} = \frac{q_u}{q_{\text{toe}}} = \frac{10.21}{2.44} = \mathbf{4.18 > 3—OK} \qquad \blacksquare$$

Example 8.2

Consider the retaining wall given in Example 8.1. Calculate the factor of safety with respect to bearing capacity using Eq. (8.21).

Solution
From Eq. (8.21)

$$q_u = c'N_{c(ei)}F_{cd} + qN_{q(ei)}F_{qd} + \tfrac{1}{2}\gamma_2 B N_{\gamma(ei)}$$

From Example 8.1, $\psi = 17.88°$. Note the change of notation. In Figures 3.26 through 3.28, $\beta = \psi$. Also, $e/B = 0.464/12.5 = 0.037$. By interpolation, for $\phi_2' = 18°$, $e/B = 0.037$ and $\psi = 17.88°$.

$$N_{c(ei)} \approx 7.5$$
$$N_{q(ei)} \approx 4$$
$$N_{\gamma(ei)} \approx 0$$

So,

$$q_u = (0.9)(7.5)(1.138) + (0.428)(4)(1.107) = 9.58 \text{ kip/ft}^2$$

$$\text{FS}_{(\text{bearing capacity})} = \frac{9.58}{2.44} = \mathbf{3.93 > 3—OK.} \qquad \blacksquare$$

Example 8.3

A concrete gravity retaining wall is shown in Figure 8.11. Determine:

a. The factor of safety against overturning
b. The factor of safety against sliding
c. The pressure on the soil at the toe and heel
   Use Rankine's earth pressure theory. (*Note:* Unit weight of concrete $= \gamma_c =$ 150 lb/ft³.)

Solution

$$H' = 15 + 2.5 = 17.5 \text{ ft}$$

$$K_a = \tan^2\left(45 - \frac{\phi'_1}{2}\right) = \tan^2\left(45 - \frac{30}{2}\right) = \frac{1}{3}$$

$$P_a = \frac{1}{2}\gamma(H')^2 K_a = \frac{1}{2}(121)(17.5)^2\left(\frac{1}{3}\right) = 6176 \text{ lb/ft}$$

$$= 6.176 \text{ kip/ft}$$

Since $\alpha = 0$

$$P_h = P_a = 6.176 \text{ kip/ft}$$

$$P_v = 0$$

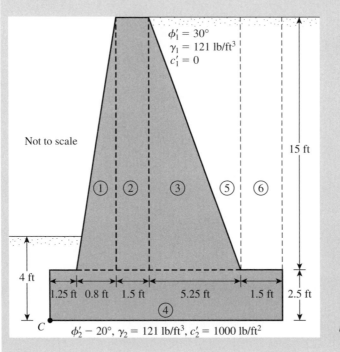

$\phi'_1 = 30°$
$\gamma_1 = 121 \text{ lb/ft}^3$
$c'_1 = 0$

Not to scale

15 ft

① ② ③ ⑤ ⑥

4 ft

1.25 ft   0.8 ft   1.5 ft   5.25 ft   1.5 ft   2.5 ft

④

C

$\phi'_2 - 20°$, $\gamma_2 = 121 \text{ lb/ft}^3$, $c'_2 = 1000 \text{ lb/ft}^2$

Figure 8.11

Part a: Factor of Safety against Overturning

The following table can now be prepared to obtain ΣM_R.

Area (from Figure 8.11)	Weight (kip/ft)	Moment arm from C (ft)	Moment about C (kip-ft/ft)
1	$\frac{1}{2}(0.8)(15)(\gamma_c) = 0.9$	$1.25 + \frac{2}{3}(0.8) = 1.783$	1.605
2	$(1.5)(15)(\gamma_c) = 3.375$	$1.25 + 0.8 + 0.75 = 2.8$	9.45
3	$\frac{1}{2}(5.25)(15)(\gamma_c) = 5.906$	$1.25 + 0.8 + 1.5 + \frac{5.25}{3} = 5.3$	31.30
4	$(10.3)(2.5)(\gamma_c) = 3.863$	$\frac{10.3}{2} = 5.15$	19.89
5	$\frac{1}{2}(5.25)(15)(0.121) = 4.764$	$1.25 + 0.8 + 1.5 + \frac{2}{3}(5.25) = 7.05$	33.59
6	$(1.5)(15)(0.121) = \underline{2.723}$ 21.531	$1.25 + 0.8 + 1.5 + 5.25 + 0.75 = 9.55$	$\underline{26.0}$ $121.84 = M_R$

The overturning moment

$$M_O = \frac{H'}{3}P_a = \left(\frac{17.5}{3}\right)(6.176) = 36.03 \text{ kip/ft}$$

$$FS_{(overturning)} = \frac{121.84}{36.03} = \mathbf{3.38}$$

Part b: Factor of Safety against Sliding

From Eq. (8.11), with $k_1 = k_2 = \frac{2}{3}$ and assuming that $P_p = 0$,

$$FS_{(sliding)} = \frac{\Sigma V \tan\left(\frac{2}{3}\right)\phi_2' + B\left(\frac{2}{3}\right)c_2'}{P_a}$$

$$= \frac{21.531 \tan\left(\frac{2 \times 20}{3}\right) + 10.3\left(\frac{2}{3}\right)(1.0)}{6.176}$$

$$= \frac{5.1 + 6.87}{6.176} = \mathbf{1.94}$$

Part c: Pressure on the Soil at the Toe and Heel

From Eqs. (8.13), (8.14), and (8.15),

$$e = \frac{B}{2} - \frac{\Sigma M_R - \Sigma M_O}{\Sigma V} = \frac{10.3}{2} - \frac{121.84 - 36.03}{21.531} = 5.15 - 3.99 = 1.16 \text{ ft}$$

$$q_{toe} = \frac{\Sigma V}{B}\left[1 + \frac{6e}{B}\right] = \frac{21.531}{10.3}\left[1 + \frac{(6)(1.16)}{10.3}\right] = \mathbf{3.5 \text{ kip/ft}^2}$$

$$q_{heel} = \frac{\Sigma V}{B}\left[1 - \frac{6e}{B}\right] = \frac{21.531}{10.3}\left[1 - \frac{(6)(1.16)}{10.3}\right] = \mathbf{0.678 \text{ kip/ft}^2} \quad \blacksquare$$

Example 8.4

A gravity retaining wall is shown in Figure 8.12. Use $\delta' = 2/3\phi'_1$ and Coulomb's active earth pressure theory. Determine

a. The factor of safety against overturning
b. The factor of safety against sliding
c. The pressure on the soil at the toe and heel

Solution
The height

$$H' = 5 + 1.5 = 6.5 \text{ m}$$

Coulomb's active force is

$$P_a = \tfrac{1}{2}\gamma_1 H'^2 K_a$$

With $\alpha = 0°$, $\beta = 75°$, $\delta' = 2/3\phi'_1$, and $\phi'_1 = 32°$, $K_a = 0.4023$. (See Table 7.4.) So,

$$P_a = \tfrac{1}{2}(18.5)\,(6.5)^2(0.4023) = 157.22 \text{ kN/m}$$

$$P_h = P_a \cos\left(15 + \tfrac{2}{3}\phi'_1\right) = 157.22 \cos 36.33 = 126.65 \text{ kN/m}$$

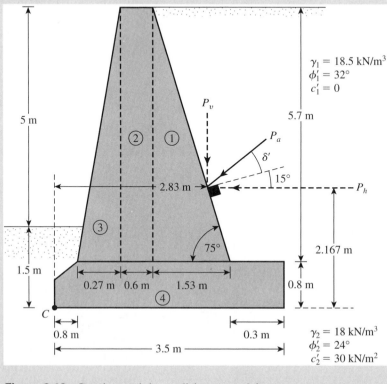

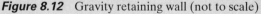

Figure 8.12 Gravity retaining wall (not to scale)

and

$$P_v = P_a \sin\left(15 + \tfrac{2}{3}\phi_1'\right) = 157.22 \sin 36.33 = 93.14 \text{ kN/m}$$

Part a: Factor of Safety against Overturning
From Figure 8.12, one can prepare the following table:

Area No.	Area (m²)	Weight* (kN/m)	Moment arm from C (m)	Moment (kN-m/m)
1	$\tfrac{1}{2}(5.7)(1.53) = 4.36$	102.81	2.18	224.13
2	$(0.6)(5.7) = 3.42$	80.64	1.37	110.48
3	$\tfrac{1}{2}(0.27)(5.7) = 0.77$	18.16	0.98	17.80
4	$\approx (3.5)(0.8) = 2.8$	66.02	1.75	115.54
		$P_v = 93.14$	2.83	263.59
		$\Sigma V = 360.77 \text{ kN/m}$		$\Sigma M_R = 731.54 \text{ kN-m/m}$

*$\gamma_{\text{concrete}} = 23.58 \text{ kN/m}^3$

Note that the weight of the soil above the back face of the wall is not taken into account in the preceding table. We have

$$\text{Overturning moment} = M_O = P_h\left(\frac{H'}{3}\right) = 126.65(2.167) = 274.45 \text{ kN-m/m}$$

Hence,

$$\text{FS}_{\text{(overturning)}} = \frac{\Sigma M_R}{\Sigma M_O} = \frac{731.54}{274.45} = \textbf{2.67} > \textbf{2, OK}$$

Part b: Factor of Safety against Sliding
We have

$$\text{FS}_{\text{(sliding)}} = \frac{(\Sigma V)\tan\left(\frac{2}{3}\phi_2'\right) + \frac{2}{3}c_2'B + P_p}{P_h}$$

$$P_p = \tfrac{1}{2}K_p\gamma_2 D^2 + 2c_2'\sqrt{K_p}D$$

and

$$K_p = \tan^2\left(45 + \frac{24}{2}\right) = 2.37$$

Hence,

$$P_p = \tfrac{1}{2}(2.37)(18)(1.5)^2 + 2(30)(1.54)(1.5) = 186.59 \text{ kN/m}$$

So

$$FS_{(sliding)} = \frac{360.77\tan\left(\dfrac{2}{3} \times 24\right) + \dfrac{2}{3}(30)(3.5) + 186.59}{126.65}$$

$$= \frac{103.45 + 70 + 186.59}{126.65} = \textbf{2.84}$$

If P_p is ignored, the factor of safety is **1.37**.

Part c: Pressure on Soil at Toe and Heel
From Eqs. (8.13), (8.14), and (8.15),

$$e = \frac{B}{2} - \frac{\Sigma M_R - \Sigma M_O}{\Sigma V} = \frac{3.5}{2} - \frac{731.54 - 274.45}{360.77} = 0.483 < \frac{B}{6} = 0.583$$

$$q_{toe} = \frac{\Sigma V}{B}\left[1 + \frac{6e}{B}\right] = \frac{360.77}{3.5}\left[1 + \frac{(6)(0.483)}{3.5}\right] = \textbf{188.43 kN/m}^2$$

and

$$q_{heel} = \frac{V}{B}\left[1 - \frac{6e}{B}\right] = \frac{360.77}{3.5}\left[1 - \frac{(6)(0.483)}{3.5}\right] = \textbf{17.73 kN/m}^2 \qquad \blacksquare$$

8.8 Construction Joints and Drainage from Backfill

Construction Joints

A retaining wall may be constructed with one or more of the following joints:

1. *Construction joints* (see Figure 8.13a) are vertical and horizontal joints that are placed between two successive pours of concrete. To increase the shear at the joints, keys may be used. If keys are not used, the surface of the first pour is cleaned and roughened before the next pour of concrete.
2. *Contraction joints* (Figure 8.13b) are vertical joints (grooves) placed in the face of a wall (from the top of the base slab to the top of the wall) that allow the concrete to shrink without noticeable harm. The grooves may be about 6 to 8 mm ($\approx$0.25 to 0.3 in.) wide and 12 to 16 mm ($\approx$0.5 to 0.6 in.) deep.
3. *Expansion joints* (Figure 8.13c) allow for the expansion of concrete caused by temperature changes; vertical expansion joints from the base to the top of the wall may also be used. These joints may be filled with flexible joint fillers. In most cases, horizontal reinforcing steel bars running across the stem are continuous through all joints. The steel is greased to allow the concrete to expand.

Drainage from the Backfill

As the result of rainfall or other wet conditions, the backfill material for a retaining wall may become saturated, thereby increasing the pressure on the wall and

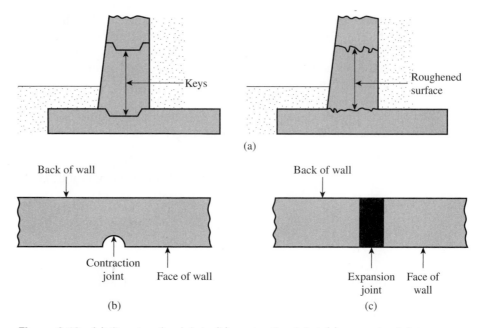

Figure 8.13   (a) Construction joints; (b) contraction joint; (c) expansion joint

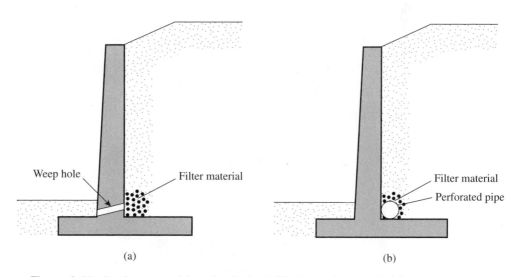

Figure 8.14   Drainage provisions for the backfill of a retaining wall: (a) by weep holes; (b) by a perforated drainage pipe

perhaps creating an unstable condition. For this reason, adequate drainage must be provided by means of *weep holes* or *perforated drainage pipes*. (See Figure 8.14.)

When provided, weep holes should have a minimum diameter of about 0.1 m (4 in.) and be adequately spaced. Note that there is always a possibility that backfill material may be washed into weep holes or drainage pipes and ultimately clog them. Thus, a filter material needs to be placed behind the weep holes or around the drainage pipes, as the case may be; geotextiles now serve that purpose.

Two main factors influence the choice of filter material: The grain-size distribution of the materials should be such that (a) the soil to be protected is not washed into the filter and (b) excessive hydrostatic pressure head is not created in the soil with a lower hydraulic conductivity (in this case, the backfill material). The preceding conditions can be satisfied if the following requirements are met (Terzaghi and Peck, 1967):

$$\frac{D_{15(F)}}{D_{85(B)}} < 5 \qquad [\text{to satisfy condition(a)}] \qquad (8.22a)$$

$$\frac{D_{15(F)}}{D_{15(B)}} > 4 \qquad [\text{to satisfy condition(b)}] \qquad (8.22b)$$

In these relations, the subscripts F and B refer to the *filter* and the *base* material (i.e., the backfill soil), respectively. Also, D_{15} and D_{85} refer to the diameters through which 15% and 85% of the soil (filter or base, as the case may be) will pass. Example 8.5 gives the procedure for designing a filter.

Example 8.5

Figure 8.15 shows the grain-size distribution of a backfill material. Using the conditions outlined in Section 8.8, determine the range of the grain-size distribution for the filter material.

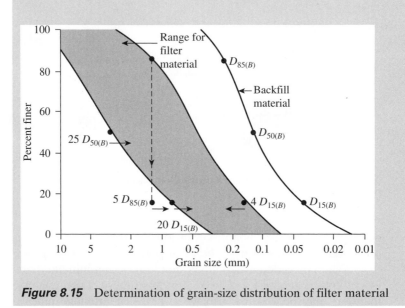

Figure 8.15   Determination of grain-size distribution of filter material

Solution

From the grain-size distribution curve given in the figure, the following values can be determined:

$$D_{15(B)} = 0.04 \text{ mm}$$

$$D_{85(B)} = 0.25 \text{ mm}$$

$$D_{50(B)} = 0.13 \text{ mm}$$

Conditions of Filter

1. $D_{15(F)}$ should be less than $5D_{85(B)}$; that is, $5 \times 0.25 = 1.25$ mm.
2. $D_{15(F)}$ should be greater than $4D_{15(B)}$; that is, $4 \times 0.04 = 0.16$ mm.
3. $D_{50(F)}$ should be less than $25D_{50(B)}$; that is, $25 \times 0.13 = 3.25$ mm.
4. $D_{15(F)}$ should be less than $20D_{15(B)}$; that is, $20 \times 0.04 = 0.8$ mm.

These limiting points are plotted in Figure 8.15. Through them, two curves can be drawn that are similar in nature to the grain-size distribution curve of the backfill material. These curves define the range of the filter material to be used. ■

8.9   *Some Comments on Design of Retaining Walls*

in Section 8.3, it was suggested that the *active earth pressure coefficient* be used to estimate the lateral force on a retaining wall due to the backfill. It is important to recognize the fact that the active state of the backfill can be established only if the wall yields sufficiently, which does not happen in all cases. The degree to which the wall yields depends on its *height* and the *section modulus*. Furthermore, the lateral force of the backfill depends on several factors identified by Casagrande (1973):

1. Effect of temperature
2. Groundwater fluctuation
3. Readjustment of the soil particles due to creep and prolonged rainfall
4. Tidal changes
5. Heavy wave action
6. Traffic vibration
7. Earthquakes

Insufficient wall yielding combined with other unforeseen factors may generate a larger lateral force on the retaining structure, compared with that obtained from the active earth-pressure theory. This is particularly true in the case of gravity retaining walls, bridge abutments, and other heavy structures that have a large section modulus. Figure 8.16 shows the lateral pressure distribution behind a bridge abutment and walls of two U-frame locks. The bridge abutment shown in Figure 8.16a is one in Germany as reported by Casagrande (1973) and walls of the two U-frames are those of the Port Allen Lock (Figure 8.16b) and the Old River Lock (Figure 8.16c) as reported by Gould (1970).

The bridge abutment shown in Figure 8.16a has a slag backfill. The effective angle of friction (ϕ') varied between 45° and 37°, depending on the degree of compaction. The backfill of walls shown in Figures 8.16b and 8.16c were clean sand with

ϕ' equal to 40°. It can be seen from these three cases that (a) the actual lateral earth-pressure distribution may not be triangular, and (b) the lateral earth pressure (at least in the top half of the wall) approached at-rest state. Hence, in the actual design, caution and judgment should be used to choose the lateral earth-pressure distribution.

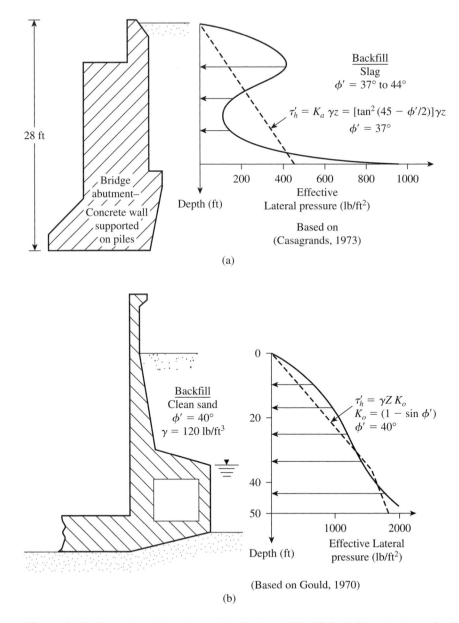

Figure 8.16 Lateral earth-pressure distribution behind (a) a bridge abutment in Germany; (b) wall of Port Allen Lock;

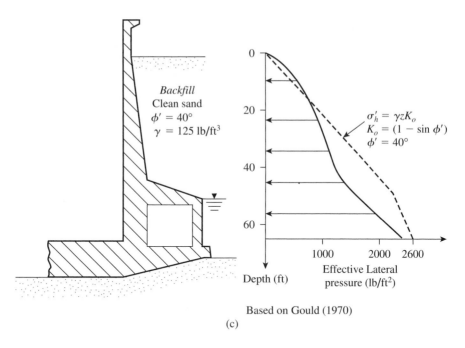

Based on Gould (1970)

(c)

Figure 8.16 (*Continued*) (c) wall of old river lock [Compiled from Casagrande (1973) and Gould (1970)]

Mechanically Stabilized Retaining Walls

More recently, soil reinforcement has been used in the construction and design of foundations, retaining walls, embankment slopes, and other structures. Depending on the type of construction, the reinforcements may be galvanized metal strips, geotextiles, geogrids, or geocomposites. Sections 8.10 and 8.11 provide a general overview of soil reinforcement and various reinforcement materials.

Reinforcement materials such as metallic strips, geotextiles, and geogrids are now being used to reinforce the backfill of retaining walls, which are generally referred to as *mechanically stabilized retaining walls*. The general principles for designing these walls are given in Sections 8.12 through 8.17.

8.10   Soil Reinforcement

The use of reinforced earth is a recent development in the design and construction of foundations and earth-retaining structures. *Reinforced earth* is a construction material made from soil that has been strengthened by tensile elements such as metal rods or strips, nonbiodegradable fabrics (geotextiles), geogrids, and the like. The fundamental idea of reinforcing soil is not new; in fact, it goes back several centuries. However, the present concept of systematic analysis and design was developed by a French engineer, H. Vidal (1966). The French Road Research Laboratory has done extensive research on the applicability and the beneficial effects of the use of reinforced earth as a construction material. This research has been documented in detail by Darbin (1970), Schlosser and Long (1974), and Schlosser and Vidal (1969). The tests that were conducted involved the use of metallic strips as reinforcing material.

Retaining walls with reinforced earth have been constructed around the world since Vidal began his work. The first reinforced-earth retaining wall with metal strips as reinforcement in the United States was constructed in 1972 in southern California.

The beneficial effects of soil reinforcement derive from (a) the soil's increased tensile strength and (b) the shear resistance developed from the friction at the soil-reinforcement interfaces. Such reinforcement is comparable to that of concrete structures. Currently, most reinforced-earth design is done with *free-draining granular soil only*. Thus, the effect of pore water development in cohesive soils, which, in turn, reduces the shear strength of the soil, is avoided.

8.11   *Considerations in Soil Reinforcement*

Metal Strips

In most instances, galvanized steel strips are used as reinforcement in soil. However, galvanized steel is subject to corrosion. The rate of corrosion depends on several environmental factors. Binquet and Lee (1975) suggested that the average rate of corrosion of galvanized steel strips varies between 0.025 and 0.050 mm/yr. So, in the actual design of reinforcement, allowance must be made for the rate of corrosion. Thus,

$$t_c = t_{\text{design}} + r \text{ (life span of structure)}$$

where

t_c = actual thickness of reinforcing strips to be used in construction
t_{design} = thickness of strips determined from design calculations
r = rate of corrosion

Further research needs to be done on corrosion-resistant materials such as fiberglass before they can be used as reinforcing strips.

Nonbiodegradable Fabrics

Nonbiodegradable fabrics are generally referred to as *geotextiles*. Since 1970, the use of geotextiles in construction has increased greatly around the world. The fabrics are usually made from petroleum products—polyester, polyethylene, and polypropylene. They may also be made from fiberglass. Geotextiles are not prepared from natural fabrics, because they decay too quickly. Geotextiles may be woven, knitted, or nonwoven.

Woven geotextiles are made of two sets of parallel filaments or strands of yarn systematically interlaced to form a planar structure. *Knitted geotextiles* are formed by interlocking a series of loops of one or more filaments or strands of yarn to form a planar structure. *Nonwoven geotextiles* are formed from filaments or short fibers arranged in an oriented or random pattern in a planar structure. These filaments or short fibers are arranged into a loose web in the beginning and then are bonded by one or a combination of the following processes:

1. *Chemical bonding*—by glue, rubber, latex, a cellulose derivative, or the like
2. *Thermal bonding*—by heat for partial melting of filaments
3. *Mechanical bonding*—by needle punching

Needle-punched nonwoven geotextiles are thick and have high in-plane permeability.
Geotextiles have four primary uses in foundation engineering:

1. *Drainage:* The fabrics can rapidly channel water from soil to various outlets, thereby providing a higher soil shear strength and hence stability.
2. *Filtration:* When placed between two soil layers, one coarse grained and the other fine grained, the fabric allows free seepage of water from one layer to the other. However, it protects the fine-grained soil from being washed into the coarse-grained soil.
3. *Separation:* Geotextiles help keep various soil layers separate after construction and during the projected service period of the structure. For example, in the construction of highways, a clayey subgrade can be kept separate from a granular base course.
4. *Reinforcement:* The tensile strength of geofabrics increases the load-bearing capacity of the soil.

Geogrids

Geogrids are high-modulus polymer materials, such as polypropylene and polyethylene, and are prepared by tensile drawing. Netlon, Ltd., of the United Kingdom was the first producer of geogrids. In 1982, the Tensar Corporation, presently Tensar Earth Technologies, Inc., introduced geogrids into the United States.

The major function of geogrids is *reinforcement*. Geogrids are relatively stiff netlike materials with openings called *apertures* that are large enough to allow interlocking with the surrounding soil or rock to perform the function of reinforcement or segregation (or both).

Geogrids generally are of two types: (a) uniaxial and (b) biaxial. Figures 8.17a and 8.17b show these two types of geogrids, which are produced by Tensar Earth

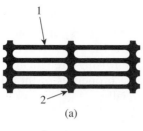

(a)

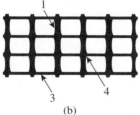

(b)

Figure 8.17 Geogrids: (a) uniaxial; (b) biaxial (*Note:* 1—longitudinal rib; 2—transverse bar; 3—transverse rib; 4—junction)

Table 8.2 Properties of TENSAR Biaxial Geogrids

Property	Geogrid		
	BX1000	**BX1100**	**BX1200**
Aperture size			
Machine direction	25 mm (1 in.) (nominal)	25 mm (1 in.) (nominal)	25 mm (1 in.) (nominal)
Cross-machine direction	33 mm (1.3 in.) (nominal)	33 mm (1.3 in.) (nominal)	33 mm (1.3 in.) (nominal)
Open area	70% (minimum)	74% (nominal)	77% (nominal)
Junction			
Thickness	2.3 mm (0.09 in.) (nominal)	2.8 mm (0.11 in.) (nominal)	4.1 mm (0.16 in.) (nominal)
Tensile modulus			
Machine direction	18.2 kN/m (12,500 lb/ft) (minimum)	204 kN/m (14,000 lb/ft) (minimum)	270 kN/m (18,500 lb/ft) (minimum)
Cross-machine direction	18.2 kN/m (12,500 lb/ft) (minimum)	292 kN/m (20,000 lb/ft) (minimum)	438 kN/m (30,000 lb/ft) (minimum)
Material			
Polypropylene	97% (minimum)	99% (nominal)	99% (nominal)
Carbon black	2% (minimum)	1% (nominal)	1% (nominal)

Technologies, Inc. Uniaxial TENSAR grids are manufactured by stretching a punched sheet of extruded high-density polyethylene in one direction under carefully controlled conditions. The process aligns the polymer's long-chain molecules in the direction of draw and results in a product with high one-directional tensile strength and a high modulus. Biaxial TENSAR grids are manufactured by stretching the punched sheet of polypropylene in two orthogonal directions. This process results in a product with high tensile strength and a high modulus in two perpendicular directions. The resulting grid apertures are either square or rectangular.

The commercial geogrids currently available for soil reinforcement have nominal rib thicknesses of about 0.5 to 1.5 mm (0.02 to 0.06 in.) and junctions of about 2.5 to 5 mm (0.1 to 0.2 in.). The grids used for soil reinforcement usually have apertures that are rectangular or elliptical. The dimensions of the apertures vary from about 25 to 150 mm (1 to 6 in.). Geogrids are manufactured so that the open areas of the grids are greater than 50% of the total area. They develop reinforcing strength at low strain levels, such as 2% (Carroll, 1988). Table 8.2 gives some properties of the TENSAR biaxial geogrids that are currently available commercially.

8.12 *General Design Considerations*

The general design procedure of any mechanically stabilized retaining wall can be divided into two parts:

1. Satisfying *internal stability* requirements
2. Checking the *external stability* of the wall

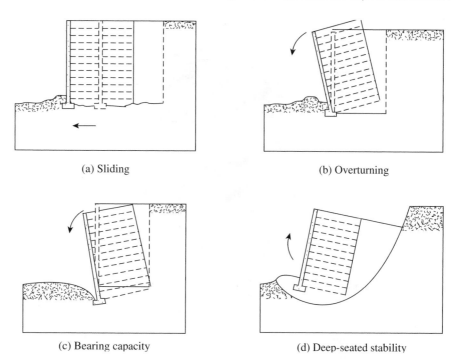

(a) Sliding (b) Overturning

(c) Bearing capacity (d) Deep-seated stability

Figure 8.18 External stability checks (After Transportation Research Board, 1995)

The internal stability checks involve determining tension and pullout resistance in the reinforcing elements and ascertaining the integrity of facing elements. The external stability checks include checks for overturning, sliding, and bearing capacity failure (Figure 8.18). The sections that follow will discuss the retaining-wall design procedures for use with metallic strips, geotextiles, and geogrids.

8.13 *Retaining Walls with Metallic Strip Reinforcement*

Reinforced-earth walls are flexible walls. Their main components are

1. *Backfill,* which is granular soil
2. *Reinforcing strips,* which are thin, wide strips placed at regular intervals, and
3. *A cover* or *skin,* on the front face of the wall

Figure 8.19 is a diagram of a reinforced-earth retaining wall. Note that, at any depth, the reinforcing strips or ties are placed with a horizontal spacing of S_H center to center; the vertical spacing of the strips or ties is S_V center to center. The skin can be constructed with sections of relatively flexible thin material. Lee et al. (1973) showed that, with a conservative design, a 5 mm-thick (≈ 0.2 in.) galvanized steel skin would be enough to hold a wall about 14 to 15 m (45 to 50 ft) high. In most cases, precast concrete slabs can also be used as skin. The slabs are grooved to fit

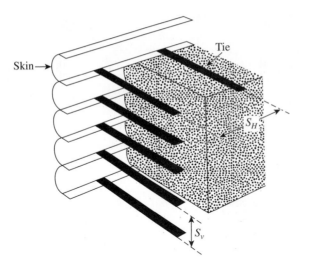

Figure 8.19 Reinforced-earth retaining wall

into each other so that soil cannot flow out between the joints. When metal skins are used, they are bolted together, and reinforcing strips are placed between the skins.

Figures 8.20 and 8.21 show a reinforced-earth retaining wall under construction; its skin (facing) is a precast concrete slab. Figure 8.22 shows a metallic reinforcement tie attached to the concrete slab.

The simplest and most common method for the design of ties is the *Rankine method*. We discuss this procedure next.

Calculation of Active Horizontal and Vertical Pressure

Figure 8.23 shows a retaining wall with a granular backfill having a unit weight of γ_1 and a friction angle of ϕ'_1. Below the base of the retaining wall, the *in situ* soil has been excavated and recompacted, with granular soil used as backfill. Below the backfill, the *in situ* soil has a unit weight of γ_2, friction angle of ϕ'_2, and cohesion of c'_2. A surcharge having an intensity of q per unit area lies atop the retaining wall, which has reinforcement ties at depths $z = 0, S_V, 2S_V, \ldots, NS_V$. The height of the wall is $NS_V = H$.

According to the Rankine active pressure theory (Section 7.3)

$$\sigma'_a = \sigma'_o K_a - 2c'\sqrt{K_a}$$

where σ'_a = Rankine active pressure at any depth z.

For dry granular soils with no surcharge at the top, $c' = 0$, $\sigma'_o = \gamma_1 z$, and $K_a = \tan^2(45 - \phi'_1/2)$. Thus,

$$\sigma'_{a(1)} = \gamma_1 z K_a \qquad (8.23)$$

Figure 8.20 Reinforced-earth retaining wall (with metallic strip) under construction

When a surcharge is added at the top, as shown in Figure 8.23,

$$\sigma_o' = \underset{\substack{\uparrow \\ = \gamma_1 z \\ \text{Due to} \\ \text{soil only}}}{\sigma_{o(1)}'} + \underset{\substack{\uparrow \\ \text{Due to the} \\ \text{surcharge}}}{\sigma_{o(2)}'} \qquad (8.24)$$

The magnitude of $\sigma_{o(2)}'$ can be calculated by using the 2:1 method of stress distribution described in Eq. (5.14) and Figure 5.5. The 2:1 method of stress distribution is shown in Figure 8.24a. According to Laba and Kennedy (1986),

$$\sigma_{o(2)}' = \frac{qa'}{a' + z} \qquad (\text{for } z \leqslant 2b') \qquad (8.25)$$

Figure 8.21   Another view of the retaining wall shown in Figure 8.20

Figure 8.22   Metallic strip attachment to the precast concrete slab used as the skin

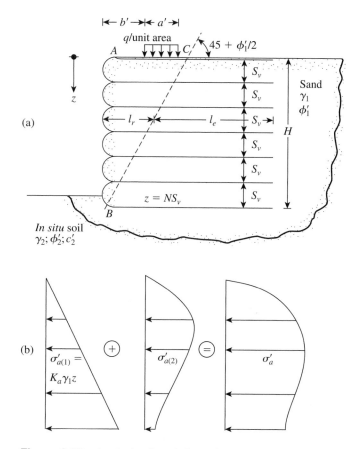

Figure 8.23 Analysis of a reinforced-earth retaining wall

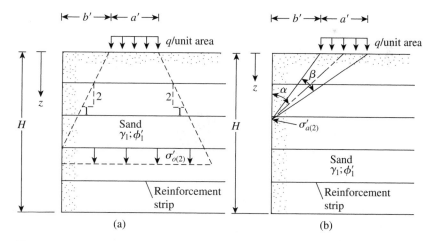

Figure 8.24 (a) Notation for the relationship of $\sigma'_{o(2)}$ in Eqs. (8.25) and (8.26); (b) notation for the relationship of $\sigma'_{a(2)}$ in Eqs. (8.28) and (8.29)

and

$$\sigma'_{o(2)} = \frac{qa'}{a' + \dfrac{z}{2} + b'} \qquad (\text{for } z > 2b') \qquad (8.26)$$

Also, when a surcharge is added at the top, the lateral pressure at any depth is

$$\sigma'_a = \underset{\uparrow}{\sigma'_{a(1)}} \qquad + \underset{\uparrow}{\sigma'_{a(2)}} \qquad (8.27)$$

$$= \underset{\substack{\text{Due to} \\ \text{soil only}}}{K_a\gamma_1 z} \quad \underset{\text{surcharge}}{\text{Due to the}}$$

According to Laba and Kennedy (1986), $\sigma'_{a(2)}$ may be expressed (see Figure 8.24b) as

$$\sigma'_{a(2)} = M\left[\frac{2q}{\pi}(\underset{\uparrow}{\beta} - \sin\beta\cos 2\alpha) \right] \qquad (8.28)$$

$$\text{(in radians)}$$

where

$$M = 1.4 - \frac{0.4b'}{0.14H} \geq 1 \qquad (8.29)$$

The net active (lateral) pressure distribution on the retaining wall calculated by using Eqs. (8.27), (8.28), and (8.29) is shown in Figure 8.23b.

Tie Force

The tie force *per unit length of the wall* developed at any depth z (see Figure 8.23) is

$$T = \text{active earth pressure at depth } z$$

$$\times \text{ area of the wall to be supported by the tie}$$

$$= (\sigma'_a)(S_V S_H) \qquad (8.30)$$

Factor of Safety against Tie Failure

The reinforcement ties at each level, and thus the walls, could fail by either (a) tie breaking or (b) tie pullout.

The factor of safety against *tie breaking* may be determined as

$$FS_{(B)} = \frac{\text{yield or breaking strength of each tie}}{\text{maximum force in any tie}}$$

$$= \frac{wtf_y}{\sigma'_a S_V S_H} \tag{8.31}$$

where

w = width of each tie
t = thickness of each tie
f_y = yield or breaking strength of the tie material

A factor of safety of about 2.5 to 3 is generally recommended for ties at all levels.

Reinforcing ties at any depth z will fail by pullout if the frictional resistance developed along the surfaces of the ties is less than the force to which the ties are being subjected. The *effective length* of the ties along which frictional resistance is developed may be conservatively taken as the length that extends *beyond the limits of the Rankine active failure zone,* which is the zone *ABC* in Figure 8.23. Line *BC* makes an angle of $45 + \phi'_1/2$ with the horizontal. Now, the maximum friction force that can be realized for a tie at depth z is

$$F_R = 2l_e w \sigma'_o \tan \phi'_\mu \tag{8.32}$$

where

l_e = effective length
σ'_o = effective vertical pressure at a depth z
ϕ'_μ = soil–tie friction angle

Thus, the factor of safety against *tie pullout* at any depth z is

$$FS_{(P)} = \frac{F_R}{T} \tag{8.33}$$

Substituting Eqs. (8.30) and (8.32) into Eq. (8.33) yields

$$FS_{(P)} = \frac{2l_e w \sigma'_o \tan \phi'_\mu}{\sigma'_a S_V S_H} \tag{8.34}$$

Total Length of Tie

The total length of ties at any depth is

$$L = l_r + l_e \tag{8.35}$$

where

l_r = length within the Rankine failure zone
l_e = effective length

For a given $FS_{(P)}$ from Eq. (8.34),

$$l_e = \frac{FS_{(P)}\sigma_a' S_V S_H}{2w\sigma_o' \tan\phi_\mu'}$$ (8.36)

Again, at any depth z,

$$l_r = \frac{(H - z)}{\tan\left(45 + \dfrac{\phi_1'}{2}\right)}$$ (8.37)

So, combining Eqs. (8.35), (8.36), and (8.37) gives

$$L = \frac{(H - z)}{\tan\left(45 + \dfrac{\phi_1'}{2}\right)} + \frac{FS_{(P)}\sigma_a' S_V S_H}{2w\sigma_o' \tan\phi_\mu'}$$ (8.38)

8.14   *Step-by-Step-Design Procedure Using Metallic Strip Reinforcement*

Following is a step-by-step procedure for the design of reinforced-earth retaining walls.

General

Step 1. Determine the height of the wall, H, and the properties of the granular backfill material, such as the unit weight (γ_1) and the angle of friction (ϕ_1').

Step 2. Obtain the soil–tie friction angle, ϕ_μ', and the required value of $FS_{(B)}$ and $FS_{(P)}$.

Internal Stability

Step 3. Assume values for horizontal and vertical tie spacing. Also, assume the width of reinforcing strip, w, to be used.

Step 4. Calculate σ_a' from Eqs. (8.27), (8.28), and (8.29).

Step 5. Calculate the tie forces at various levels from Eq. (8.30).

Step 6. For the known values of $FS_{(B)}$, calculate the thickness of ties, t, required to resist the tie breakout:

$$T = \sigma_a' S_V S_H = \frac{wt f_y}{FS_{(B)}}$$

or

$$t = \frac{(\sigma_a' S_V S_H)[FS_{(B)}]}{w f_y}$$ (8.39)

The convention is to keep the magnitude of t the same at all levels, so σ'_a in Eq. (8.39) should equal $\sigma'_{a(max)}$.

Step 7.   For the known values of ϕ'_μ and $FS_{(P)}$, determine the length L of the ties at various levels from Eq. (8.38).

Step 8.   The magnitudes of S_V, S_H, t, w, and L may be changed to obtain the most economical design.

External Stability

Step 9.   Check for *overturning,* using Figure 8.25 as a guide. Taking the moment about B yields the overturning moment for the unit length of the wall:

$$M_o = P_a z' \tag{8.40}$$

Here,

$$P_a = \text{active force} = \int_0^H \sigma'_a dz$$

The resisting moment per unit length of the wall is

$$M_R = W_1 x_1 + W_2 x_2 + \cdots + qa'\left(b' + \frac{a'}{2}\right) \tag{8.41}$$

where

$W_1 = (\text{area } AFEGI)\,(1)\,(\gamma_1)$
$W_2 = (\text{area } FBDE)\,(1)\,(\gamma_1)$
$\vdots$

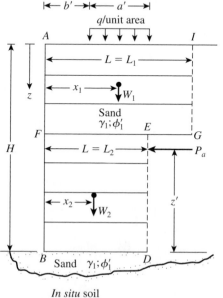

In situ soil
$\gamma_2; \phi'_2; c'_2$

Figure 8.25   Stability check for the retaining wall

So,

$$FS_{(overturning)} = \frac{M_R}{M_o}$$

$$= \frac{W_1 x_1 + W_2 x_2 + \cdots + qa'\left(b' + \frac{a'}{2}\right)}{\left(\int_0^H \sigma_a' \, dz\right) z'}$$

(8.42)

Step 10. The check for *sliding* can be done by using Eq. (8.11), or

$$FS_{(sliding)} = \frac{(W_1 + W_2 + \cdots + qa')[\tan(k\phi_1')]}{P_a}$$

(8.43)

where $k \approx \frac{2}{3}$.

Step 11. Check for ultimate bearing capacity failure, which can be given as

$$q_u = c_2' N_c + \frac{1}{2}\gamma_2 L_2' N_\gamma$$

(8.44a)

The bearing capacity factors N_c and N_γ correspond to the soil friction angle ϕ_2'. (See Table 3.3.) In Eq. (8.44a), L_2' is the effective length; that is,

$$L_2' = L_2 - 2e$$

(8.44b)

where $e =$ eccentricity given by

$$e = \frac{L_2}{2} - \frac{M_R - M_O}{\Sigma V}$$

(8.44c)

in which $\Sigma V = W_1 + W_2 \ldots + qa'$.
From Eq. 8.24, the vertical stress at $z = H$ is

$$\sigma_{o(H)}' = \gamma_1 H + \sigma_{o(2)}'$$

(8.45)

So the factor of safety against bearing capacity failure is

$$FS_{(bearing\ capacity)} = \frac{q_{ult}}{\sigma_{o(H)}'}$$

(8.46)

Generally, minimum values of $FS_{(overturning)} = 3$, $FS_{(sliding)} = 3$, and $FS_{(bearing\ capacity\ failure)} = 3$ to 5 are recommended.

Example 8.6

A 30-ft-high retaining wall with galvanized steel-strip reinforcement in a granular backfill has to be constructed. Referring to Figure 8.23, given:

Granular backfill: $\phi'_1 = 36°$
$\gamma_1 = 105 \text{ lb/ft}^3$

Foundation soil: $\phi'_2 = 28°$
$\gamma_2 = 110 \text{ lb/ft}^3$
$c'_2 = 1000 \text{ lb/ft}^2$

Galvanized steel reinforcement:

Width of strip, $w = 3 \text{ in.}$
$S_V = 2 \text{ ft center-to-center}$
$S_H = 3 \text{ ft center-to-center}$
$f_y = 35{,}000 \text{ lb/in}^2$
$\phi'_\mu = 20°$

Required $\text{FS}_{(B)} = 3$

Required $\text{FS}_{(P)} = 3$

Check for the external and internal stability. Assume the corrosion rate of the galvanized steel to be 0.001 in./year and the life span of the structure to be 50 years.

Solution
Internal Stability Check

Tie thickness: Maximum tie force, $T_{\max} = \sigma'_{a(\max)} S_V S_H$

$$\sigma_{a(\max)} = \gamma_1 H K_a = \gamma H \tan^2\left(45 - \frac{\phi'_1}{2}\right)$$

so

$$T_{\max} = \gamma_1 H \tan^2\left(45 - \frac{\phi'_1}{2}\right) S_V S_H$$

From Eq. (8.39), for *tie break*,

$$t = \frac{(\sigma'_a S_V S_H)[\text{FS}_{(B)}]}{w f_y} = \frac{\left[\gamma_1 H \tan^2\left(45 - \frac{\phi'_1}{2}\right) S_V S_H\right]}{w f_y}$$

or

$$t = \frac{\left[(105)(30)\tan^2\left(45 - \frac{36}{2}\right)(2)(3)\right](3)}{\left(\frac{3}{12}\text{ ft}\right)(35{,}000 \times 144 \text{ lb/ft}^2)} = 0.0117 \text{ ft} = 0.14 \text{ in.}$$

If the rate of corrosion is 0.001 in./yr and the life span of the structure is 50 yr, then the actual thickness, t, of the ties will be

$$t = 0.14 + (0.001)(50) = 0.19 \text{ in.}$$

So a **tie thickness of 0.2 in.** would be enough.

Tie length: Refer to Eq. (8.38). For this case, $\sigma_a' = \gamma_1 z K_a$ and $\sigma_O' = \gamma_1 z$, so

$$L = \frac{(H - z)}{\tan\left(45 + \dfrac{\phi_1'}{2}\right)} + \frac{FS_{(P)}\gamma_1 z K_a S_V S_H}{2w\gamma_1 z \tan\phi_\mu'}$$

Now the following table can be prepared. (Note: $FS_{(P)} = 3$, $H = 30$ ft, $w = 3$ in., and $\phi_\mu' = 20°$.)

z(ft)	Tie length L (ft) [Eq. (8.38)]
5	38.45
10	35.89
15	33.34
20	30.79
25	28.25
30	25.7

So use a **tie length of L = 40 ft.**

External Stability Check
Check for overturning: Refer to Figure 8.26. For this case, using Eq. (8.42)

$$FS_{(overturning)} = \frac{W_1 x_1}{\left[\displaystyle\int_0^H \sigma_a' \, dz\right] z'}$$

$$W_1 = \gamma_1 HL = (105)(30)(40) = 126,000 \text{ lb}$$

$$x_1 = 20 \text{ ft}$$

$$P_a = \int_0^H \sigma_a' \, dz = \tfrac{1}{2}\gamma_1 K_a H^2 = \left(\tfrac{1}{2}\right)(105)(0.26)(30)^2 = 12,285 \text{ lb/ft}$$

$$z' = \frac{30}{3} = 10 \text{ ft}$$

$$FS_{(overturning)} = \frac{(126,000)(20)}{(12,285)(10)} = \textbf{20.5} > 3\textbf{—OK}$$

Check for sliding: From Eq. (8.43)

$$FS_{(sliding)} = \frac{W_1 \tan(k\phi_1')}{P_a} = \frac{126,000 \tan\left[\left(\dfrac{2}{3}\right)(36)\right]}{12,285} = \textbf{4.57} > 3\textbf{—OK}$$

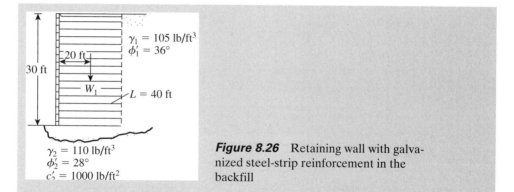

Figure 8.26 Retaining wall with galvanized steel-strip reinforcement in the backfill

Check for bearing capacity: For $\phi_2' = 28°$, $N_c = 25.8$, $N_\gamma = 16.78$ (Table 3.3). From Eq. (8.44a),

$$q_{ult} = c_2' N_c + \tfrac{1}{2}\gamma_2 L' N_\gamma$$

$$e = \frac{L}{2} - \frac{M_R - M_O}{\Sigma V} = \frac{40}{2} - \left[\frac{(126{,}000 \times 20) - (12{,}285 \times 10)}{126{,}000} \right] = 0.975 \text{ ft}$$

$$L' = 40 - (2 \times 0.975) = 38.05 \text{ ft}$$

$$q_{ult} = (1000)(25.8) + (\tfrac{1}{2})(110)(38.05)(16.72) = 60{,}791 \text{ lb/ft}^2$$

From Eq. (8.45),

$$\sigma_{O(H)}' = \gamma_1 H = (105)(30) = 3150 \text{ lb/ft}^2$$

$$FS_{(bearing\ capacity)} = \frac{q_{ult}}{\sigma_{O(H)}'} = \frac{60{,}791}{3150} = 19.3 > 5 \text{—OK} \qquad \blacksquare$$

8.15 *Retaining Walls with Geotextile Reinforcement*

Figure 8.27 shows a retaining wall in which layers of geotextile have been used as reinforcement. As in Figure 8.25, the backfill is a granular soil. In this type of retaining wall, the facing of the wall is formed by lapping the sheets as shown with a lap length of l_l. When construction is finished, the exposed face of the wall must be covered; otherwise, the geotextile will deteriorate from exposure to ultraviolet light. *Bitumen emulsion* or *Gunite* is sprayed on the wall face. A wire mesh anchored to the geotextile facing may be necessary to keep the coating on.

The design of this type of retaining wall is similar to that presented in Section 8.14. Following is a step-by-step procedure for design based on the recommendations of Bell et al. (1975) and Koerner (1990):

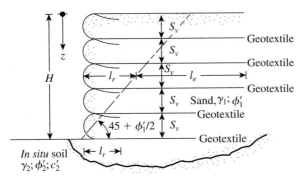

Figure 8.27 Retaining wall with geotextile reinforcement

Internal Stability

Step 1. Determine the active pressure distribution on the wall from the formula

$$\sigma'_a = K_a \sigma'_o = K_a \gamma_1 z \tag{8.47}$$

where

K_a = Rankine active pressure coefficient = $\tan^2(45 - \phi'_1/2)$
γ_1 = unit weight of the granular backfill
ϕ'_1 = friction angle of the granular backfill

Step 2. Select a geotextile fabric with an allowable strength of σ_G (lb/ft or kN/m).
Step 3. Determine the vertical spacing of the layers at any depth z from the formula

$$S_V = \frac{\sigma_G}{\sigma'_a FS_{(B)}} = \frac{\sigma_G}{(\gamma_1 z K_a)[FS_{(B)}]} \tag{8.48}$$

Note that Eq. (8.48) is similar to Eq. (8.31). The magnitude of $FS_{(B)}$ is generally 1.3 to 1.5.

Step 4. Determine the length of each layer of geotextile from the formula

$$L = l_r + l_e \tag{8.49}$$

where

$$l_r = \frac{H - z}{\tan\left(45 + \dfrac{\phi'_1}{2}\right)} \tag{8.50}$$

and

$$l_e = \frac{S_V \sigma'_a [FS_{(P)}]}{2\sigma'_o \tan \phi'_F} \tag{8.51}$$

in which

$$\sigma'_a = \gamma_1 z K_a$$

$$\sigma'_o = \gamma_1 z$$

$$FS_{(P)} = 1.3 \text{ to } 1.5$$

$$\phi'_F = \text{friction angle at geotextile–soil interface}$$

$$\approx \tfrac{2}{3}\phi'_1$$

Note that Eqs. (8.49), (8.50), and (8.51) are similar to Eqs. (8.35), (8.37), and (8.36), respectively.

Based on the published results, the assumption of $\phi'_F/\phi'_1 \approx \tfrac{2}{3}$ is reasonable and appears to be conservative. Martin et al. (1984) presented the following laboratory test results for ϕ'_F/ϕ'_1 between various types of geotextiles and sand.

Type	ϕ'_F/ϕ'_1
Woven—monofilament/concrete sand	0.87
Woven—silt film/concrete sand	0.8
Woven—silt film/rounded sand	0.86
Woven—silt film/silty sand	0.92
Nonwoven—melt-bonded/concrete sand	0.87
Nonwoven—needle-punched concrete sand	1.0
Nonwoven—needle-punched/rounded sand	0.93
Nonwoven—needle-punched/silty sand	0.91

Step 5. Determine the lap length, l_l, from

$$l_l = \frac{S_V \sigma'_a FS_{(P)}}{4\sigma'_o \tan\phi'_F} \tag{8.52}$$

The minimum lap length should be 1 m (3 ft).

External Stability

Step 6. Check the factors of safety against overturning, sliding, and bearing capacity failure as described in Section 8.14 (Steps 9, 10, and 11).

Example 8.7

A geotextile-reinforced retaining wall 16 ft high is shown in Figure 8.28. For the granular backfill, $\gamma_1 = 110 \text{ lb/ft}^3$ and $\phi'_1 = 36°$. For the geotextile, $\sigma_G = 80 \text{ lb/in.}$ For the design of the wall, determine S_V, L, and l_l.

Solution
We have

$$K_a = \tan^2\!\left(45 - \frac{\phi'_1}{2}\right) = 0.26$$

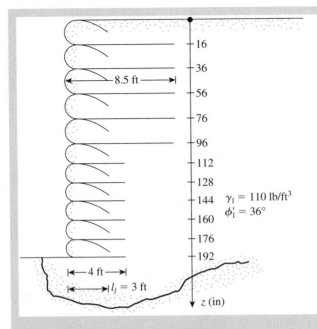

Figure 8.28 Geotextile-reinforced retaining wall

Determination of S_V

To find S_V, we make a few trials. From Eq. (8.48),

$$S_V = \frac{\sigma_G}{(\gamma_1 z K_a)[FS_{(B)}]}$$

With $FS_{(B)} = 1.5$ at $z = 8$ ft,

$$S_V = \frac{(80 \times 12 \text{ lb/ft})}{(110)\,(8)\,(0.26)\,(1.5)} = 2.8 \text{ ft} \approx 33.6 \text{ in.}$$

At $z = 12$ ft,

$$S_V = \frac{(80 \times 12 \text{ lb/ft})}{(110)\,(12)\,(0.26)\,(1.5)} = 1.87 \text{ ft} \approx 22 \text{ in.}$$

At $z = 16$ ft,

$$S_V = \frac{(80 \times 12 \text{ lb/ft})}{(110)\,(16)\,(0.26)\,(1.5)} = 1.4 \text{ ft} \approx 16.8 \text{ in.}$$

So, use $S_V = 20$ in. for $z = 0$ to $z = 8$ ft and $S_V = 16$ in. for $z > 8$ ft. (See Figure 8.27.)

Determination of L

From Eqs. (8.49), (8.50), and (8.51),

$$L = \frac{(H - z)}{\tan\left(45 + \dfrac{\phi_1'}{2}\right)} + \frac{S_V K_a [FS_{(P)}]}{2 \tan \phi_F'}$$

For $FS_{(P)} = 1.5$, $\tan\phi_F' = \tan[(\frac{2}{3})(36)] = 0.445$, and it follows that

$$L = (0.51)(H - z) + 0.438S_V$$

Now the following table can be prepared:

z (in.)	(ft)	S_V (ft)	(0.51)($H - z$) (ft)	0.438S_V (ft)	L (ft)
16	1.33	1.67	7.48	0.731	8.21
56	4.67	1.67	5.78	0.731	6.51
76	6.34	1.67	4.93	0.731	5.66
96	8.0	1.67	4.08	0.731	4.81
112	9.34	1.33	3.40	0.582	3.982
144	12.0	1.33	2.04	0.582	2.662
176	14.67	1.33	0.68	0.582	1.262

On the basis of the preceding calculations, **use $L = 8.5$ ft for $z \leqslant 8$ ft and $L = 4$ ft for $z > 8$ ft.**

Determination of l_l
From Eq. (8.52),

$$l_l = \frac{S_V \sigma_a'[FS_{(P)}]}{4\sigma_o' \tan\phi_F'}$$

With $\sigma_a' = \gamma_1 z K_a$, $FS_{(P)} = 1.5$; with $\sigma_o' = \gamma_1 z$, $\phi_F' = \frac{2}{3}\phi_1'$. So

$$l_l = \frac{S_V K_a[FS_{(P)}]}{4\tan\phi_F'} = \frac{S_V(0.26)(1.5)}{4\tan[(\frac{2}{3})(36)]} = 0.219S_V$$

At $z = 16$ in.,

$$l_l = 0.219S_V = (0.219)\left(\frac{20}{12}\right) = 0.365 \text{ ft} \leqslant 3 \text{ ft}$$

So, use $l_l = 3$ ft.   ∎

8.16   *Retaining Walls with Geogrid Reinforcement*

Geogrids can also be used as reinforcement in granular backfill for the construction of retaining walls. Figure 8.29 shows typical schematic diagrams of retaining walls with geogrid reinforcement.

   Relatively few field measurements are available for lateral earth pressure on retaining walls constructed with geogrid reinforcement. Figure 8.30 shows a comparison of measured and design lateral pressures (Berg et al., 1986) for two retaining walls constructed with precast panel facing. The figure indicates that the measured earth pressures were substantially smaller than those calculated for the Rankine active case.

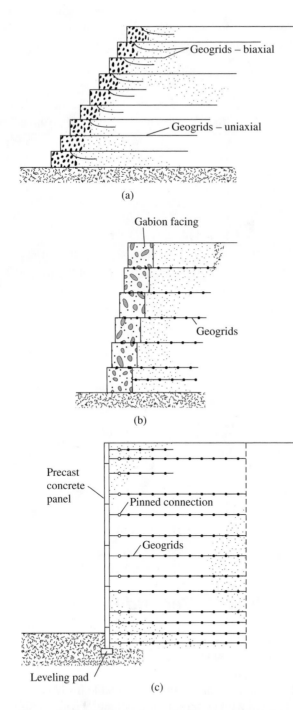

(a)

(b)

(c)

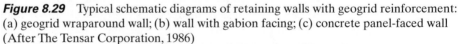

Figure 8.29 Typical schematic diagrams of retaining walls with geogrid reinforcement: (a) geogrid wraparound wall; (b) wall with gabion facing; (c) concrete panel-faced wall (After The Tensar Corporation, 1986)

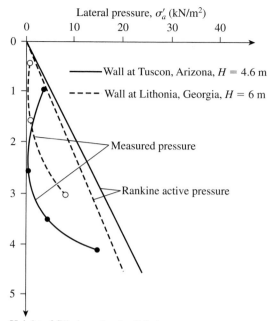

Figure 8.30 Comparison of theoretical and measured lateral pressures in geogrid reinforced retaining walls (Based on Berg et al., 1986)

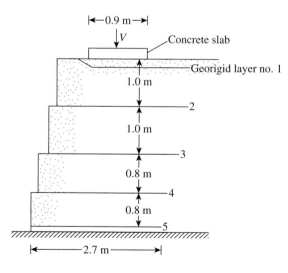

Figure 8.31 Schematic diagram of the retaining wall tested by Thamm et al. (1990)

The results of another interesting full-scale test on a retaining wall with geogrid reinforcement, granular backfill, and a height of 3.6 m was reported by Thamm et al. (1990). The main reinforcement for the wall was TENSAR SR2 geogrid. Figure 8.31 shows a schematic diagram of the retaining wall. Failure in the wall was caused by applying load to a concrete slab measuring 2.4 m $\times$ 0.9 m. The wall failed when the vertical load, V, on the concrete slab reached 1065 kN. Figure 8.32 shows the variation of the wall face displacement and the distribution of lateral pressure as the loading progressed, from which the following conclusions can be drawn:

1. The shape of the lateral earth pressure distribution on the wall face is similar to that shown in Figure 8.23b.
2. At failure load, the magnitude of $\Delta L/H$ (ΔL = facing displacement) at the top of the wall was about 1.7%, which is considerably higher than may be encountered for a rigid retaining wall.

8.17   *General Comments*

Great progress is being made in the development of rational design procedures for mechanically stabilized earth (MSE) retaining walls. Readers are directed to Transportation Research Circular No. 444 (1995) and Federal Highway Administration Publication No. FHWA-SA-96-071 (1996) for further information. However, following is a summary of a few recent developments:

1. In this chapter, we have used Rankine's active pressure in the design of MSE retaining walls. The appropriate value of the earth-pressure coefficient depends, however, on the degree of restraint that the reinforcing elements impose on the soil. If the wall can yield substantially, the Rankine active earth pressure may be appropriate, which is not the case for all types of MSE walls. Figure 8.33 shows the recommended design values for the lateral earth-pressure coefficient K. Note that

$$\sigma'_h = K\sigma'_o = K\gamma_1 z$$

where

σ'_h = effective lateral earth pressure
σ'_o = effective vertical stress
γ_1 = unit weight of granular backfill

In the figure, $K_a = \tan^2(45 - \phi'_1/2)$, where ϕ'_1 is the effective angle of friction of the backfill.

2. In Sections 8.13 and 8.15, the effective length l_e against tie pullout was calculated behind the Rankine failure surface (e.g., see Figure 8.23a). Recent field measurements and theoretical analysis show that the potential failure surface may depend on the type of reinforcement. Figure 8.34a shows the potential failure plane locations for walls with inextensible reinforcement in the granular backfill, and Figure 8.34b shows the locations for such walls with extensible reinforcement.

New developments in the design of MSE walls will be incorporated into future editions of this text.

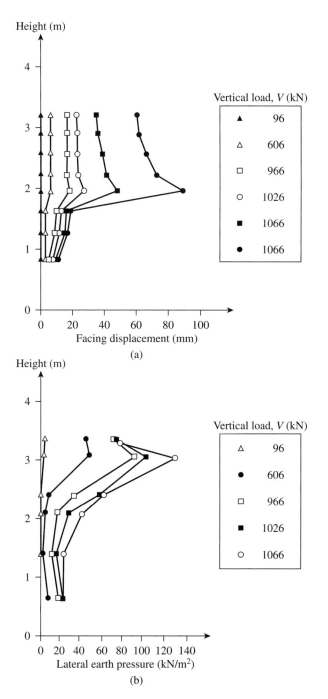

Figure 8.32 Observations from tests on the retaining wall shown in Figure 8.31: (a) facing displacement with loading; (b) lateral earth pressure with loading (Based on Thamm et al., 1990)

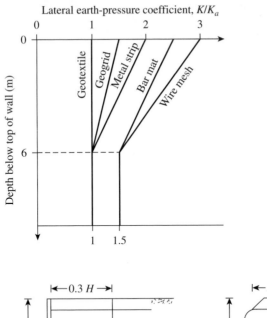

Figure 8.33 Recommended design values for lateral earth-pressure coefficient K (After Transportation Research Board, 1995)

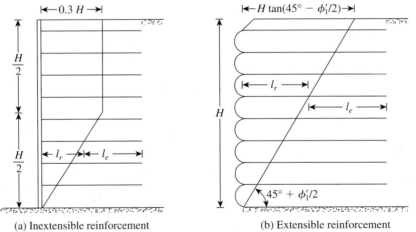

(a) Inextensible reinforcement   (b) Extensible reinforcement

Figure 8.34 Location of potential failure surface (After Transportation Research Board, 1995)

Problems

In problems 8.1 through 8.6, use $\gamma_{concrete} = 23.58$ kN/m³ (150 lb/ft³). Also, in Eq. (8.11), use $k_1 = k_2 = 2/3$ and $P_p = 0$.

8.1 For the cantilever retaining wall shown in Figure P8.1, let the following data be given:

Wall dimensions: $H = 18$ ft, $x_1 = 18$ in., $x_2 = 30$ in., $x_3 = 4$ ft, $x_4 = 6$ ft, $x_5 = 2.75$ ft, $D = 4$ ft, $\alpha = 10°$

Soil properties: $\gamma_1 = 117$ lb/ft³, $\phi'_1 = 34°$, $\gamma_2 = 110$ lb/ft³, $\phi'_2 = 18°$, $c'_2 = 800$ lb/ft²

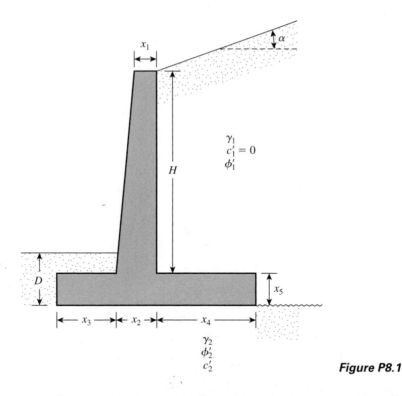

Figure P8.1

Calculate the factor of safety with respect to overturning, sliding, and bearing capacity.

8.2 Repeat Problem 8.1 with the following:

Wall dimensions: $H = 22$ ft, $x_1 = 12$ in., $x_2 = 27$ in., $x_3 = 4.5$ ft, $x_4 = 8$ ft, $x_5 = 2.75$ ft, $D = 4$ ft, $\alpha = 5°$

Soil properties: $\gamma_1 = 110$ lb/ft^3, $\phi_1' = 36°$, $\gamma_2 = 120$ lb/ft^3, $\phi_2' = 15°$, $c_2' = 1000$ lb/ft^2

8.3 Repeat Problem 8.1 with the following:

Wall dimensions: $H = 6.5$ m, $x_1 = 0.3$ m, $x_2 = 0.6$ m, $x_3 = 0.8$ m, $x_4 = 2$ m, $x_5 = 0.8$ m, $D = 1.5$ m, $\alpha = 0°$

Soil properties: $\gamma_1 = 18.08$ kN/m^3, $\phi_1' = 36°$, $\gamma_2 = 19.65$ kN/m^3, $\phi_2' = 15°$, $c_2' = 30$ kN/m^2

8.4 A gravity retaining wall is shown in Figure P8.4. Calculate the factor of safety with respect to overturning and sliding, given the following data:

Wall dimensions: $H = 6$ m, $x_1 = 0.6$ m, $x_2 = 2$ m, $x_3 = 2$ m, $x_4 = 0.5$ m, $x_5 = 0.75$ m, $x_6 = 0.8$ m, $D = 1.5$ m

Soil properties: $\gamma_1 = 16.5$ kN/m^3, $\phi_1' = 32°$, $\gamma_2 = 18$ kN/m^3, $\phi_2' = 22°$, $c_2' = 40$ kN/m^2

Use the Rankine active earth pressure in your calculation.

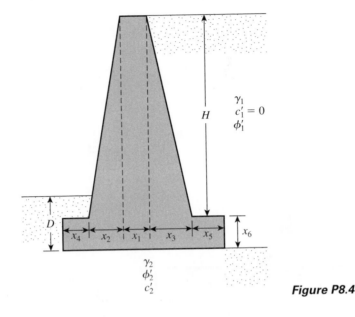

$$\gamma_1$$
$$c_1' = 0$$
$$\phi_1'$$

H

D

$x_4 \quad x_2 \quad x_1 \quad x_3 \quad x_5$

x_6

$$\gamma_2$$
$$\phi_2'$$
$$c_2'$$

Figure P8.4

8.5   Repeat Problem 8.4 using Coulomb's active earth pressure in your calculation and letting $\delta' = 2/3\ \phi_1'$.

8.6   Repeat Problem 8.4 with the following:

Wall dimensions: $H = 12$ ft, $x_1 = 1.5$ ft, $x_2 = 1$ ft, $x_3 = 5$ ft, $x_4 = 1.5$ ft, $x_5 = 2$ ft,
$x_6 = 2$ ft, $D = 3.5$ ft

Soil properties: $\gamma_1 = 115$ lb/ft³, $\phi_1' = 35°$, $\gamma_2 = 120$ lb/ft³, $\phi_2' = 25°$, $c_2' = 800$ lb/ft²

Use the Rankine active pressure in your calculation.

8.7   In Figure 8.23a, use the following parameters:

Wall: $H = 20$ ft

Soil: $\gamma_1 = 105$ lb/ft³, $\phi_1' = 30°$

Reinforcement: $S_V = 3$ ft and $S_H = 4$ ft

Surcharge: $q = 2000$ lb/ft², $a' = 5$ ft, and $b' = 6$ ft

Calculate the vertical stress σ_o' [Eqs. (8.24), (8.25) and (8.26)] at $z = 5$ ft, 10 ft,
15 ft and 20 ft.

8.8   For the data given in Problem 8.7, calculate the lateral pressure σ_a' at $z = 5$ ft,
10 ft, 15 ft and 20 ft. Use Eqs. (8.27), (8.28) and (8.29).

8.9   A reinforced earth retaining wall (Figure 8.23) is to be 30 ft. high. Here,

Backfill: unit weight, $\gamma_1 = 119$ lb/ft³ and soil friction angle, $\sigma_a' = 34°$

Reinforcement: vertical spacing, $S_V = 3$ ft; horizontal spacing, $S_H = 4$ ft; width
of reinforcement $= 4.75$ in., $f_\gamma = 38{,}000$ lb/in.² $\phi_\mu = 25°$;
factor of safety against tie pullout $= 3$; and factor of safety
against tie breaking $= 3$

Determine:
a. The required thickness of ties
b. The required maximum length of ties

8.10 In Problem 8.9 assume that the ties at all depths are the length determined in Part b. For the *in situ* soil, $\phi_2' = 25°$, $\gamma_2 = 116$ lb/ft^3, $c_2' = 650$ lb/ft^2. Calculate the factor of safety against (a) overturning, (b) sliding, and (c) bearing capacity failure.

8.11 Redo Problem 8.9 for a retaining wall with a height of 24 ft.

8.12 A retaining wall with geotextile reinforcement is 6-m high. For the granular backfill, $\gamma_1 = 15.9$ kN/m^3 and $\phi_1' = 30°$. For the geotextile, $\sigma_G = 16$ kN/m. For the design of the wall, determine S_V, L, and l_l. Use $FS_{(B)} = FS_{(P)} = 1.5$.

8.13 The S_V, L, and l_l determined in Problem 8.12, check the overall stability (i.e., factor of safety overturning, sliding, and bearing capacity failure) of the wall. For the *in situ* soil, $\gamma_2 = 16.8$ kN/m^3, $\phi_2' = 20°$, and $c_2' = 55$ kN/m^2.

References

Bell, J. R., Stilley, A. N., and Vandre, B. (1975). "Fabric Retaining Earth Walls," *Proceedings, Thirteenth Engineering Geology and Soils Engineering Symposium,* Moscow, ID.

Berg, R. R., Bonaparte, R., Anderson, R. P., and Chouery, V. E. (1986). "Design Construction and Performance of Two Tensar Geogrid Reinforced Walls," *Proceedings, Third International Conference on Geotextiles,* Vienna, pp. 401–406.

Binquet, J., and Lee, K. L. (1975). "Bearing Capacity Analysis of Reinforced Earth Slabs," *Journal of the Geotechnical Engineering Division,* American Society of Civil Engineers, Vol. 101, No. GT12, pp. 1257–1276.

Carroll, R., Jr. (1988). "Specifying Geogrids," *Geotechnical Fabric Report,* Industrial Fabric Association International, St. Paul, March/April.

Casagrande, L. (1973). "Comments on Conventional Design of Retaining Structure," *Journal of the Soil Mechanics and Foundations Division,* ASCE, Vol. 99, No. SM2, pp. 181–198.

Darbin, M. (1970). "Reinforced Earth for Construction of Freeways" (in French), *Revue Générale des Routes et Aerodromes,* No. 457, September.

Federal Highway Administration (1996). *Mechanically Stabilized Earth Walls and Reinforced Soil Slopes Design and Construction Guidelines,* Publication No. FHWA-SA-96-071, Washington, DC.

Gould, J. P. (1970). "Lateral Pressures on Rigid Permanent Structures," *Proceedings,* Specialty Conference on Lateral Stresses in the Ground and Design of Earth Retaining Structures, American Society of Civil Engineers, pp. 219–270.

Koerner, R. B. (1990). *Design with Geosynthetics,* 2d ed., Prentice Hall, Englewood Cliffs, NJ.

Laba, J. T., and Kennedy, J. B. (1986). "Reinforced Earth Retaining Wall Analysis and Design," *Canadian Geotechnical Journal,* Vol. 23, No. 3, pp. 317–326.

Lee, K. L., Adams, B. D., and Vagneron, J. J. (1973). "Reinforced Earth Retaining Walls," *Journal of the Soil Mechanics and Foundations Division,* American Society of Civil Engineers, Vol. 99, No. SM10, pp. 745–763.

Martin, J. P., Koerner, R. M., and Whitty, J. E. (1984). "Experimental Friction Evaluation of Slippage Between Geomembranes, Geotextiles, and Soils," *Proceedings,* International Conference on Geomembranes, Denver, pp. 191–196.

Schlosser, F., and Long, N. (1974). "Recent Results in French Research on Reinforced Earth," *Journal of the Construction Division,* American Society of Civil Engineers, Vol. 100, No. CO3, pp. 113–237.

Schlosser, F., and Vidal, H. (1969). "Reinforced Earth" (in French), *Bulletin de Liaison des Laboratoires Routier,* Ponts et Chassées, Paris, France, November, pp. 101–144.

Tensar Corporation (1986). Tensar Technical Note. No. TTN:RW1, August.

Terzaghi, K., and Peck, R. B. (1967). *Soil Mechanics in Engineering Practice,* Wiley, New York.

Thamm, B. R., Krieger, B., and Krieger, J. (1990). "Full-Scale Test on a Geotextile-Reinforced Retaining Structure," *Proceedings,* Fourth International Conference on Geotextiles Geomembranes, and Related Products, The Hague, Vol. 1, pp. 3–8.

Transportation Research Board (1995). Transportation Research Circular No. 444, National Research Council, Washington, DC.

Vidal, H. (1966). "La terre Armée," *Annales de l'Institut Technique du Bâtiment et des Travaux Publiques,* France, July–August, pp. 888–938.

9

Sheet Pile Walls

9.1 Introduction

Connected or semiconnected sheet piles are often used to build continuous walls for waterfront structures that range from small waterfront pleasure boat launching facilities to large dock facilities. (See Figure 9.1.) In contrast to the construction of other types of retaining wall, the building of sheet pile walls does not usually require dewatering of the site. Sheet piles are also used for some temporary structures, such as braced cuts. (See Chapter 10.) The principles of sheet-pile wall design are discussed in the current chapter.

Several types of sheet pile are commonly used in construction: (a) wooden sheet piles, (b) precast concrete sheet piles, and (c) steel sheet piles. Aluminum sheet piles are also marketed.

Wooden sheet piles are used only for temporary, light structures that are above the water table. The most common types are ordinary wooden planks and *Wakefield piles*. The wooden planks are about 50 mm × 300 mm (2 in. × 12 in.) in cross section and are driven edge to edge (Figure 9.2a). Wakefield piles are made by nailing three planks together, with the middle plank offset by 50 to 75 mm (2 to 3 in.) (Figure 9.2b). Wooden planks can also be milled to form *tongue-and-groove piles,* as shown in

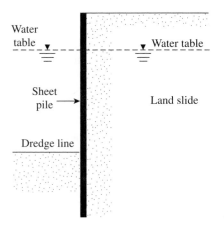

Figure 9.1 Example of waterfront sheet-pile wall

409

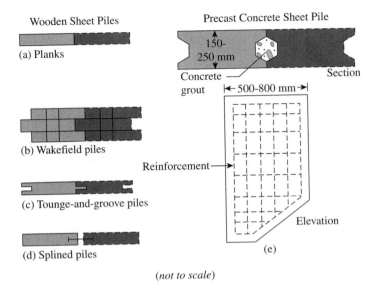

Figure 9.2   Various types of wooden and concrete sheet pile

Figure 9.2c. Figure 9.2d shows another type of wooden sheet pile that has precut grooves. Metal *splines* are driven into the grooves of the adjacent sheetings to hold them together after they are sunk into the ground.

Precast concrete sheet piles are heavy and are designed with reinforcements to withstand the permanent stresses to which the structure will be subjected after construction and also to handle the stresses produced during construction. In cross section, these piles are about 500 to 800 mm (20 to 32 in.) wide and 150 to 250 mm (6 to 10 in.) thick. Figure 9.2e is a schematic diagram of the elevation and the cross section of a reinforced concrete sheet pile.

Steel sheet piles in the United States are about 10 to 13 mm (0.4 to 0.5 in.) thick. European sections may be thinner and wider. Sheet-pile sections may be *Z*, *deep arch, low arch,* or *straight web* sections. The interlocks of the sheet-pile sections are shaped like a *thumb-and-finger* or *ball-and-socket* joint for watertight connections. Figure 9.3a is a schematic diagram of the thumb-and-finger type of interlocking for straight web sections. The ball-and-socket type of interlocking for *Z* section piles is shown in Figure 9.3b. Figure 9.3c shows a sheet pile wall. Table 9.1 lists the properties of the steel sheet pile sections produced by the Bethlehem Steel Corporation. The allowable design flexural stress for the steel sheet piles is as follows:

Type of steel	Allowable stress	
ASTM A-328	170 MN/m^2	(25,000 lb/in^2)
ASTM A-572	210 MN/m^2	(30,000 lb/in^2)
ASTM A-690	210 MN/m^2	(30,000 lb/in^2)

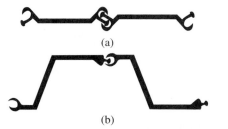

(a)

(b)

(c)

Figure 9.3 (a) Thumb-and-finger type sheet pile connection; (b) ball-and-socket type sheet-pile connection; (c) steel sheet-pile wall

Table 9.1 Properties of Some Sheet-Pile Sections Produced by Bethlehem Steel Corporation

Section designation	Sketch of section	Section modulus		Moment of inertia	
		m³/m of wall	in³/ft of wall	m⁴/m of wall	in⁴/ft of wall
PZ-40		326.4×10^{-5}	60.7	670.5×10^{-6}	490.8

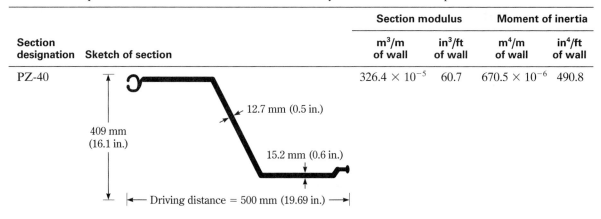

(Continued)

Table 9.1 (Continued)

Section designation	Sketch of section	Section modulus		Moment of inertia	
		m³/m of wall	in³/ft of wall	m⁴/m of wall	in⁴/ft of wall
PZ-35		260.5×10^{-5}	48.5	493.4×10^{-6}	361.2
PZ-27		162.3×10^{-5}	30.2	251.5×10^{-6}	184.2
PZ-22		97×10^{-5}	18.1	115.2×10^{-6}	84.4
PSA-31		10.8×10^{-5}	2.01	4.41×10^{-6}	3.23
PSA-23		12.8×10^{-5}	2.4	5.63×10^{-6}	4.13

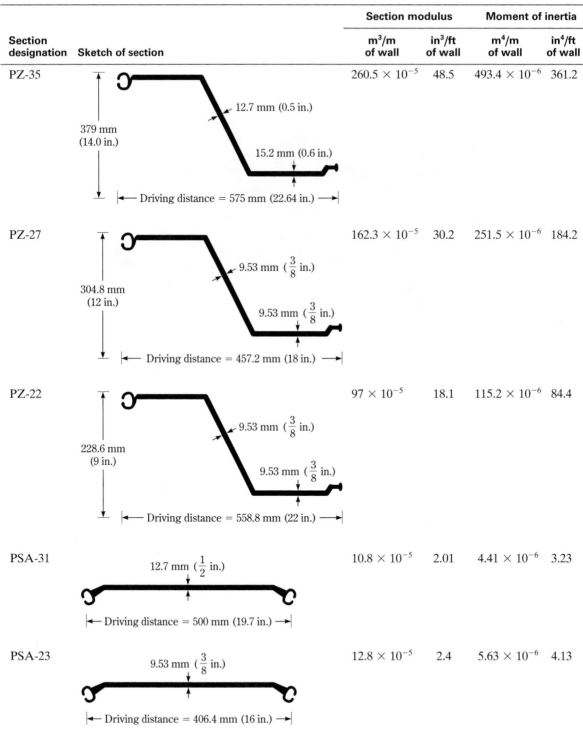

Steel sheet piles are convenient to use because of their resistance to the high driving stress that is developed when they are being driven into hard soils. Steel sheet piles are also lightweight and reusable.

9.2 *Construction Methods*

Sheet pile walls may be divided into two basic categories: (a) cantilever and (b) anchored.

In the construction of sheet pile walls, the sheet pile may be driven into the ground and then the backfill placed on the land side, or the sheet pile may first be driven into the ground and the soil in front of the sheet pile dredged. In either case, the soil used for backfill behind the sheet pile wall is usually granular. The soil below the dredge line may be sandy or clayey. The surface of soil on the water side is referred to as the *mud line* or *dredge line*.

Thus, construction methods generally can be divided into two categories (Tsinker, 1983):

1. Backfilled structure
2. Dredged structure

The sequence of construction for a *backfilled structure* is as follows (see Figure 9.4):

Step 1. Dredge the *in situ* soil in front and back of the proposed structure.
Step 2. Drive the sheet piles.
Step 3. Backfill up to the level of the anchor, and place the anchor system.
Step 4. Backfill up to the top of the wall.

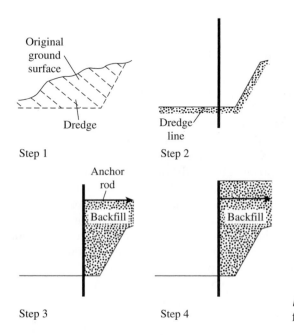

Figure 9.4 Sequence of construction for a backfilled structure

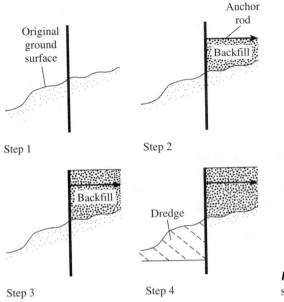

Figure 9.5 Sequence of construction for a dredged structure

For a cantilever type of wall, only Steps 1, 2, and 4 apply. The sequence of construction for a *dredged structure* is as follows (see Figure 9.5):

Step 1. Drive the sheet piles.
Step 2. Backfill up to the anchor level, and place the anchor system.
Step 3. Backfill up to the top of the wall.
Step 4. Dredge the front side of the wall.

With cantilever sheet pile walls, Step 2 is not required.

9.3 *Cantilever Sheet Pile Walls*

Cantilever sheet pile walls are usually recommended for walls of moderate height—about 6 m ($\approx$20 ft) or less, measured above the dredge line. In such walls, the sheet piles act as a wide cantilever beam above the dredge line. The basic principles for estimating net lateral pressure distribution on a cantilever sheet-pile wall can be explained with the aid of Figure 9.6. The figure shows the nature of lateral yielding of a cantilever wall penetrating a sand layer below the dredge line. The wall rotates about point O. Because the hydrostatic pressures at any depth from both sides of the wall will cancel each other, we consider only the effective lateral soil pressures. In zone A, the lateral pressure is just the active pressure from the land side. In zone B, because of the nature of yielding of the wall, there will be active pressure from the land side and passive pressure from the water side. The condition is reversed in zone C—that is, below the point of rotation, O. The net actual pressure distribution on the wall is like that shown in Figure 9.6b. However, for design purposes, Figure 9.6c shows a simplified version.

Sections 9.4 through 9.7 present the mathematical formulation of the analysis of cantilever sheet pile walls. Note that, in some waterfront structures, the water

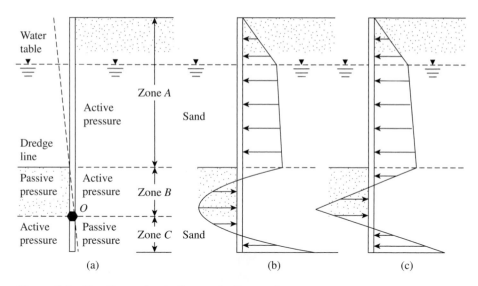

Figure 9.6 Cantilever sheet pile penetrating sand

level may fluctuate as the result of tidal effects. Care should be taken in determining the water level that will affect the net pressure diagram.

9.4 *Cantilever Sheet Piling Penetrating Sandy Soils*

To develop the relationships for the proper depth of embedment of sheet piles driven into a granular soil, examine Figure 9.7a. The soil retained by the sheet piling above the dredge line also is sand. The water table is at a depth L_1 below the top of the wall. Let the effective angle of friction of the sand be ϕ'. The intensity of the active pressure at a depth $z = L_1$ is

$$\sigma'_1 = \gamma L_1 K_a \tag{9.1}$$

where

K_a = Rankine active pressure coefficient = $\tan^2(45 - \phi'/2)$
γ = unit weight of soil above the water table

Similarly, the active pressure at a depth $z = L_1 + L_2$ (i.e., at the level of the dredge line) is

$$\sigma'_2 = (\gamma L_1 + \gamma' L_2) K_a \tag{9.2}$$

where γ' = effective unit weight of soil = $\gamma_{sat} - \gamma_w$.
Note that, at the level of the dredge line, the hydrostatic pressures from both sides of the wall are the same magnitude and cancel each other.
To determine the net lateral pressure below the dredge line up to the point of rotation, O, as shown in Figure 9.6a, an engineer has to consider the passive pressure acting from the left side (the water side) toward the right side (the land side) of the

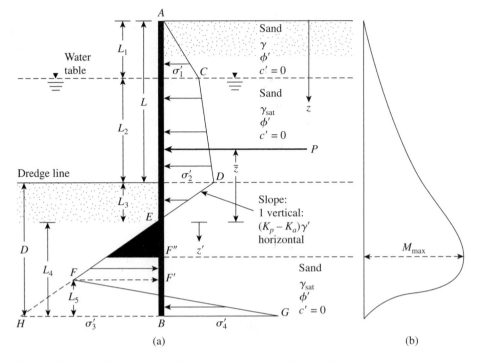

Figure 9.7 Cantilever sheet pile penetrating sand: (a) variation of net pressure diagram; (b) variation of moment

wall and also the active pressure acting from the right side toward the left side of the wall. For such cases, ignoring the hydrostatic pressure from both sides of the wall, the active pressure at depth z is

$$\sigma'_a = [\gamma L_1 + \gamma' L_2 + \gamma'(z - L_1 - L_2)]K_a \tag{9.3}$$

Also, the passive pressure at depth z is

$$\sigma'_p = \gamma'(z - L_1 - L_2)K_p \tag{9.4}$$

where K_p = Rankine passive pressure coefficient = $\tan^2(45 + \phi'/2)$.

Combining Eqs. (9.3) and (9.4) yields the net lateral pressure, namely,

$$\sigma' = \sigma'_a - \sigma'_p = (\gamma L_1 + \gamma' L_2)K_a - \gamma'(z - L_1 - L_2)(K_p - K_a)$$

$$= \sigma'_2 - \gamma'(z - L)(K_p - K_a) \tag{9.5}$$

where $L = L_1 + L_2$.

The net pressure, σ' equals zero at a depth L_3 below the dredge line, so

$$\sigma'_2 - \gamma'(z - L)(K_p - K_a) = 0$$

or

$$(z - L) = L_3 = \frac{\sigma'_2}{\gamma'(K_p - K_a)} \tag{9.6}$$

Equation (9.6) indicates that the slope of the net pressure distribution line DEF is 1 vertical to $(K_p - K_a)\gamma'$ horizontal, so, in the pressure diagram,

$$\overline{HB} = \sigma_3' = L_4(K_p - K_a)\gamma' \tag{9.7}$$

At the bottom of the sheet pile, passive pressure, σ_p', acts from the right toward the left side, and active pressure acts from the left toward the right side of the sheet pile, so, at $z = L + D$,

$$\sigma_p' = (\gamma L_1 + \gamma' L_2 + \gamma' D)K_p \tag{9.8}$$

At the same depth,

$$\sigma_a' = \gamma' D K_a \tag{9.9}$$

Hence, the net lateral pressure at the bottom of the sheet pile is

$$\begin{aligned}
\sigma_p' - \sigma_a' = \sigma_4' &= (\gamma L_1 + \gamma' L_2)K_p + \gamma' D(K_p - K_a) \\
&= (\gamma L_1 + \gamma' L_2)K_p + \gamma' L_3(K_p - K_a) + \gamma' L_4(K_p - K_a) \\
&= \sigma_5' + \gamma' L_4(K_p - K_a) \tag{9.10}
\end{aligned}$$

where

$$\sigma_5' = (\gamma L_1 + \gamma' L_2)K_p + \gamma' L_3(K_p - K_a) \tag{9.11}$$
$$D = L_3 + L_4 \tag{9.12}$$

For the stability of the wall, the principles of statics can now be applied:

$$\Sigma \text{ horizontal forces per unit length of wall} = 0$$

and

$$\Sigma \text{ moment of the forces per unit length of wall about point } B = 0$$

For the summation of the horizontal forces, we have

Area of the pressure diagram $ACDE$ − area of $EFHB$ + area of $FHBG = 0$

or

$$P - \tfrac{1}{2}\sigma_3' L_4 + \tfrac{1}{2}L_5(\sigma_3' + \sigma_4') = 0 \tag{9.13}$$

where P = area of the pressure diagram $ACDE$.

Summing the moment of all the forces about point B yields

$$P(L_4 + \bar{z}) - \left(\frac{1}{2}L_4\sigma_3'\right)\left(\frac{L_4}{3}\right) + \frac{1}{2}L_5(\sigma_3' + \sigma_4')\left(\frac{L_5}{3}\right) = 0 \tag{9.14}$$

From Eq. (9.13),

$$L_5 = \frac{\sigma_3' L_4 - 2P}{\sigma_3' + \sigma_4'} \tag{9.15}$$

Combining Eqs. (9.7), (9.10), (9.14), and (9.15) and simplifying them further, we obtain the following fourth-degree equation in terms of L_4:

$$L_4^4 + A_1 L_4^3 - A_2 L_4^2 - A_3 L_4 - A_4 = 0 \tag{9.16}$$

In this equation,

$$A_1 = \frac{\sigma_5'}{\gamma'(K_p - K_a)} \tag{9.17}$$

$$A_2 = \frac{8P}{\gamma'(K_p - K_a)} \tag{9.18}$$

$$A_3 = \frac{6P[2\bar{z}\gamma'(K_p - K_a) + \sigma_5']}{\gamma'^2(K_p - K_a)^2} \tag{9.19}$$

$$A_4 = \frac{P(6\bar{z}\sigma_5' + 4P)}{\gamma'^2(K_p - K_a)^2} \tag{9.20}$$

Step-by-Step Procedure for Obtaining the Pressure Diagram

Based on the preceding theory, a step-by-step procedure for obtaining the pressure diagram for a cantilever sheet pile wall penetrating a granular soil is as follows:

Step 1. Calculate K_a and K_p.
Step 2. Calculate σ_1' [Eq. (9.1)] and σ_2' [Eq. (9.2)]. (*Note:* L_1 and L_2 will be given.)
Step 3. Calculate L_3 [Eq. (9.6)].
Step 4. Calculate P.
Step 5. Calculate $\bar{z}$ (i.e., the center of pressure for the area $ACDE$) by taking the moment about E.
Step 6. Calculate σ_5' [Eq. (9.11)].
Step 7. Calculate A_1, A_2, A_3, and A_4 [Eqs. (9.17) through (9.20)].
Step 8. Solve Eq. (9.16) by trial and error to determine L_4.
Step 9. Calculate σ_4' [Eq. (9.10)].
Step 10. Calculate σ_3' [Eq. (9.7)].
Step 11. Obtain L_5 from Eq. (9.15).
Step 12. Draw a pressure distribution diagram like the one shown in Figure 9.7a.
Step 13. Obtain the theoretical depth [see Eq. (9.12)] of penetration as $L_3 + L_4$. The actual depth of penetration is increased by about 20 to 30%.

Note that some designers prefer to use a factor of safety on the passive earth pressure coefficient at the beginning. In that case, in Step 1,

$$K_{p(\text{design})} = \frac{K_p}{FS}$$

where FS = factor of safety (usually between 1.5 and 2).

For this type of analysis, follow Steps 1 through 12 with the value of $K_a = \tan^2(45 - \phi'/2)$ and $K_{p(\text{design})}$ (instead of K_p). The actual depth of penetration can now be determined by adding L_3, obtained from Step 3, and L_4, obtained from Step 8.

Calculation of Maximum Bending Moment

The nature of the variation of the moment diagram for a cantilever sheet pile wall is shown in Figure 9.7b. The maximum moment will occur between points E and F'. Obtaining the maximum moment ($M_{\max}$) per unit length of the wall requires determining the point of zero shear. For a new axis z' (with origin at point E) for zero shear,

$$P = \tfrac{1}{2}(z')^2(K_p - K_a)\gamma'$$

or

$$z' = \sqrt{\frac{2P}{(K_p - K_a)\gamma'}} \tag{9.21}$$

Once the point of zero shear force is determined (point F'' in Figure 9.7a), the magnitude of the maximum moment can be obtained as

$$M_{\max} = P(\bar{z} + z') - [\tfrac{1}{2}\gamma'z'^2(K_p - K_a)](\tfrac{1}{3})z' \tag{9.22}$$

The necessary profile of the sheet piling is then sized according to the allowable flexural stress of the sheet pile material, or

$$S = \frac{M_{\max}}{\sigma_{\text{all}}} \tag{9.23}$$

where

S = section modulus of the sheet pile required per unit length of the structure
σ_{all} = allowable flexural stress of the sheet pile

Example 9.1

Figure 9.8 shows a cantilever sheet-pile wall penetrating a granular soil.

a. What is the theoretical depth of embedment, D?
b. For a 30% increase in D, what should be the total length of the sheet piles?
c. What should be the minimum section modulus of the sheet piles? Use $\sigma_{\text{all}} = 30$ kip/ft^2.

Solution
Part a
The step-by-step procedure given in Section 9.4 will be followed here.

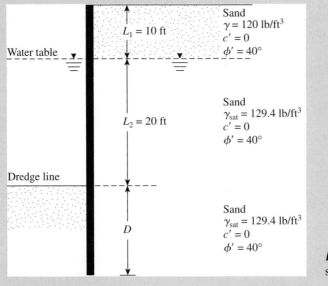

Figure 9.8 Cantilever sheet-pile wall

Step 1.

$$K_a = \tan^2\left(45 - \frac{\phi'}{2}\right) = \tan^2\left(45 - \frac{40}{2}\right) = 0.217$$

$$K_p = \tan^2\left(45 + \frac{\phi'}{2}\right) = 4.599$$

Step 2.

$$\sigma_1' = \gamma L_1 K_a = (0.12)(10)(0.217) = 0.26 \text{ kip/ft}^2$$

$$\sigma_2' = (\gamma L_1 + \gamma' L_2) K_a = [(0.12)(10) + (0.1294 - 0.0624)(20)]0.217 = 0.551 \text{ kip/ft}^2$$

Step 3.

$$L_3 = \frac{\sigma_2'}{\gamma'(K_p - K_a)} = \frac{0.551}{(0.1294 - 0.0624)(4.599 - 0.217)} = 1.88 \text{ ft}$$

Step 4.

$$P = \frac{1}{2}\sigma_1' L_1 + \sigma_1' L_2 + \frac{1}{2}(\sigma_2' - \sigma_1')L_2 + \frac{1}{2}\sigma_2' L_3$$

$$= \frac{1}{2}(0.26)(10) + (0.26)(20) + \frac{1}{2}(0.551 - 0.26)(20) + \frac{1}{2}(0.551)(1.88)$$

$$= 1.3 + 5.2 + 2.91 + 0.518 = 9.928 \text{ kip/ft}$$

Step 5.

Taking the moment about E (see Figure 9.7)yields

$$\frac{(1.3)\left(\dfrac{10}{3} + 20 + 1.88\right) + (5.2)\left(1.88 + \dfrac{20}{2}\right) + (2.91)\left(1.88 + \dfrac{20}{3}\right) + (0.518)\left(\dfrac{2}{3}\right)(1.88)}{9.928} = 12.1\text{ft}$$

Step 6.

$$\sigma'_5 = (\gamma L_1 + \gamma' L_2)K_p + \gamma' L_3(K_p - K_a)$$

$$= [(0.12)(10) + (0.1294 - 0.0624)(20)]4.599 + (0.067)(1.88)(4.599 - 0.217)$$

$$= 12.233 \text{ kip/ft}^2$$

Step 7.

$$A_1 = \frac{\sigma'_5}{\gamma'(K_p - K_a)} = \frac{12.233}{(0.1294 - 0.0624)(4.382)} = 41.7$$

$$A_2 = \frac{8P}{\gamma'(K_p - K_a)} = \frac{(8)(9.928)}{(0.1294 - 0.0624)(4.382)} = 270.5$$

$$A_3 = \frac{6P[2\bar{z}\gamma'(K_p - K_a) + \sigma'_5]}{\gamma'^2(K_p - K_a)^2} = \frac{(6)(9.928)[(2)(12.1)(0.067)(4.382) + 12.233]}{(0.067)^2(4.382)^2} = 13,363$$

$$A_4 = \frac{P(6\bar{z}\sigma'_5 + 4P)}{\gamma'^2(K_p - K_a)^2} = \frac{(9.928)[(6)(12.1)(12.233) + (4)(9.928)]}{(0.067)^2(4.382)^2} = 106,863$$

Step 8.

From Eq. (9.16),

$$L_4^4 + 41.7L_4^3 - 270.5L_4^2 - 13,363L_4 - 106,863 = 0$$

By trial and error, $L_4 \approx 20$ ft and

$$D = 1.88 + 20 \approx \mathbf{21.88 \text{ ft}}$$

Part b

Total length of the sheet piles $= 10 + 20 + (1.3)(21.88) = \mathbf{58.4 \text{ ft}}$

Part c
Using Eq. (9.21) gives

$$z' = \sqrt{\frac{2P}{\gamma'(K_p - K_a)}} = \sqrt{\frac{(2)(9.928)}{(0.067)(4.382)}} = 8.23 \text{ ft}$$

From Eq. (9.22),

$$M_{max} = P(\bar{z} + z') - \left[\frac{1}{2}\gamma'z'^2(K_p - K_a)\right]\left(\frac{z'}{3}\right)$$

$$= (9.928)(12.1 + 8.23) - \left[\frac{1}{6}(0.067)(8.23)^3(4.382)\right] = 174.7 \text{ kip-ft/ft}$$

$$= 2097 \text{ kip-in./ft}$$

$$S = \frac{2097}{\sigma_{all}} = \frac{2097}{30} = 69.9 \text{ in}^3/\text{ft}$$

∎

9.5   *Special Cases for Cantilever Walls Penetrating a Sandy Soil*

In the absence of the water table, the net pressure diagram on the cantilever sheet-pile wall will be as shown in Figure 9.9, which is a modified version of Figure 9.7. In this case,

$$\sigma_2' = \gamma L K_a \tag{9.24}$$

$$\sigma_3' = L_4(K_p - K_a)\gamma \tag{9.25}$$

$$\sigma_4' = \sigma_5' + \gamma L_4(K_p - K_a) \tag{9.26}$$

$$\sigma_5' = \gamma L K_p + \gamma L_3(K_p - K_a) \tag{9.27}$$

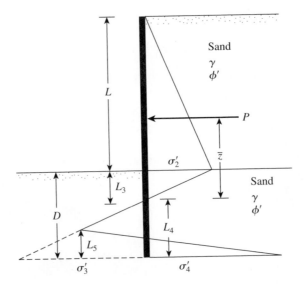

Figure 9.9 Sheet piling penetrating a sandy soil in the absence of the water table

$$L_3 = \frac{\sigma_2'}{\gamma(K_p - K_a)} = \frac{LK_a}{(K_p - K_a)} \tag{9.28}$$

$$P = \tfrac{1}{2}\sigma_2'L + \tfrac{1}{2}\sigma_2'L_3 \tag{9.29}$$

$$\bar{z} = L_3 + \frac{L}{3} = \frac{LK_a}{K_p - K_a} + \frac{L}{3} = \frac{L(2K_a + K_p)}{3(K_p - K_a)} \tag{9.30}$$

and Eq. (9.16) transforms to

$$L_4^4 + A_1'L_4^3 - A_2'L_4^2 - A_3'L_4 - A_4' = 0 \tag{9.31}$$

where

$$A_1' = \frac{\sigma_5'}{\gamma(K_p - K_a)} \tag{9.32}$$

$$A_2' = \frac{8P}{\gamma(K_p - K_a)} \tag{9.33}$$

$$A_3' = \frac{6P[2\bar{z}\gamma(K_p - K_a) + \sigma_5']}{\gamma^2(K_p - K_a)^2} \tag{9.34}$$

$$A_4' = \frac{P(6\bar{z}\sigma_5' + 4P)}{\gamma^2(K_p - K_a)^2} \tag{9.35}$$

9.6   *Cantilever Sheet Piling Penetrating Clay*

At times, cantilever sheet piles must be driven into a clay layer possessing an undrained cohesion $c(\phi = 0)$. The net pressure diagram will be somewhat different from that shown in Figure 9.7a. Figure 9.10 shows a cantilever sheet-pile wall driven into clay with a backfill of granular soil above the level of the dredge line. The water table is at a depth L_1 below the top of the wall. As before, Eqs. (9.1) and (9.2) give the intensity of the net pressures σ_1' and σ_2', and the diagram for pressure distribution above the level of the dredge line can be drawn. The diagram for net pressure distribution below the dredge line can now be determined as follows.

At any depth greater than $L_1 + L_2$, for $\phi = 0$, the Rankine active earth-pressure coefficient $K_a = 1$. Similarly, for $\phi = 0$, the Rankine passive earth-pressure coefficient $(K_p) = 1$. Thus, above the point of rotation (point O in Figure 9.6a), the active pressure, from right to left is

$$\sigma_a = [\gamma L_1 + \gamma'L_2 + \gamma_{\text{sat}}(z - L_1 - L_2)] - 2c \tag{9.36}$$

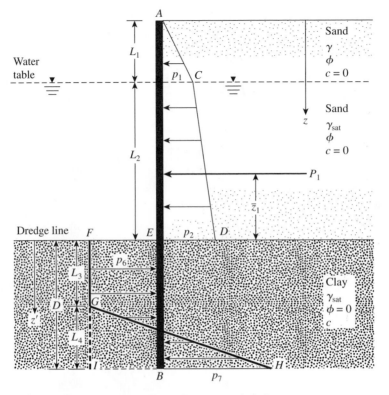

Figure 9.10 Cantilever sheet pile penetrating clay

Similarly, the passive pressure from left to right may be expressed as

$$\sigma_p = \gamma_{sat}(z - L_1 - L_2) + 2c \tag{9.37}$$

Thus, the net pressure is

$$\begin{aligned}
\sigma_6 = \sigma_p - \sigma_a &= [\gamma_{sat}(z - L_1 - L_2) + 2c] \\
&\quad - [\gamma L_1 + \gamma' L_2 + \gamma_{sat}(z - L_1 - L_2)] + 2c \\
&= 4c - (\gamma L_1 + \gamma' L_2)
\end{aligned} \tag{9.38}$$

At the bottom of the sheet pile, the passive pressure from right to left is

$$\sigma_p = (\gamma L_1 + \gamma' L_2 + \gamma_{sat}D) + 2c \tag{9.39}$$

Similarly, the active pressure from left to right is

$$\sigma_a = \gamma_{sat}D - 2c \tag{9.40}$$

Hence, the net pressure is

$$\sigma_7 = \sigma_p - \sigma_a = 4c + (\gamma L_1 + \gamma' L_2) \tag{9.41}$$

For equilibrium analysis, $\Sigma F_H = 0$; that is, the area of the pressure diagram $ACDE$ minus the area of $EFIB$ plus the area of $GIH = 0$, or

$$P_1 - [4c - (\gamma L_1 + \gamma' L_2)]D + \tfrac{1}{2}L_4[4c - (\gamma L_1 + \gamma' L_2) + 4c + (\gamma L_1 + \gamma' L_2)] = 0$$

where P_1 = area of the pressure diagram $ACDE$.

Simplifying the preceding equation produces

$$L_4 = \frac{D[4c - (\gamma L_1 + \gamma' L_2)] - P_1}{4c} \qquad (9.42)$$

Now, taking the moment about point B ($\Sigma M_B = 0$) yields

$$P_1(D + \bar{z}_1) - [4c - (\gamma L_1 + \gamma' L_2)]\frac{D^2}{2} + \frac{1}{2}L_4(8c)\left(\frac{L_4}{3}\right) = 0 \qquad (9.43)$$

where $\bar{z}_1$ = distance of the center of pressure of the pressure diagram ACDE, measured from the level of the dredge line.
    Combining Eqs. (9.42) and (9.43) yields

$$D^2[4c - (\gamma L_1 + \gamma' L_2)] - 2DP_1 - \frac{P_1(P_1 + 12c\bar{z}_1)}{(\gamma L_1 + \gamma' L_2) + 2c} = 0 \qquad (9.44)$$

Equation (9.44) may be solved to obtain D, the theoretical depth of penetration of the clay layer by the sheet pile.

Step-by-Step Procedure for Obtaining the Pressure Diagram

Step 1. Calculate $K_a = \tan^2(45 - \phi'/2)$ for the granular soil (backfill).
Step 2. Obtain σ_1' and σ_2'. [See Eqs. (9.1) and (9.2).]
Step 3. Calculate P_1 and $\bar{z}_1$.
Step 4. Use Eq. (9.44) to obtain the theoretical value of D.
Step 5. Using Eq. (9.42), calculate L_4.
Step 6. Calculate σ_6 and σ_7. [See Eqs. (9.38) and (9.41).]
Step 7. Draw the pressure distribution diagram as shown in Figure 9.10.
Step 8. The actual depth of penetration is

$$D_{actual} = 1.4 \text{ to } 1.6(D_{theoretical})$$

Maximum Bending Moment

According to Figure 9.10, the maximum moment (zero shear) will be between $L_1 + L_2 < z < L_1 + L_2 + L_3$. Using a new coordinate system z' (with $z' = 0$ at the dredge line) for zero shear gives

$$P_1 - \sigma_6 z' = 0$$

or

$$z' = \frac{P_1}{\sigma_6} \qquad (9.45)$$

The magnitude of the maximum moment may now be obtained:

$$M_{max} = P_1(z' + \bar{z}_1) - \frac{\sigma_6 z'^2}{2} \tag{9.46}$$

Knowing the maximum bending moment, we determine the section modulus of the sheet pile section from Eq. (9.23).

Example 9.2

In Figure 9.11, for the sheet pile wall, determine

a. The theoretical and actual depth of penetration. Use $D_{actual} = 1.5 D_{theory}$.
b. The minimum size of sheet pile section necessary. Use $\sigma_{all} = 172.5 \text{ MN/m}^2$.

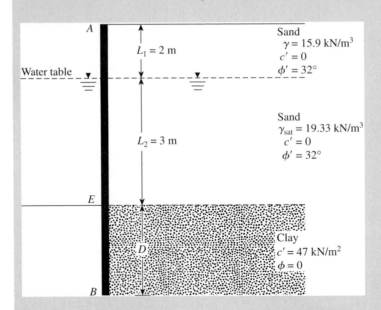

Figure 9.11 Cantilever sheet pile penetrating into saturated clay

Solution

We will follow the step-by-step procedure given in Section 9.6:

Step 1.

$$K_a = \tan^2\left(45 - \frac{\phi'}{2}\right) = \tan^2\left(45 - \frac{32}{2}\right) = 0.307$$

Step 2.

$$\sigma_1' = \gamma L_1 K_a = (15.9)(2)(0.307) = 9.763 \text{ kN/m}^2$$

$$\sigma_2' = (\gamma L_1 + \gamma' L_2) K_a = [(15.9)(2) + (19.33 - 9.81)3]0.307$$

$$= 18.53 \text{ kN/m}^2$$

Step 3. From the net pressure distribution diagram given in Figure 9.10, we have

$$P_1 = \frac{1}{2}\sigma_1' L_1 + \sigma_1' L_2 + \frac{1}{2}(\sigma_2' - \sigma_1') L_2$$

$$= 9.763 + 29.289 + 13.151 = 52.2 \text{ kN/m}$$

and

$$\bar{z}_1 = \frac{1}{52.2}\left[9.763\left(3 + \frac{2}{3}\right) + 29.289\left(\frac{3}{2}\right) + 13.151\left(\frac{3}{3}\right)\right]$$

$$= 1.78 \text{ m}$$

Step 4. From Eq. (9.44),

$$D^2[4c - (\gamma L_1 + \gamma' L_2)] - 2DP_1 - \frac{P_1(P_1 + 12c\bar{z}_1)}{(\gamma L_1 + \gamma' L_2) + 2c} = 0$$

Substituting proper values yields

$$D^2\{(4)(47) - [(2)(15.9) + (19.33 - 9.81)3]\} - 2D(52.2)$$

$$- \frac{52.2[52.2 + (12)(47)(1.78)]}{[(15.9)(2) + (19.33 - 9.81)3] + (2)(47)} = 0$$

or

$$127.64D^2 - 104.4D - 357.15 = 0$$

Solving the preceding equation, we obtain $D = 2.13$ m.

Step 5. From Eq. (9.42),

$$L_4 = \frac{D[4c - (\gamma L_1 + \gamma' L_2)] - P_1}{4c}$$

and

$$4c - (\gamma L_1 + \gamma' L_2) = (4)(47) - [(15.9)(2) + (19.33 - 9.81)3]$$

$$= 127.64 \text{ kN/m}^2$$

So,

$$L_4 = \frac{2.13(127.64) - 52.2}{(4)(47)} = 1.17 \text{ m}$$

Step 6.

$$\sigma_6 = 4c - (\gamma L_1 + \gamma' L_2) = 127.64 \text{ kN/m}^2$$

$$\sigma_7 = 4c + (\gamma L_1 + \gamma' L_2) = 248.36 \text{ kN/m}^2$$

Step 7.   The net pressure distribution diagram can now be drawn, as shown in Figure 9.10.

Step 8. $D_{\text{actual}} \approx 1.5 D_{\text{theoretical}} = 1.5(2.13) \approx$ **3.2 m**

Maximum-Moment Calculation
From Eq. (9.45),

$$z' = \frac{P_1}{\sigma_6} = \frac{52.2}{127.64} \approx 0.41 \text{ m}$$

Again, from Eq. (9.46),

$$M_{\text{max}} = P_1(z' + \bar{z}_1) - \frac{\sigma_6 z'^2}{2}$$

So

$$M_{\text{max}} = 52.2(0.41 + 1.78) - \frac{127.64(0.41)^2}{2}$$

$$= 114.32 - 10.73 = 103.59 \text{ kN-m/m}$$

The minimum required section modulus (assuming that $\sigma_{\text{all}} = 172.5 \text{ MN/m}^2$) is

$$S = \frac{103.59 \text{ kN-m/m}}{172.5 \times 10^3 \text{ kN/m}^2} = \textbf{0.6} \times \textbf{10}^{-3} \textbf{ m}^3\textbf{/m of the wall} \qquad \blacksquare$$

9.7   *Special Cases for Cantilever Walls Penetrating Clay*

As in Section 9.5, relationships for special cases for cantilever walls penetrating clay may also be derived. Referring to Figure 9.12, we can write

$$\sigma_2' = \gamma L K_a \qquad (9.47)$$

$$\sigma_6 = 4c - \gamma L \qquad (9.48)$$

$$\sigma_7 = 4c + \gamma L \qquad (9.49)$$

$$P_1 = \tfrac{1}{2} L \sigma_2' = \tfrac{1}{2} \gamma L^2 K_a \qquad (9.50)$$

and

$$L_4 = \frac{D(4c - \gamma L) - \tfrac{1}{2}\gamma L^2 K_a}{4c} \qquad (9.51)$$

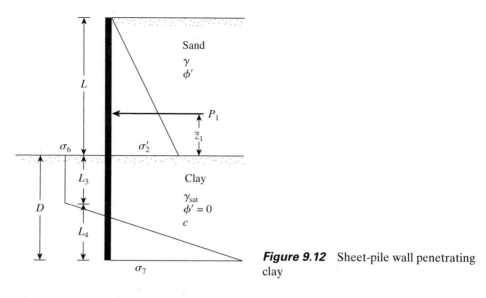

Figure 9.12 Sheet-pile wall penetrating clay

The theoretical depth of penetration, D, can be calculated [in a manner similar to the calculation of Eq. (9.44)] as

$$D^2(4c - \gamma L) - 2DP_1 - \frac{P_1(P_1 + 12c\bar{z}_1)}{\gamma L + 2c} = 0 \qquad (9.52)$$

where $\bar{z}_1 = \dfrac{L}{3}$. (9.53)

The magnitude of the maximum moment in the wall is

$$M_{max} = P_1(z' + \bar{z}_1) - \frac{\sigma_6 z'^2}{2} \qquad (9.54)$$

where $z' = \dfrac{P_1}{\sigma_6} = \dfrac{\frac{1}{2}\gamma L^2 K_a}{4c - \gamma L}$. (9.55)

9.8   *Anchored Sheet-Pile Walls*

When the height of the backfill material behind a cantilever sheet-pile wall exceeds about 6 m ($\approx$20 ft), tying the wall near the top to anchor plates, anchor walls, or anchor piles becomes more economical. This type of construction is referred to as *anchored sheet-pile wall* or an *anchored bulkhead*. Anchors minimize the depth of penetration required by the sheet piles and also reduce the cross-sectional area and weight of the sheet piles needed for construction. However, the tie rods and anchors must be carefully designed.

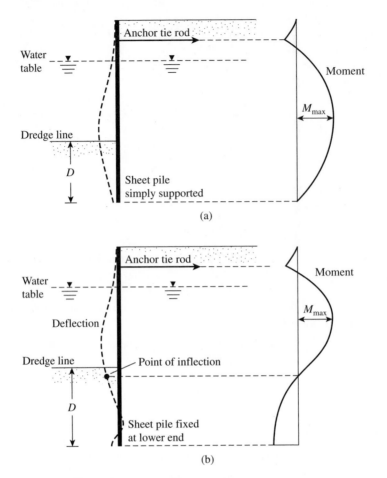

Figure 9.13 Nature of variation of deflection and moment for anchored sheet piles: (a) free earth supprt method; (b) fixed earth support method

The two basic methods of designing anchored sheet-pile walls are (a) the *free earth support* method and (b) the *fixed earth support* method. Figure 9.13 shows the assumed nature of deflection of the sheet piles for the two methods.

The free earth support method involves a minimum penetration depth. Below the dredge line, no pivot point exists for the static system. The nature of the variation of the bending moment with depth for both methods is also shown in Figure 9.13. Note that

$$D_{\text{free earth}} < D_{\text{fixed earth}}$$

9.9 Free Earth Support Method for Penetration of Sandy Soil

Figure 9.14 shows an anchor sheet-pile wall with a granular soil backfill; the wall has been driven into a granular soil. The tie rod connecting the sheet pile and the anchor is located at a depth l_1 below the top of the sheet-pile wall.

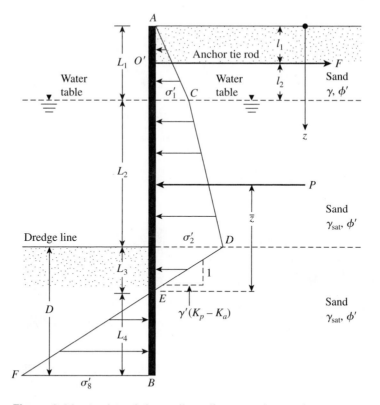

Figure 9.14 Anchored sheet-pile wall penetrating sand

The diagram of the net pressure distribution above the dredge line is similar to that shown in Figure 9.7. At depth $z = L_1$, $\sigma_1' = \gamma L_1 K_a$, and at $z = L_1 + L_2$, $\sigma_2' = (\gamma L_1 + \gamma' L_2) K_a$. Below the dredge line, the net pressure will be zero at $z = L_1 + L_2 + L_3$. The relation for L_3 is given by Eq. (9.6), or

$$L_3 = \frac{\sigma_2'}{\gamma'(K_p - K_a)}$$

At $z = L_1 + L_2 + L_3 + L_4$, the net pressure is given by

$$\sigma_8' = \gamma'(K_p - K_a)L_4 \tag{9.56}$$

Note that the slope of the line DEF is 1 vertical to $\gamma'(K_p - K_a)$ horizontal.

For equilibrium of the sheet pile, Σ horizontal forces $= 0$, and Σ moment about $O' = 0$. (*Note:* Point O' is located at the level of the tie rod.)

Summing the forces in the horizontal direction (per unit length of the wall) gives

Area of the pressure diagram $ACDE$ − area of EBF − $F = 0$

where F = tension in the tie rod/unit length of the wall, or.

$$P - \tfrac{1}{2}\sigma_8' L_4 - F = 0$$

or

$$F = P - \tfrac{1}{2}[\gamma'(K_p - K_a)]L_4^2 \tag{9.57}$$

where P = area of the pressure diagram $ACDE$. Now, taking the moment about point O' gives

$$-P[(L_1 + L_2 + L_3) - (\bar{z} + l_1)] + \tfrac{1}{2}[\gamma'(K_p - K_a)]L_4^2(l_2 + L_2 + L_3 + \tfrac{2}{3}L_4) = 0$$

or

$$L_4^3 + 1.5L_4^2(l_2 + L_2 + L_3) - \frac{3P[(L_1 + L_2 + L_3) - (\bar{z} + l_1)]}{\gamma'(K_p - K_a)} = 0 \tag{9.58}$$

Equation (9.58) may be solved by trial and error to determine the theoretical depth, L_4:

$$D_{\text{theoretical}} = L_3 + L_4$$

The theoretical depth is increased by about 30–40% for actual construction, or

$$D_{\text{actual}} = 1.3 \text{ to } 1.4 D_{\text{theoretical}} \tag{9.59}$$

The step-by-step procedure in Section 9.4 indicated that a factor of safety can be applied to K_p at the beginning [i.e., $K_{p(\text{design})} = K_p/\text{FS}$]. If that is done, there is no need to increase the theoretical depth by 30 to 40%. This approach is often more conservative.

The maximum theoretical moment to which the sheet pile will be subjected occurs at a depth between $z = L_1$ and $z = L_1 + L_2$. The depth z for zero shear and hence maximum moment may be evaluated from

$$\tfrac{1}{2}\sigma_1' L_1 - F + \sigma_1'(z - L_1) + \tfrac{1}{2}K_a\gamma'(z - L_1)^2 = 0 \tag{9.60}$$

Once the value of z is determined, the magnitude of the maximum moment is easily obtained.

Sometimes, the dredge line slopes at an angle β with respect to the horizontal, as shown in Figure 9.15a. In that case, the passive pressure coefficient will not be equal to $\tan^2(45 + \phi'/2)$. The variations of K_p (Coulomb's passive earth-pressure analysis for a wall friction angle of zero) with β for $\phi' = 30°$ and $35°$ are shown in Figure 9.15b. With these values of K_p, the procedure described in this section may be used to determine the depth of penetration, D.

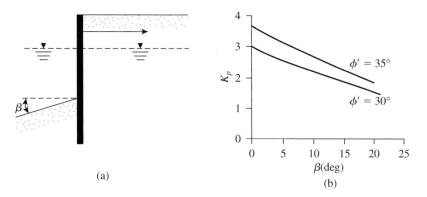

Figure 9.15 (a) Anchored sheet-pile wall with sloping dredge line; (b) variation of K_p with β and ϕ'

Example 9.3

Refer to Figure 9.14. Here $L_1 = 2$ m, $L_2 = 3$ m, $l_1 = l_2 = 1$ m, $c' = 0$, $\phi' = 32°$, $\gamma = 15.9$ kN/m³, and $\gamma_{sat} = 19.33$ kN/m³.

  a. Determine the theoretical and actual depths of penetration
  b. Find the anchor force per unit length of the wall
  c. Determine M_{max}

Solution
Part a: Depths of Penetration

$$K_a = \tan^2\left(45 - \frac{\phi'}{2}\right) = \tan^2\left(45 - \frac{32}{2}\right) = 0.307$$

$$K_p = \tan^2\left(45 + \frac{\phi'}{2}\right) = 3.25$$

$$\sigma_1' = \gamma L_1 K_a = (15.9)(2)(0.307) = 9.763 \text{ kN/m}^2$$

$$\sigma_2' = (\gamma L_1 + \gamma' L_2)K_a = [(15.9)(2) + (19.33 - 9.81)3]0.307$$

$$= 18.53 \text{ kN/m}^2$$

$$L_3 = \frac{\sigma_2'}{\gamma'(K_p - K_a)} = \frac{18.53}{(19.33 - 9.81)(3.25 - 0.307)} = 0.66 \text{ m}$$

$$P = \frac{1}{2}\sigma_1' L_1 + \sigma_1' L_2 + \frac{1}{2}(\sigma_2' - \sigma_1')L_2 + \frac{1}{2}\sigma_2' L_3$$

$$= \frac{1}{2}(9.763)(2) + (9.763)(3) + \frac{1}{2}(18.53 - 9.763)3 + \frac{1}{2}(18.53)(0.66)$$

$$= 9.763 + 29.289 + 13.151 + 6.115 = 58.32 \text{ kN/m}$$

Taking the moment about E yields

$$\bar{z} = \frac{1}{58.32}\left[9.763\left(0.66 + 3 + \frac{2}{3}\right) + 29.289\left(0.66 + \frac{3}{2}\right)\right.$$

$$\left. + 13.151\left(0.66 + \frac{3}{3}\right) + 6.115\left(0.66 \times \frac{2}{3}\right)\right] = 2.23 \text{ m}$$

Into Eq. (9.58),

$$L_3^4 + 1.5L_4^2(l_2 + L_2 + L_3) - \frac{3P[(L_1 + L_2 + L_3) - (\bar{z} + l_1)]}{\gamma'(K_p - K_a)} = 0$$

we substitute $l_1 = 1$ m, $l_2 = 1$ m, $K_p = 3.25$, and $K_a = 0.307$ to get

$$L_3^4 + 1.5L_4^2(1 + 3 + 0.66) - \frac{3(58.32)[(2 + 3 + 0.66) - (2.23 + 1)]}{9.52(3.25 - 0.307)} = 0$$

or

$$L_4^3 + 6.99L_4^2 - 14.55 = 0 \tag{a}$$

The magnitude of L_4 is obtained by trial and error:

Assumed L_4 (m)	Left-hand side of Eq. (a)
2.0	+21.41
1.5	+4.55
1.4	+1.89
1.3	−0.54

Hence $L_4 \approx 1.4$ m and

$$D_{\text{theoretical}} = L_3 + L_4 = 0.66 + 1.4 = 2.06 \text{ m}$$

$$D_{\text{actual}} \approx 1.4 D_{\text{theory}} = (1.4)(2.06) = 2.88 \text{ m (rounded to } \textbf{2.9 m})$$

Part b: Anchor Force
From Eq. (9.57),

$$F = P - \frac{1}{2}[\gamma'(K_p - K_a)]L_4^2$$

$$= 58.32 - \frac{1}{2}[9.52(3.25 - 0.307)](1.4)^2 = \textbf{30.86 kN/m}$$

Part c: Maximum Moment (M_{max})
From Eq. (9.60) for zero shear,

$$\frac{1}{2}\sigma_1' L_1 - F + \sigma_1'(z - L_1) + \frac{1}{2}K_a\gamma'(z - L_1)^2 = 0$$

or

$$\frac{1}{2}(9.763)(2) - 30.86 + (9.763)(z - 2) + \frac{1}{2}(0.307)(9.52)(z - 2)^2 = 0$$

Let $z - 2 = x$. So,

$$9.763 - 30.86 + 9.763x + 1.461x^2 = 0$$
$$x^2 + 6.682x - 14.44 = 0$$
$$x = 1.72 \text{ m}$$

or

$$z = x + 2 = 1.72 + 2 = 3.72 \text{ m} \qquad (L_1 + L_2 < z < L_1 \text{—checks})$$

Taking the moment about the point of zero shear force ($z = 3.72$ m or $x = 1.72$ m) gives

$$M_{max} = -\left(\frac{1}{2}\sigma_1' L_1\right)\left[x + \left(\frac{1}{3}\right)(2)\right] + F(x + 1) - (\sigma_1' x)\left(\frac{x}{2}\right) - \frac{1}{2}K_a \gamma'(x)^2\left(\frac{x}{3}\right)$$

or

$$M_{max} = -(9.763)(2.387) + (30.86)(2.72) - \frac{9.763(1.72)^2}{2} - \frac{(0.307)(9.52)(1.72)^3}{6}$$
$$= -23.3 + 83.94 - 14.44 - 2.48 = \textbf{43.72 kN} \cdot \textbf{m/m.} \qquad \blacksquare$$

9.10   *Design Charts for Free Earth Support Method (Penetration into Sandy Soil)*

Using the free earth support method, Hagerty and Nofal (1992) provided simplified design charts for quick estimation of the depth of penetration, D, anchor force, F, and maximum moment, M_{max}, for anchored sheet-pile walls penetrating into sandy soil, as shown in Figure 9.14. They made the following assumptions for their analysis.

a. The soil friction angle, ϕ', above and below the dredge line is the same.
b. The angle of friction between the sheet-pile wall and the soil is $\phi'/2$.
c. The passive earth pressure below the dredge line has a logarithmic spiral failure surface.
d. For active earth-pressure calculation, Coulomb's theory is valid.

The magnitudes of D, F, and M_{max} may be calculated from the following relationships:

$$\frac{D}{L_1 + L_2} = (GD)(CDL_1) \tag{9.61}$$

$$\frac{F}{\gamma_a(L_1 + L_2)^2} = (GF)(CFL_1) \tag{9.62}$$

$$\frac{M_{max}}{\gamma_a(L_1 + L_2)^3} = (GM)(CML_1) \tag{9.63}$$

where

γ_a = average unit weight of soil

$$= \frac{\gamma L_1^2 + (\gamma_{sat} - \gamma_w)L_2^2 + 2\gamma L_1 L_2}{(L_1 + L_2)^2} \tag{9.64}$$

GD = generalized nondimensional embedment

$$= \frac{D}{L_1 + L_2} \quad \text{(for } L_1 = 0 \text{ and } L_2 = L_1 + L_2)$$

GF = generalized nondimensional anchor force

$$= \frac{F}{\gamma_a(L_1 + L_2)^2} \quad \text{(for } L_1 = 0 \text{ and } L_2 = L_1 + L_2)$$

GM = generalized nondimensional moment

$$= \frac{M_{max}}{\gamma_a(L_1 + L_2)^3} \quad \text{(for } L_1 = 0 \text{ and } L_2 = L_1 + L_2)$$

CDL_1, CFL_1, CML_1 = correction factors for $L_1 \neq 0$

The variations of GD, GF, GM, CDL_1, CFL_1, and CML_1 are shown in Figures 9.16, 9.17, 9.18, 9.19, 9.20, and 9.21, respectively.

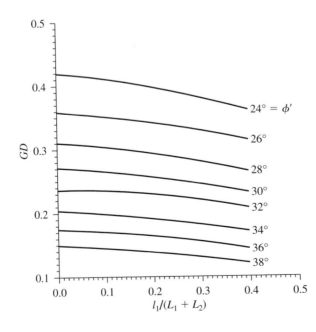

Figure 9.16 Variation of GD with $l_1/(L_1 + L_2)$ and ϕ' (After Hagerty and Nofal, 1992)

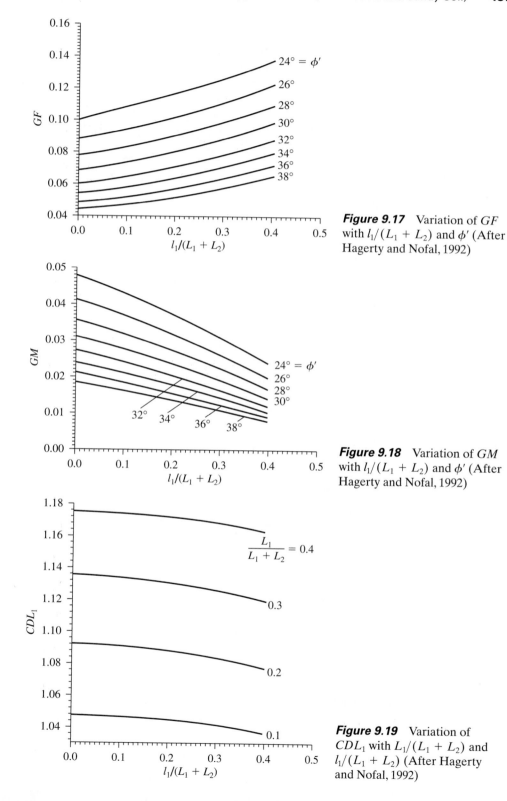

Figure 9.17 Variation of GF with $l_1/(L_1 + L_2)$ and ϕ' (After Hagerty and Nofal, 1992)

Figure 9.18 Variation of GM with $l_1/(L_1 + L_2)$ and ϕ' (After Hagerty and Nofal, 1992)

Figure 9.19 Variation of CDL_1 with $L_1/(L_1 + L_2)$ and $l_1/(L_1 + L_2)$ (After Hagerty and Nofal, 1992)

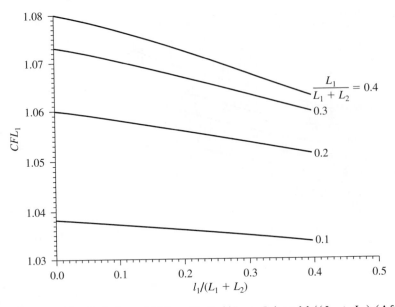

Figure 9.20 Variation of CFL_1 with $L_1/(L_1 + L_2)$ and $l_1/(L_1 + L_2)$ (After Hagerty and Nofal, 1992)

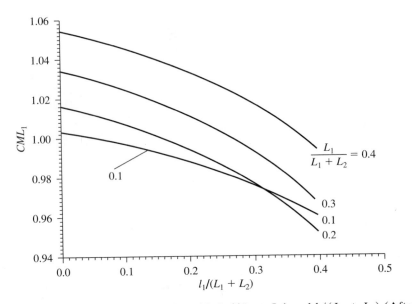

Figure 9.21 Variation of CML_1 with $L_1/(L_1 + L_2)$ and $l_1/(L_1 + L_2)$ (After Hagerty and Nofal, 1992)

Example 9.4

Use the charts just presented to redo Example 9.3.

Solution

Part a
From Eq. (9.61),

$$\frac{D}{L_1 + L_2} = (GD)(CDL_1)$$

For Problem 9.3,

$$\frac{l_1}{L_1 + L_2} = \frac{1}{2 + 3} = 0.2$$

From Figure 9.16 for $l_1/(L_1 + L_2) = 0.2$ and $\phi' = 32°$, $GD = 0.23$. From Figure 9.19 for

$$\frac{L_1}{L_1 + L_2} = \frac{2}{2 + 3} = 0.4 \quad \text{and} \quad \frac{l_1}{L_1 + L_2} = 0.2$$

$CDL_1 \approx 1.172$. So,

$$D_{\text{theoretical}} = (L_1 + L_2)(GD)(CDL_1) = (5)(0.23)(1.172) \approx 1.35$$

$$D_{\text{actual}} \approx (1.4)(1.35) = 1.89 \approx \textbf{2 m}$$

Part b
From Figure 9.17 for $l_1/(L_1 + L_2) = 0.2$ and $\phi' = 32°$, $GF \approx 0.07$. Also, from Figure 9.20, for

$$\frac{L_1}{L_1 + L_2} = \frac{2}{2 + 3} = 0.4, \quad \frac{l_1}{L_1 + L_2} = 0.2, \quad \text{and} \quad \phi' = 32°$$

$CFL_1 = 1.073$. From Eq. (9.64),

$$\gamma_a = \frac{\gamma L_1^2 + \gamma' L_2^2 + 2\gamma L_1 L_2}{(L_1 + L_2)^2}$$

$$= \frac{(15.9)(2)^2 + (19.33 - 9.81)(3)^2 + (2)(15.9)(2)(3)}{(2 + 3)^2} = 13.6 \text{ kN/m}^3$$

Using Eq. (9.62) yields

$$F = \gamma_a(L_1 + L_2)^2(GF)(CFL_1) = (13.6)(5)^2(0.07)(1.073) \approx \textbf{25.54 kn/m}$$

Part c
From Figure 9.18, for $l_1/(L_1 + L_2) = 0.2$ and $\phi' = 32°$, $GM = 0.021$. Also, from Figure 9.21, for

$$\frac{L_1}{L_1 + L_2} = \frac{2}{2 + 3} = 0.4, \quad \frac{l_1}{L_1 + L_2} = 0.2, \quad \text{and} \quad \phi' = 32°$$

$CML_1 = 1.036$. Hence from Eq. (9.63),

$$M_{max} = \gamma_a (L_1 + L_2)^3 (GM)(CML_1) = (13.6)(5)^3(0.021)(1.036) = \mathbf{36.99\ kN \cdot m/m}$$

(*Note:* The difference between the results in Examples 9.3 and 9.4 is primarily due to the wall friction angle assumed and the method used to calculate passive earth pressure.)   ∎

9.11   *Moment Reduction for Anchored Sheet-Pile Walls*

Sheet piles are flexible, and hence sheet-pile walls yield (i.e., become displaced laterally), which redistributes the lateral earth pressure. This change tends to reduce the maximum bending moment, M_{max}, as calculated by the procedure outlined in Section 9.9. For that reason, Rowe (1952, 1957) suggested a procedure for reducing the maximum design moment on the sheet pile walls *obtained from the free earth support method.* This section discusses the procedure of moment reduction for sheet piles *penetrating into sand.*

In Figure 9.22, which is valid for the case of a sheet pile penetrating sand, the following notation is used:

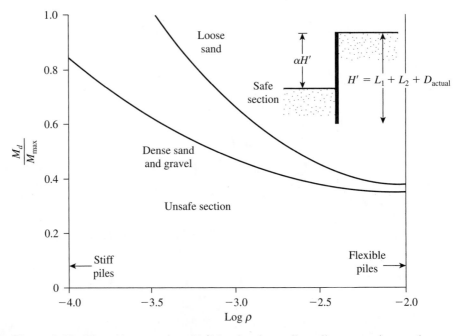

Figure 9.22   Plot of log ρ against M_d/M_{max} for sheet-pile walls penetrating sand (After Rowe, 1952)

1. H' = total height of pile driven (i.e., $L_1 + L_2 + D_{actual}$)

2. Relative flexibility of pile = $\rho = 10.91 \times 10^{-7} \left(\dfrac{H'^4}{EI} \right)$ (9.65)

where

H' is in meters
E = modulus of elasticity of the pile material (MN/m^2)
I = moment of inertia of the pile section per meter of the wall (m^4/m of wall)

3. M_d = design moment
4. M_{max} = maximum theoretical moment

In English units, Eq. (9.65) takes the form

$$\rho = \frac{H'^4}{EI}$$ (9.66)

where H' is in ft, E is in lb/in^2, and I is in in^4/ft of the wall.

The procedure for the use of the moment reduction diagram (see Figure 9.22) is as follows:

Step 1. Choose a sheet pile section (e.g., from among those given in Table 9.1).
Step 2. Find the modulus S of the selected section (Step 1) per unit length of the wall.
Step 3. Determine the moment of inertia of the section (Step 1) per unit length of the wall.
Step 4. Obtain H' and calculate ρ [see Eq. (9.65) or Eq. (9.66)].
Step 5. Find $\log \rho$.
Step 6. Find the moment capacity of the pile section chosen in Step 1 as $M_d = \sigma_{all} S$.
Step 7. Determine M_d/M_{max}. Note that M_{max} is the maximum theoretical moment determined before.
Step 8. Plot $\log \rho$ (Step 5) and M_d/M_{max} in Figure 9.22.
Step 9. Repeat Steps 1 through 8 for several sections. The points that fall above the curve (in loose sand or dense sand, as the case may be) are *safe sections.* The points that fall below the curve are *unsafe sections.* The cheapest section may now be chosen from those points which fall above the proper curve. Note that the section chosen will have an $M_d < M_{max}$.

For anchor sheet-pile walls penetrating into sand with a sloping dredge line (see Figure 9.15), a moment reduction procedure similar to that just outlined may be adopted. For this procedure, Figure 9.23 (which was developed by Schroeder and Roumillac, 1983) should be used.

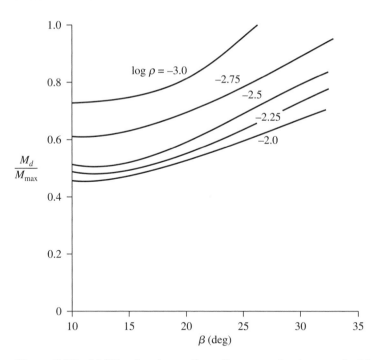

Figure 9.23 M_d/M_{max} for sheet-pile walls penetrating into sand with a sloping dredge line

Example 9.5

Refer to Example 9.3. Use Rowe's moment reduction diagram (Figure 9.22) to find an appropriate sheet pile section. For the sheet pile, use $E = 207 \times 10^3$ MN/m^2 and $\sigma_{all} = 172{,}500$ kN/m^2.

Solution

$$H' = L_1 + L_2 + D_{actual} = 2 + 3 + 2.9 = 7.9 \text{ m}$$

$M_{max} = 43.72$ kN · m/m. Now the following table can be prepared.

Sheet pile section	I (m^4/m)	H' (m)	$\rho = 10.91 \times 10^{-7}\left(\dfrac{H'^4}{EI}\right)$ Log ρ	S (m^3/m)	$M_d = S\sigma_{all}$ (kN · m/m)	$\dfrac{M_d}{M_{max}}$
PSA-31	4.41×10^{-6}	7.9	0.00466	-2.33	10.8×10^{-5}	18.63	0.426
PSA-23	5.63×10^{-6}	7.9	0.00365	-2.44	12.8×10^{-5}	22.08	0.505

Figure 9.24 shows a plot of M_d/M_{max} versus log ρ. It can be seen that **PSA-31** will be sufficient.

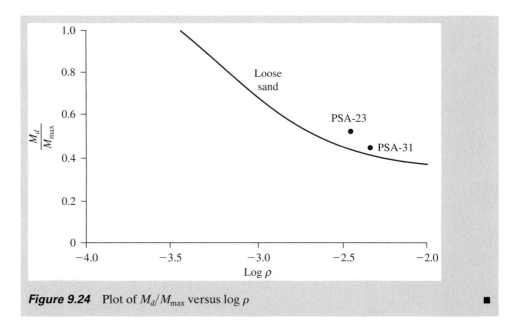

Figure 9.24 Plot of M_d/M_{max} versus log ρ

9.12 *Computational Pressure Diagram Method for Penetration into Sandy Soil*

The computational pressure diagram (CPD) method for sheet pile penetrating a sandy soil is a simplified method of design and an alternative to the free earth method described in Sections 9.9 and 9.11 (Nataraj and Hoadley, 1984). In this method, the net pressure diagram shown in Figure 9.14 is replaced by rectangular pressure diagrams, as in Figure 9.25. Note that $\overline{\sigma}'_a$ is the width of the net active pressure diagram above the dredge line and $\overline{\sigma}'_p$ is the width of the net passive pressure diagram below the dredge line. The magnitudes of $\overline{\sigma}'_a$ and $\overline{\sigma}'_p$ may respectively be expressed as

$$\overline{\sigma}'_a = C K_a \gamma'_{av} L \tag{9.67}$$

and

$$\overline{\sigma}'_p = R C K_a \gamma'_{av} L = R\overline{\sigma}'_a \tag{9.68}$$

where

γ'_{av} = average effective unit weight of sand

$$\approx \frac{\gamma L_1 + \gamma' L_2}{L_1 + L_2} \tag{9.69}$$

C = coefficient

$$R = \text{coefficient} = \frac{L(L - 2l_1)}{D(2L + D - 2l_1)} \tag{9.70}$$

The range of values for C and R is given in Table 9.2.

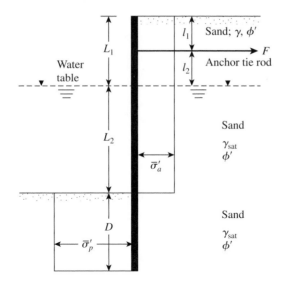

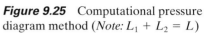

Figure 9.25 Computational pressure diagram method (*Note:* $L_1 + L_2 = L$)

Table 9.2 Range of Values for C and R [from Eqs. (9.67) and (9.68)]

Soil type	C[a]	R
Loose sand	0.8–0.85	0.3–0.5
Medium sand	0.7–0.75	0.55–0.65
Dense sand	0.55–0.65	0.60–0.75

[a] Valid for the case in which there is no surcharge above the granular backfill (i.e., on the right side of the wall, as shown in Figure 9.25)

The depth of penetration, D, anchor force per unit length of the wall, F, and maximum moment in the wall, M_{max}, are obtained from the following relationships.

Depth of Penetration

For the depth of penetration, we have

$$D^2 + 2DL\left[1 - \left(\frac{l_1}{L}\right)\right] - \left(\frac{L^2}{R}\right)\left[1 - 2\left(\frac{l_1}{L}\right)\right] = 0 \qquad (9.71)$$

Anchor Force

The anchor force is

$$F = \overline{\sigma}'_a(L - RD) \qquad (9.72)$$

Maximum Moment

The maximum moment is calculated from

$$M_{\max} = 0.5\overline{\sigma}_a' L^2 \left[\left(1 - \frac{RD}{L} \right)^2 - \left(\frac{2l_1}{L} \right)\left(1 - \frac{RD}{L} \right) \right] \tag{9.73}$$

Note the following qualifications:

1. The magnitude of D obtained from Eq. (9.71) is about 1.25 to 1.5 times the value of D_{theory} obtained by the conventional free earth support method (see Section 9.9), so

$$D \approx D_{\text{actual}}$$

$$\uparrow \qquad \uparrow$$

Eq. (9.71)   Eq. (9.59)

2. The magnitude of F obtained by using Eq. (9.72) is about 1.2 to 1.6 times the value obtained by using Eq. (9.57). Thus, an additional factor of safety for the actual design of anchors need not be used.
3. The magnitude of $M_{\max}$ obtained from Eq. (9.73) is about 0.6 to 0.75 times the value of $M_{\max}$ obtained by the conventional free earth support method. Hence, the former value of $M_{\max}$ can be used as the actual design value, and Rowe's moment reduction need not be applied.

Example 9.6

For the anchored sheet pile wall shown in Figure 9.26, determine (a) D, (b) F, and (c) $M_{\max}$. Use the CPD method; assume that $C = 0.68$ and $R = 0.6$.

Solution

Part a

$$\gamma' = \gamma_{\text{sat}} - \gamma_w = 122.4 - 62.4 = 60 \text{ lb/ft}^3$$

From Eq. (9.69)

$$\gamma_{\text{av}}' = \frac{\gamma L_1 + \gamma' L_2}{L_1 + L_2} = \frac{(110)(10) + (60)(20)}{10 + 20} = 76.67 \text{ lb/ft}^3$$

$$K_a = \tan^2\left(45 - \frac{\phi'}{2} \right) = \tan^2\left(45 - \frac{35}{2} \right) = 0.271$$

$$\overline{\sigma}_a' = CK_a\gamma_{\text{av}}' L = (0.68)(0.271)(76.67)(30) = 423.9 \text{ lb/ft}^2$$

$$\overline{\sigma}_p' = R\overline{\sigma}_a' = (0.6)(423.9) = 254.3 \text{ lb/ft}^2$$

From Eq. (9.71)

$$D^2 + 2DL\left[1 - \left(\frac{l_1}{L} \right) \right] - \frac{L^2}{R}\left[1 - 2\left(\frac{l_1}{L} \right) \right] = 0$$

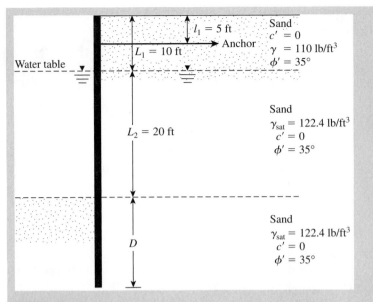

Figure 9.26

or

$$D^2 + 2(D)(30)\left[1 - \left(\frac{5}{30}\right)\right] - \frac{(30)^2}{0.6}\left[1 - 2\left(\frac{5}{30}\right)\right] = D^2 + 50D - 1000 = 0$$

Hence $D \approx$ **15.3 ft.**
Check for the assumption of R:

$$R = \frac{L(L - 2l_1)}{D(2L + D - 2l_1)} = \frac{30[30 - (2)(5)]}{15.3[(2)(30) + 15.3 - (2)(5)]} \approx \mathbf{0.6—OK}$$

Part b
From Eq. (9.72)

$$F = \overline{\sigma}_a'(L - RD) = 423.9[30 - (0.6)(15.3)] = \textbf{8825 lb/ft}$$

Part c
From Eq. (9.73)

$$M_{max} = 0.5\overline{\sigma}_a'L^2\left[\left(1 - \frac{RD}{L}\right)^2 - \left(\frac{2l_1}{L}\right)\left(1 - \frac{RD}{L}\right)\right]$$

$$1 - \frac{RD}{L} = 1 - \frac{(0.6)(15.3)}{30} = 0.694$$

So,

$$M_{max} = (0.5)(423.9)(30)^2\left[(0.694)^2 - \frac{(2)(5)(0.694)}{30}\right] = \textbf{47,753 lb-ft/ft} \quad \blacksquare$$

Field Observations of an Anchored Sheet Pile Wall

In Section 9.9, a large factor of safety was recommended for the depth of penetration, D. In most cases, designers use smaller magnitude of soil friction angle, ϕ', thereby ensuring a built-in factor of safety for the active earth pressure. This procedure is followed primarily because of the uncertainties involved in predicting the actual earth pressure to which a sheet-pile wall will be subjected in the field. In addition, Casagrande (1973) observed that, if the soil behind the sheet pile has grain sizes that are predominantly smaller than those of coarse sand, the active earth pressure may increase to an at-rest earth-pressure condition after construction. Such an increase causes a large increase in the anchor force, F. Following is a summary of the observations made by Casagrande (1973) relating to the distribution of lateral earth pressure on the Pier C bulkhead in the Long Beach, California Harbor during its construction period (May through August 1949).

Figure 9.27 shows a schematic diagram of the cross section of the bulkhead along with distributions of lateral earth pressure on May 24, June 3, and August 6, 1949. Except for a rockfill dike that was constructed with 76 mm (3 in.) maximum size quarry wastes, the backfill consisted of fine sand. The fine sand backfill reached

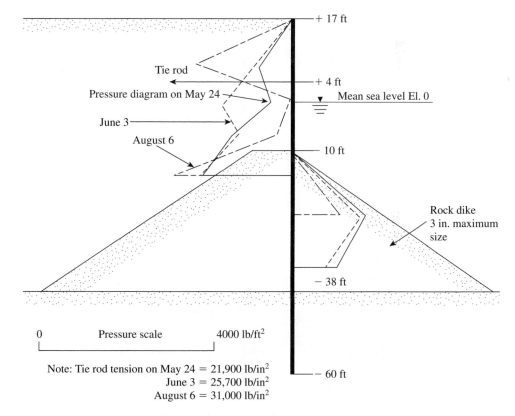

Figure 9.27   Pressure distribution on Pier C bulkhead, Long Beach harbor (Adapted after Casagrande, 1973)

design grade on May 24, 1949. On that day, due to the yielding of the wall, the earth pressure reached an active state. Between May 24 and June 3, the anchor resisted further yielding and the lateral earth pressure increased to an at-rest state.

It was explained in Section 9.11 that sheet piles are flexible and, hence, sheet pile walls yield. This process redistributes the lateral earth pressure and ultimately reduces the maximum bending moment, M_{max}. This can be seen by observing the distribution of lateral earth pressure on August 6, 1949 (Figure 9.27). Based on these field observations, it is important to realize that the actual lateral pressure diagram may not be identical to that used for design.

9.14   *Free Earth Support Method for Penetration of Clay*

Figure 9.28 shows an anchored sheet-pile wall penetrating a clay soil and with a granular soil backfill. The diagram of pressure distribution above the dredge line is similar to that shown in Figure 9.10. From Eq. (9.38), the net pressure distribution below the dredge line (from $z = L_1 + L_2$ to $z = L_1 + L_2 + D$) is

$$\sigma_6 = 4c - (\gamma L_1 + \gamma' L_2)$$

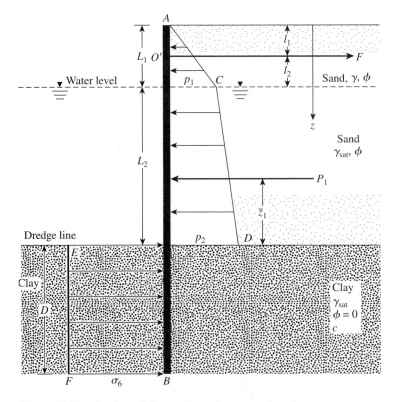

Figure 9.28 Anchored sheet-pile wall penetrating clay

For static equilibrium, the sum of the forces in the horizontal direction is

$$P_1 - \sigma_6 D = F \tag{9.74}$$

where

P_1 = area of the pressure diagram ACD
F = anchor force per unit length of the sheet pile wall

Again, taking the moment about O' produces

$$P_1(L_1 + L_2 - l_1 - \bar{z}_1) - \sigma_6 D\left(l_2 + L_2 + \frac{D}{2}\right) = 0$$

Simplification yields

$$\sigma_6 D^2 + 2\sigma_6 D(L_1 + L_2 - l_1) - 2P_1(L_1 + L_2 - l_1 - \bar{z}_1) = 0 \tag{9.75}$$

Equation (9.75) gives the theoretical depth of penetration, D.

As in Section 9.9, the maximum moment in this case occurs at a depth $L_1 < z < L_1 + L_2$. The depth of zero shear (and thus the maximum moment) may be determined from Eq. (9.60).

A moment reduction technique similar to that in Section 9.11 for anchored sheet piles penetrating into clay has also been developed by Rowe (1952, 1957). This technique is presented in Figure 9.29, in which the following notation is used:

1. The stability number is

$$S_n = 1.25\frac{c}{(\gamma L_1 + \gamma' L_2)} \tag{9.76}$$

where c = undrained cohesion ($\phi = 0$).

For the definition of γ, γ', L_1, and L_2, see Figure 9.28.
2. The nondimensional wall height is

$$\alpha = \frac{L_1 + L_2}{L_1 + L_2 + D_{\text{actual}}} \tag{9.77}$$

3. The flexibility number is ρ [see Eq. (9.65) or Eq. (9.66)]
4. M_d = design moment
M_{max} = maximum theoretical moment

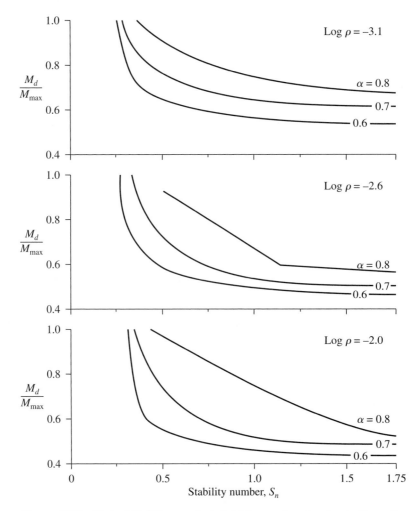

Figure 9.29 Plot of M_d/M_{max} against stability number for sheet-pile wall penetrating clay (After Rowe, 1957)

The procedure for moment reduction, using Figure 9.29, is as follows:

Step 1. Obtain $H' = L_1 + L_2 + D_{actual}$.

Step 2. Determine $\alpha = (L_1 + L_2)/H'$.

Step 3. Determine S_n [from Eq. (9.76)].

Step 4. For the magnitudes of α and S_n obtained in Steps 2 and 3, determine M_d/M_{max} for various values of log ρ from Figure 9.29, and plot M_d/M_{max} against log ρ.

Step 5. Follow Steps 1 through 9 as outlined for the case of moment reduction of sheet-pile walls penetrating granular soil. (See Section 9.11.)

Example 9.7

In Figure 9.28, let $L_1 = 3$ m, $L_2 = 6$ m, and $l_1 = 1.5$ m. Also, let $\gamma = 17$ kN/m³, $\gamma_{sat} = 20$ kN/m³, $\phi' = 35°$, and $c = 41$ kN/m².

a. Determine the theoretical depth of embedment of the sheet-pile wall.
b. Calculate the anchor force per unit length of the wall.

Solution

Part a
We have

$$K_a = \tan^2\left(45 - \frac{\phi'}{2}\right) = \tan^2\left(45 - \frac{35}{2}\right) = 0.271$$

and

$$K_p = \tan^2\left(45 + \frac{\phi'}{2}\right) = \tan^2\left(45 + \frac{35}{2}\right) = 3.69$$

From the pressure diagram in Figure 9.30,

$\sigma'_1 = \gamma L_1 K_a = (17)(3)(0.271) = 13.82$ kN/m²
$\sigma'_2 = (\gamma L_1 + \gamma' L_2)K_a = [(17)(3) + (20 - 9.81)(6)](0.271) = 30.39$ kN/m²
$P_1 = $ areas $1 + 2 + 3 = 1/2(3)(13.82) + (13.82)(6) + 1/2(30.39 - 13.82)(6)$
$\quad = 20.73 + 82.92 + 49.71 = 153.36$ kN/m

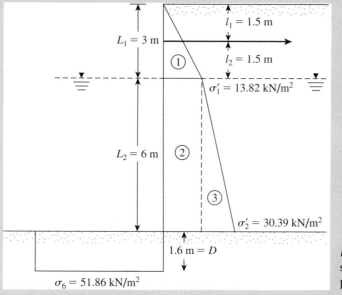

$\sigma_6 = 51.86$ kN/m²

Figure 9.30   Free earth support method, sheet pile penetrating into clay

and

$$\bar{z}_1 = \frac{(20.73)\left(6 + \dfrac{3}{3}\right) + (82.92)\left(\dfrac{6}{2}\right) + (49.71)\left(\dfrac{6}{3}\right)}{153.36} = 3.2 \text{ m}$$

From Eq. (9.75),

$$\sigma_6 D^2 + 2\sigma_6 D(L_1 + L_2 - l_1) - 2P_1(L_1 + L_2 - l_1 - \bar{z}_1) = 0$$

or

$$\sigma_6 = 4c - (\gamma L_1 + \gamma' L_2) = (4)(41) - [(17)(3)$$
$$+ (20 - 9.81)(6)] = 51.86 \text{ kN/m}^2$$

So,

$$(51.86)D^2 + (2)(51.86)(D)(3 + 6 - 1.5)$$
$$- (2)(153.36)(3 + 6 - 1.5 - 3.2) = 0$$

or

$$D^2 + 15D - 25.43 = 0$$

Hence,

$$D \approx \textbf{1.6 m}$$

Part b
From Eq. (9.74),

$$F = P_1 - \sigma_6 D = 153.36 - (51.86)(1.6) = \textbf{70.38 kN/m} \qquad \blacksquare$$

9.15   *Anchors*

Sections 9.9 through 9.14 gave an analysis of anchored sheet-pile walls and discussed how to obtain the force F per unit length of the sheet-pile wall that has to be sustained by the anchors. The current section covers in more detail the various types of anchor generally used and the procedures for evaluating their ultimate holding capacities.

The general types of anchor used in sheet-pile walls are as follows:

1. Anchor plates and beams (deadman)
2. Tie backs
3. Vertical anchor piles
4. Anchor beams supported by batter (compression and tension) piles

Anchor plates and beams are generally made of cast concrete blocks. (See Figure 9.31a.) The anchors are attached to the sheet pile by *tie-rods*. A *wale* is placed at the front or back face of a sheet pile for the purpose of conveniently attaching the tie-rod to the wall. To protect the tie rod from corrosion, it is generally coated with paint or asphaltic materials.

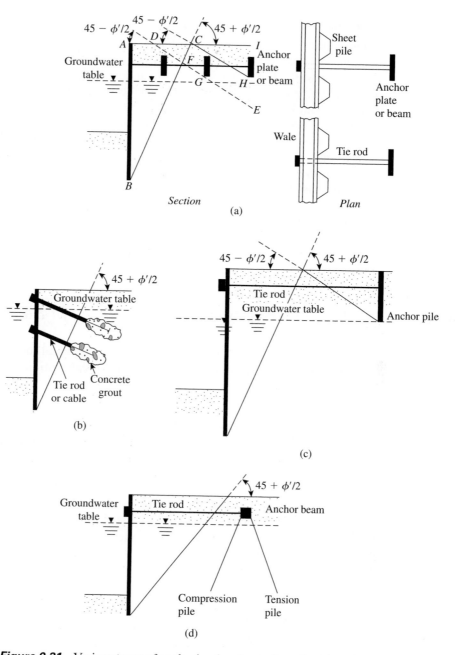

Figure 9.31 Various types of anchoring for sheet-pile walls: (a) anchor plate or beam; (b) tieback; (c) vertical anchor pile; (d) anchor beam with batter piles

In the construction of *tiebacks,* bars or cables are placed in predrilled holes (see Figure 9.31b) with concrete grout (cables are commonly high-strength, pre-stressed steel tendons). Figures 9.31c and 9.31d show a vertical anchor pile and an anchor beam with batter piles.

Placement of Anchors

The resistance offered by anchor plates and beams is derived primarily from the passive force of the soil located in front of them. Figure 9.31a, in which AB is the sheet-pile wall, shows the best location for maximum efficiency of an anchor plate. If the anchor is placed inside wedge ABC, which is the Rankine active zone, it would not provide any resistance to failure. Alternatively, the anchor could be placed in zone $CFEH$. Note that line DFG is the slip line for the Rankine passive pressure. If part of the passive wedge is located inside the active wedge ABC, full passive resistance of the anchor cannot be realized upon failure of the sheet-pile wall. However, if the anchor is placed in zone ICH, the Rankine passive zone in front of the anchor slab or plate is located completely outside the Rankine active zone ABC. In this case, full passive resistance from the anchor can be realized.

Figures 9.31b, 9.31c, and 9.31d also show the proper locations for the placement of tiebacks, vertical anchor piles, and anchor beams supported by batter piles.

9.16   Holding Capacity of Anchor Plates in Sand

Ovesen and Stromann (1972) proposed a semi-empirical method for determining the ultimate resistance of anchors in sand. Their calculations, made in three steps, are carried out as follows:

Step 1.   **Basic Case.** Determine the depth of embedment, H. Assume that the anchor slab has height H and is continuous (i.e., B = length of anchor slab perpendicular to the cross section = ∞), as shown in Figure 9.32, in which the following notation is used:

P_p = passive force per unit length of anchor
P_a = active force per unit length of anchor
ϕ' = effective soil friction angle
δ' = friction angle between anchor slab and soil
P'_u = ultimate resistance per unit length of anchor
W = effective weight per unit length of anchor slab

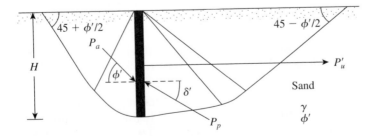

Figure 9.32   Basic case: continuous vertical anchor in granular soil

Also,

$$P'_u = \tfrac{1}{2}\gamma H^2 K_p \cos \delta' - P_a \cos \phi' = \tfrac{1}{2}\gamma H^2 K_p \cos \delta' - \tfrac{1}{2}\gamma H^2 K_a \cos \phi'$$
$$= \tfrac{1}{2}\gamma H^2 (K_p \cos \delta' - K_a \cos \phi')$$

(9.78)

where

K_a = active pressure coefficient with $\delta' = \phi'$
(see Figure 9.33a)
K_p = passive pressure coefficient

To obtain $K_p \cos \delta'$, first calculate

$$K_p \sin \delta' = \frac{W + P_a \sin \phi'}{\tfrac{1}{2}\gamma H^2} = \frac{W + \tfrac{1}{2}\gamma H^2 K_a \sin \phi'}{\tfrac{1}{2}\gamma H^2}$$

(9.79)

Then use the magnitude of $K_p \sin \delta'$ obtained from Eq. (9.79) to estimate the magnitude of $K_p \cos \delta'$ from the plots given in Figure 9.33b.

Step 2. **Strip Case.** Determine the actual height h of the anchor to be constructed. If a continuous anchor (i.e., an anchor for which $B = \infty$) of
height h is placed in the soil so that its depth of embedment is H, as
shown in Figure 9.34, the ultimate resistance per unit length is

$$P'_{us} = \left[\frac{C_{ov} + 1}{C_{ov} + \left(\dfrac{H}{h}\right)} \right] \underset{\text{Eq. 9.78}}{\overset{\uparrow}{P'_u}}$$

(9.80)

where

P'_{us} = ultimate resistance for the *strip case*
C_{ov} = 19 for dense sand and 14 for loose sand

Step 3. **Actual Case.** In practice, the anchor plates are placed in a row with
center-to-center spacing S', as shown in Figure 9.35a. The ultimate
resistance of each anchor is

$$P_u = P'_{us} B_e$$

(9.81)

where B_e = equivalent length.

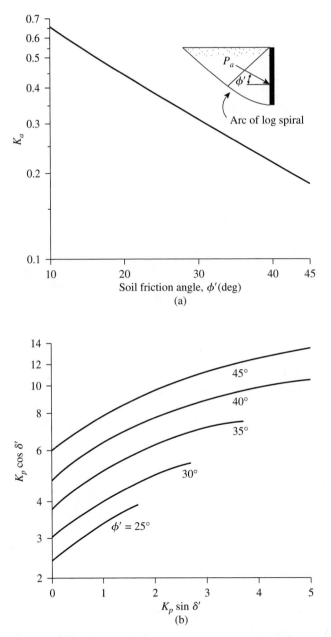

Figure 9.33   (a) Variation of K_a for $\delta' = \phi'$, (b) variation of $K_p \cos \delta'$ with $K_p \sin \delta'$ (Based on Ovesen and Stromann, 1972)

The equivalent length is a function of S', B, H, and h. Figure 9.35b shows a plot of $(B_e - B)/(H + h)$ against $(S' - B)/(H + h)$ for the cases of loose and dense sand. With known values of S', B, H, and h, the value of B_e can be calculated and used in Eq. (9.81) to obtain P_u.

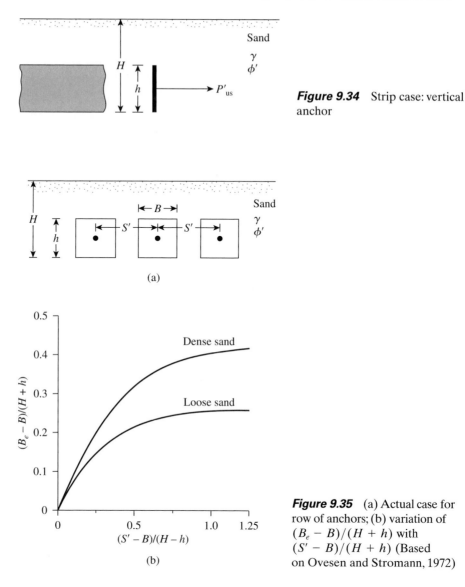

Figure 9.34 Strip case: vertical anchor

(a)

(b)

Figure 9.35 (a) Actual case for row of anchors; (b) variation of $(B_e - B)/(H + h)$ with $(S' - B)/(H + h)$ (Based on Ovesen and Stromann, 1972)

Empirical Correlation Based on Model Tests

Ghaly (1997) used the results of 104 laboratory tests, 15 centrifugal model tests, and 9 field tests to propose an empirical correlation for the ultimate resistance of single anchors. The correlation can be written as

$$P_u = \frac{5.4}{\tan \phi'} \left(\frac{H^2}{A} \right)^{0.28} \gamma A H \tag{9.82}$$

where A = area of the anchor = Bh.

Ghaly also used the model test results of Das and Seeley (1975) to develop a load–displacement relationship for single anchors. The relationship can be given as

$$\frac{P}{P_u} = 2.2\left(\frac{u}{H}\right)^{0.3} \tag{9.83}$$

where u = horizontal displacement of the anchor at a load level P.

Equations (9.82) and (9.83) apply to single anchors (i.e., anchors for which $S'/B = \infty$). For all practical purposes, when $S'/B \approx 2$ the anchors behave as single anchors.

Factor of Safety for Anchor Plates

The allowable resistance per anchor plate may be given as

$$P_{all} = \frac{P_u}{FS}$$

where FS = factor of safety.

Generally, a factor of safety of 2 is suggested when the method of Ovesen and Stromann is used. A factor of safety of 3 is suggested for P_u calculated by Eq. (9.82).

Spacing of Anchor Plates

The center-to-center spacing of anchors, S', may be obtained from

$$S' = \frac{P_{all}}{F}$$

where F = force per unit length of the sheet pile.

Example 9.8

A row of vertical anchors embedded in sand is shown in Figure 9.36. The anchor plates are made of 6-in. thick concrete. The design parameters are $B = h = 15$ in., $S' = 48$ in., $H = 37.5$ in., $\gamma = 105$ lb/ft³, $\phi' = 35°$, and the unit weight of concrete = 150 lb/ft³. Determine the ultimate resistance of each anchor plate.

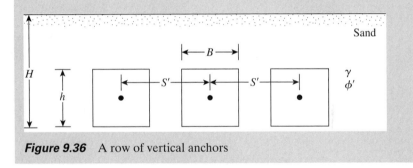

Figure 9.36   A row of vertical anchors

Solution

From Figure 9.33a for $\phi' = 35°$, the magnitude of K_a is about 0.26. Also,

$$W = Ht\gamma_{\text{concrete}} = \left(\frac{37.5}{12}\right)\left(\frac{6}{12}\right)(150) = 234.4 \text{ lb/ft}$$

From Eq. (9.79),

$$K_p \sin \delta' = \frac{W + \dfrac{1}{2}\gamma H^2 K_a \sin \phi'}{\dfrac{1}{2}\gamma H^2}$$

$$= \frac{234.4 + \dfrac{1}{2}(105)\left(\dfrac{37.5}{12}\right)^2 (0.26)(\sin 35)}{\dfrac{1}{2}(10.5)\left(\dfrac{37.5}{12}\right)^2} = 0.606$$

From Figure 9.33b with $\phi' = 35°$ and $K_p \sin \delta' = 0.606$, the magnitude of $K_p \cos \delta'$ is about 4.5. Now, from Eq. (9.78),

$$P'_u = \frac{1}{2}\gamma H^2 (K_p \cos \delta' - K_a \cos \phi')$$

$$= \frac{1}{2}(105)\left(\frac{37.5}{12}\right)^2 [4.5 - (0.26)(\cos 35)] = 2198 \text{ lb/ft}$$

To calculate P'_{us}, we assume the sand to be loose. So, C_{ov} in Eq. (9.80) is 14. Hence

$$P'_{us} = \left[\frac{C_{ov} + 1}{C_{ov} + \left(\dfrac{H}{h}\right)}\right] P'_u = \left[\frac{14 + 1}{14 + \left(\dfrac{37.5}{15}\right)}\right](2198) = 1998 \text{ lb/ft}$$

$$\frac{S' - B}{H + h} = \frac{48 - 15}{37.5 + 15} = 0.63$$

For $(S' - B)/(H + h) = 0.63$ and loose sand, Figure 9.35b yields

$$\frac{B_e - B}{H - h} = 0.227$$

So,

$$B_e = (0.227)(H + h) + B = \frac{(0.227)(37.5 + 15) + 15}{12} = 2.24 \text{ ft}$$

Hence, from Eq. (9.81),

$$P_u = P'_{us} B_e = (1998)(2.24) = \textbf{4476 lb}$$ ∎

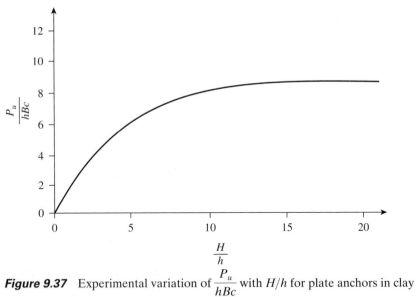

Figure 9.37 Experimental variation of $\dfrac{P_u}{hBc}$ with H/h for plate anchors in clay
[Based on Mackenzie (1955) and Tschebotarioff (1973)]

9.17   Holding Capacity of Anchor Plates in Clay ($\phi = 0$ Condition)

Relatively few studies have been conducted on the ultimate resistance of anchor plates in clayey soils ($\phi = 0$). Mackenzie (1955) and Tschebotarioff (1973) identified the nature of variation of the ultimate resistance of strip anchors and beams as a function of H, h, and c (undrained cohesion based on $\phi = 0$) in a nondimensional form based on laboratory model test results. This is shown in the form of a nondimensional plot in Figure 9.37 (P_u/hBc versus H/h) and can be used to estimate the ultimate resistance of anchor plates in saturated clay ($\phi = 0$).

9.18   Ultimate Resistance of Tiebacks

According to Figure 9.38, the ultimate resistance offered by a tieback in sand is

$$P_u = \pi dl\overline{\sigma}'_o K \tan \phi' \tag{9.84}$$

where

ϕ' = effective angle of friction of soil
$\overline{\sigma}'_o$ = average effective vertical stress ($=\gamma z$ in dry sand)
K = earth pressure coefficient

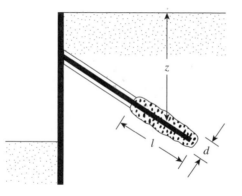

Figure 9.38 Parameters for defining the ultimate resistance of tiebacks

The magnitude of K can be taken to be equal to the earth pressure coefficient at rest (K_o) if the concrete grout is placed under pressure (Littlejohn, 1970). The lower limit of K can be taken to be equal to the Rankine active earth pressure coefficient. In clays, the ultimate resistance of tiebacks may be approximated as

$$P_u = \pi dlc_a \tag{9.85}$$

where c_a = adhesion.

The value of c_a may be approximated as $\frac{2}{3}c_u$ (where c_u = undrained cohesion). A factor of safety of 1.5 to 2 may be used over the ultimate resistance to obtain the allowable resistance offered by each tieback.

Problems

9.1   Figure P9.1 shows a cantilever sheet pile wall penetrating a granular soil. Here, $L_1 = 4$ m, $L_2 = 8$ m, $\gamma = 16.1$ kN/m³, $\gamma_{sat} = 18.2$ kN/m³, and $\phi' = 32°$.
   a. What is the theoretical depth of embedment, D?
   b. For a 30% increase in D, what should be the total length of the sheet piles?
   c. Determine the theoretical maximum moment of the sheet pile.
9.2   Redo Problem 9.1 with the following: $L_1 = 3$ m, $L_2 = 6$ m, $\gamma = 17.3$ kN/m³, $\gamma_{sat} = 19.4$ kN/m³, and $\phi' = 30°$.
9.3   Refer to Figure 9.9. Given: $L = 15$ ft, $\gamma = 108$ lb/ft³, and $\phi' = 35°$. Calculate the theoretical depth of penetration, D, and the maximum moment.
9.4   Repeat Problem 9.3 with the following: $L = 3$ m, $\gamma = 16.7$ kN/m³, and $\phi' = 30°$.
9.5   Refer to Figure P9.5, for which $L_1 = 2.4$ m, $L_2 = 4.6$ m, $\gamma = 15.7$ kN/m³, $\gamma_{sat} = 17.3$ kN/m³, and $\phi' = 30°$, and $c = 29$ kN/m².
   a. What is the theoretical depth of embedment, D?
   b. Increase D by 40%. What length of sheet piles is needed?
   c. Determine the theoretical maximum moment in the sheet pile.
9.6   Solve Problem 9.5 with the following: $L_1 = 5$ ft, $L_2 = 10$ ft, $\gamma = 108$ lb/ft³, $\gamma_{sat} = 122.4$ lb/ft³, $\phi' = 36°$, and $c = 800$ lb/ft².

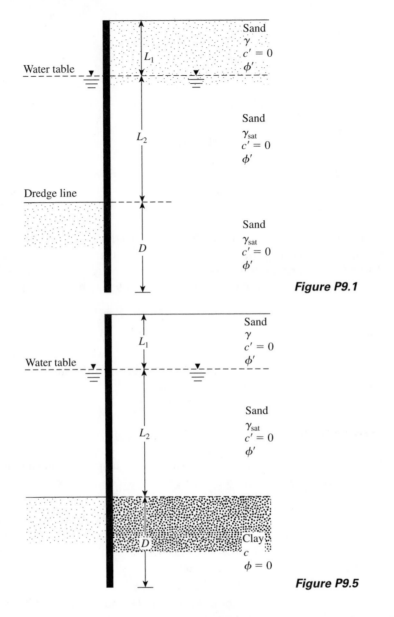

Figure P9.1

Figure P9.5

9.7 Refer to Figure 9.12. Given: $L = 4$ m; for sand, $\gamma = 16$ kN/m³; $\phi' = 35°$; and, for clay, $\gamma_{sat} = 19.2$ kN/m³ and $c = 45$ kN/m². Determine the theoretical value of D and the maximum moment.

9.8 An anchored sheet pile bulkhead is shown in Figure P9.8. Let $L_1 = 4$ m, $L_2 = 9$ m, $l_1 = 2$ m, $\gamma = 17$ kN/m³, $\gamma_{sat} = 19$ kN/m³, and $\phi' = 34°$.
a. Calculate the theoretical value of the depth of embedment, D.
b. Draw the pressure distribution diagram.
c. Determine the anchor force per unit length of the wall.
Use the free earth support method.

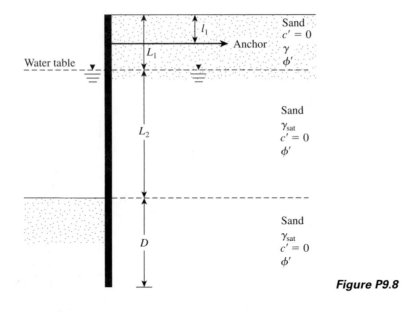

Figure P9.8

9.9 In Problem 9.8, assume that $D_{actual} = 1.3D_{theory}$.
   a. Determining the theoretical maximum moment.
   b. Using Rowe's moment reduction technique, choose a sheet pile section. Take $E = 210 \times 10^3$ MN/m^2 and $\sigma_{all} = 210,000$ kN/m^2.

9.10 Redo Problem 9.8 with the following: $L_1 = 10$ ft, $L_2 = 25$ ft, $l_1 = 4$ ft, $\gamma = 120$ lb/ft^3, $\gamma_{sat} = 129.4$ lb/ft^3, and $\phi' = 40°$.

9.11 Refer to Problem 9.10. Assume that $D_{actual} = 1.3D_{theory}$. Using Rowe's moment reduction method, choose a sheet pile section. Take $E = 29 \times 10^6$ lb/in.^{2} and $\sigma_{all} = 25$ kip/in^2.

9.12 Refer to Figure P9.8. Given: $L_1 = 4$ m, $L_2 = 8$ m, $l_1 = l_2 = 2$ m, $\gamma = 16$ kN/m^3, $\gamma_{sat} = 18.5$ kN/m^3, and $\phi' = 35°$. Use the charts presented in Section 9.10 and determine:
   a. Theoretical depth of penetration
   b. Anchor force per unit length
   c. Maximum moment in the sheet pile.

9.13 Refer to Figure P9.8, for which $L_1 = 3$ m, $L_2 = 6$ m, $l_1 = 1.5$ m, $\gamma = 17.5$ kN/m^3, $\gamma_{sat} = 19.5$ kN/m^3, and $\phi' = 35°$. Use the computational diagram method (Section 9.12) to determine D, F, and M_{max}. Assume that $C = 0.68$ and $R = 0.6$.

9.14 An anchored sheet-pile bulkhead is shown in Figure P9.14. Let $L_1 = 3$ m, $L_2 = 8$ m, $l_1 = 1.5$ m, $\gamma = 17$ kN/m^3, $\gamma_{sat} = 19.5$ kN/m^3, $\phi' = 36°$, and $c = 40$ kN/m^2.
   a. Determine the theoretical depth of embedment, D.
   b. Calculate the anchor force per unit length of the sheet-pile wall.
   Use the free earth support method.

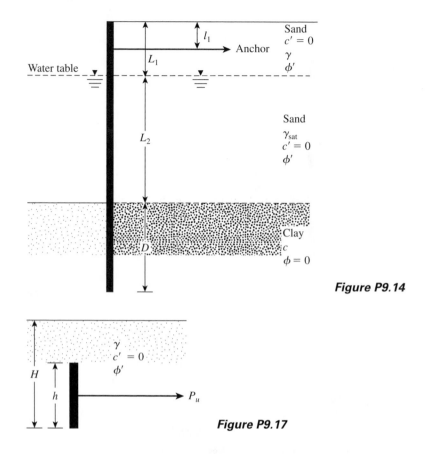

Figure P9.14

Figure P9.17

9.15 Repeat Problem 9.14 with the following: $L_1 = 8$ ft, $L_2 = 20$ ft, $l_1 = 4$ ft, $\gamma = 115$ lb/ft^3, $\gamma_{sat} = 128$ lb/ft^3, $\phi' = 40°$, and $c = 1500$ lb/ft^2.

9.16 In Figure 9.35a, for the anchor slab in sand, $H = 5$ ft, $h = 3$ ft, $B = 4$ ft, $S' = 7$ ft, $\phi' = 30°$, and $\gamma = 110$ lb/ft^3. The anchor plates are made of concrete and have a thickness of 3 in. Using Ovesen and Stromann's method, calculate the ultimate holding capacity of each anchor. Take $\gamma_{concrete} = 150$ lb/ft^3.

9.17 A single anchor slab is shown in Figure P9.17. Here, $H = 0.9$ m, $h = 0.3$ m, $\gamma = 17$ kN/m^3, and $\phi' = 32°$. Calculate the ultimate holding capacity of the anchor slab if the width B is (a) 0.3 m, (b) 0.6 m, and (c) = 0.9 m. (*Note*: center-to-center spacing, $S' = \infty$.) Use the empirical correlation given in Section 9.15 [Eq. (9.82)].

References

Casagrande, L. (1973). "Comments on Conventional Design of Retaining Structures," *Journal of the Soil Mechanics and Foundations Division,* ASCE, Vol. 99, No. SM2, pp. 181–198.

Das, B. M., and Seeley, G. R. (1975). "Load–Displacement Relationships for Vertical Anchor Plates," *Journal of the Geotechnical Engineering Division,* American Society of Civil Engineers, Vol. 101, No, GT7, pp. 711–715.

Ghaly, A. M. (1997). "Load–Displacement Prediction for Horizontally Loaded Vertical Plates." *Journal of Geotechnical and Geoenvironmental Engineering,* ASCE, Vol. 123, No. 1, pp. 74–76.

Hagerty, D. J., and Nofal, M. M. (1992). "Design Aids: Anchored Bulkheads in Sand," *Canadian Geotechnical Journal,* Vol. 29, No. 5, pp. 789–795.

Littlejohn, G. S. (1970). "Soil Anchors," *Proceedings, Conference on Ground Engineering,* Institute of Civil Engineers, London, pp. 33–44.

Mackenzie, T. R. (1955). *Strength of Deadman Anchors in Clay,* M.S. Thesis, Princeton University, Princeton, N. J.

Nataraj, M. S., and Hoadley, P. G. (1984). "Design of Anchored Bulkheads in Sand," *Journal of Geotechnical Engineering,* American Society of Civil Engineers, Vol. 110, No. GT4, pp. 505–515.

Ovesen, N. K., and Stromann, H. (1972). "Design Methods for Vertical Anchor Slabs in Sand," *Proceedings, Specialty Conference on Performance of Earth and Earth-Supported Structures.* American Society of Civil Engineers, Vol. 2.1, pp. 1481–1500.

Rowe, P. W. (1952). "Anchored Sheet Pile Walls," *Proceedings, Institute of Civil Engineers,* Vol. 1, Part 1, pp. 27–70.

Rowe, P. W. (1957). "Sheet Pile Walls in Clay," *Proceedings, Institute of Civil Engineers,* Vol. 7, pp. 654–692.

Schroeder, W. L., and Roumillac, P. (1983). "Anchored Bulkheads with Sloping Dredge Lines," *Journal of Geotechnical Engineering,* American Society of Civil Engineers, Vol. 109. No. 6, pp. 845–851.

Tschebotarioff, G. P. (1973). *Foundations, Retaining and Earth Structures,* 2nd ed., McGraw-Hill, New York.

Tsinker, G. P. (1983). "Anchored Street Pile Bulkheads: Design Practice," *Journal of Geotechnical Engineering,* American Society of Civil Engineers, Vol. 109, No. GT8, pp. 1021–1038.

10

Braced Cuts

10.1   *Introduction*

Sometimes construction work requires ground excavations with vertical or near-vertical faces—for example, basements of buildings in developed areas or underground transportation facilities at shallow depths below the ground surface (a cut-and-cover type of construction). The vertical faces of the cuts need to be protected by temporary bracing systems to avoid failure that may be accompanied by considerable settlement or by bearing capacity failure of nearby foundations.

Figure 10.1 shows two types of braced cut commonly used in construction work. One type uses the *soldier beam* (Figure 10.1a), which is driven into the ground before excavation and is a vertical steel or timber beam. *Laggings*, which are horizontal timber planks, are placed between soldier beams as the excavation proceeds. When the excavation reaches the desired depth, *wales* and *struts* (horizontal steel beams) are installed. The struts are compression members. Figure 10.1b shows another type of braced excavation. In this case, interlocking *sheet piles* are driven into the soil before excavation. Wales and struts are inserted immediately after excavation reaches the appropriate depth.

Figure 10.2 shows the braced-cut construction used for the Chicago subway in 1940. Timber lagging, timber struts, and steel wales were used. Figure 10.3 shows a braced cut made during the construction of the Washington, DC, metro in 1974. In this cut, timber lagging, steel H-soldier piles, steel wales, and pipe struts were used.

To design braced excavations (i.e., to select wales, struts, sheet piles, and soldier beams), an engineer must estimate the lateral earth pressure to which the braced cuts will be subjected. The theoretical aspects of the lateral earth pressure on a braced cut were discussed in Section 7.7. The total active force per unit length of the wall (P_a) was calculated using the general wedge theory. However, that analysis does not provide the relationships required for estimating the variation of lateral pressure with depth, which is a function of several factors, such as the type of soil, the experience of the construction crew, the type of construction equipment used, and so forth. For that reason, empirical pressure envelopes developed from field observations are used for the design of braced cuts. This procedure is discussed in the next section.

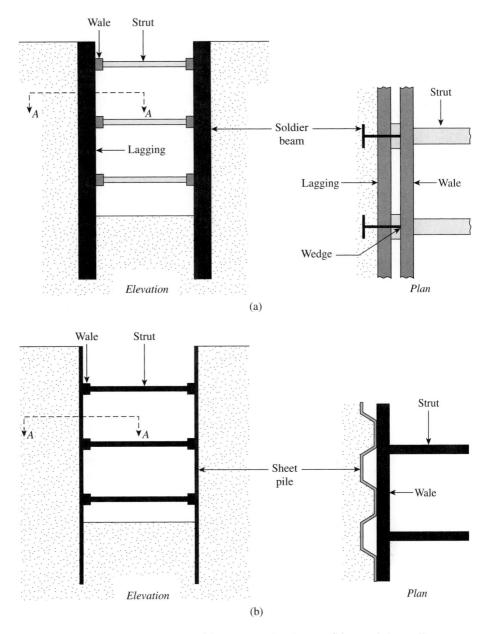

Figure 10.1   Types of braced cut: (a) use of soldier beams; (b) use of sheet piles

10.2   *Pressure Envelope for Braced-Cut Design*

As mentioned in Section 10.1, the lateral earth pressure in a braced cut is dependent on the type of soil, construction method, and type of equipment used. The lateral earth pressure changes from place to place. Each strut should also be designed for

Figure 10.2 Braced cut in Chicago subway construction, January 1940 (Courtesy of Ralph B. Peck)

the maximum load to which it may be subjected. Therefore, the braced cuts should be designed using apparent-pressure diagrams that are envelopes of all the pressure diagrams determined from measured strut loads in the field. Figure 10.4 shows the method for obtaining the apparent-pressure diagram at a section from strut loads. In this figure, let P_1, P_2, P_3, P_4, ... be the measured strut loads. The apparent horizontal pressure can then be calculated as

$$\sigma_1 = \frac{P_1}{(s)\left(d_1 + \dfrac{d_2}{2}\right)}$$

$$\sigma_2 = \frac{P_2}{(s)\left(\dfrac{d_2}{2} + \dfrac{d_3}{2}\right)}$$

$$\sigma_3 = \frac{P_3}{(s)\left(\dfrac{d_3}{2} + \dfrac{d_4}{2}\right)}$$

$$\sigma_4 = \frac{P_4}{(s)\left(\dfrac{d_4}{2} + \dfrac{d_5}{2}\right)}$$

Figure 10.3 Braced cut in the construction of Washington, DC, metro, May 1974 (Courtesy of Ralph B. Peck)

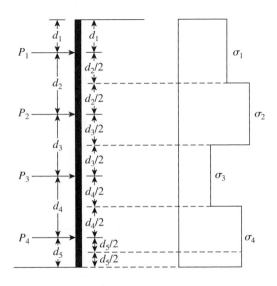

Figure 10.4 Procedure for calculating apparent-pressure diagram from measured strut loads

where

$$\sigma_1, \sigma_2, \sigma_3, \sigma_4 = \text{apparent pressures}$$
$$s = \text{center-to-center spacing of the struts}$$

Using the procedure just described for strut loads observed from the Berlin subway cut, Munich subway cut, and New York subway cut, Peck (1969) provided the envelope of apparent-lateral-pressure diagrams for design of cuts in *sand*. This envelope is illustrated in Figure 10.5, in which

$$\sigma_a = 0.65\gamma H K_a \tag{10.1}$$

where

$$\gamma = \text{unit weight}$$
$$H = \text{height of the cut}$$
$$K_a = \text{Rankine active pressure coefficient} = \tan^2(45 - \phi'/2)$$
$$\phi' = \text{effective friction angle of sand}$$

Cuts in Clay

In a similar manner, Peck (1969) also provided the envelopes of apparent-lateral-pressure diagrams for cuts in *soft to medium clay* and in *stiff clay*. The pressure envelope for soft to medium clay is shown in Figure 10.6 and is applicable to the condition

$$\frac{\gamma H}{c} > 4$$

where c = undrained cohesion ($\phi = 0$).

The pressure, σ_a, is the larger of

$$\sigma_a = \gamma H \left[1 - \left(\frac{4c}{\gamma H}\right)\right]$$
$$\text{and} \tag{10.2}$$
$$\sigma_a = 0.3\gamma H$$

where γ = unit weight of clay.

The pressure envelope for cuts in stiff clay is shown in Figure 10.7, in which

$$\sigma_a = 0.2\gamma H \text{ to } 0.4\gamma H \qquad \text{(with an average of } 0.3\gamma H) \tag{10.3}$$

is applicable to the condition $\gamma H/c \leq 4$.

When using the pressure envelopes just described, keep the following points in mind:

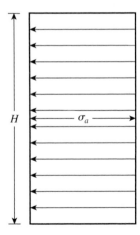

Figure 10.5 Peck's (1969) apparent-pressure envelope for cuts in sand

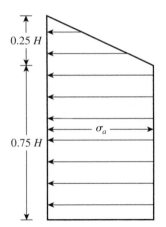

Figure 10.6 Peck's (1969) apparent-pressure envelope for cuts in soft to medium clay

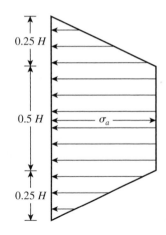

Figure 10.7 Peck's (1969) apparent-pressure envelope for cuts in stiff clay

1. They apply to excavations having depths greater than about 6 m ($\approx$20 ft).
2. They are based on the assumption that the water table is below the bottom of the cut.
3. Sand is assumed to be drained with zero pore water pressure.
4. Clay is assumed to be undrained and pore water pressure is not considered.

10.3 *Pressure Envelope for Cuts in Layered Soil*

Sometimes, layers of both sand and clay are encountered when a braced cut is being constructed. In this case, Peck (1943) proposed that an equivalent value of cohesion ($\phi = 0$) should be determined according to the formula (see Figure 10.8a).

$$c_{av} = \frac{1}{2H}[\gamma_s K_s H_s^2 \tan \phi_s' + (H - H_s)n'q_u] \tag{10.4}$$

where

H = total height of the cut
γ_s = unit weight of sand
H_s = height of the sand layer
K_s = a lateral earth pressure coefficient for the sand layer ($\approx$1)
ϕ_s' = effective angle of friction of sand
q_u = unconfined compression strength of clay
n' = a coefficient of progressive failure (ranging from 0.5 to 1.0; average value 0.75)

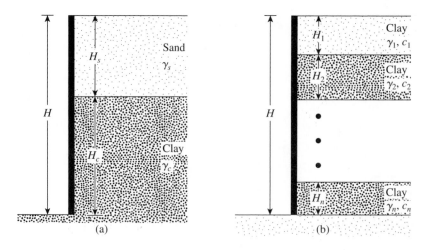

Figure 10.8   Layered soils in braced cuts

The average unit weight of the layers may be expressed as

$$\gamma_a = \frac{1}{H}\left[\gamma_s H_s + (H - H_s)\gamma_c\right] \qquad (10.5)$$

where γ_c = saturated unit weight of clay layer.

Once the average values of cohesion and unit weight are determined, the pressure envelopes in clay can be used to design the cuts.

Similarly, when several clay layers are encountered in the cut (Figure 10.8b), the average undrained cohesion becomes

$$c_{av} = \frac{1}{H}(c_1 H_1 + c_2 H_2 + \cdots + c_n H_n) \qquad (10.6)$$

where

$c_1, c_2, \ldots, c_n$ = undrained cohesion in layers $1, 2, \ldots, n$
$H_1, H_2, \ldots, H_n$ = thickness of layers $1, 2, \ldots, n$

The average unit weight is now

$$\gamma_a = \frac{1}{H}(\gamma_1 H_1 + \gamma_2 H_2 + \gamma_3 H_3 + \cdots + \gamma_n H_n) \qquad (10.7)$$

10.4   *Design of Various Components of a Braced Cut*

Struts

In construction work, struts should have a minimum vertical spacing of about 2.75 m (9 ft) or more. Struts are horizontal columns subject to bending. The load-carrying capacity of columns depends on their *slenderness ratio,* which can be reduced by providing vertical and horizontal supports at intermediate points. For wide cuts, splicing the struts may be necessary. For braced cuts in clayey soils, the depth of the first strut below the ground surface should be less than the depth of tensile crack, z_c. From Eq. (7.8),

$$\sigma'_a = \gamma z K_a - 2c'\sqrt{K_a}$$

where K_a = coefficient of Rankine active pressure.
For determining the depth of tensile crack,

$$\sigma'_a = 0 = \gamma z_c K_a - 2c'\sqrt{K_a}$$

or

$$z_c = \frac{2c'}{\sqrt{K_a}\gamma}$$

With $\phi = 0$, $K_a = \tan^2(45 - \phi/2) = 1$, so

$$z_c = \frac{2c}{\gamma}$$

A simplified conservative procedure may be used to determine the strut loads. Although this procedure will vary, depending on the engineers involved in the project, the following is a step-by-step outline of the general methodology (see Figure 10.9):

Step 1.   Draw the pressure envelope for the braced cut. (See Figures 10.5, 10.6, and 10.7.) Also, show the proposed strut levels. Figure 10.9a shows a pressure envelope for a sandy soil; however, it could also be for a clay. The strut levels are marked A, B, C, and D. The sheet piles (or soldier beams) are assumed to be hinged at the strut levels, except for the top and bottom ones. In Figure 10.9a, the hinges are at the level of struts B and C. (Many designers also assume the sheet piles or soldier beams to be hinged at all strut levels except for the top.)

Step 2.   Determine the reactions for the two simple cantilever beams (top and bottom) and all the simple beams between. In Figure 10.9b, these reactions are A, B_1, B_2, C_1, C_2, and D.

Step 3.   The strut loads in the figure may be calculated via the formulas

$$P_A = (A)(s)$$
$$P_B = (B_1 + B_2)(s) \qquad\qquad (10.8)$$
$$P_C = (C_1 + C_2)(s)$$

and

$$P_D = (D)(s)$$

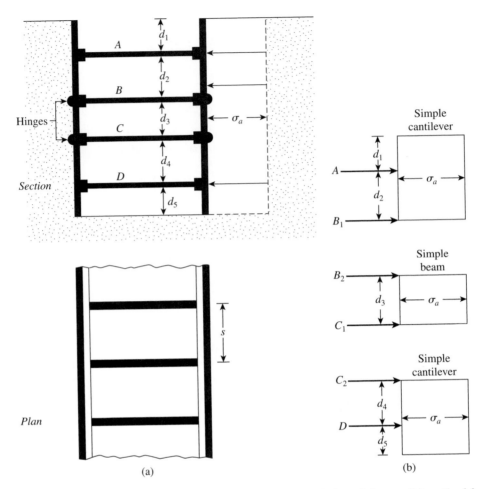

Figure 10.9 Determination of strut loads: (a) section and plan of the cut; (b) method for determining strut loads

where

P_A, P_B, P_C, P_D = loads to be taken by the individual struts at levels $A, B, C,$ and D, respectively

A, B_1, B_2, C_1, C_2, D = reactions calculated in Step 2 (note the unit: force/unit length of the braced cut)

s = horizontal spacing of the struts (see plan in Figure 10.9a)

Step 4. Knowing the strut loads at each level and the intermediate bracing conditions allows selection of the proper sections from the steel construction manual.

Sheet Piles

The following steps are involved in designing the sheet piles:

Step 1. For each of the sections shown in Figure 10.9b, determine the maximum bending moment.

Step 2. Determine the maximum value of the maximum bending moments (M_{max}) obtained in Step 1. Note that the unit of this moment will be, for example, kN · m/m (lb-ft/ft) length of the wall.

Step 3. Obtain the required section modulus of the sheet piles, namely,

$$S = \frac{M_{max}}{\sigma_{all}} \tag{10.9}$$

where σ_{all} = allowable flexural stress of the sheet pile material.

Step 4. Choose a sheet pile having a section modulus greater than or equal to the required section modulus from a table such as Table 9.1.

Wales

Wales may be treated as continuous horizontal members if they are spliced properly. Conservatively, they may also be treated as though they are pinned at the struts. For the section shown in Figure 10.9a, the maximum moments for the wales (assuming that they are pinned at the struts) are,

$$\text{At level } A, \quad M_{max} = \frac{(A)(s^2)}{8}$$

$$\text{At level } B, \quad M_{max} = \frac{(B_1 + B_2)s^2}{8}$$

$$\text{At level } C, \quad M_{max} = \frac{(C_1 + C_2)s^2}{8}$$

and

$$\text{At level } D, \quad M_{max} = \frac{(D)(s^2)}{8}$$

where A, B_1, B_2, C_1, C_2, and D are the reactions under the struts per unit length of the wall (see Step 2 of strut design).

Now determine the section modulus of the wales:

$$S = \frac{M_{max}}{\sigma_{all}}$$

The wales are sometimes fastened to the sheet piles at points that satisfy the lateral support requirements.

Example 10.1

The cross section of a long braced cut is shown in Figure 10.10a.

a. Draw the earth-pressure envelope.
b. Determine the strut loads at levels *A, B*, and *C*.
c. Determine the section modulus of the sheet pile section required.
d. Determine a design section modulus for the wales at level *B*.

(*Note:* The struts are placed at 3 m, center to center, in the plan.) Use

$$\sigma_{all} = 170 \times 10^3 \text{ kN/m}^2$$

Solution

Part a
We are given that $\gamma = 18$ kN/m^2, $c = 35$ kN/m^2, and $H = 7$ m. So,

$$\frac{\gamma H}{c} = \frac{(18)(7)}{35} = 3.6 < 4$$

Thus, the pressure envelope will be like the one in Figure 10.7. The envelope is plotted in Figure 10.10a with maximum pressure intensity, σ_a, equal to $0.3\gamma H = 0.3(18)(7) = $ **37.8 kN/m^2.**

Part b
To calculate the strut loads, examine Figure 10.10b. Taking the moment about B_1, we have $\Sigma M_{B_1} = 0$, and

$$A(2.5) - \left(\frac{1}{2}\right)(37.8)(1.75)\left(1.75 + \frac{1.75}{3}\right) - (1.75)(37.8)\left(\frac{1.75}{2}\right) = 0$$

or

$$A = 54.02 \text{ kN/m}$$

Also, Σ vertical forces $= 0$. Thus,

$$\tfrac{1}{2}(1.75)(37.8) + (37.8)(1.75) = A + B_1$$

or

$$33.08 + 66.15 - A = B_1$$

So,

$$B_1 = 45.2 \text{ kN/m}$$

Due to symmetry,

$$B_2 = 45.2 \text{ kN/m}$$

and

$$C = 54.02 \text{ kN/m}$$

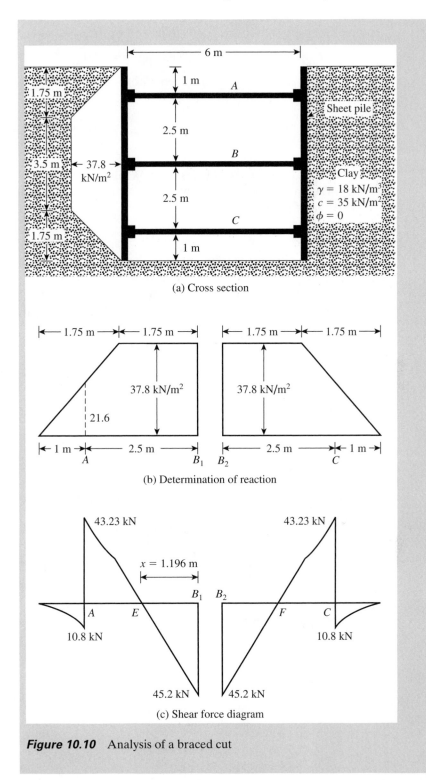

(a) Cross section

(b) Determination of reaction

(c) Shear force diagram

Figure 10.10   Analysis of a braced cut

Hence, the strut loads at the levels indicated by the subscripts are

$$P_a = 54.02 \times \text{horizontal spacing}, s = 54.02 \times 3 = \textbf{162.06 kN}$$

$$P_B = (B_1 + B_2)3 = (45.2 + 45.2)3 = \textbf{271.2 kN}$$

and

$$P_C = 54.02 \times 3 = \textbf{162.06 kN}$$

Part c

At the left side of Figure 10.10b, for the maximum moment, the shear force should be zero. The nature of the variation of the shear force is shown in Figure 10.10c. The location of point E can be given as

$$x = \frac{\text{reaction at } B_1}{37.8} = \frac{45.2}{37.8} = 1.196 \text{ m}$$

Also,

$$\text{Magnitude of moment at } A = \frac{1}{2}(1)\left(\frac{37.8}{1.75} \times 1\right)\left(\frac{1}{3}\right)$$

$$= 3.6 \text{ kN-m/meter of wall}$$

and

$$\text{Magnitude of moment at } E = (45.2 \times 1.196) - (37.8 \times 1.196)\left(\frac{1.196}{2}\right)$$

$$= 54.06 - 27.03 = 27.03 \text{ kN-m/meter of wall}$$

Because the loading on the left and right sections of Figure 10.10b are the same, the magnitudes of the moments at F and C (see Figure 10.10c) will be the same as those at E and A, respectively. Hence, the maximum moment is 27.03 kN-m/meter of wall.

The section modulus of the sheet piles is thus

$$S = \frac{M_{max}}{\sigma_{all}} = \frac{27.03 \text{ kN-m}}{170 \times 10^3 \text{ kN/m}^2} = \textbf{15.9} \times \textbf{10}^{-5}\textbf{m}^3\textbf{/m of the wall}$$

Part d

The reaction at level B has been calculated in Part b. Hence,

$$M_{max} = \frac{(B_1 + B_2)s^2}{8} = \frac{(45.2 + 45.2)3^2}{8} = 101.7 \text{ kN-m}$$

and

$$\text{Section modulus, } S = \frac{101.7}{\sigma_{all}} = \frac{101.7}{(170 \times 1000)}$$

$$= \textbf{0.598} \times \textbf{10}^{-3} \textbf{ m}^3 \qquad \blacksquare$$

Example 10.2

Refer to the braced cut shown in Figure 10.11, for which $\gamma = 112 \text{ lb/ft}^3$, $\phi' = 32°$, and $c' = 0$. The struts are located 12 ft on center in the plan. Draw the earth-pressure envelope and determine the strut loads at levels A, B, and C.

Solution

For this case, the earth-pressure envelope shown in Figure 10.5 is applicable. Hence,

$$K_a = \tan^2\left(45 - \frac{\phi'}{2}\right) = \tan^2\left(45 - \frac{32}{2}\right) = 0.307$$

From Equation (10.1)

$$\sigma_a = 0.65\,\gamma H K_a = (0.65)(112)(27)(0.307) = 603.44 \text{ lb/ft}^2$$

Figure 10.12a shows the pressure envelope. Refer to Figure 10.12b and calculate B_1:

$$\sum M_{B_1} = 0$$

$$A = \frac{(603.44)(15)\left(\dfrac{15}{2}\right)}{9} = 7543 \text{ lb/ft}$$

$$B_1 = (603.44)(15) - 7543 = 1508.6 \text{ lb/ft} \approx 1509 \text{ lb/ft}$$

Now, refer to Figure 10.12c and calculate B_2:

$$\sum M_{B_2} = 0$$

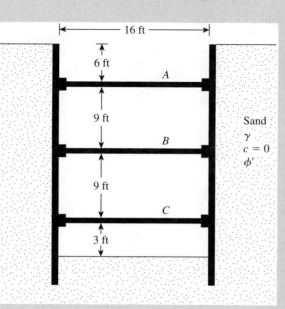

Figure 10.11

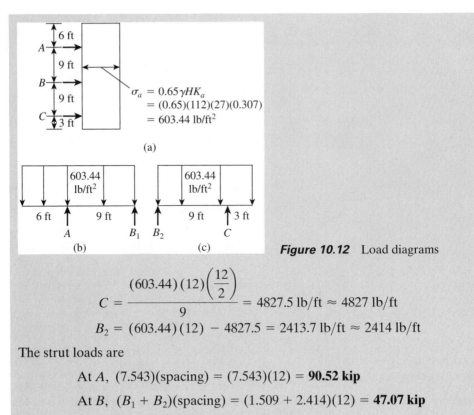

$$\sigma_a = 0.65\gamma H K_a$$
$$= (0.65)(112)(27)(0.307)$$
$$= 603.44 \text{ lb/ft}^2$$

(a)

603.44 lb/ft²

6 ft   9 ft
A   B₁

603.44 lb/ft²

9 ft   3 ft
B₂   C

(b)   (c)

Figure 10.12   Load diagrams

$$C = \frac{(603.44)(12)\left(\dfrac{12}{2}\right)}{9} = 4827.5 \text{ lb/ft} \approx 4827 \text{ lb/ft}$$

$$B_2 = (603.44)(12) - 4827.5 = 2413.7 \text{ lb/ft} \approx 2414 \text{ lb/ft}$$

The strut loads are

At A, $(7.543)(\text{spacing}) = (7.543)(12) = \mathbf{90.52 \text{ kip}}$

At B, $(B_1 + B_2)(\text{spacing}) = (1.509 + 2.414)(12) = \mathbf{47.07 \text{ kip}}$

At C, $(4.827)(\text{spacing}) = (4.827)(12) = \mathbf{57.93 \text{ kip}}$ ∎

Example 10.3

For the braced cut described in Example 10.2, determine:

a. The sheet-pile section modulus
b. The required section modulus of the wales at level A; assume that $\sigma_{\text{all}} = 24 \text{ kip/in}^2$

Solution

Part a
Refer to the load diagrams shown in Figure 10.12b and 10.12c. Figure 10.13 shows the shear force diagrams based on the load diagrams. First, determine x_1 and x_2:

$$x_1 = \frac{3.923}{0.603} = 6.5 \text{ ft}$$

$$x_2 = \frac{3.017}{0.603} = 5 \text{ ft}$$

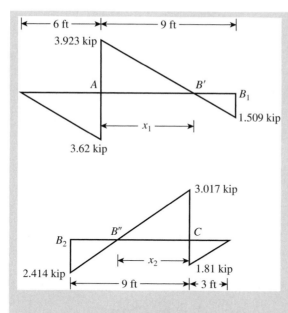

Figure 10.13   Shear force diagrams

Then the moments are

$$\text{At } A, \quad \frac{1}{2}(3.62)(6) = 10.86 \text{ kip-ft}$$

$$\text{At } C, \quad \frac{1}{2}(1.81)(3) = 2.715 \text{ kip-ft}$$

$$\text{At } B', \quad \frac{1}{2}(1.509)(2.5) = 1.89 \text{ kip-ft}$$

$$\text{At } B'', \quad \frac{1}{2}(2.414)(4) = 4.828 \text{ kip-ft}$$

M_A is maximum, so

$$S = \frac{M_{\text{max}}}{\sigma_{\text{all}}} = \frac{(10.86 \text{ kip-ft})(12)}{24 \text{ kip/in}^2} = \textbf{5.43 in}^3\textbf{/ft}$$

Part b
For the wale at level A,

$$M_{\text{max}} = \frac{A(s^2)}{8}$$

$A = 7543$ lb/ft-(from Example 10.2). So,

$$M_{\text{max}} = \frac{(7.543)(12^2)}{8} = 135.77 \text{ kip-ft/ft}$$

$$S = \frac{M_{\text{max}}}{\sigma_{\text{all}}} = \frac{(135.77)(12)}{24 \text{ kip/in}^2} = \textbf{67.9 in}^3\textbf{/ft of wall}$$ ■

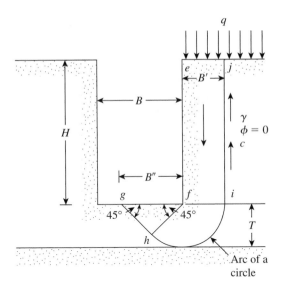

Figure 10.14   Heaving in braced cuts in clay

10.5   *Bottom Heave of a Cut in Clay*

Braced cuts in clay may become unstable as a result of heaving of the bottom of the excavation. Terzaghi (1943) analyzed the factor of safety of long braced excavations against bottom heave. The failure surface for such a case in a homogeneous soil is shown in Figure 10.14. In the figure, the following notations are used: B = width of the cut, H = depth of the cut, T = thickness of the clay below the base of excavation, and q = uniform surcharge adjacent to the excavation.

The ultimate bearing capacity at the base of a soil column with a width of B' can be given as

$$q_u = cN_c \tag{10.10}$$

where $N_c = 5.7$ (for a perfectly rough foundation).

The vertical load per unit area along fi is

$$q = \gamma H + q - \frac{cH}{B'} \tag{10.11}$$

Hence, the factor of safety against bottom heave is

$$\text{FS} = \frac{q_u}{q} = \frac{cN_c}{\gamma H + q - \dfrac{cH}{B'}} = \frac{cN_c}{\left(\gamma + \dfrac{q}{H} - \dfrac{c}{B'}\right)H} \tag{10.12}$$

For excavations of limited length L, the factor of safety can be modified to

$$\text{FS} = \frac{cN_c\left(1 + 0.2\dfrac{B'}{L}\right)}{\left(\gamma + \dfrac{q}{H} - \dfrac{c}{B'}\right)H} \tag{10.13}$$

where $B' = T$ or $B/\sqrt{2}$ (whichever is smaller).

In 2000, Chang suggested a revision of Eq. (10.13) with the following changes:

1. The shearing resistance along ij may be considered as an increase in resistance rather than a reduction in loading.
2. In Figure 10.14, fg with a width of B'' at the base of the excavation may be treated as a negatively loaded footing.
3. The value of the bearing capacity factor N_c should be 5.14 (not 5.7) for a perfectly smooth footing, because of the restraint-free surface at the base of the excavation.

With the foregoing modifications, Eq. (10.13) takes the form

$$FS = \frac{5.14c\left(1 + \dfrac{0.2B''}{L}\right) + \dfrac{cH}{B'}}{\gamma H + q} \qquad (10.14)$$

where

$$B' = T \text{ if } T \leq B/\sqrt{2}$$
$$B' = B/\sqrt{2} \text{ if } T > B/\sqrt{2}$$
$$B'' = \sqrt{2}B'$$

Bjerrum and Eide (1956) compiled a number of case records for the bottom heave of cuts in clay. Chang (2000) used those records to calculate FS by means of Eq. (10.14); his findings are summarized in Table 10.1. It can be seen from this table that the actual field observations agree well with the calculated factors of safety.

Table 10.1   Calculated Factors of Safety for Selected Case Records Compiled by Bjerrum and Eide (1956) and Calculated by Chang (2000)

Site	B (m)	B/L	H (m)	H/B	γ (kN/m³)	c (kN/m²)	q (kN/m²)	FS [Eq. (10.14)]	Type of failure
Pumping station, Fornebu, Oslo	5.0	1.0	3.0	0.6	17.5	7.5	0	1.05	Total failure
Storehouse, Drammen	4.8	0	2.4	0.5	19.0	12	15	1.05	Total failure
Sewerage tank, Drammen	5.5	0.69	3.5	0.64	18.0	10	10	0.92	Total failure
Excavation, Grey Wedels Plass, Oslo	5.8	0.72	4.5	0.78	18.0	14	10	1.07	Total failure
Pumping station, Jernbanetorget, Oslo	8.5	0.70	6.3	0.74	19.0	22	0	1.26	Partial failure
Storehouse, Freia, Oslo	5.0	0	5.0	1.00	19.0	16	0	1.10	Partial failure
Subway, Chicago	16	0	11.3	0.70	19.0	35	0	1.00	Near failure

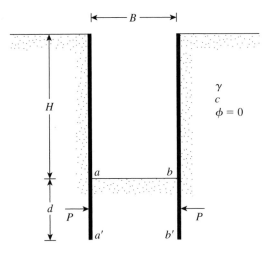

Figure 10.15 Force on the buried length of sheet pile

Equation (10.14) is recommended for use in this test. In most cases, a factor of safety of about 1.5 is recommended.

In homogeneous clay, if FS becomes less than 1.5, the sheet pile is driven deeper. (See Figure 10.15.) Usually, the depth d is kept less than or equal to $B/2$, in which case the force P per unit length of the buried sheet pile (aa' and bb') may be expressed as (U.S. Department of the Navy, 1971)

$$P = 0.7(\gamma HB - 1.4cH - \pi cB) \qquad \text{for } d > 0.47B \qquad (10.15)$$

and

$$P = 1.5d\left(\gamma H - \frac{1.4cH}{B} - \pi c\right). \qquad \text{for } d < 0.47B \qquad (10.16)$$

Example 10.4

Refer to the braced cut described in Example 10.1 (see Figure 10.10a). Assume that the length of the cut is 25 m and a hard stratum is located at a depth of 4 m below the bottom of the cut. Calculate the factor of safety against bottom heave. Use Eq. (10.14)

Solution
From Eq. (10.14),

$$FS = \frac{5.14c\left(1 + \dfrac{0.2B''}{L}\right) + \dfrac{cH}{B'}}{\gamma H + q}$$

with $T = 4$ m,

$$\frac{B}{\sqrt{2}} = \frac{6}{\sqrt{2}} = 4.24 \text{ m}$$

So,

$$T \leqslant \frac{B}{\sqrt{2}}$$

Hence, $B' = T = 4$ m, and it follows that

$$B'' = \sqrt{2}B' = (\sqrt{2})(4) = 5.66 \text{ m}$$

and

$$\text{FS} = \frac{(5.14)(35)\left[1 + \dfrac{(0.2)(5.66)}{25}\right] + \dfrac{(35)(7)}{4}}{(18)(7)} = \mathbf{1.98} \qquad \blacksquare$$

10.6 *Stability of the Bottom of a Cut in Sand*

The bottom of a cut in sand is generally stable. When the water table is encountered, the bottom of the cut is stable as long as the water level inside the excavation is higher than the groundwater level. In case dewatering is needed (see Figure 10.16), the factor of safety against piping should be checked. [*Piping* is another term for failure by heave, as defined in Section 1.10; see Eq. (1.37).] Piping may occur when a high hydraulic gradient is created by water flowing into the excavation. To check the

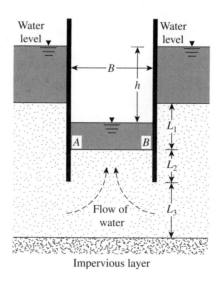

Figure 10.16 Stability of the bottom of a cut in sand

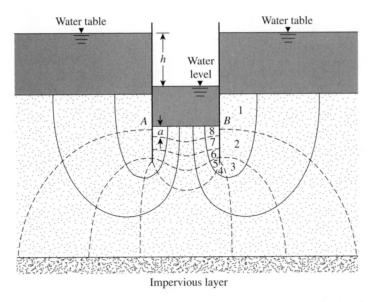

Figure 10.17 Determining the factor of safety against piping by drawing a flow net

factor of safety, draw flow nets and determine the maximum exit gradient $[i_{max(exit)}]$ that will occur at points A and B. Figure 10.17 shows such a flow net, for which the maximum exit gradient is

$$i_{max(exit)} = \frac{\dfrac{h}{N_d}}{a} = \frac{h}{N_d a} \tag{10.17}$$

where

a = length of the flow element at A (or B)
N_d = number of drops (*Note:* in Figure 10.17, $N_d = 8$; see also Section 1.9)

The factor of safety against piping may be expressed as

$$FS = \frac{i_{cr}}{i_{max(exit)}} \tag{10.18}$$

where i_{cr} = critical hydraulic gradient.
    The relationship for i_{cr} was given in Chapter 1 as

$$i_{cr} = \frac{G_s - 1}{e + 1}$$

The magnitude of i_{cr} varies between 0.9 and 1.1 in most soils, with an average of about 1. A factor of safety of about 1.5 is desirable.

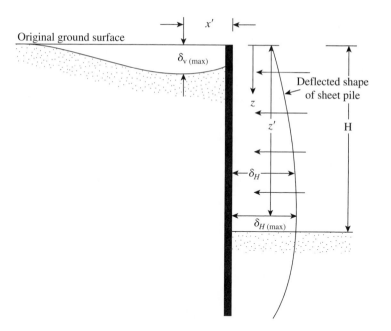

Figure 10.18 Lateral yielding of sheet pile and ground settlement

10.7 *Lateral Yielding of Sheet Piles and Ground Settlement*

In braced cuts, some lateral movement of sheet pile walls may be expected. (See Figure 10.18.) The amount of lateral yield (δ_H) depends on several factors, the most important of which is the elapsed time between excavation and the placement of wales and struts. As discussed before, in several instances the sheet piles (or the soldier, piles as the case may be) are driven to a certain depth below the bottom of the excavation. The reason is to reduce the lateral yielding of the walls during the last stages of excavation. Lateral yielding of the walls will cause the ground surface surrounding the cut to settle. The degree of lateral yielding, however, depends mostly on the type of soil below the bottom of the cut. If clay below the cut extends to a great depth and $\gamma H/c$ is less than about 6, extension of the sheet piles or soldier piles below the bottom of the cut will help considerably in reducing the lateral yield of the walls.

However, under similar circumstances, if $\gamma H/c$ is about 8, the extension of sheet piles into the clay below the cut does not help greatly. In such circumstances, we may expect a great degree of wall yielding that could result in the total collapse of the bracing systems. If a hard layer of soil lies below a clay layer at the bottom of the cut, the piles should be embedded in the stiffer layer. This action will greatly reduce lateral yield.

The lateral yielding of walls will generally induce ground settlement, δ_V, around a braced cut. Such settlement is generally referred to as *ground loss*. On the basis of several field observations, Peck (1969) provided curves for predicting ground settlement in various types of soil. (See Figure 10.19.) The magnitude of ground loss varies extensively; however, the figure may be used as a general guide.

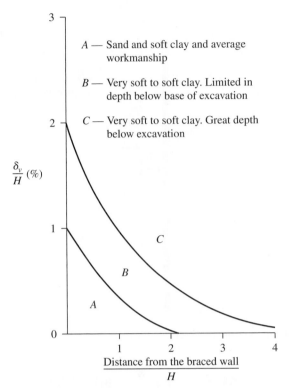

Figure 10.19   Variation of ground settlement with distance (After Peck, 1969)

Moormann (2004) recently analyzed about 153 case histories dealing mainly with the excavation in soft clay (that is, undrained shear strength, $c \leqslant 75 \text{ kN/m}^2$). Following is a summary of his analysis relating to $\delta_{V(max)}$, x', $\delta_{H(max)}$, and z' (see Figure 10.18).

- Maximum Vertical Movement $[\delta_{V(max)}]$

  $\delta_{V(max)}/H \approx 0.1$ to 10.1% with an average of 1.07% (soft clay)
  $\delta_{V(max)}/H \approx 0$ to 0.9% with an average of 0.18% (stiff clay)
  $\delta_{V(max)}/H \approx 0$ to 2.43% with an average of 0.33% (non-cohesive soils)

- Location of $\delta_{V(max)}$, that is x' (Figure 10.18)

  For 70% of all case histories considered, $x' \leqslant 0.5H$.
  However, in soft clays, x' may be as much as $2H$.

- Maximum Horizontal Deflection of Sheet Piles, $\delta_{H(max)}$

  For 40% of excavation in soft clay, $0.5\% \leqslant \delta_{H(max)}/H \leqslant 1\%$.
  The average value of $\delta_{H(max)}/H$ is about 0.87%.
  In stiff clays, the average value of $\delta_{H(max)}/H$ is about 0.25%.
  In non-cohesive soils, $\delta_{H(max)}/H$ is about 0.27% of the average.

- Location of $\delta_{H(max)}$, that is z' (Figure 10.18)

  For deep excavation of soft and stiff cohesive soils, z'/H is about 0.5 to 1.0.

Problems

10.1 Refer to the braced cut shown in Figure P10.1. Given: $\gamma = 17 \text{ kN/m}^3$. $\phi' = 35°$, and $c' = 0$. The struts are located at 3 m center-to-center in the plan. Draw the earth-pressure envelope and determine the strut loads at levels A, B, and C.

10.2 For the braced cut described in Problem 10.1, determine the following:
a. The sheet-pile section modulus
b. The section modulus of the wales at level B
Assume that $\sigma_{all} = 170 \text{ MN/m}^2$.

10.3 Redo Problem 10.1 with $\gamma = 18 \text{ kN/m}^3$, $\phi' = 40°$, $c' = 0$, and the center-to-center strut spacing in the plan $= 4$ m.

10.4 Determine the sheet-pile section modulus for the braced cut described in Problem 10.3. Given: $\sigma_{all} = 170 \text{ MN/m}^2$.

10.5 Refer to Figure 10.8a. For the braced cut, given $H = 8$ m; $H_s = 3$ m; $\gamma_s = 17.5 \text{ kN/m}^3$; angle of friction of sand, $\phi'_s = 34°$; $H_c = 5$ m; $\gamma_c = 18.2 \text{ kN/m}^3$; and unconfined compression strength of clay layer, $q_u = 55 \text{ kN/m}^2$.
a. Estimate the average cohesion (c_{av}) and average unit weight (γ_{av}) for the construction of the earth-pressure envelope.
b. Plot the earth-pressure envelope.

10.6 Refer to Figure 10.8b, which shows a braced cut in clay. Given: $H = 25$ ft, $H_1 = 5$ ft, $c_1 = 2125 \text{ lb/ft}^2$, $\gamma_1 = 111 \text{ lb/ft}^3$, $H_2 = 10$ ft, $c_2 = 1565 \text{ lb/ft}^2$, $\gamma_2 = 107 \text{ lb/ft}^3$, $H_3 = 10$ ft, $c_3 = 1670 \text{ lb/ft}^2$, and $\gamma_3 = 109 \text{ lb/ft}^3$.
a. Determine the average cohesion (c_{av}) and average unit weight (γ_{av}) for the construction of the earth-pressure envelope.
b. Plot the earth-pressure envelope.

10.7 Refer to Figure P10.7. Given: $\gamma = 118 \text{ lb/ft}^3$, $c = 800 \text{ lb/ft}^2$, and center-to-center spacing of struts in the plan $= 12$ ft. Draw the earth-pressure envelope and determine the strut loads at levels A, B, and C.

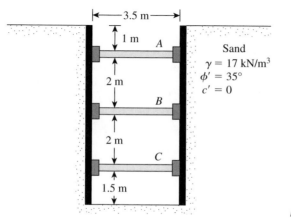

Figure P10.1

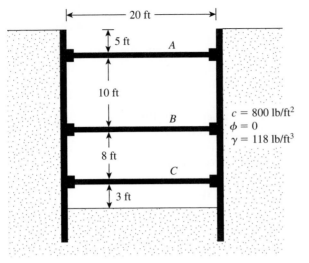

Figure P10.7

10.8 Determine the sheet-pile section modulus for the braced cut described in Problem 10.7. Use $\sigma_{all} = 25{,}000$ lb/in.^{2}.

10.9 Redo Problem 10.7 assuming that $c = 600$ lb/ft^2.

10.10 Determine the factor of safety against bottom heave for the braced cut described in Problem 10.7. Use Eq. (10.14) and assume the length of the cut, $L = 30$ ft.

10.11 Determine the factor of safety against bottom heave for the braced cut described in Problem 10.9. Use Eq. (10.13). The length of the cut is 40 ft.

References

Bjerrum, L, and Eide, O. (1956). "Stability of Strutted Excavation in Clay," *Geotechnique,* Vol. 6, No. 1, pp. 32–47.

Chang, M. F. (2000). "Basal Stability Analysis of Braced Cuts in Clay," *Journal of Geotechnical and Geoenvironmental Engineering,* ASCE, Vol. 126, No. 3, pp. 276–279.

Moormann, C. (2004). "Analysis of Wall and Ground Movements Due to Deep Excavations in Soft Soil Based on New Worldwide Data Base," *Soils and Foundations,* Vol. 44, No. 1, pp. 87–98.

Peck, R. B. (1943). "Earth Pressure Measurements in Open Cuts, Chicago (ILL.) Subway," *Transactions,* American Society of Civil Engineers, Vol. 108, pp. 1008–1058.

Peck, R. B. (1969). "Deep Excavation and Tunneling in Soft Ground," *Proceedings Seventh International Conference on Soil Mechanics and Foundation Engineering,* Mexico City, State-of-the-Art Volume, pp. 225–290.

Terzaghi, K. (1943). *Theoretical Soil Mechanics,* Wiley, New York.

11

Pile Foundations

11.1 ## Introduction

Piles are structural members that are made of steel, concrete, or timber. They are used to build pile foundations, which are deep and which cost more than shallow foundations. (See Chapters 3, 4, and 5.) Despite the cost, the use of piles often is necessary to ensure structural safety. The following list identifies some of the conditions that require pile foundations (Vesic, 1977):

1. When one or more upper soil layers are highly compressible and too weak to support the load transmitted by the superstructure, piles are used to transmit the load to underlying bedrock or a stronger soil layer, as shown in Figure 11.1a. When bedrock is not encountered at a reasonable depth below the ground surface, piles are used to transmit the structural load to the soil gradually. The resistance to the applied structural load is derived mainly from the frictional resistance developed at the soil–pile interface. (See Figure 11.1b.)

2. When subjected to horizontal forces (see Figure 11.1c), pile foundations resist by bending, while still supporting the vertical load transmitted by the superstructure. This type of situation is generally encountered in the design and construction of earth-retaining structures and foundations of tall structures that are subjected to high wind or to earthquake forces.

3. In many cases, expansive and collapsible soils may be present at the site of a proposed structure. These soils may extend to a great depth below the ground surface. Expansive soils swell and shrink as their moisture content increases and decreases, and the pressure of the swelling can be considerable. If shallow foundations are used in such circumstances, the structure may suffer considerable damage. However, pile foundations may be considered as an alternative when piles are extended beyond the active zone, which is where swelling and shrinking occur. (See Figure 11.1d)

   Soils such as loess are collapsible in nature. When the moisture content of these soils increases, their structures may break down. A sudden decrease in the void ratio of soil induces large settlements of structures supported by shallow foundations. In such cases, pile foundations may be used in which the piles are extended into stable soil layers beyond the zone where moisture will change.

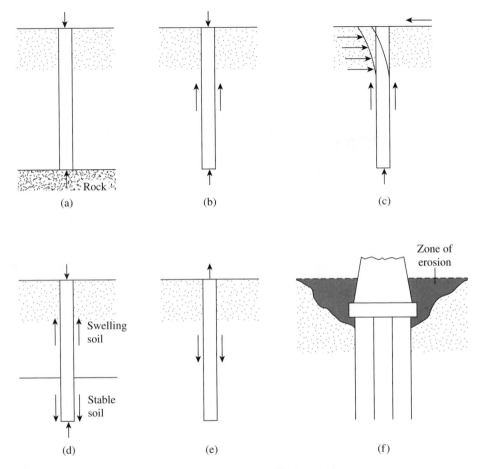

Figure 11.1 Conditions that require the use of pile foundations

4. The foundations of some structures, such as transmission towers, offshore platforms, and basement mats below the water table, are subjected to uplifting
forces. Piles are sometimes used for these foundations to resist the uplifting
force. (See Figure 11.1e.)
5. Bridge abutments and piers are usually constructed over pile foundations to
avoid the loss of bearing capacity that a shallow foundation might suffer because of soil erosion at the ground surface. (See Figure 11.1f.)

Although numerous investigations, both theoretical and experimental, have
been conducted in the past to predict the behavior and the load-bearing capacity
of piles in granular and cohesive soils, the mechanisms are not yet entirely understood and may never be. The design and analysis of pile foundations may thus be
considered somewhat of an art as a result of the uncertainties involved in
working with some subsoil conditions. This chapter discusses the present state of
the art.

Types of Piles and Their Structural Characteristics

Different types of piles are used in construction work, depending on the type of load to be carried, the subsoil conditions, and the location of the water table. Piles can be divided into the following categories: (a) steel piles, (b) concrete piles, (c) wooden (timber) piles, and (d) composite piles.

Steel Piles

Steel piles generally are either *pipe piles* or *rolled steel* H-*section piles*. Pipe piles can be driven into the ground with their ends open or closed. Wide-flange and I-section steel beams can also be used as piles. However, H-section piles are usually preferred because their web and flange thicknesses are equal. (In wide-flange and I-section beams, the web thicknesses are smaller than the thicknesses of the flange.) Table 11.1 gives the dimensions of some standard H-section steel piles used in the United States. Table 11.2 shows selected pipe sections frequency used for piling purposes. In many cases, the pipe piles are filled with concrete after they have been driven.

The allowable structural capacity for steel piles is

$$Q_{all} = A_s f_s \tag{11.1}$$

where

A_s = cross-sectional area of the steel
f_s = allowable stress of steel (≈ 0.33–$0.5 f_y$)

Once the design load for a pile is fixed, one should determine, on the basis of geotechnical considerations, whether $Q_{(design)}$ is within the allowable range as defined by Eq. 11.1.

When necessary, steel piles are spliced by welding or by riveting. Figure 11.2a shows a typical splice by welding for an H-pile. A typical splice by welding for a pipe pile is shown in Figure 11.2b. Figure 11.2c is a diagram of a splice of an H-pile by rivets or bolts.

When hard driving conditions are expected, such as driving through dense gravel, shale, or soft rock, steel piles can be fitted with driving points or shoes. Figures 11.2d and 11.2e are diagrams of two types of shoe used for pipe piles.

Steel piles may be subject to corrosion. For example, swamps, peats, and other organic soils are corrosive. Soils that have a pH greater than 7 are not so corrosive. To offset the effect of corrosion, an additional thickness of steel (over the actual designed cross-sectional area) is generally recommended. In many circumstances factory-applied epoxy coatings on piles work satisfactorily against corrosion. These coatings are not easily damaged by pile driving. Concrete encasement of steel piles in most corrosive zones also protects against corrosion.

Following are some general facts about steel piles:

- Usual length: 15 m to 60 m (50 ft to 200 ft)
- Usual load: 300 kN to 1200 kN (67 kip to 265 kip)

- Advantages:
  - **a.** Easy to handle with respect to cutoff and extension to the desired length
  - **b.** Can stand high driving stresses
  - **c.** Can penetrate hard layers such as dense gravel and soft rock
  - **d.** High load-carrying capacity
- Disadvantages:
  - **a.** Relatively costly
  - **b.** High level of noise during pile driving
  - **c.** Subject to corrosion
  - **d.** H-piles may be damaged or deflected from the vertical during driving through hard layers or past major obstructions

Table 11.1a Common H-Pile Sections used in the United States (SI Units)

Designation, size (mm) × weight (kg/m)	Depth d_1 (mm)	Section area ($m^2 \times 10^{-3}$)	Flange and web thickness w (mm)	Flange width d_2 (mm)	Moment of inertia ($m^4 \times 10^{-6}$)	
					I_{xx}	I_{yy}
HP 200 × 53	204	6.84	11.3	207	49.4	16.8
HP 250 × 85	254	10.8	14.4	260	123	42
× 62	246	8.0	10.6	256	87.5	24
HP 310 × 125	312	15.9	17.5	312	271	89
× 110	308	14.1	15.49	310	237	77.5
× 93	303	11.9	13.1	308	197	63.7
× 79	299	10.0	11.05	306	164	62.9
HP 330 × 149	334	19.0	19.45	335	370	123
× 129	329	16.5	16.9	333	314	104
× 109	324	13.9	14.5	330	263	86
× 89	319	11.3	11.7	328	210	69
HP 360 × 174	361	22.2	20.45	378	508	184
× 152	356	19.4	17.91	376	437	158
× 132	351	16.8	15.62	373	374	136
× 108	346	13.8	12.82	371	303	109

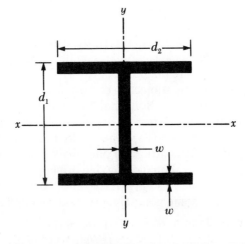

Table 11.1b Common H-Pile Sections used in the United States (English Units)

Designation size (in.) × weight (lb/ft)	Depth d_1 (in.)	Section area (in²)	Flange and web thickness w (in.)	Flange width d_2 (in.)	Moment of inertia (in⁴)	
					I_{xx}	I_{yy}
HP 8 × 36	8.02	10.6	0.445	8.155	119	40.3
HP 10 × 57	9.99	16.8	0.565	10.225	294	101
× 42	9.70	12.4	0.420	10.075	210	71.7
HP 12 × 84	12.28	24.6	0.685	12.295	650	213
× 74	12.13	21.8	0.610	12.215	570	186
× 63	11.94	18.4	0.515	12.125	472	153
× 53	11.78	15.5	0.435	12.045	394	127
HP 13 × 100	13.15	29.4	0.766	13.21	886	294
× 87	12.95	25.5	0.665	13.11	755	250
× 73	12.74	21.6	0.565	13.01	630	207
× 60	12.54	17.5	0.460	12.90	503	165
HP 14 × 117	14.21	34.4	0.805	14.89	1220	443
× 102	14.01	30.0	0.705	14.78	1050	380
× 89	13.84	26.1	0.615	14.70	904	326
× 73	13.61	21.4	0.505	14.59	729	262

Table 11.2a Selected Pipe Pile Sections (SI Units)

Outside diameter (mm)	Wall thickness (mm)	Area of steel (cm²)
219	3.17	21.5
	4.78	32.1
	5.56	37.3
	7.92	52.7
254	4.78	37.5
	5.56	43.6
	6.35	49.4
305	4.78	44.9
	5.56	52.3
	6.35	59.7
406	4.78	60.3
	5.56	70.1
	6.35	79.8
457	5.56	80
	6.35	90
	7.92	112
508	5.56	88
	6.35	100
	7.92	125
610	6.35	121
	7.92	150
	9.53	179
	12.70	238

Table 11.2b Selected Pipe Pile Sections (English Units)

Outside diameter (in.)	Wall thickness (in.)	Area of steel (in²)
$8\frac{5}{8}$	0.125	3.34
	0.188	4.98
	0.219	5.78
	0.312	8.17
10	0.188	5.81
	0.219	6.75
	0.250	7.66
12	0.188	6.96
	0.219	8.11
	0.250	9.25
16	0.188	9.34
	0.219	10.86
	0.250	12.37
18	0.219	12.23
	0.250	13.94
	0.312	17.34
20	0.219	13.62
	0.250	15.51
	0.312	19.30
24	0.250	18.7
	0.312	23.2
	0.375	27.8
	0.500	36.9

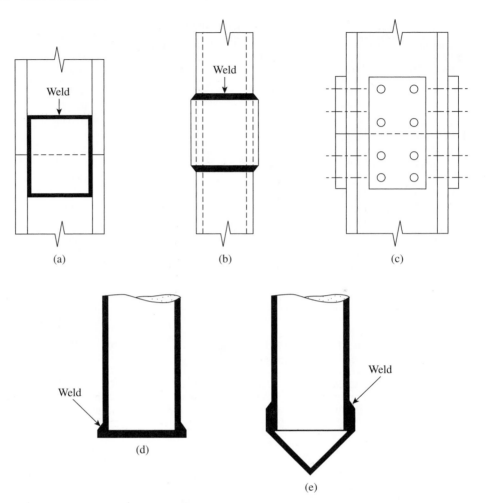

Figure 11.2 Steel piles: (a) splicing of H-pile by welding; (b) splicing of pipe pile by welding; (c) splicing of H-pile by rivets and bolts; (d) flat driving point of pipe pile; (e) conical driving point of pipe pile

Concrete Piles

Concrete piles may be divided into two basic categories: (a) precast piles and (b) cast-*in-situ* piles. *Precast piles* can be prepared by using ordinary reinforcement, and they can be square or octagonal in cross section. (See Figure 11.3.) Reinforcement is provided to enable the pile to resist the bending moment developed during pickup and transportation, the vertical load, and the bending moment caused by a lateral load. The piles are cast to desired lengths and cured before being transported to the work sites.

Some general facts about concrete piles are as follows:

- Usual length: 10 m to 15 m (30 ft to 50 ft)
- Usual load: 300 kN to 3000 kN (67 kip to 675 kip)

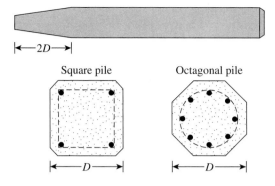

Figure 11.3 Precast piles with ordinary reinforcement

- Advantages:
  a. Can be subjected to hard driving
  b. Corrosion resistant
  c. Can be easily combined with a concrete superstructure
- Disadvantages:
  a. Difficult to achieve proper cutoff
  b. Difficult to transport

Precast piles can also be prestressed by the use of high-strength steel pre-stressing cables. The ultimate strength of these cables is about $1800\ \text{MN/m}^2$ (≈ 260 ksi). During casting of the piles, the cables are pretensioned to about 900 to $1300\ \text{MN/m}^2$ (≈ 130 to 190 ksi), and concrete is poured around them. After curing, the cables are cut, producing a compressive force on the pile section. Table 11.3 gives additional information about prestressed concrete piles with square and octagonal cross sections.

Some general facts about precast prestressed piles are as follows:

- Usual length: 10 m to 45 m (30 ft to 150 ft)
- Maximum length: 60 m (200 ft)
- Maximum load: 7500 kN to 8500 kN (1700 kip to 1900 kip)

The advantages and disadvantages are the same as those of precast piles.

Cast-in-situ, or *cast-in-place, piles* are built by making a hole in the ground and then filling it with concrete. Various types of cast-in-place concrete piles are currently used in construction, and most of them have been patented by their manufacturers. These piles may be divided into two broad categories: (a) cased and (b) uncased. Both types may have a pedestal at the bottom.

Cased piles are made by driving a steel casing into the ground with the help of a mandrel placed inside the casing. When the pile reaches the proper depth the mandrel is withdrawn and the casing is filled with concrete. Figures 11.4a, 11.4b, 11.4c, and 11.4d show some examples of cased piles without a pedestal. Figure 11.4e

Table 11.3a   Typical Prestressed Concrete Pile in Use (SI Units)

Pile shape[a]	D (mm)	Area of cross section (cm²)	Perimeter (mm)	Number of strands 12.7-mm diameter	11.1-mm diameter	Minimum effective prestress force (kN)	Section modulus (m³ × 10⁻³)	Design bearing capacity (kN) Strength of concrete (MN/m²) 34.5	41.4
S	254	645	1016	4	4	312	2.737	556	778
O	254	536	838	4	4	258	1.786	462	555
S	305	929	1219	5	6	449	4.719	801	962
O	305	768	1016	4	5	369	3.097	662	795
S	356	1265	1422	6	8	610	7.489	1091	1310
O	356	1045	1168	5	7	503	4.916	901	1082
S	406	1652	1626	8	11	796	11.192	1425	1710
O	406	1368	1346	7	9	658	7.341	1180	1416
S	457	2090	1829	10	13	1010	15.928	1803	2163
O	457	1729	1524	8	11	836	10.455	1491	1790
S	508	2581	2032	12	16	1245	21.844	2226	2672
O	508	2136	1677	10	14	1032	14.355	1842	2239
S	559	3123	2235	15	20	1508	29.087	2694	3232
O	559	2587	1854	12	16	1250	19.107	2231	2678
S	610	3658	2438	18	23	1793	37.756	3155	3786
O	610	3078	2032	15	19	1486	34.794	2655	3186

[a]S = square section; O = octagonal section

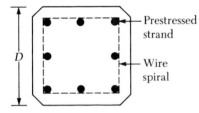

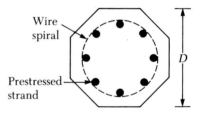

shows a cased pile with a pedestal. The pedestal is an expanded concrete bulb that is formed by dropping a hammer on fresh concrete.

Some general facts about cased cast-in-place piles are as follows:

- Usual length: 5 m to 15 m (15 ft to 50 ft)
- Maximum length: 30 m to 40 m (100 ft to 130 ft)
- Usual load: 200 kN to 500 kN (45 kip to 115 kip)
- Approximate maximum load: 800 kN (180 kip)
- Advantages:
  a. Relatively cheap
  b. Allow for inspection before pouring concrete
  c. Easy to extend

Table 11.3b   Typical Prestressed Concrete Pile in Use (English Units)

Pile shape[a]	D (in.)	Area of cross section (in²)	Perimeter (in.)	Number of strands		Minimum effective prestress force (kip)	Section modulus (in³)	Design bearing capacity (kip) Strength of Concrete	
				$\frac{1}{2}$-in diameter	$\frac{7}{16}$-in diameter			5000 psi	6000 psi
S	10	100	40	4	4	70	167	125	175
O	10	83	33	4	4	58	109	104	125
S	12	144	48	5	6	101	288	180	216
O	12	119	40	4	5	83	189	149	178
S	14	196	56	6	8	137	457	245	295
O	14	162	46	5	7	113	300	203	243
S	16	256	64	8	11	179	683	320	385
O	16	212	53	7	9	148	448	265	318
S	18	324	72	10	13	227	972	405	486
O	18	268	60	8	11	188	638	336	402
S	20	400	80	12	16	280	1333	500	600
O	20	331	66	10	14	234	876	414	503
S	22	484	88	15	20	339	1775	605	727
O	22	401	73	12	16	281	1166	502	602
S	24	576	96	18	23	403	2304	710	851
O	24	477	80	15	19	334	2123	596	716

[a]S = square section; O = octagonal section

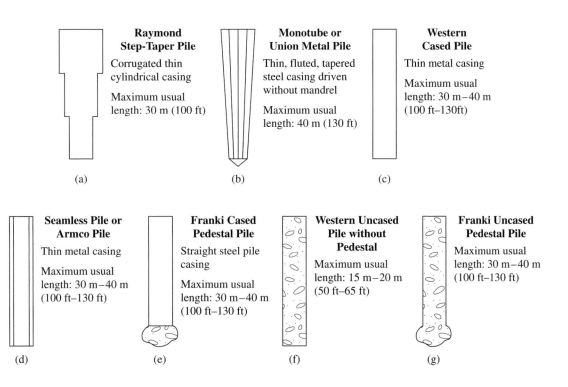

Figure 11.4   Cast-in-place concrete piles

- Disadvantages:
  a. Difficult to splice after concreting
  b. Thin casings may be damaged during driving
- Allowable load:

$$Q_{all} = A_s f_s + A_c f_c \qquad (11.2)$$

where

A_s = area of cross section of steel
A_c = area of cross section of concrete
f_s = allowable stress of steel
f_c = allowable stress of concrete

Figures 11.4f and 11.4g are two types of uncased pile, one with a pedestal and the other without. The uncased piles are made by first driving the casing to the desired depth and then filling it with fresh concrete. The casing is then gradually withdrawn. Following are some general facts about uncased cast-in-place concrete piles:

- Usual length: 5 m to 15 m (15 ft to 50 ft)
- Maximum length: 30 m to 40 m (100 ft to 130 ft)
- Usual load: 300 kN to 500 kN (67 kip to 115 kip)
- Approximate maximum load: 700 kN (160 kip)
- Advantages:
  a. Initially economical
  b. Can be finished at any elevation
- Disadvantages:
  a. Voids may be created if concrete is placed rapidly
  b. Difficult to splice after concreting
  c. In soft soils, the sides of the hole may cave in, squeezing the concrete
- Allowable load:

$$Q_{all} = A_c f_c \qquad (11.3)$$

where

A_c = area of cross section of concrete
f_c = allowable stress of concrete

Timber Piles

Timber piles are tree trunks that have had their branches and bark carefully trimmed off. The maximum length of most timber piles is 10 to 20 m (30 to 65 ft). To qualify for use as a pile, the timber should be straight, sound, and without any defects. The American Society of Civil Engineers' *Manual of Practice*, No. 17 (1959), divided timber piles into three classes:

1. *Class A piles* carry heavy loads. The minimum diameter of the butt should be 356 mm (14 in.).
2. *Class B piles* are used to carry medium loads. The minimum butt diameter should be 305 to 330 mm (12 to 13 in.).
3. *Class C piles* are used in temporary construction work. They can be used permanently for structures when the entire pile is below the water table. The minimum butt diameter should be 305 mm (12 in.).

In any case, a pile tip should not have a diameter less than 150 mm (6 in.).

Timber piles cannot withstand hard driving stress; therefore, the pile capacity is generally limited. Steel shoes may be used to avoid damage at the pile tip (bottom). The tops of timber piles may also be damaged during the driving operation. The crushing of the wooden fibers caused by the impact of the hammer is referred to as *brooming*. To avoid damage to the top of the pile, a metal band or a cap may be used.

Splicing of timber piles should be avoided, particularly when they are expected to carry a tensile load or a lateral load. However, if splicing is necessary, it can be done by using *pipe sleeves* (see Figure 11.5a) or *metal straps and bolts* (see Figure 11.5b). The length of the sleeve should be at least five times the diameter of the pile. The butting ends should be cut square so that full contact can be maintained. The spliced portions should be carefully trimmed so that they fit tightly to the inside of the pipe sleeve. In the case of metal straps and bolts, the butting ends should also be cut square. The sides of the spliced portion should be trimmed plane for putting the straps on.

Timber piles can stay undamaged indefinitely if they are surrounded by saturated soil. However, in a marine environment, timber piles are subject to attack by various organisms and can be damaged extensively in a few months. When located above the water table, the piles are subject to attack by insects. The life of the piles may be increased by treating them with preservatives such as creosote.

The allowable load-carrying capacity of wooden piles is

$$Q_{all} = A_p f_w \tag{11.4}$$

where

A_p = average area of cross section of the pile
f_w = allowable stress on the timber

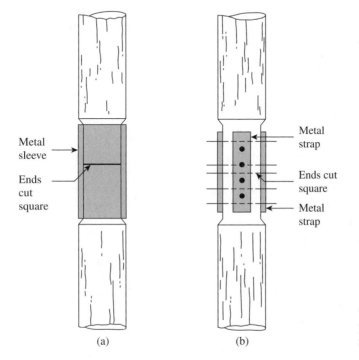

Metal
sleeve

Ends
cut
square

Metal
strap

Ends cut
square

Metal
strap

(a)   (b)

Figure 11.5 Splicing of timber piles: (a) use of pipe sleeves; (b) use of metal straps and bolts

The following allowable stresses are for pressure-treated round timber piles made from Pacific Coast Douglas fir and Southern pine used in hydraulic structures (ASCE, 1993):

Pacific Coast Douglas Fir

- Compression parallel to grain: 6.04 MN/m^2 (875 lb/in.^{2})
- Bending: 11.7 MN/m^2 (1700 lb/in.^{2})
- Horizontal shear: 0.66 MN/m^2 (95 lb/in.^{2})
- Compression perpendicular to grain: 1.31 MN/m^2 (190 lb/in.^{2})

Southern Pine

- Compression parallel to grain: 5.7 MN/m^2 (825 lb/in.^{2})
- Bending: 11.4 MN/m^2 (1650 lb/in.^{2})
- Horizontal shear: 0.62 MN/m^2 (90 lb/in.^{2})
- Compression perpendicular to grain: 1.41 MN/m^2 (205 lb/in.^{2})

The usual length of wooden piles is 5 m to 15 m (15 ft to 50 ft). The maximum length is about 30 m to 40 m (100 ft to 130 ft). The usual load carried by wooden piles is 300 kN to 500 kN (67 kip to 115 kip).

Composite Piles

The upper and lower portions of *composite piles* are made of different materials. For example, composite piles may be made of steel and concrete or timber and concrete. Steel-and-concrete piles consist of a lower portion of steel and an upper portion of cast-in-place concrete. This type of pile is used when the length of the pile required for adequate bearing exceeds the capacity of simple cast-in-place concrete piles. Timber-and-concrete piles usually consist of a lower portion of timber pile below the permanent water table and an upper portion of concrete. In any case, forming proper joints between two dissimilar materials is difficult, and for that reason, composite piles are not widely used.

11.3   *Estimating Pile Length*

Selecting the type of pile to be used and estimating its necessary length are fairly difficult tasks that require good judgment. In addition to being broken down into the classification given in Section 11.2, piles can be divided into three major categories, depending on their lengths and the mechanisms of load transfer to the soil: (a) point bearing piles, (b) friction piles, and (c) compaction piles.

Point Bearing Piles

If soil-boring records establish the presence of bedrock or rocklike material at a site within a reasonable depth, piles can be extended to the rock surface. (See Figure 11.6a.) In this case, the ultimate capacity of the piles depends entirely on the load-bearing capacity of the underlying material; thus, the piles are called *point bearing piles*. In most of these cases, the necessary length of the pile can be fairly well established.

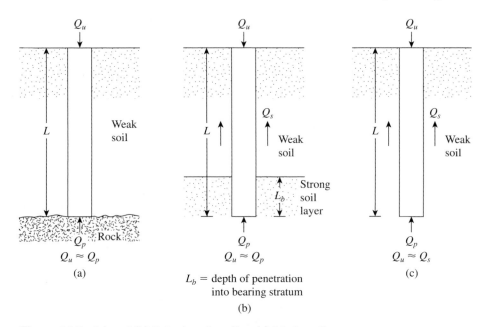

Figure 11.6 (a) and (b) Point bearing piles; (c) friction piles

If, instead of bedrock, a fairly compact and hard stratum of soil is encountered at a reasonable depth, piles can be extended a few meters into the hard stratum. (See Figure 11.6b.) Piles with pedestals can be constructed on the bed of the hard stratum, and the ultimate pile load may be expressed as

$$Q_u = Q_p + Q_s \tag{11.5}$$

where

Q_p = load carried at the pile point
Q_s = load carried by skin friction developed at the side of the pile (caused by shearing resistance between the soil and the pile)

If Q_s is very small,

$$Q_s \approx Q_p \tag{11.6}$$

In this case, the required pile length may be estimated accurately if proper subsoil exploration records are available.

Friction Piles

When no layer of rock or rocklike material is present at a reasonable depth at a site, point bearing piles become very long and uneconomical. In this type of subsoil, piles are driven through the softer material to specified depths. (See Figure 11.6c.) The ultimate load of the piles may be expressed by Eq. (11.5). However, if the value of Q_p is relatively small, then

$$Q_u \approx Q_s \tag{11.7}$$

These piles are called *friction piles,* because most of their resistance is derived from skin friction. However, the term *friction pile,* although used often in the literature, is a misnomer: In clayey soils, the resistance to applied load is also caused by *adhesion.*

The lengths of friction piles depend on the shear strength of the soil, the applied load, and the pile size. To determine the necessary lengths of these piles, an engineer needs a good understanding of soil–pile interaction, good judgment, and experience. Theoretical procedures for calculating the load-bearing capacity of piles are presented later in the chapter.

Compaction Piles

Under certain circumstances, piles are driven in granular soils to achieve proper compaction of soil close to the ground surface. These piles are called *compaction piles.* The lengths of compaction piles depend on factors such as (a) the relative density of the soil before compaction, (b) the desired relative density of the soil after compaction, and (c) the required depth of compaction. These piles are generally short; however, some field tests are necessary to determine a reasonable length.

11.4   *Installation of Piles*

Most piles are driven into the ground by means of *hammers* or *vibratory drivers.* In special circumstances, piles can also be inserted by *jetting* or *partial augering.* The types of hammer used for pile driving include (a) the drop hammer, (b) the single-acting air or steam hammer, (c) the double-acting and differential air or steam hammer, and (d) the diesel hammer. In the driving operation, a cap is attached to the top of the pile. A cushion may be used between the pile and the cap. The cushion has the effect of reducing the impact force and spreading it over a longer time; however, the use of the cushion is optional. A hammer cushion is placed on the pile cap. The hammer drops on the cushion.

Figure 11.7 illustrates various hammers. A drop hammer (see Figure 11.7a) is raised by a winch and allowed to drop from a certain height *H.* It is the oldest type of hammer used for pile driving. The main disadvantage of the drop hammer is its slow rate of blows. The principle of the single-acting air or steam hammer is shown in Figure 11.7b. The striking part, or ram, is raised by air or steam pressure and then drops by gravity. Figure 11.7c shows the operation of the double-acting and differential air or steam hammer. Air or steam is used both to raise the ram and to push it downward, thereby increasing the impact velocity of the ram. The diesel hammer (see Figure 11.7d) consists essentially of a ram, an anvil block, and a fuel-injection system. First the ram is raised and fuel is injected near the anvil. Then the ram is released. When the ram drops, it compresses the air–fuel mixture, which ignites. This action, in effect, pushes the pile downward and raises the ram. Diesel hammers work well under hard driving conditions. In soft soils, the downward movement of the pile is rather large, and the upward movement of the ram is small. This differential may not be sufficient to ignite the air–fuel system, so the ram may have to be lifted manually.

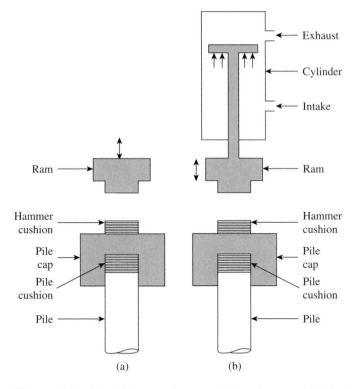

Figure 11.7 Pile-driving equipment: (a) drop hammer; (b) single-acting air or steam hammer.

The principles of operation of a vibratory pile driver are shown in Figure 11.7e. This driver consists essentially of two counterrotating weights. The horizontal components of the centrifugal force generated as a result of rotating masses cancel each other. As a result, a sinusoidal dynamic vertical force is produced on the pile and helps drive the pile downward.

Figure 11.7f is a photograph of a vibratory pile driver. Figure 11.8 shows a pile-driving operation in the field.

Jetting is a technique that is sometimes used in pile driving when the pile needs to penetrate a thin layer of hard soil (such as sand and gravel) overlying a layer of softer soil. In this technique, water is discharged at the pile point by means of a pipe 50 to 75 mm (2 to 3 in.) in diameter to wash and loosen the sand and gravel.

Piles driven at an angle to the vertical, typically 14 to 20°, are referred to as *batter piles*. Batter piles are used in group piles when higher lateral load-bearing capacity is required. Piles also may be advanced by partial augering, with power augers (see Chapter 2) used to predrill holes part of the way. The piles can then be inserted into the holes and driven to the desired depth.

Piles may be divided into two categories based on the nature of their placement: *displacement piles* and *nondisplacement piles*. Driven piles are displacement piles, because they move some soil laterally; hence, there is a tendency for

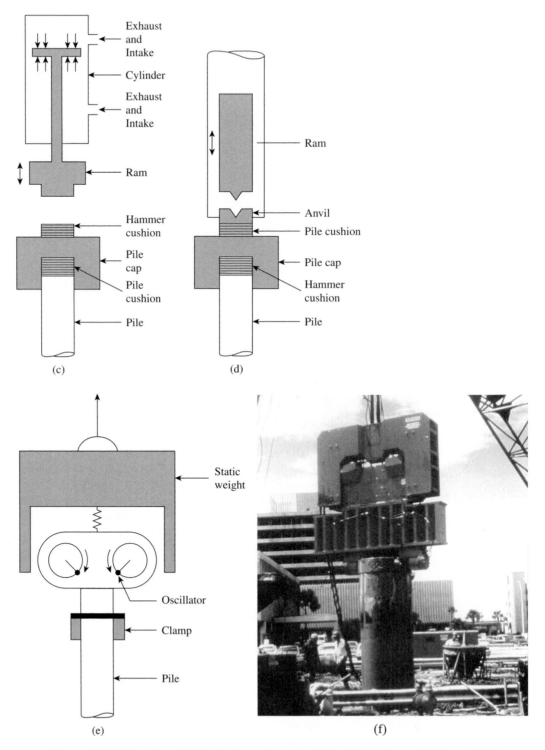

Figure 11.7 (*continued*) Pile-driving equipment: (c) double-acting and differential air or steam hammer; (d) diesel hammer; (e) vibratory pile driver; (f) photograph of a vibratory pile driver (Courtesy of Michael W. O'Neill, University of Houston)

Figure 11.8 A pile-driving operation in the field (Courtesy of E. C. Shin, University of Incheon, Korea)

densification of soil surrounding them. Concrete piles and closed-ended pipe piles are high-displacement piles. However, steel H-piles displace less soil laterally during driving, so they are low-displacement piles. In contrast, bored piles are nondisplacement piles because their placement causes very little change in the state of stress in the soil.

11.5 *Load Transfer Mechanism*

The load transfer mechanism from a pile to the soil is complicated. To understand it, consider a pile of length L, as shown in Figure 11.9a. The load on the pile is gradually increased from zero to $Q_{(z=0)}$ at the ground surface. Part of this load will be resisted by the side friction developed along the shaft, Q_1, and part by the soil below the tip of

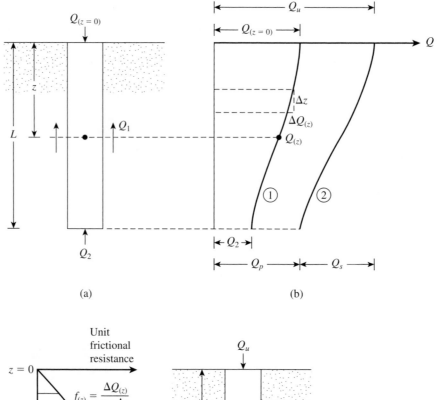

(a)

(b)

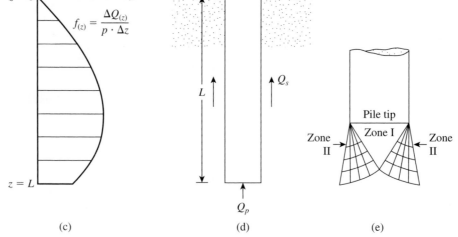

(c)

(d)

(e)

Figure 11.9 Load transfer mechanism for piles

the pile, Q_2. Now, how are Q_1 and Q_2 related to the total load? If measurements are made to obtain the load carried by the pile shaft, $Q_{(z)}$, at any depth z, the nature of the variation found will be like that shown in curve 1 of Figure 11.9b. The *frictional resistance per unit area* at any depth z may be determined as

$$f_{(z)} = \frac{\Delta Q_{(z)}}{(p)(\Delta z)} \tag{11.8}$$

where p = perimeter of the cross section of the pile. Figure 11.9c shows the variation of $f_{(z)}$ with depth.

   If the load Q at the ground surface is gradually increased, maximum frictional resistance along the pile shaft will be fully mobilized when the relative displacement between the soil and the pile is about 5 to 10 mm (0.2 to 0.3 in.), irrespective of the pile size and length L. However, the maximum point resistance $Q_2 = Q_p$ will not be mobilized until the tip of the pile has moved about 10 to 25% of the pile width (or diameter). (The lower limit applies to driven piles and the upper limit to bored piles). At ultimate load (Figure 11.9d and curve 2 in Figure 11.9b), $Q_{(z=0)} = Q_u$. Thus,

$$Q_1 = Q_s$$

and

$$Q_2 = Q_p$$

The preceding explanation indicates that Q_s (or the unit skin friction, f, along the pile shaft) is developed at a *much smaller pile displacement compared with the point resistance, Q_p*.

   At ultimate load, the failure surface in the soil at the pile tip (a bearing capacity failure caused by Q_p) is like that shown in Figure 11.9e. Note that pile foundations are deep foundations and that the soil fails mostly in a *punching mode*, as illustrated previously in Figures 3.1c and 3.3. That is, a *triangular zone*, I, is developed at the pile tip, which is pushed downward without producing any other visible slip surface. In dense sands and stiff clayey soils, a *radial shear zone*, II, may partially develop. Hence, the load displacement curves of piles will resemble those shown in Figure 3.1c.

11.6   *Equations for Estimating Pile Capacity*

The ultimate load-carrying capacity Q_u of a pile is given by the equation

$$Q_u = Q_p + Q_s \tag{11.9}$$

where

Q_p = load-carrying capacity of the pile point
Q_s = frictional resistance (skin friction) derived from the soil–pile interface (see Figure 11.10)

Numerous published studies cover the determination of the values of Q_p and Q_s. Excellent reviews of many of these investigations have been provided by Vesic (1977), Meyerhof (1976), and Coyle and Castello (1981). These studies afford an insight into the problem of determining the ultimate pile capacity.

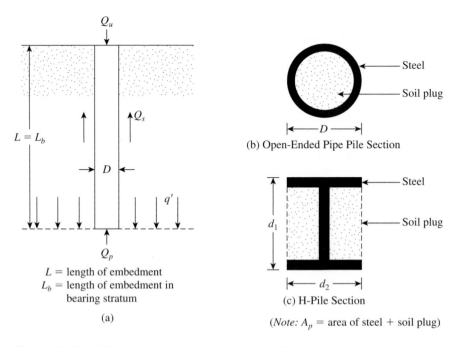

L = length of embedment
L_b = length of embedment in
bearing stratum

(a)

(b) Open-Ended Pipe Pile Section

(c) H-Pile Section

(*Note:* A_p = area of steel + soil plug)

Figure 11.10   Ultimate load-carrying capacity of pile

Point Bearing Capacity, Q_p

The ultimate bearing capacity of shallow foundations was discussed in Chapter 3. According to Terzaghi's equations,

$$q_u = 1.3c'N_c + qN_q + 0.4\gamma BN_\gamma \qquad \text{(for shallow square foundations)}$$

and

$$q_u = 1.3c'N_c + qN_q + 0.3\gamma BN_\gamma \qquad \text{(for shallow circular foundations)}$$

Similarly, the general bearing capacity equation for shallow foundations was given in Chapter 3 (for vertical loading) as

$$q_u = c'N_cF_{cs}F_{cd} + qN_qF_{qs}F_{qd} + \tfrac{1}{2}\gamma BN_\gamma F_{\gamma s}F_{\gamma d}$$

Hence, in general, the ultimate load-bearing capacity may be expressed as

$$q_u = c'N_c^* + qN_q^* + \gamma BN_\gamma^* \qquad (11.10)$$

where N_c^*, N_q^*, and N_γ^* are the bearing capacity factors that include the necessary shape and depth factors.

Pile foundations are deep. However, the ultimate resistance per unit area developed at the pile tip, q_p, may be expressed by an equation similar in form to Eq. (11.10), although the values of N_c^*, N_q^*, and N_γ^* will change. The notation used in this chapter for the width of a pile is D. Hence, substituting D for B in Eq. (11.10) gives

$$q_u = q_p = c'N_c^* + qN_q^* + \gamma DN_\gamma^* \qquad (11.11)$$

Because the width D of a pile is relatively small, the term $\gamma D N_\gamma^*$ may be dropped from the right side of the preceding equation without introducing a serious error; thus, we have

$$q_p = c'N_c^* + q'N_q^* \tag{11.12}$$

Note that the term q has been replaced by q' in Eq. (11.12), to signify effective vertical stress. Thus, the point bearing of piles is

$$Q_p = A_p q_p = A_p(c'N_c^* + q'N_q^*) \tag{11.13}$$

where

$$
\begin{aligned}
A_p &= \text{area of pile tip} \\
c' &= \text{cohesion of the soil supporting the pile tip} \\
q_p &= \text{unit point resistance} \\
q' &= \text{effective vertical stress at the level of the pile tip} \\
N_c^*, N_q^* &= \text{the bearing capacity factors}
\end{aligned}
$$

Frictional Resistance, Q_s

The frictional, or skin, resistance of a pile may be written as

$$Q_s = \Sigma\, p\, \Delta L f \tag{11.14}$$

where

$$
\begin{aligned}
p &= \text{perimeter of the pile section} \\
\Delta L &= \text{incremental pile length over which } p \text{ and } f \text{ are taken to be constant} \\
f &= \text{unit friction resistance at any depth } z
\end{aligned}
$$

The various methods for estimating Q_p and Q_s are discussed in the next several sections. It needs to be reemphasized that, in the field, for full mobilization of the point resistance (Q_p), the pile tip must go through a displacement of 10 to 25% of the pile width (or diameter).

Allowable Load, Q_{all}

After the total ultimate load-carrying capacity of a pile has been determined by summing the point bearing capacity and the frictional (or skin) resistance, a reasonable factor of safety should be used to obtain the total allowable load for each pile, or

$$Q_{all} = \frac{Q_u}{FS}$$

where

$$
\begin{aligned}
Q_{all} &= \text{allowable load-carrying capacity for each pile} \\
FS &= \text{factor of safety}
\end{aligned}
$$

The factor of safety generally used ranges from 2.5 to 4, depending on the uncertainties surrounding the calculation of ultimate load.

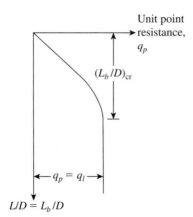

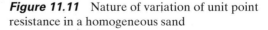

Figure 11.11   Nature of variation of unit point resistance in a homogeneous sand

Meyerhof's Method for Estimating Q_p

Sand

The point bearing capacity, q_p, of a pile in sand generally increases with the depth of embedment in the bearing stratum and reaches a maximum value at an embedment ratio of $L_b/D = (L_b/D)_{cr}$. Note that in a homogeneous soil L_b is equal to the actual embedment length of the pile, L. (See Figure 11.10a.) However, where a pile has penetrated into a bearing stratum, $L_b < L$. (See Figure 11.6b.) Beyond the critical embedment ratio, $(L_b/D)_{cr}$, the value of q_p remains constant $(q_p = q_l)$. That is, as shown in Figure 11.11 for the case of a homogeneous soil, $L = L_b$.

For piles in sand, $c' = 0$, and Eq. (11.13) simpifies to

$$Q_p = A_p q_p = A_p q' N_q^* \qquad (11.15)$$

The variation of N_q^* with soil friction angle ϕ' is shown in Figure 11.12. However, Q_p should not exceed the limiting value $A_p q_l$; that is,

$$Q_p = A_p q' N_q^* \leq A_p q_l \qquad (11.16)$$

The limiting point resistance is

$$q_l = 0.5 \, p_a N_q^* \tan \phi' \qquad (11.17)$$

where

p_a = atmospheric pressure ($=100 \text{ kN/m}^2$ or 2000 lb/ft^2)
ϕ' = effective soil friction angle of the bearing stratum

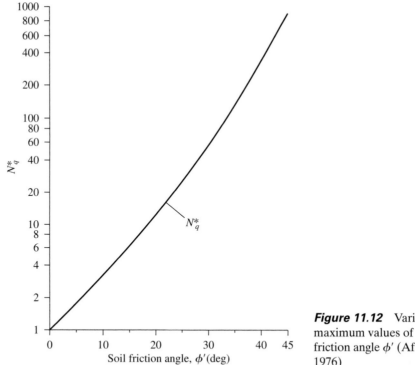

Figure 11.12 Variation of the maximum values of N_q^* with soil friction angle ϕ' (After Meyerhof, 1976)

On the basis of field observations, Meyerhof (1976) also suggested that the ultimate point resistance q_p in a homogeneous granular soil ($L = L_b$) may be obtained from standard penetration numbers as

$$q_l = 0.4\, p_a N_{60} \frac{L}{D} \leqslant 4\, p_a N_{60} \tag{11.18}$$

where

N_{60} = the average value of the standard penetration number near the pile point (about $10D$ above and $4D$ below the pile point)

p_a = atmospheric pressure ($\approx 100 \text{ kN/m}^2$ or 2000 lb/ft^2)

A good example of the concept of the critical embedment ratio can be found from the field load tests on a pile in sand at the Ogeechee River site reported by Vesic (1970). The pile tested was a steel pile with a diameter of 457 mm (18 in.). Table 11.4 shows the ultimate resistance at various depths. Figure 11.13 shows the plot of q_p with depth obtained from the field tests along with the range of standard penetration resistance at the site. From the figure, the following observations can be made.

1. There is a limiting value of q_p. For the tests under consideration, it is about $12,000 \text{ kN/m}^2$.

2. The $(L/D)_{cr}$ value is about 16 to 18.

Table 11.4 Ultimate Point Resistance, q_p, of Test Pile at the
Ogeechee River Site as Reported by Vesic (1970)

Pile diameter, D (m)	Depth of embedment, L (m)	L/D	q_p (kN/m²)
0.457	3.02	6.61	3,304
0.457	6.12	13.39	9,365
0.457	8.87	19.4	11,472
0.457	12.0	26.26	11,587
0.457	15.00	32.82	13,971

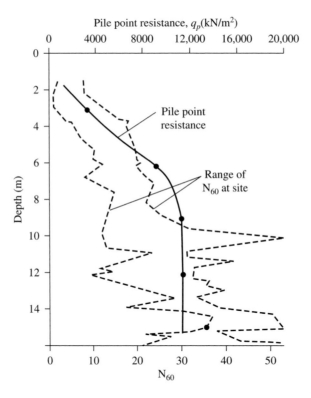

Figure 11.13 Vesic's pile test (1970) result—variation of q_p and N_{60} with depth

3. The average N_{60} value is about 30 for $L/D \geqslant (L/D)_{cr}$. Using Eq. (11.18), the limiting point resistance is $4p_a N_{60} = (4)(100)(30) = 12,000$. This value is generally consistent with the field observation.

Clay ($\phi = 0$)

For piles in *saturated clays* under undrained conditions ($\phi = 0$),

$$Q_p = N_c^* c_u A_p = 9c_u A_p \qquad (11.19)$$

where c_u = undrained cohesion of the soil below the tip of the pile.

11.8 *Vesic's Method for Estimating Q_p*

Vesic (1977) proposed a method for estimating the pile point bearing capacity based on the theory of *expansion of cavities.* According to this theory, on the basis of effective stress parameters, we may write

$$Q_p = A_p q_p = A_p(c'N_c^* + \overline{\sigma}_o' N_\sigma^*) \qquad (11.20)$$

where

$\overline{\sigma}_o'$ = mean effective normal ground stress at the level of the pile point

$$= \left(\frac{1 + 2K_o}{3}\right)q' \qquad (11.21)$$

K_o = earth pressure coefficient at rest = $1 - \sin \phi'$ \qquad (11.22)

and

$$N_c^*, N_\sigma^* = \text{bearing capacity factors}$$

Note that Eq. (11.20) is a modification of Eq. (11.13) with

$$N_\sigma^* = \frac{3N_q^*}{(1 + 2K_o)} \qquad (11.23)$$

Note also that N_c^* in Eq. (11.20) may be expressed as

$$N_c^* = (N_q^* - 1)\cot \phi' \qquad (11.24)$$

According to Vesic's theory,

$$N_\sigma^* = f(I_{rr}) \qquad (11.25)$$

where I_{rr} = reduced rigidity index for the soil. However,

$$I_{rr} = \frac{I_r}{1 + I_r\Delta} \qquad (11.26)$$

where

$$I_r = \text{rigidity index} = \frac{E_s}{2(1 + \mu_s)(c' + q' \tan \phi')} = \frac{G_s}{c' + q' \tan \phi'} \qquad (11.27)$$

E_s = modulus of elasticity of soil
μ_s = Poisson's ratio of soil
G_s = shear modulus of soil
Δ = average volumatic strain in the plastic zone below the pile point

When the volume does not change (e.g., for dense sand or saturated clay), $\Delta = 0$, so

$$I_r = I_{rr} \qquad (11.28)$$

Table 11.5 gives the values of N_c^* and N_σ^* for various values of the soil friction angle ϕ' and I_{rr}. For $\phi = 0$ (an undrained condition),

$$N_c^* = \frac{4}{3}(\ln I_{rr} + 1) + \frac{\pi}{2} + 1 \qquad (11.29)$$

The values of I_r can be estimated from laboratory consolidation and triaxial tests corresponding to the proper stress levels. However, for preliminary use, the following values are recommended:

Type of soil	I_r
Sand	70–150
Silts and clays (drained condition)	50–100
Clays (undrained condition)	100–200

On the basis of cone penetration tests in the field, Baldi et al. (1981) gave the following correlations for I_r:

$$I_r = \frac{300}{F_r(\%)} \quad \text{(for mechanical cone penetration)} \qquad (11.30a)$$

and

$$I_r = \frac{170}{F_r(\%)} \quad \text{(for electric cone penetration)} \qquad (11.30b)$$

For the definition of F_r, see Eq. (2.41).

11.9   *Janbu's Method for Estimating Q_p*

Janbu (1976) proposed calculating Q_p as follows:

$$Q_p = A_p(c'N_c^* + q'N_q^*) \qquad (11.31)$$

Note that Eq. (11.31) has the same form as Eq. (11.13). The bearing capacity factors N_c^* and N_q^* are calculated by assuming a failure surface in soil at the pile tip similar to that shown in Figure 11.14. The bearing capacity relationships are then

$$N_q^* = (\tan \phi' + \sqrt{1 + \tan^2 \phi'})^2 (e^{2\eta' \tan \phi'}) \qquad (11.32a)$$

(the angle η' is defined in the figure) and

$$N_c^* = (N_q^* - 1) \cot \phi' \qquad (11.32b)$$

$$\uparrow$$

from Eq. (11.32a)

Table 11.5 Bearing Capacity Factors N_c^* and N_σ^* Based on the Theory of Expansion of Cavities

ϕ'		10	20	40	60	80	100	200	300	400	500
0		6.97	7.90	8.82	9.36	9.75	10.04	10.97	11.51	11.89	12.19
		1.00	1.00	1.00	1.00	1.00	1.00	1.00	1.00	1.00	1.00
1		7.34	8.37	9.42	10.04	10.49	10.83	11.92	12.57	13.03	13.39
		1.13	1.15	1.16	1.18	1.18	1.19	1.21	1.22	1.23	1.23
2		7.72	8.87	10.06	10.77	11.28	11.69	12.96	13.73	14.28	14.71
		1.27	1.31	1.35	1.38	1.39	1.41	1.45	1.48	1.50	1.51
3		8.12	9.40	10.74	11.55	12.14	12.61	14.10	15.00	15.66	16.18
		1.43	1.49	1.56	1.61	1.64	1.66	1.74	1.79	1.82	1.85
4		8.54	9.96	11.47	12.40	13.07	13.61	15.34	16.40	17.18	17.80
		1.60	1.70	1.80	1.87	1.91	1.95	2.07	2.15	2.20	2.24
5		8.99	10.56	12.25	13.30	14.07	14.69	16.69	17.94	18.86	19.59
		1.79	1.92	2.07	2.16	2.23	2.28	2.46	2.57	2.65	2.71
6		9.45	11.19	13.08	14.26	15.14	15.85	18.17	19.62	20.70	21.56
		1.99	2.18	2.37	2.50	2.59	2.67	2.91	3.06	3.18	3.27
7		9.94	11.85	13.96	15.30	16.30	17.10	19.77	21.46	22.71	23.73
		2.22	2.46	2.71	2.88	3.00	3.10	3.43	3.63	3.79	3.91
8		10.45	12.55	14.90	16.41	17.54	18.45	21.51	23.46	24.93	26.11
		2.47	2.76	3.09	3.31	3.46	3.59	4.02	4.30	4.50	4.67
9		10.99	13.29	15.91	17.59	18.87	19.90	23.39	25.64	27.35	28.73
		2.74	3.11	3.52	3.79	3.99	4.15	4.70	5.06	5.33	5.55
10		11.55	14.08	16.97	18.86	20.29	21.46	25.43	28.02	29.99	31.59
		3.04	3.48	3.99	4.32	4.58	4.78	5.48	5.94	6.29	6.57
11		12.14	14.90	18.10	20.20	21.81	23.13	27.64	30.61	32.87	34.73
		3.36	3.90	4.52	4.93	5.24	5.50	6.37	6.95	7.39	7.75
12		12.76	15.77	19.30	21.64	23.44	24.92	30.03	33.41	36.02	38.16
		3.71	4.35	5.10	5.60	5.98	6.30	7.38	8.10	8.66	9.11
13		13.41	16.69	20.57	23.17	25.18	26.84	32.60	36.46	39.44	41.89
		4.09	4.85	5.75	6.35	6.81	7.20	8.53	9.42	10.10	10.67
14		14.08	17.65	21.92	24.80	27.04	28.89	35.38	39.75	43.15	45.96
		4.51	5.40	6.47	7.18	7.74	8.20	9.82	10.91	11.76	12.46
15		14.79	18.66	23.35	26.53	29.02	31.08	38.37	43.32	47.18	50.39
		4.96	6.00	7.26	8.11	8.78	9.33	11.28	12.61	13.64	14.50
16		15.53	19.73	24.86	28.37	31.13	33.43	41.58	47.17	51.55	55.20
		5.45	6.66	8.13	9.14	9.93	10.58	12.92	14.53	15.78	16.83

(Continued)

Table 11.5 (Continued)

I_{rr}

φ'	10	20	40	60	80	100	200	300	400	500
17	16.30	20.85	26.46	30.33	33.37	35.92	45.04	51.32	56.27	60.42
	5.98	7.37	9.09	10.27	11.20	11.98	14.77	16.99	18.20	19.47
18	17.11	22.03	28.15	32.40	35.76	38.59	48.74	55.80	61.38	66.07
	6.56	8.16	10.15	11.53	12.62	13.54	16.84	19.13	20.94	22.47
19	17.95	23.26	29.93	34.59	38.30	41.42	52.71	60.61	66.89	72.18
	7.18	9.01	11.31	12.91	14.19	15.26	19.15	21.87	24.03	25.85
20	18.83	24.56	31.81	36.92	40.99	44.43	56.97	65.79	72.82	78.78
	7.85	9.94	12.58	14.44	15.92	17.17	21.73	24.94	27.51	29.67
21	19.75	25.92	33.80	39.38	43.85	47.64	61.51	71.34	79.22	85.90
	8.58	10.95	13.97	16.12	17.83	19.29	24.61	28.39	31.41	33.97
22	20.71	27.35	35.89	41.98	46.88	51.04	66.37	77.30	86.09	93.57
	9.37	12.05	15.50	17.96	19.94	21.62	27.82	32.23	35.78	38.81
23	21.71	28.84	38.09	44.73	50.08	54.66	71.56	83.68	93.47	101.83
	10.21	13.24	17.17	19.99	22.26	24.20	31.37	36.52	40.68	44.22
24	22.75	30.41	40.41	47.63	53.48	58.49	77.09	90.51	101.39	110.70
	11.13	14.54	18.99	22.21	24.81	27.04	35.32	41.30	46.14	50.29
25	23.84	32.05	42.85	50.69	57.07	62.54	82.98	97.81	109.88	120.23
	12.12	15.95	20.98	24.64	27.61	30.16	39.70	46.61	52.24	57.06
26	24.98	33.77	45.42	53.93	60.87	66.84	89.25	105.61	118.96	130.44
	13.18	17.47	23.15	27.30	30.69	33.60	44.53	52.51	59.02	64.62
27	26.16	35.57	48.13	57.34	64.88	71.39	95.02	113.92	128.67	141.39
	14.33	19.12	25.52	30.21	34.06	37.37	49.88	59.05	66.56	73.04
28	27.40	37.45	50.96	60.93	69.12	76.20	103.01	122.79	139.04	153.10
	15.57	20.91	28.10	33.40	37.75	41.51	55.77	66.29	74.93	82.40
29	28.69	39.42	53.95	64.71	73.58	81.28	110.54	132.23	150.11	165.61
	16.90	22.85	30.90	36.87	41.79	46.05	62.27	74.30	84.21	92.80
30	30.03	41.49	57.08	68.69	78.30	86.64	118.53	142.27	161.91	178.98
	18.24	24.95	33.95	40.66	46.21	51.02	69.43	83.14	94.48	104.33
31	31.43	43.64	60.37	72.88	83.27	92.31	126.99	152.95	174.49	193.23
	19.88	27.22	37.27	44.79	51.03	56.46	77.31	92.90	105.84	117.11
32	32.89	45.90	63.82	77.29	88.50	98.28	135.96	164.29	187.87	208.43
	21.55	29.68	40.88	49.30	56.30	62.41	85.96	103.66	118.39	131.24
33	34.41	48.26	67.44	81.92	94.01	104.58	145.46	176.33	202.09	224.62
	23.34	32.34	44.80	54.20	62.05	68.92	95.46	115.51	132.24	146.87
34	35.99	50.72	71.24	86.80	99.82	111.22	155.51	189.11	217.21	241.84
	25.28	35.21	49.05	59.54	68.33	76.02	105.90	128.55	147.51	164.12

Table 11.5 (Continued)

ϕ'		I_{rr}								
	10	20	40	60	80	100	200	300	400	500
35	37.65	53.30	75.22	91.91	105.92	118.22	166.14	202.64	233.27	260.15
	27.36	38.32	53.67	65.36	75.17	83.78	117.33	142.89	164.33	183.16
36	39.37	55.99	79.39	97.29	112.34	125.59	177.38	216.98	250.30	279.60
	29.60	41.68	58.68	71.69	82.62	92.24	129.87	158.65	182.85	204.14
37	41.17	58.81	83.77	102.94	119.10	133.34	189.25	232.17	268.36	300.26
	32.02	45.31	64.13	78.57	90.75	101.48	143.61	175.95	203.23	227.26
38	43.04	61.75	88.36	108.86	126.20	141.50	201.78	248.23	287.50	322.17
	34.63	49.24	70.03	86.05	99.60	111.56	158.65	194.94	225.62	252.71
39	44.99	64.83	93.17	115.09	133.66	150.09	215.01	265.23	307.78	345.41
	37.44	53.50	76.45	94.20	109.24	122.54	175.11	215.78	250.23	280.71
40	47.03	68.04	98.21	121.62	141.51	159.13	228.97	283.19	329.24	370.04
	40.47	58.10	83.40	103.05	119.74	134.52	193.13	238.62	277.26	311.50
41	49.16	71.41	103.49	128.48	149.75	168.63	243.69	302.17	351.95	396.12
	43.74	63.07	90.96	112.68	131.18	147.59	212.84	263.67	306.94	345.34
42	51.38	74.92	109.02	135.68	158.41	178.62	259.22	322.22	375.97	423.74
	47.27	68.46	99.16	123.16	143.64	161.83	234.40	291.13	339.52	382.53
43	53.70	78.60	114.82	143.23	167.51	189.13	275.59	343.40	401.36	452.96
	51.08	74.30	108.08	134.56	157.21	177.36	257.99	321.22	375.28	423.39
44	56.13	82.45	120.91	151.16	177.07	200.17	292.85	365.75	428.21	483.88
	55.20	80.62	117.76	146.97	172.00	194.31	283.80	354.20	414.51	468.28
45	58.66	86.48	127.28	159.48	187.12	211.79	311.04	389.35	456.57	516.58
	59.66	87.48	128.28	160.48	188.12	212.79	312.03	390.35	457.57	517.58
46	61.30	90.70	133.97	168.22	197.67	224.00	330.20	414.26	486.54	551.16
	64.48	94.92	139.73	175.20	205.70	232.96	342.94	429.98	504.82	571.74
47	64.07	95.12	140.99	177.40	208.77	236.85	350.41	440.54	518.20	587.72
	69.71	103.00	152.19	191.24	224.88	254.99	376.77	473.42	556.70	631.25
48	66.97	99.75	148.35	187.04	220.43	250.36	371.70	468.28	551.64	626.36
	75.38	111.78	165.76	208.73	245.81	279.06	413.82	521.08	613.65	696.64
49	70.01	104.60	156.09	197.17	232.70	264.58	394.15	497.56	586.96	667.21
	81.54	121.33	180.56	227.82	268.69	305.37	454.42	573.38	676.22	768.53
50	73.19	109.70	164.21	207.83	245.60	279.55	417.82	528.46	624.28	710.39
	88.23	131.73	196.70	248.68	293.70	334.15	498.94	630.80	744.99	847.61

From "Design of Pile Foundations," by A. S. Vesic, in *NCHRP Synthesis of Highway Practice 42*, Transportation Research Board, 1977. Reprinted by permission.

(*Note*: Upper number N_c^*, lower number N_σ^*.)

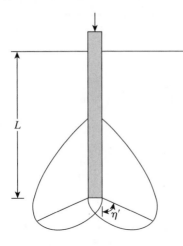

Figure 11.14 Failure surface at the pile tip

Table 11.6 Janbu's Bearing Capacity Factors

ϕ'°	$\eta' = 60°$		$\eta' = 75°$		$\eta' = 90°$	
	N_c^*	N_q^*	N_c^*	N_q^*	N_c^*	N_q^*
0	5.74	1.0	5.74	1.0	5.74	1.0
10	5.95	2.05	7.11	2.25	8.34	2.47
20	9.26	4.37	11.78	5.29	14.83	6.40
30	19.43	10.05	21.82	13.60	30.14	18.40
40	30.58	26.66	48.11	41.37	75.31	64.20
45	46.32	47.32	78.90	79.90	133.87	134.87

The angle η' varies from 60° for soft clays to about 105° for dense sandy soils. It is recommended that, for practical use,

$$60° \leqslant \eta' \leqslant 90°$$

Table 11.6 gives the variation of N_c^* and N_q^* for $\eta' = 60°, 75°$, and $90°$.

11.10 *Coyle and Castello's Method for Estimating Q_p in Sand*

Coyle and Castello (1981) analyzed 24 large-scale field load tests of driven piles in sand. On the basis of the test results, they suggested that, in sand,

$$Q_p = q'N_q^*A_p \qquad (11.33)$$

where

q' = effective vertical stress at the pile tip
N_q^* = bearing capacity factor

Figure 11.15 shows the variation of N_q^* with L/D and the soil friction angle ϕ'.

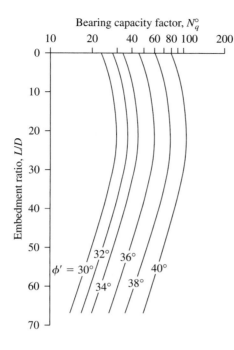

Figure 11.15 Variation of N_q^* with L/D (Redrawn after Coyle and Castello, 1981)

11.11 Other Correlations for Calculating Q_p with SPT and CPT Results

There are two major methods for estimating the magnitude of q_p using the cone penetration resistance q_c:

1. The LCPC method, developed by Laboratoire Central des Ponts at Chaussées (Bustamante and Gianeselli, 1982); and
2. The Dutch method (DeRuiter and Beringen, 1979).

LCPC Method

According to the LCPC method,

$$q_p = q_{c(eq)}k_b \qquad (11.34)$$

where

$q_{c(eq)}$ = equivalent average cone resistance
k_b = empirical bearing capacity factor

The magnitude of $q_{c(eq)}$ is calculated in the following manner:

Step 1. Consider the cone tip resistance q_c within a range of 1.5D below the pile tip to 1.5D above the pile tip, as shown in Figure 11.16.

Step 2. Calculate the average value of $q_c[q_{c(av)}]$ within the zone shown in Figure 11.16.

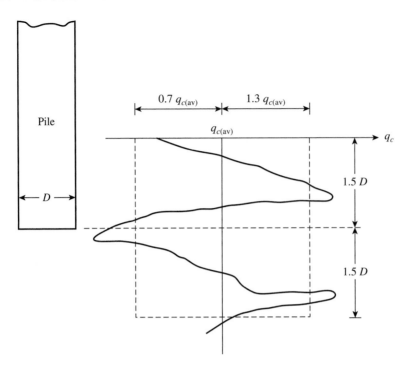

Figure 11.16 LCPC method

Step 3. Eliminate the q_c values that are higher than $1.3q_{c(av)}$ and the q_c values that are lower than $0.7q_{c(av)}$.

Step 4. Calculate $q_{c(eq)}$ by averaging the remaining q_c values.

Briaud and Miran (1991) suggested that

$$k_b = 0.6 \text{ (for clays and silts)}$$

and

$$k_b = 0.375 \text{ (for sands and gravels)}$$

Dutch Method

According to the Dutch method, one considers the variation of q_c in the range of $4D$ below the pile tip to $8D$ above the pile tip, as shown in Figure 11.17. Then one conducts the following operations:

Step 1. Average the q_c values over a distance yD below the pile tip. This is path *a–b–c.* Sum q_c values along the downward path *a–b* (i.e., the actual path *a*) and the upward path *b–c* (i.e., the minimum path). Determine the minimum value q_{c1} = average value of q_c for $0.7 < y < 4$.

Step 2. Average the q_c values (q_{c2}) between the pile tip and $8D$ above the pile tip along the path *c–d–e–f–g,* using the minimum path and ignoring minor peak depressions.

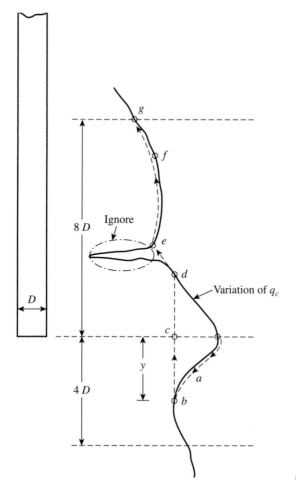

Figure 11.17 Dutch method

Step 3. Calculate

$$q_p = \frac{(q_{c1} + q_{c2})}{2} k_b' \leq 150 p_a \qquad (11.35)$$

where p_a = atmospheric pressure (≈ 100 kN/m^2, or 2000 lb/ft^2)

DeRuiter and Beringen (1979) recommended the following values for k_b' for sand:

- 1.0 for OCR (overconsolidation ratio) = 1
- 0.67 for OCR = 2 to 4

Nottingham and Schmertmann (1975) and Schmertmann (1978) recommended the following relationship for q_p in clay:

$$q_p = R_1 R_2 \frac{(q_{c1} + q_{c2})}{2} k_b' \leq 150 p_a \qquad (11.36)$$

In this equation,

R_1 = reduction factor, which is a function of the undrained shear strength c_u
R_2 = 1 for electric cone penetrometer; = 0.6 for mechanical cone penetrometer

The interpolated values of R_1 with c_u provided by Schmertmann (1978) are as follows:

$\dfrac{c_u}{p_a}$	R_1
⩾0.5	1
0.75	0.64
1.0	0.53
1.25	0.42
1.5	0.36
1.75	0.33
2.0	0.30

11.12   *Frictional Resistance (Q_s) in Sand*

According to Eq. (11.14), the frictional resistance

$$Q_s = \Sigma p\,\Delta Lf$$

The unit frictional resistance, f, is hard to estimate. In making an estimation of f, several important factors must be kept in mind:

1. The nature of the pile installation. For driven piles in sand, the vibration caused during pile driving helps densify the soil around the pile. The zone of sand densification may be as much as 2.5 times the pile diameter, in the sand surrounding the pile.
2. It has been observed that the nature of variation of f in the field is approximately as shown in Figure 11.18. The unit skin friction increases with depth more or less

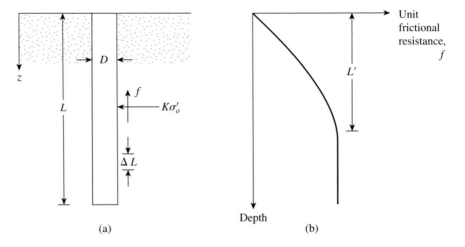

(a)   (b)

Figure 11.18   Unit frictional resistance for piles in sand

linearly to a depth of L' and remains constant thereafter. The magnitude of the critical depth L' may be 15 to 20 pile diameters. A conservative estimate would be

$$L' \approx 15D \tag{11.37}$$

3. At similar depths, the unit skin friction in loose sand is higher for a high-displacement pile, compared with a low-displacement pile.
4. At similar depths, bored, or jetted, piles will have a lower unit skin friction, compared with driven piles.

Taking into account the preceding factors, we can give the following approximate relationship for f (see Figure 11.18):
For $z = 0$ to L',

$$f = K\sigma_o'\tan \delta' \tag{11.38}$$

and for $z = L'$ to L,

$$f = f_{z=L'} \tag{11.39}$$

In these equations,

K = effective earth pressure coefficient
σ_o' = effective vertical stress at the depth under consideration
δ' = soil-pile friction angle

In reality, the magnitude of K varies with depth; it is approximately equal to the Rankine passive earth pressure coefficient, K_p, at the top of the pile and may be less than the at-rest pressure coefficient, K_o, at a greater depth. Based on presently available results, the following average values of K are recommended for use in Eq. (11.38):

Pile type	K
Bored or jetted	$\approx K_o = 1 - \sin \phi'$
Low-displacement driven	$\approx K_o = 1 - \sin \phi'$ to $1.4K_o = 1.4(1 - \sin \phi')$
High-displacement driven	$\approx K_o = 1 - \sin \phi'$ to $1.8K_o = 1.8(1 - \sin \phi')$

The values of δ' from various investigations appear to be in the range from $0.5\phi'$ to $0.8\phi'$.

Coyle and Castello (1981), in conjunction with the material presented in Section 11.10, proposed that

$$Q_s = f_{av}pL = (K\overline{\sigma}_o' \tan \delta')pL \tag{11.40}$$

where

$\overline{\sigma}_o'$ = average effective overburden pressure
δ' = soil–pile friction angle = $0.8\phi'$

The lateral earth pressure coefficient K, which was determined from field observations, is shown in Figure 11.19. Thus, if that figure is used,

$$Q_s = K\overline{\sigma}_o'\tan(0.8\phi')pL \tag{11.41}$$

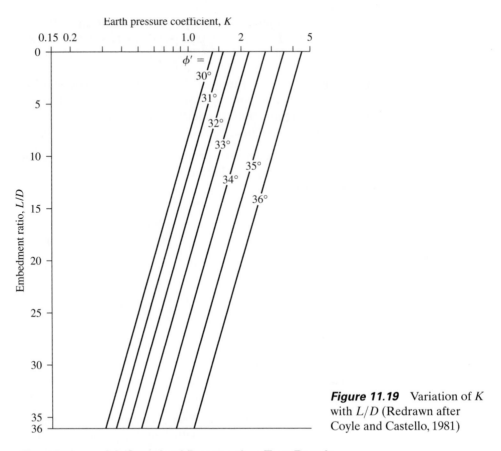

Figure 11.19   Variation of K with L/D (Redrawn after Coyle and Castello, 1981)

Correlation with Standard Penetration Test Results

Meyerhof (1976) indicated that the average unit frictional resistance, f_{av}, for high-displacement driven piles may be obtained from average corrected standard penetration resistance values as

$$f_{av} = 0.02 p_a (\overline{N_{60}}) \tag{11.42}$$

where

$(\overline{N_{60}})$ = average value of standard penetration resistance
p_a = atmospheric pressure (≈ 100 kN/m^2 or 2000 lb/ft^2)

For low-displacement driven piles

$$f_{av} = 0.01 p_a (\overline{N_{60}}) \tag{11.43}$$

Thus,

$$Q_s = p L f_{av} \tag{11.44}$$

Correlation with Cone Penetration Test Results

In Section 11.11, the Dutch method for calculating pile tip capacity Q_p using cone penetration test results was described. In conjunction with using that method, Nottingham and Schmertmann (1975) and Schmertmann (1978) provided correlations

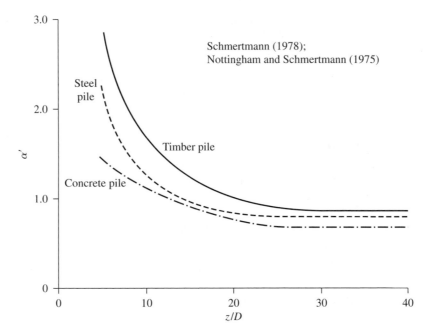

Figure 11.20 Variation of α' with embedment ratio for pile in sand: electric cone penetrometer

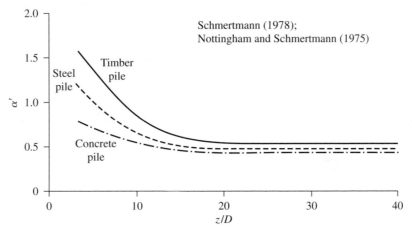

Figure 11.21 Variation of α' with embedment ratio for piles in sand: mechanical cone penetrometer

for estimating Q_s using the frictional resistance (f_c) obtained during cone penetration tests. According to this method

$$f = \alpha' f_c \qquad (11.45)$$

The variations of α' with z/D for electric cone and mechanical cone penetrometers are shown in Figures 11.20 and 11.21, respectively. We have

$$Q_s = \Sigma p(\Delta L)f = \Sigma p(\Delta L)\alpha' f_c \qquad (11.46)$$

11.13   *Frictional (Skin) Resistance in Clay*

Estimating the frictional (or skin) resistance of piles in clay is almost as difficult a task as estimating that in sand (see Section 11.12), due to the presence of several variables that cannot easily be quantified. Several methods for obtaining the unit frictional resistance of piles are described in the literature. We examine some of them next.

λ *Method*

This method, proposed by Vijayvergiya and Focht (1972), is based on the assumption that the displacement of soil caused by pile driving results in a passive lateral pressure at any depth and that the average unit skin resistance is

$$f_{\text{av}} = \lambda(\overline{\sigma}_o' + 2c_u) \tag{11.47}$$

where

$\overline{\sigma}_o'$ = mean effective vertical stress for the entire embedment length
c_u = mean undrained shear strength ($\phi = 0$)

The value of λ changes with the depth of penetration of the pile. (See Table 11.7.) Thus, the total frictional resistance may be calculated as

$$Q_s = pLf_{\text{av}}$$

Care should be taken in obtaining the values of $\overline{\sigma}_o'$ and c_u in layered soil. Figure 11.22 helps explain the reason. Figure 11.22a shows a pile penetrating three layers of clay. According to Figure 11.22b, the mean value of c_u is $(c_{u(1)}L_1 + c_{u(2)}L_2 + \cdots)/L$.

Table 11.7 Variation of λ with pile embedment length, L

Embedment Length, L (m)	λ
0	0.5
5	0.336
10	0.245
15	0.200
20	0.173
25	0.150
30	0.136
35	0.132
40	0.127
50	0.118
60	0.113
70	0.110
80	0.110
90	0.110

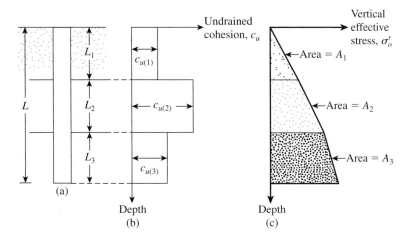

Figure 11.22 Application of λ method in layered soil

Similarly, Figure 11.22c shows the plot of the variation of effective stress with depth. The mean effective stress is

$$\overline{\sigma}'_o = \frac{A_1 + A_2 + A_3 + \cdots}{L} \tag{11.48}$$

where $A_1, A_2, A_3, \ldots$ = areas of the vertical effective stress diagrams.

α Method

According to the α method, the unit skin resistance in clayey soils can be represented by the equation

$$f = \alpha c_u \tag{11.49}$$

where α = empirical adhesion factor. The approximate variation of the value of α is shown in Figure 11.23, where σ'_o is the vertical effective stress. This variation of α with c_u/σ'_o was obtained by Randolph and Murphy (1985). With it, we have

$$Q_s = \Sigma f p \, \Delta L = \Sigma \alpha c_u p \, \Delta L \tag{11.50}$$

β Method

When piles are driven into saturated clays, the pore water pressure in the soil around the piles increases. The excess pore water pressure in normally consolidated clays may be four to six times c_u. However, within a month or so, this pressure gradually dissipates.

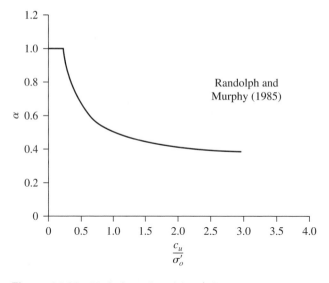

Figure 11.23   Variation of α with c_u/σ'_o

Hence, the unit frictional resistance for the pile can be determined on the basis of the effective stress parameters of the clay in a remolded state $(c' = 0)$. Thus, at any depth,

$$f = \beta\sigma'_o \qquad (11.51)$$

where

σ'_o = vertical effective stress
$\beta = K \tan \phi'_R$ (11.52)
ϕ'_R = drained friction angle of remolded clay
K = earth pressure coefficient

Conservatively, the magnitude of K is the earth pressure coefficient at rest, or

$$K = 1 - \sin \phi'_R \qquad \text{(for normally consolidated clays)} \qquad (11.53)$$

and

$$K = (1 - \sin \phi'_R)\sqrt{\text{OCR}} \qquad \text{(for overconsolidated clays)} \qquad (11.54)$$

where OCR = overconsolidation ratio.

Combining Eqs. (11.51), (11.52), (11.53), and (11.54), for normally consolidated clays yields

$$f = (1 - \sin \phi'_R) \tan \phi'_R\sigma'_o \qquad (11.55)$$

and for overconsolidated clays,

$$f = (1 - \sin \phi'_R)\tan \phi'_R\sqrt{\text{OCR}}\ \sigma'_o \qquad (11.56)$$

With the value of f determined, the total frictional resistance may be evaluated as

$$Q_s = \Sigma f p\ \Delta L$$

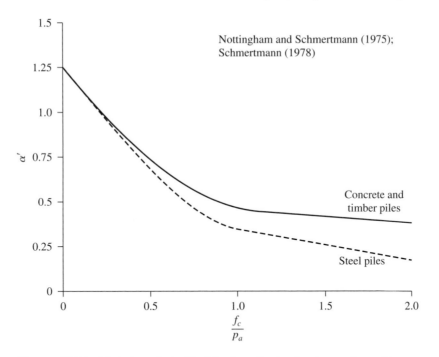

Figure 11.24   Variation of α' with f_c/p_a for piles in clay (p_a = atmosphic pressure $\approx 100 \text{ kN/m}^2$ or 2000 lb/ft^2)

Correlation with Cone Penetration Test Results

Nottingham and Schmertmann (1975) and Schmertmann (1978) found the correlation for unit skin friction in clay (with $\phi = 0$) to be

$$f = \alpha' f_c \tag{11.57}$$

The variation of α' with the frictional resistance f_c is shown in Figure 11.24. Thus,

$$Q_s = \Sigma f p(\Delta L) = \Sigma \alpha' f_c p(\Delta L) \tag{11.58}$$

11.14   *Point Bearing Capacity of Piles Resting on Rock*

Sometimes piles are driven to an underlying layer of rock. In such cases, the engineer must evaluate the bearing capacity of the rock. The ultimate unit point resistance in rock (Goodman, 1980) is approximately

$$q_p = q_u(N_\phi + 1) \tag{11.59}$$

where

$N_\phi = \tan^2(45 + \phi'/2)$
q_u = unconfined compression strength of rock
ϕ' = drained angle of friction

Table 11.8 Typical Unconfined Compressive Strength of Rocks

Type of rock	MN/m²	lb/in²
	q_u	q_u
Sandstone	70–140	10,000–20,000
Limestone	105–210	15,000–30,000
Shale	35–70	5000–10,000
Granite	140–210	20,000–30,000
Marble	60–70	8500–10,000

Table 11.9 Typical Values of Angle of Friction ϕ' of Rocks

Type of rock	Angle of friction, ϕ' (deg)
Sandstone	27–45
Limestone	30–40
Shale	10–20
Granite	40–50
Marble	25–30

The unconfined compression strength of rock can be determined by laboratory tests on rock specimens collected during field investigation. However, extreme caution should be used in obtaining the proper value of q_u, because laboratory specimens usually are small in diameter. As the diameter of the specimen increases, the unconfined compression strength decreases—a phenomenon referred to as the *scale effect*. For specimens larger than about 1 m (3 ft) in diameter, the value of q_u remains approximately constant. There appears to be a fourfold to fivefold reduction of the magnitude of q_u in this process. The scale effect in rock is caused primarily by randomly distributed large and small fractures and also by progressive ruptures along the slip lines. Hence, we always recommend that

$$q_{u(\text{design})} = \frac{q_{u(\text{lab})}}{5} \tag{11.60}$$

Table 11.8 lists some representative values of (laboratory) unconfined compression strengths of rock. Representative values of the rock friction angle ϕ' are given in Table 11.9.

A factor of safety of at least 3 should be used to determine the allowable point bearing capacity of piles. Thus,

$$Q_{p(\text{all})} = \frac{[q_{u(\text{design})}(N_\phi + 1)]A_p}{\text{FS}} \tag{11.61}$$

Example 11.1

A concrete pile is 50 ft (L) long and 16 in. × 16 in. in cross section. The pile is fully embedded in sand for which $\gamma = 110$ lb/ft³ and $\phi' = 30°$. Calculate the ultimate point load, Q_p, by using

a. Meyerhof's method (Section 11.7)
b. Vesic's method (Section 11.8). Use $I_r = I_{rr} = 50$
c. Janbu's method (Section 11.9). Use $\eta' = 90°$

Solution

Part a
From Eq. (11.15),

$$Q_p = A_p q' N_q^* = A_p \gamma L N_q^*$$

For $\phi' = 30°$, $N_q^* \approx 55$ (Figure 11.12), so

$$Q_p = \left(\frac{16 \times 16}{16 \times 12} \text{ ft}^2 \right) \left(\frac{110 \times 50}{1000} \text{ kip/ft}^2 \right)(55) = 537.8 \text{ kip}$$

Again, from Eq. (11.17)

$$q_p = (0.5 p_a N_q^* \tan \phi') A_p$$

$$= (0.4)(2000)(55)(\tan 30)\left(\frac{16 \times 16}{16 \times 12} \right) = 56,452 \text{ lb} \approx \textbf{56.45 kip}$$

Part b
From Eqs. (11.20), (11.21), and (11.22), with $c' = 0$,

$$Q_p = A_p \sigma_o' N_\sigma^* = A_p \left[\frac{1 + 2(1 - \sin \phi')}{3} \right] q' N_\sigma^*$$

For $\phi' = 30°$ and $I_{rr} = 50$, the value of N_σ^* is about 36 (Table 11.5), so

$$Q_p = \left(\frac{16 \times 16}{12 \times 12} \text{ ft}^2 \right) \left[\frac{1 + 2(1 - \sin 30)}{3} \right] \left(\frac{110 \times 50}{1000} \text{ kip/ft}^2 \right)(36) = \textbf{234.7 kip}$$

Part c
From Eq. (11.31) with $c' = 0$,

$$Q_p = A_p q' N_q^*$$

For $\phi' = 30°$ and $\eta' = 90°$, the value of $N_q^* \approx 18.4$ (Table 11.6)

$$Q_p = \left(\frac{16 \times 16}{12 \times 12} \text{ ft}^2 \right) \left(\frac{110 \times 50}{1000} \text{ kip/ft}^2 \right)(18.4) = \textbf{179.9 kip} \qquad ■$$

Example 11.2

For the pile described in Example 11.1

a. Determine the frictional resistance, Q_s. Use Eqs. (11.14), (11.38), and (11.39). Given: $K = 1.3$ and $\delta' = 0.8\phi'$.
b. Using the results of Example 11.1 and Part a of this problem, estimate the allowable load-carrying capacity of the pile. Given: FS = 4.

Solution

Part a

From Eq. (11.37),

$$L \approx 15D = 15\left(\frac{16}{12}\text{ ft}\right) = 20\text{ ft}$$

From Eq. (11.38), at $z = 0$, $\sigma_o' = 0$, so $f = 0$. Again, at $z = L' = 20$ ft,

$$\sigma_o' = \gamma L' = \frac{(110)(20)}{1000} = 2.2\text{ kip/ft}^2$$

So,

$$f = K\sigma_o' \tan \delta' = (1.3)(2.2)[\tan(0.8 \times 30)] = 1.273\text{ kip/ft}^2$$

Thus,

$$Q_s = \left(\frac{f_{z=0} + f_{z=20\text{ ft}}}{2}\right)pL' + f_{z=20\text{ ft}}p(L - L')$$

$$= \left(\frac{0 + 1.273}{2}\right)\left(4 \times \frac{16}{12}\right)(20) + (1.273)\left(4 \times \frac{16}{12}\right)(50 - 20)$$

$$= 67.9 + 203.7 = \mathbf{271.6\text{ kip}}$$

Part b

$$Q_u = Q_p + Q_s$$

Average value of Q_p from Example 11.1 is

$$\frac{56.45 + 234.7 + 179.9}{3} \approx 157\text{ kip}$$

So,

$$Q_{\text{all}} = \frac{Q_u}{\text{FS}} = \frac{1}{4}\left(157 + 271.6\right) = \mathbf{107.15\text{ kip}}$$ ∎

Example 11.3

For the pile described in Example 11.1, estimate the Q_{all} using Coyle and Castello's method [Sections 11.10 and Eq. (11.41)].

Solution

From Eqs. (11.33) and (11.41),

$$Q_u = Q_p + Q_s = q'N_q^* A_p + K\sigma_o' \tan(0.8\phi')pL$$

$$\frac{L}{D} = \frac{50}{\left(\dfrac{16}{12}\right)} = 37.5$$

For $\phi' = 30°$ and $L/D = 37.5$, $N_q^* = 25$ (Figure 11.15) and $K = 0.25$ (Figure 11.19). Thus,

$$Q_u = \left(\frac{110 \times 50}{1000} \text{ kip/ft}^2\right)(25)\left(\frac{16 \times 16}{12 \times 12} \text{ ft}^2\right)$$

$$+ (0.25)\left(\frac{110 \times 50}{1000 \times 2}\right)\tan(0.8 \times 30)\left(\frac{4 \times 16}{12}\right)(50)$$

$$= 244.4 + 81.62 = 326.02 \text{ kip}$$

$$Q_{\text{all}} = \frac{Q_u}{\text{FS}} = \frac{326.02}{4} = \textbf{81.5 kip}$$

■

Example 11.4

A driven-pipe pile in clay is shown in Figure 11.25a. The pipe has an outside diameter of 406 mm and a wall thickness of 6.35 mm.

a. Calculate the net point bearing capacity. Use Eq. (11.19).
b. Calculate the skin resistance (1) by using Eqs. (11.49) and (11.50) (α method), (2) by using Eq. (11.47) (λ method), and (3) by using Eq. (11.51) (β method). For all clay layers, $\phi_R' = 30°$. The top 10 m of clay is normally consolidated. The bottom clay layer has an OCR of 2.
c. Estimate the net allowable pile capacity. Use FS = 4.

Solution

The area of cross section of the pile, including the soil inside the pile, is

$$A_p = \frac{\pi}{4}D^2 = \frac{\pi}{4}(0.406)^2 = 0.1295 \text{ m}^2$$

Part a: Calculation of Net Point Bearing Capacity
From Eq. (11.19),

$$Q_p = A_p q_p = A_p N_c^* c_{u(2)} = (0.1295)(9)(100) = \textbf{116.55 kN}$$

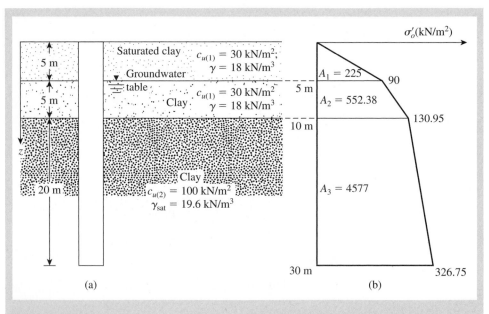

Figure 11.25 Estimation of the load bearing capacity of a driven-pipe pile

Part b: Calculation of Skin Resistance
(1) Using Eqs. (11.49) and (11.50), we have

$$Q_s = \Sigma \alpha c_u p \Delta L$$

The variation of vertical effective stress with depth is shown in Figure 11.25b. Now the following table can be prepared:

Depth (m)	Average depth (m)	Average vertical effective stress, $\overline{\sigma}'_o$ (kN/m²)	c_u (kN/m²)	$\dfrac{c_u}{\overline{\sigma}'_o}$ (kN/m²)	α (Figure 11.23)
0–5	2.5	$\dfrac{0 + 90}{2} = 45$	30	0.67	0.6
5–10	7.5	$\dfrac{90 + 130.95}{2} = 110.5$	30	0.27	0.9
10–30	20	$\dfrac{130.95 + 326.75}{2} = 228.85$	100	0.44	0.725

Thus,

$$Q_s = [\alpha_1 c_{u(1)} L_1 + \alpha_2 c_{u(1)} L_2 + \alpha_3 c_{u(2)} L_3]p$$
$$= [(0.6)(30)(5) + (0.9)(30)(5)$$
$$+ (0.725)(100)(20)](\pi \times 0.406) = \mathbf{2136\ kN}$$

(2) From Eq. 11.47, $f_{av} = \lambda(\overline{\sigma}'_o + 2c_u)$. Now, the average value of c_u is

$$\frac{c_{u(1)}(10) + c_{u(2)}(20)}{30} = \frac{(30)(10) + (100)(20)}{30} = 76.7 \text{ kN/m}^2$$

To obtain the average value of $\overline{\sigma}'_o$, the diagram for vertical effective stress variation with depth is plotted in Figure 11.25b. From Eq. (11.48),

$$\overline{\sigma}'_o = \frac{A_1 + A_2 + A_3}{L} = \frac{225 + 552.38 + 4577}{30} = 178.48 \text{ kN/m}^2$$

From Table 11.7, the magnitude of λ is 0.136. So

$$f_{av} = 0.136[178.48 + (2)(76.7)] = 45.14 \text{ kN/m}^2$$

Hence,

$$Q_s = pLf_{av} = \pi(0.406)(30)(45.14) = \mathbf{1727 \text{ kN}}$$

(3) The top layer of clay (10 m) is normally consolidated, and $\phi'_R = 30°$. For $z = 0\text{--}5$ m, from Eq. (11.55), we have

$$f_{av(1)} = (1 - \sin \phi'_R) \tan \phi'_R \, \overline{\sigma}'_o$$

$$= (1 - \sin 30°)(\tan 30°)\left(\frac{0 + 90}{2}\right) = 13.0 \text{ kN/m}^2$$

Similarly, for $z = 5\text{--}10$ m.

$$f_{av(2)} = (1 - \sin 30°)(\tan 30°)\left(\frac{90 + 130.95}{2}\right) = 31.9 \text{ kN/m}^2$$

For $z = 10\text{--}30$ m from Eq. (11.56),

$$f_{av} = (1 - \sin \phi'_R)\tan \phi'_R \sqrt{OCR} \, \overline{\sigma}'_o$$

For OCR = 2,

$$f_{av(3)} = (1 - \sin 30°)(\tan 30°)\sqrt{2}\left(\frac{130.95 + 326.75}{2}\right) = 93.43 \text{ kN/m}^2$$

So,

$$Q_s = p[f_{av(1)}(5) + f_{av(2)}(5) + f_{av(3)}(20)]$$
$$= (\pi)(0.406)[(13)(5) + (31.9)(5) + (93.43)(20)] = \mathbf{2670 \text{ kN}}$$

Part c: Calculation of Net Ultimate Capacity, Q_u
We have

$$Q_{s(average)} = \frac{2136 + 1727 + 2670}{3} \approx 2178 \text{ kN}$$

Thus,

$$Q_u = Q_p + Q_s = 116.55 + 2178 = 2294.55 \text{ kN}$$

and

$$Q_{\text{all}} = \frac{Q_u}{\text{FS}} = \frac{2294.55}{4} \approx \mathbf{574 \text{ kN}} \qquad \blacksquare$$

Example 11.5

A concrete pile 305 mm × 305 mm in cross section is driven to a depth of 20 m below the ground surface in a saturated clay soil. A summary of the variation of frictional resistance f_c obtained from a cone penetration test is as follows:

Depth (m)	Friction resistance, f_c (kg/cm²)
0–6	0.35
6–12	0.56
12–20	0.72

Estimate the frictional resistance Q_s for the pile.

Solution
We can prepare the following table:

Depth (m)	f_c (kN/m²)	α' (Figure 11.24)	ΔL (m)	$\alpha' f_c p(\Delta L)$ [Eq. (11.58)] (kN)
0–6	34.34	0.84	6	211.5
6–12	54.94	0.71	6	285.5
12–20	70.63	0.63	8	434.2

(*Note:* $p = (4)(0.305) = 1.22$ m)

Thus,

$$Q_s = \Sigma \alpha' f_c p(\Delta L) = \mathbf{931 \text{ kN}} \qquad \blacksquare$$

11.15　*Pile Load Tests*

In most large projects, a specific number of load tests must be conducted on piles. The primary reason is the unreliability of prediction methods. The vertical and lateral load-bearing capacity of a pile can be tested in the field. Figure 11.26a shows a schematic

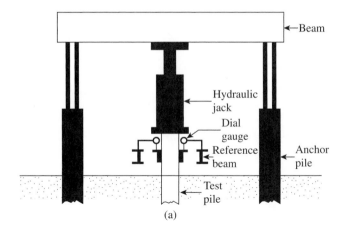

(a)

(b)

Figure 11.26 (a) Schematic diagram of pile load test arrangement; (b) a load test in progress (Courtesy of E. C. Shin, University of Incheon, Korea)

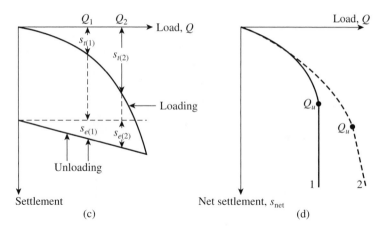

Figure 11.26   *(Continued)* (c) plot of load against total settlement; (d) plot of load against net settlement

diagram of the pile load arrangement for testing *axial compression* in the field. The load is applied to the pile by a hydraulic jack. Step loads are applied to the pile, and sufficient time is allowed to elapse after each load so that a small amount of settlement occurs. The settlement of the pile is measured by dial gauges. The amount of load to be applied for each step will vary, depending on local building codes. Most building codes require that each step load be about one-fourth of the proposed working load. The load test should be carried out to at least a total load of two times the proposed working load. After the desired pile load is reached, the pile is gradually unloaded.

Figure 11.26b shows a pile load test in progress.

Figure 11.26c shows a load–settlement diagram obtained from field loading and unloading. For any load Q, the net pile settlement can be calculated as follows:

When $Q = Q_1$,

$$\text{Net settlement, } s_{\text{net}(1)} = s_{t(1)} - s_{e(1)}$$

When $Q = Q_2$,

$$\text{Net settlement, } s_{\text{net}(2)} = s_{t(2)} - s_{e(2)}$$

$$\vdots$$

where

s_{net} = net settlement
s_e = elastic settlement of the pile itself
s_t = total settlement

These values of Q can be plotted in a graph against the corresponding net settlement, s_{net}, as shown in Figure 11.26d. The ultimate load of the pile can then be determined from the graph. Pile settlement may increase with load to a certain point, beyond which the load–settlement curve becomes vertical. The load corresponding to the point where the curve of Q versus s_{net} becomes vertical is the ultimate load, Q_u, for the pile; it is shown by curve 1 in Figure 11.26d. In many cases, the latter stage of the load–settlement curve is almost linear, showing a large degree of settlement for a small increment of load; this is shown by curve 2 in the figure. The ultimate load, Q_u, for such a case is determined from the point of the curve of Q versus s_{net} where this steep linear portion starts.

The load test procedure just described requires the application of step loads on the piles and the measurement of settlement and is called a *load-controlled* test. Another technique used for a pile load test is the *constant-rate-of-penetration* test, wherein the load on the pile is continuously increased to maintain a constant rate of penetration, which can vary from 0.25 to 2.5 mm/min (0.01 to 0.1 in./min). This test gives a load–settlement plot similar to that obtained from the load-controlled test. Another type of pile load test is *cyclic loading,* in which an incremental load is repeatedly applied and removed.

In order to conduct a load test on piles, it is important to take into account the time lapse after the end of driving (EOD). When piles are driven into soft clay, a certain zone surrounding the clay becomes remolded or compressed, as shown in Figure 11.27a. This results in a reduction of undrained shear strength, c_u (Figure 11.27b). With time, the loss of undrained shear strength is partially or fully regained. The time lapse may range from 30 to 60 days.

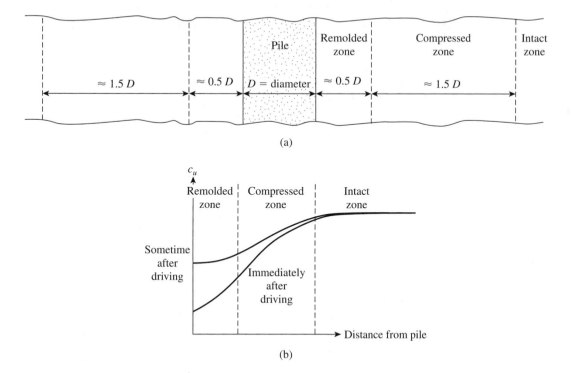

Figure 11.27 (a) Remolded or compacted zone around a pile driven into soft clay; (b) Nature of variation of undrained shear strength (c_u) with time around a pile driven into soft clay

For piles driven in dilative (dense to very dense) saturated fine sands, relaxation is possible. Negative pore water pressure, if developed during pile driving, will dissipate over time, resulting in a reduction in pile capacity with time after the driving operation is completed. At the same time, excess pore water pressure may be generated in contractive fine sands during pile driving. The excess pore water pressure will dissipate over time, which will result in greater pile capacity.

Several empirical relationships have been developed to predict changes in pile capacity with time. For further information, readers may refer to Skov and Denver (1988), Guang-Yu (1988), and Svinkin (1996).

11.16   *Comparison of Theory with Field Load Test Results*

Details of many field studies related to the estimation of the ultimate load-carrying capacity of various types of piles are available in the literature. In some cases, the results agree reasonably well with the theoretical predictions, and in others, they vary widely. The discrepancy between theory and field-test results may be attributed to factors such as improper interpretation of subsoil properties, incorrect theoretical assumptions, erroneous acquisition of field-test results, and other factors.

We saw from Example 11.1 that, for similar soil properties, the ultimate point load Q_p can vary over 400% or more, depending on which theory and equation are used. Also, from the calculation of part (a) of the example, it is easy to see that, in most cases, for long piles embedded in sand, the limiting point resistance q_l [Eq. 11.17] controls the unit point resistance q_p.

Briaud et al. (1989) reported the results of 28 axial load tests performed by the U.S. Army Engineering District (St. Louis) on impact-driven H-piles and pipe piles in sand during the construction of the New Lock and Dam No. 26 on the Mississippi River. The results of load tests on four of the H-piles are summarized in Table 11.10.

Briaud and his colleagues made a statistical analysis to determine the ratio of theoretical ultimate load to measured ultimate load. The results of this analysis are summarized in Table 11.11 for the plugged case. (See Figure 11.10.) Note that a perfect prediction would have a mean of 1.0, a standard deviation of 0, and a coefficient of variation of 0. The table indicates that no method gave a perfect prediction; in general, Q_p was overestimated and Q_s was underestimated. Again, this shows the uncertainty in predicting the load-bearing capacity of piles.

Table 11.10   Pile Load Test Results

Pile no.	Pile type	Batter	Q_p (ton)	Q_s (ton)	Q_u (ton)	Pile length (ft)
1–3A	HP14 × 73	Vertical	152	161	313	54
1–6	HP14 × 73	Vertical	75	353	428	53
1–9	HP14 × 73	1:2.5	85	252	337	58
2–5	HP14 × 73	1:2.5	46	179	225	59

Table 11.11 Summary of Briaud et al.'s Statistical Analysis for H-Piles, Plugged Case

Theoretical method	Q_p			Q_s			Q_u		
	Mean	Standard deviation	Coefficient of variation	Mean	Standard deviation	Coefficient of variation	Mean	Standard deviation	Coefficient of variation
Coyle and Castello (1981)	2.38	1.31	0.55	0.87	0.36	0.41	1.17	0.44	0.38
Briaud and Tucker (1984)	1.79	1.02	0.59	0.81	0.32	0.40	0.97	0.39	0.40
Meyerhof (1976)	4.37	2.76	0.63	0.92	0.43	0.46	1.68	0.76	0.45
API (1984)	1.62	1.00	0.62	0.59	0.25	0.43	0.79	0.34	0.43

Lessons from the preceding case studies and others available in the literature show that previous experience and good practical judgment, along with a knowledge of theoretical developments, are required to design safe pile foundations.

11.17 Elastic Settlement of Piles

The total settlement of a pile under a vertical working load Q_w is given by

$$s_e = s_{e(1)} + s_{e(2)} + s_{e(3)} \tag{11.62}$$

where

$s_{e(1)}$ = elastic settlement of pile
$s_{e(2)}$ = settlement of pile caused by the load at the pile tip
$s_{e(3)}$ = settlement of pile caused by the load transmitted along the pile shaft

If the pile material is assumed to be elastic, the deformation of the pile shaft can be evaluated, in accordance with the fundamental principles of mechanics of materials, as

$$s_{e(1)} = \frac{(Q_{wp} + \xi Q_{ws})L}{A_p E_p} \tag{11.63}$$

where

Q_{wp} = load carried at the pile point under working load condition
Q_{ws} = load carried by frictional (skin) resistance under working load condition
A_p = area of cross section of pile
L = length of pile
E_p = modulus of elasticity of the pile material

The magnitude of ξ will depend on the nature of the distribution of the unit friction (skin) resistance f along the pile shaft. If the distribution of f is uniform or parabolic, as shown in Figures 11.28a and 11.28b, then $\xi = 0.5$. However, for a triangular distribution of f (Figure 11.28c), the magnitude of ξ is about 0.67 (Vesic, 1977).

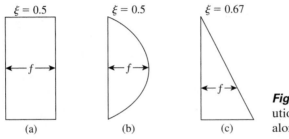

Figure 11.28 Various types of distribution of unit friction (skin) resistance along the pile shaft

The settlement of a pile caused by the load carried at the pile point may be expressed in the form:

$$s_{e(2)} = \frac{q_{\text{wp}} D}{E_s} (1 - \mu_s^2) I_{\text{wp}}$$

(11.64)

where

D = width or diameter of pile
q_{wp} = point load per unit area at the pile point = Q_{wp}/A_p
E_s = modulus of elasticity of soil at or below the pile point
μ_s = Poisson's ratio of soil
I_{wp} = influence factor ≈ 0.85

Vesic (1977) also proposed a semi-empirical method for obtaining the magnitude of the settlement of $s_{e(2)}$. His equation is

$$s_{e(2)} = \frac{Q_{\text{wp}} C_p}{D q_p}$$

(11.65)

where

q_p = ultimate point resistance of the pile
C_p = an empirical coefficient

Representative values of C_p for various soils are given in Table 11.12.
The settlement of a pile caused by the load carried by the pile shaft is given by a relation similar to Eq. (11.64), namely,

$$s_{e(3)} = \left(\frac{Q_{\text{ws}}}{pL} \right) \frac{D}{E_s} (1 - \mu_s^2) I_{\text{ws}}$$

(11.66)

where

p = perimeter of the pile
L = embedded length of pile
I_{ws} = influence factor

Table 11.12 Typical Values of C_p [from Eq. (11.65)]

Type of soil	Driven pile	Bored pile
Sand (dense to loose)	0.02–0.04	0.09–0.18
Clay (stiff to soft)	0.02–0.03	0.03–0.06
Silt (dense to loose)	0.03–0.05	0.09–0.12

From "Design of Pile Foundations," by A. S. Vesic, in *NCHRP Synthesis of Highway Practice 42*, Transportation Research Board, 1977. Reprinted by permission.

Note that the term Q_{ws}/pL in Eq. (11.66) is the average value of f along the pile shaft. The influence factor, I_{ws}, has a simple empirical relation (Vesic, 1977):

$$I_{ws} = 2 + 0.35\sqrt{\frac{L}{D}} \tag{11.67}$$

Vesic (1977) also proposed a simple empirical relation similar to Eq. (11.65) for obtaining $s_{e(3)}$:

$$s_{e(3)} = \frac{Q_{ws}C_s}{Lq_p} \tag{11.68}$$

In this equation, C_s = an empirical constant = $(0.93 + 0.16\sqrt{L/D})C_p$ (11.69) The values of C_p for use in Eq. (11.65) may be estimated from Table 11.12.

Example 11.6

The allowable working load on a prestressed concrete pile 21-m long that has been driven into sand is 502 kN. The pile is octagonal in shape with $D = 356$ mm (see Table 11.3a). Skin resistance carries 350 kN of the allowable load, and point bearing carries the rest. Use $E_p = 21 \times 10^6$ kN/m², $E_s = 25 \times 10^3$ kN/m², μ_s 0.35, and $\xi = 0.62$.. Determine the settlement of the pile.

Solution
From Eq. (11.63),

$$S_{e(1)} = \frac{(Q_{wp} + \xi Q_{ws})L}{A_p E_p}$$

From Table 11.3a for $D = 356$ mm, the area of pile cross section. $A_p = 1045$ cm², Also, perimeter $p = 1.168$ m. Given: $Q_{ws} = 350$ kN, so

$$Q_{wp} = 502 - 350 = 152 \text{ kN}$$

$$S_{e(1)} = \frac{[152 + 0.62(350)](21)}{(0.1045 \text{ m}^2)(21 \times 10^6)} = 0.00353 \text{ m} = 3.35 \text{ mm}$$

From Eq. (11.64),

$$s_{e(2)} = \frac{q_{wp}D}{E_s}(1 - \mu_s^2)I_{wp} = \left(\frac{152}{0.1045}\right)\left(\frac{0.356}{25 \times 10^3}\right)(1 - 0.35^2)(0.85)$$

$$= 0.0155 \text{ m} = 15.5 \text{ mm}$$

Again, from Eq. (11.66),

$$s_{e(3)} = \left(\frac{Q_{ws}}{pL}\right)\left(\frac{D}{E_s}\right)(1 - \mu_s^2)I_{ws}$$

$$I_{ws} = 2 + 0.35\sqrt{\frac{L}{D}} = 2 + 0.35\sqrt{\frac{21}{0.356}} = 4.69$$

$$s_{e(3)} = \left[\frac{350}{(1.168)(21)}\right]\left(\frac{0.356}{25 \times 10^3}\right)(1 - 0.35^2)(4.69)$$

$$= 0.00084 \text{ m} = 0.84 \text{ mm}$$

Hence, total settlement is

$$s_e = s_{e(1)} + s_{e(2)} + s_{e(3)} = 3.35 + 15.5 + 0.84 = \textbf{19.69 mm} \qquad \blacksquare$$

11.18   *Laterally Loaded Piles*

A vertical pile resists a lateral load by mobilizing passive pressure in the soil surrounding it. (See Figure 11.1c.) The degree of distribution of the soil's reaction depends on (a) the stiffness of the pile, (b) the stiffness of the soil, and (c) the fixity of the ends of the pile. In general, laterally loaded piles can be divided into two major categories: (1) short or rigid piles and (2) long or elastic piles. Figures 11.29a and 11.29b show the nature of the variation of the pile deflection and the distribution of the moment and shear force along the pile length when the pile is subjected to lateral loading. We next summarize the current solutions for laterally loaded piles.

Elastic Solution

A general method for determining moments and displacements of a vertical pile embedded in a *granular soil* and subjected to lateral load and moment at the ground surface was given by Matlock and Reese (1960). Consider a pile of length L subjected to a lateral force Q_g and a moment M_g at the ground surface ($z = 0$), as shown in Figure 11.30a. Figure 11.30b shows the general deflected shape of the pile and the soil resistance caused by the applied load and the moment.

According to a simpler Winkler's model, an elastic medium (soil in this case) can be replaced by a series of infinitely close independent elastic springs. Based on this assumption,

$$k = \frac{p'(\text{kN/m or lb/ft})}{x(\text{m or ft})} \qquad (11.70)$$

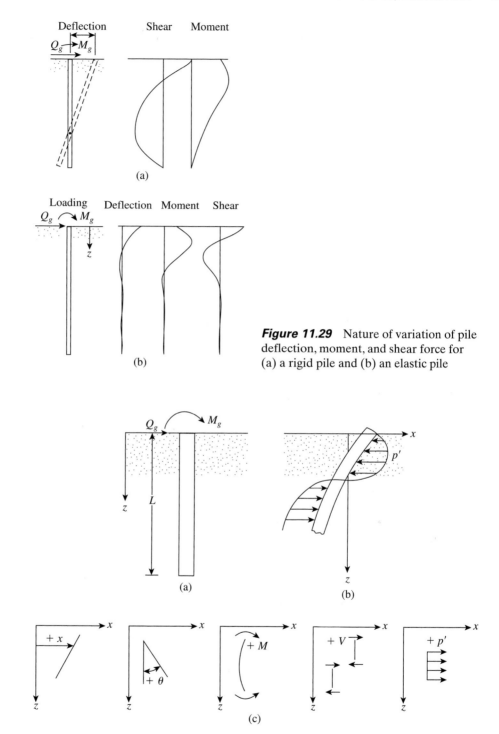

Figure 11.29 Nature of variation of pile deflection, moment, and shear force for (a) a rigid pile and (b) an elastic pile

Figure 11.30 (a) Laterally loaded pile; (b) soil resistance on pile caused by lateral load; (c) sign conventions for displacement, slope, moment, shear, and soil reaction

where

k = modulus of subgrade reaction
p' = pressure on soil
x = deflection

The subgrade modulus for *granular soils* at a depth z is defined as

$$k_z = n_h z \tag{11.71}$$

where n_h = constant of modulus of horizontal subgrade reaction.

Referring to Figure 11.30b and using the theory of beams on an elastic foundation, we can write

$$E_p I_p \frac{d^4 x}{dz^4} = p' \tag{11.72}$$

where

E_p = modulus of elasticity in the pile material
I_p = moment of inertia of the pile section

Based on Winkler's model

$$p' = -kx \tag{11.73}$$

The sign in Eq. (11.73) is negative because the soil reaction is in the direction opposite that of the pile deflection.

Combining Eqs. (11.72) and (11.73) gives

$$E_p I_p \frac{d^4 x}{dz^4} + kx = 0 \tag{11.74}$$

The solution of Eq. (11.74) results in the following expressions:

Pile Deflection at Any Depth $[x_z(z)]$

$$x_z(z) = A_x \frac{Q_g T^3}{E_p I_p} + B_x \frac{M_g T^2}{E_p I_p} \tag{11.75}$$

Slope of Pile at Any Depth $[\theta_z(z)]$

$$\theta_z(z) = A_\theta \frac{Q_g T^2}{E_p I_p} + B_\theta \frac{M_g T}{E_p I_p} \tag{11.76}$$

Moment of Pile at Any Depth $[M_z(z)]$

$$M_z(z) = A_m Q_g T + B_m M_g \tag{11.77}$$

Shear Force on Pile at Any Depth [$V_z(z)$]

$$V_z(z) = A_v Q_g + B_v \frac{M_g}{T} \qquad (11.78)$$

Soil Reaction at Any Depth [$p_z'(z)$]

$$p_z'(z) = A_{p'} \frac{Q_g}{T} + B_{p'} \frac{M_g}{T^2} \qquad (11.79)$$

where

A_x, B_x, A_θ, B_θ, A_m, B_m, A_v, B_v, $A_{p'}$, and $B_{p'}$ are coefficients
T = characteristic length of the soil–pile system

$$= \sqrt[5]{\frac{E_p I_p}{n_h}} \qquad (11.80)$$

n_h has been defined in Eq. (11.71)

When $L \geqslant 5T$, the pile is considered to be a *long pile*. For $L \leqslant 2T$, the pile is considered to be a *rigid pile*. Table 11.13 gives the values of the coefficients for long piles ($L/T \geqslant 5$) in Eqs. (11.75) through (11.79). Note that, in the first column of the table,

$$Z = \frac{z}{T} \qquad (11.81)$$

is the nondimensional depth.

The positive sign conventions for $x_z(z)$, $\theta_z(z)$, $M_z(z)$, $V_z(z)$, and $p_z'(z)$ assumed in the derivations in Table 11.13 are shown in Figure 11.30c. Figure 11.31 shows the variation of A_x, B_x, A_m, and B_m for various values of $L/T = Z_{\max}$. It indicates that, when L/T is greater than about 5, the coefficients do not change, which is true of long piles only.

Calculating the characteristic length T for the pile requires assuming a proper value of n_h. Table 11.14 gives some representative values.

Elastic solutions similar to those given in Eqs. 11.75 through 11.79 for piles embedded in *cohesive soil* were developed by Davisson and Gill (1963). Their equations are

$$x_z(z) = A_x' \frac{Q_g R^3}{E_p I_p} + B_x' \frac{M_g R^2}{E_p I_p} \qquad (11.82)$$

and

$$M_z(z) = A_m' Q_g R + B_m' M_g \qquad (11.83)$$

Table 11.13 Coefficients for Long Piles, $k_z = n_h z$

Z	A_x	A_θ	A_m	A_v	A'_p	B_x	B_θ	B_m	B_v	B'_p
0.0	2.435	−1.623	0.000	1.000	0.000	1.623	−1.750	1.000	0.000	0.000
0.1	2.273	−1.618	0.100	0.989	−0.227	1.453	−1.650	1.000	−0.007	−0.145
0.2	2.112	−1.603	0.198	0.956	−0.422	1.293	−1.550	0.999	−0.028	−0.259
0.3	1.952	−1.578	0.291	0.906	−0.586	1.143	−1.450	0.994	−0.058	−0.343
0.4	1.796	−1.545	0.379	0.840	−0.718	1.003	−1.351	0.987	−0.095	−0.401
0.5	1.644	−1.503	0.459	0.764	−0.822	0.873	−1.253	0.976	−0.137	−0.436
0.6	1.496	−1.454	0.532	0.677	−0.897	0.752	−1.156	0.960	−0.181	−0.451
0.7	1.353	−1.397	0.595	0.585	−0.947	0.642	−1.061	0.939	−0.226	−0.449
0.8	1.216	−1.335	0.649	0.489	−0.973	0.540	−0.968	0.914	−0.270	−0.432
0.9	1.086	−1.268	0.693	0.392	−0.977	0.448	−0.878	0.885	−0.312	−0.403
1.0	0.962	−1.197	0.727	0.295	−0.962	0.364	−0.792	0.852	−0.350	−0.364
1.2	0.738	−1.047	0.767	0.109	−0.885	0.223	−0.629	0.775	−0.414	−0.268
1.4	0.544	−0.893	0.772	−0.056	−0.761	0.112	−0.482	0.688	−0.456	−0.157
1.6	0.381	−0.741	0.746	−0.193	−0.609	0.029	−0.354	0.594	−0.477	−0.047
1.8	0.247	−0.596	0.696	−0.298	−0.445	−0.030	−0.245	0.498	−0.476	0.054
2.0	0.142	−0.464	0.628	−0.371	−0.283	−0.070	−0.155	0.404	−0.456	0.140
3.0	−0.075	−0.040	0.225	−0.349	0.226	−0.089	0.057	0.059	−0.213	0.268
4.0	−0.050	0.052	0.000	−0.106	0.201	−0.028	0.049	−0.042	0.017	0.112
5.0	−0.009	0.025	−0.033	0.015	0.046	0.000	−0.011	−0.026	0.029	−0.002

From *Drilled Pier Foundations,* by R. J. Woodwood, W. S. Gardner, and D. M. Greer. Copyright 1972 by McGraw-Hill. Used with the permission of McGraw-Hill Book Company.

Table 11.14 Representative Values of n_h

Soil	n_h kN/m³	n_h lb/in³
Dry or moist sand		
Loose	1800–2200	6.5–8.0
Medium	5500–7000	20–25
Dense	15,000–18,000	55–65
Submerged sand		
Loose	1000–1400	3.5–5.0
Medium	3500–4500	12–18
Dense	9000–12,000	32–45

where A'_x, B_x, A'_m, and B'_m are coefficients.
and

$$R = \sqrt[4]{\frac{E_p I_p}{k}} \qquad (11.84)$$

The values of the coefficients A' and B' are given in Figure 11.32. Note that

$$Z = \frac{z}{R} \qquad (11.85)$$

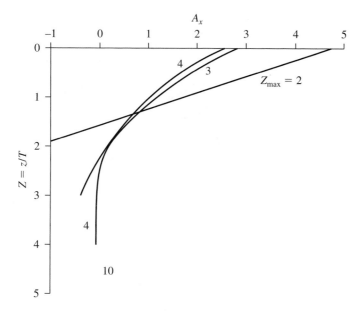

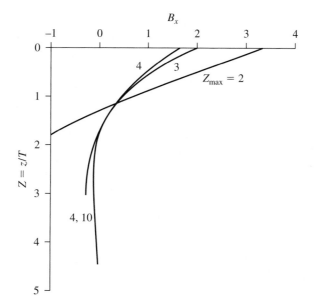

Figure 11.31 Variation of A_x, B_x, A_m, and B_m with Z (After Matlock and Reese, 1960)

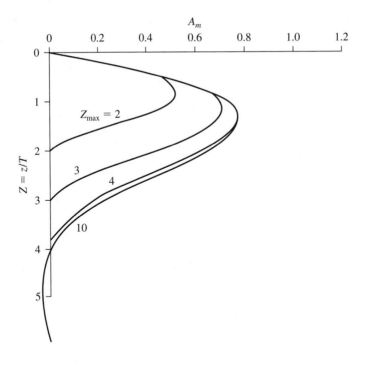

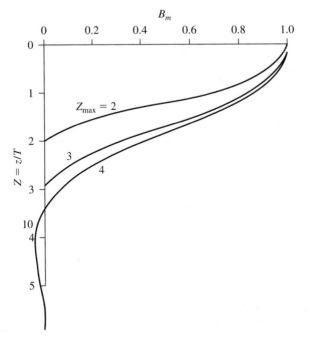

Figure 11.31 (Continued)

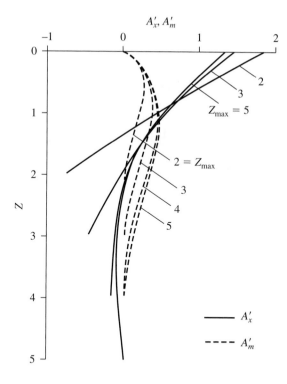

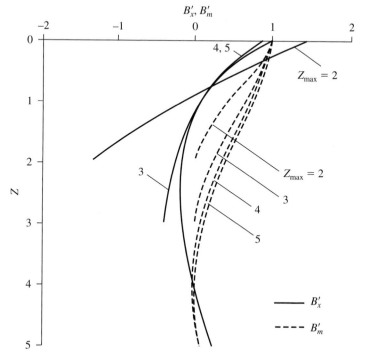

Figure 11.32 Variation of A_x, B'_x, A'_m, and B'_m with Z
(After Davisson and Gill, 1963)

and

$$Z_{max} = \frac{L}{R} \tag{11.86}$$

The use of Eqs. (11.82) and (11.83) requires knowing the magnitude of the characteristic length, R. This can be calculated from Eq. (11.84), provided that the coefficient of the subgrade reaction is known. For sands, the coefficient of the subgrade reaction was given by Eq. (11.71), which showed a linear variation with depth. However, in cohesive soils, the subgrade reaction may be assumed to be approximately constant with depth. Vesic (1961) proposed the following equation to estimate the value of k:

$$k = 0.65 \sqrt[12]{\frac{E_s D^4}{E_p I_p}} \frac{E_s}{1 - \mu_s^2} \tag{11.87}$$

Here,

E_s = modulus of elasticity of soil
D = pile width (or diameter)
μ_s = Poisson's ratio for the soil

Ultimate Load Analysis: Broms's Method

For laterally loaded piles, Broms (1965) developed a simplified solution based on the assumptions of (a) shear failure in soil, which is the case for short piles, and (b) bending of the pile, governed by the plastic yield resistance of the pile section, which is applicable to long piles. Broms's solution for calculating the ultimate load resistance, $Q_{u(g)}$, for *short piles* is given in Figure 11.33a. A similar solution for piles embedded in cohesive soil is shown in Figure 11.33b. In Figure 11.33a, note that

$$K_p = \text{Rankine passive earth pressure coefficient} = \tan^2\left(45 + \frac{\phi'}{2}\right) \tag{11.88}$$

Similarly, in Figure 11.33b,

$$c_u = \text{undrained cohesion} \approx \frac{0.75q_u}{FS} = \frac{0.75q_u}{2} = 0.375q_u \tag{11.89}$$

where

FS = factor of safety($=2$)
q_u = unconfined compression strength

Figure 11.34 shows Broms's analysis of long piles. In the figure, the yield moment for the pile is

$$M_y = SF_Y \tag{11.90}$$

where

S = section modulus of the pile section
F_Y = yield stress of the pile material

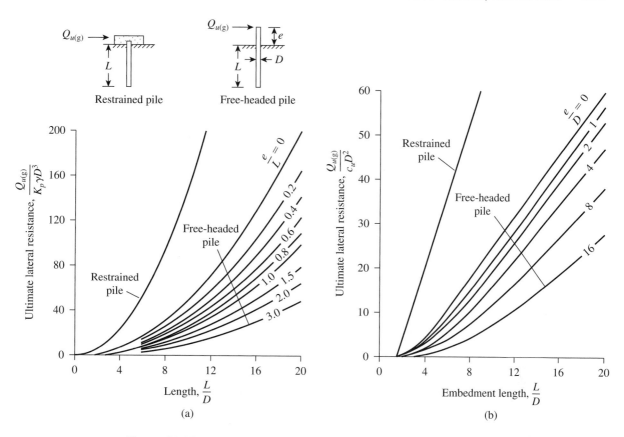

Figure 11.33 Broms's solution for ultimate lateral resistance of short piles (a) in sand and (b) in clay

In solving a given problem, both cases (i.e., Figure 11.33 and Figure 11.34) should be checked.

The deflection of the pile head, $x_z(z = 0)$, under working load conditions can be estimated from Figure 11.35. In Figure 11.35a, the term η can be expressed as

$$\eta = \sqrt[5]{\frac{n_h}{E_p I_p}} \tag{11.91}$$

The range of n_h for granular soil is given in Table 11.14. Similarly, in Figure 11.35b, which is for clay, the term K is the horizontal soil modulus and can be defined as

$$K = \frac{\text{pressure (kN/m}^2 \text{ or lb/in}^2)}{\text{displacement (m or in.)}} \tag{11.92}$$

Also, the term β can be defined as

$$\beta = \sqrt[4]{\frac{KD}{4E_p I_p}} \tag{11.93}$$

Note that, in Figure 11.35, Q_g is the working load.

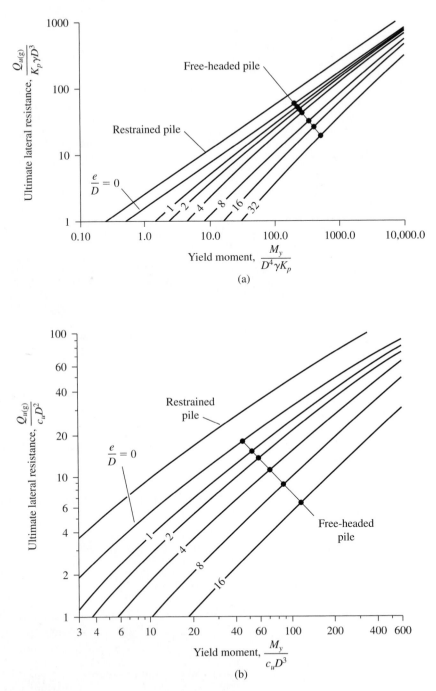

Figure 11.34 Broms's solution for ultimate lateral resistance of long piles (a) in sand (b) in clay

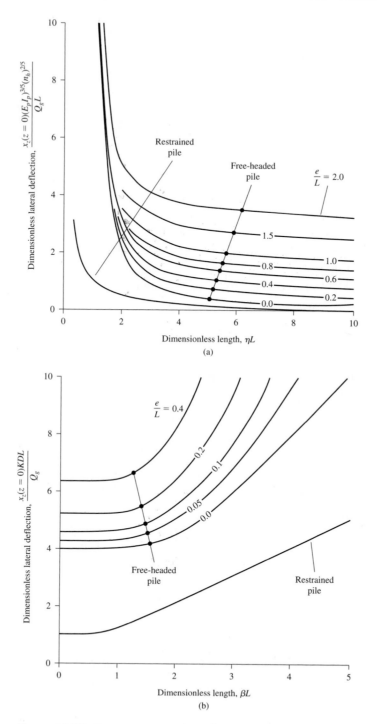

Figure 11.35 Broms's solution for estimating deflection of pile head (a) in sand and (b) in clay

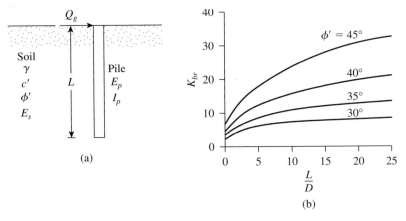

Figure 11.36 (a) Pile with lateral loading at ground level; (b) Variation of resultant net soil pressure coefficient K_{br}

Ultimate Load Analysis: Meyerhof's Method

In 1995, Meyerhof offered solutions for laterally loaded rigid and flexible piles. (See Figure 11.36a.) According to Meyerhof's method, a pile can be defined as flexible if

$$K_r = \text{relative stiffness of pile} = \frac{E_p I_p}{E_s L^4} < 0.01 \tag{11.94}$$

where E_s = average horizontal soil modulus of elasticity.

Piles in Sand For short (rigid) piles in *sand*, the ultimate load resistance can be given as

$$Q_{u(g)} = 0.12 \gamma D L^2 K_{br} \leqslant 0.4 p_l D L \tag{11.95}$$

where

γ = unit weight of soil
K_{br} = resultant net soil pressure coefficient (Figure 11.36b)
p_l = limit pressure obtained from pressuremeter tests (see Chapter 2)

The limit pressure can be given as

$$p_l = 0.4 p_a N_q \tan \phi' \quad \text{(for a Menard pressuremeter)} \tag{11.96}$$

and

$$p_l = 0.6 p_a N_q \tan \phi' \tag{11.97}$$

(for self-boring and full-displacement pressuremeters)

where

N_q = bearing capacity factor (see Table 3.3)
p_a = atmospheric pressure (≈ 100 kN/m^2 or 2000 lb/ft^2)

The maximum moment in the pile due to the lateral load, $Q_{u(g)}$, is

$$M_{max} = 0.35 Q_{u(g)} L \leqslant M_y$$
$$\uparrow$$
$$\text{Eq. (11.90)} \tag{11.98}$$

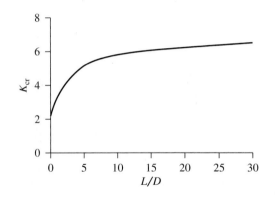

Figure 11.37 Variation of K_{cr}

For long (flexible) piles in sand, the ultimate lateral load, $Q_{u(g)}$, can be estimated from Eq. (11.95) by substituting an effective length L_e for L, where

$$\frac{L_e}{L} = 1.65K_r^{-0.12} \leqslant 1 \tag{11.99}$$

The maximum moment in a flexible pile due to a working lateral load Q_g applied at the ground surface is

$$M_{\max} = 0.3K_r^{0.2}Q_gL \leqslant 0.3Q_gL \tag{11.100}$$

Piles in Clay The ultimate lateral load applied at the ground surface for short (rigid) piles embedded in clay can be given as

$$Q_{u(g)} = 0.4c_uK_{cr}DL \leqslant 0.4p_lDL \tag{11.101}$$

where

p_l = limit pressure from pressuremeter test
K_{cr} = net soil pressure coefficient (see Figure 11.37)

The limit pressure in clay is

$$p_l \approx 6c_u \qquad \text{(for a Menard pressuremeter)} \tag{11.102}$$

and

$$p_l \approx 8c_u \qquad \text{(for self-boring and full-displacement pressuremeters)} \tag{11.103}$$

The maximum bending moment in the pile due to $Q_{u(g)}$ is

$$M_{\max} = 0.22Q_{u(g)}L \leqslant M_y \tag{11.104}$$
$$\uparrow$$
$$\text{Eq. (11.90)}$$

For long (flexible) piles, Eq. (11.101) can be used to estimate $Q_{u(g)}$ by substituting the effective length L_e in place of L, where

$$\frac{L_e}{L} = 1.5K_r^{0.12} \leqslant 1 \tag{11.105}$$

The maximum moment in a flexible pile due to a working lateral load Q_g applied at the ground surface is

$$M_{\text{max}} = 0.3K_r^{0.2}Q_gL \leqslant 0.15Q_gL \tag{11.106}$$

Example 11.7

Consider a steel H-pile (HP 250 × 85) 25 m long, embedded fully in a granular soil. Assume that $n_h = 12,000 \text{ kN/m}^3$. The allowable displacement at the top of the pile is 8 mm. Determine the allowable lateral load, Q_g. Let $M_g = 0$. Use the elastic solution.

Solution
From Table 11.1a, for an HP 250 × 85 pile,

$$I_p = 123 \times 10^{-6} \text{ m}^4 \qquad \text{(about the strong axis)}$$

and let

$$E_p = 207 \times 10^6 \text{ kN/m}^2$$

From Eq. (11.80),

$$T = \sqrt[5]{\frac{E_pI_p}{n_h}} = \sqrt[5]{\frac{(207 \times 10^6)(123 \times 10^{-6})}{12,000}} = 1.16 \text{ m}$$

Here, $L/T = 25/1.16 = 21.55 > 5$, so the pile is a long one. Because $M_g = 0$, Eq. (11.75) takes the form

$$x_z(z) = A_x\frac{Q_gT^3}{E_pI_p}$$

and it follows that

$$Q_g = \frac{x_z(z)E_pI_p}{A_xT^3}$$

At $z = 0$, $x_z = 8 \text{ mm} = 0.008 \text{ m}$ and $A_x = 2.435$ (see Table 11.13), so

$$Q_g = \frac{(0.008)(207 \times 10^6)(123 \times 10^{-6})}{(2.435)(1.16^3)} = 53.59 \text{ kN}$$

This magnitude of Q_g is based on the *limiting displacement condition only*. However, the magnitude of Q_g based on the *moment capacity* of the pile also needs to be determined. For $M_g = 0$, Eq. (11.77) becomes

$$M_z(z) = A_m Q_g T$$

According to Table 11.13, the maximum value of A_m at any depth is 0.772. The maximum allowable moment that the pile can carry is

$$M_{z(max)} = F_Y \frac{I_p}{\dfrac{d_1}{2}}$$

Let $F_Y = 248{,}000 \text{ kN/m}^2$. From Table 11.1a, $I_p = 123 \times 10^{-6} \text{ m}^4$ and $d_1 = 0.254$ m, so

$$\frac{I_p}{\left(\dfrac{d_1}{2}\right)} = \frac{123 \times 10^{-6}}{\left(\dfrac{0.254}{2}\right)} = 968.5 \times 10^{-6} \text{ m}^3$$

Now,

$$Q_g = \frac{M_{z(max)}}{A_m T} = \frac{(968.5 \times 10^{-6})(248.000)}{(0.772)(1.16)} = 268.2 \text{ kN}$$

Because $Q_g = 268.2 \text{ kN} > 53.59 \text{ kN}$, the deflection criteria apply. Hence, $Q_g = \textbf{53.59 kN}$. ∎

Example 11.8

Solve Example 11.7 by Broms's method. Assume that the pile is flexible and is free headed. Let the yield stress of the pile material, $F_y = 248 \text{ MN/m}^2$; the unit weight of soil, $\gamma = 18 \text{ kN/m}^3$; and the soil friction angle $\phi' = 35°$.

Solution
We check for bending failure. From Eq. (11.90),

$$M_y = S F_y$$

From Table 11.1a,

$$S = \frac{I_p}{\dfrac{d_1}{2}} = \frac{123 \times 10^{-6}}{\dfrac{0.254}{2}}$$

Also,

$$M_y = \left[\frac{123 \times 10^{-6}}{\dfrac{0.254}{2}} \right] (248 \times 10^3) = 240.2 \text{ kN-m}$$

and

$$\frac{M_y}{D^4\gamma K_p} = \frac{M_y}{D^4\gamma \tan^2\left(45 + \dfrac{\phi'}{2}\right)} = \frac{240.2}{(0.254)^4(18)\tan^2\left(45 + \dfrac{35}{2}\right)} = 868.8$$

From Figure 11.34a, for $M_y/D^4\gamma K_p = 868.8$, the magnitude of $Q_{u(g)}/K_p D^3\gamma$ (for a free-headed pile with $e/D = 0$) is about 140, so

$$Q_{u(g)} = 140K_p D^3\gamma = 140\tan^2\left(45 + \frac{35}{2}\right)(0.254)^3(18) = 152.4 \text{ kN}$$

Next, we check for pile head deflection. From Eq. (11.91),

$$\eta = \sqrt[5]{\frac{n_h}{E_p I_p}} = \sqrt[5]{\frac{12,000}{(207 \times 10^6)(123 \times 10^{-6})}} = 0.86 \text{ m}^{-1}$$

so

$$\eta L = (0.86)(25) = 21.5$$

From Figure 11.35a, for $\eta L = 21.5$, $e/L = 0$ (free-headed pile): thus,

$$\frac{x_o(E_p I_p)^{3/5}(n_h)^{2/5}}{Q_g L} \approx 0.15 \qquad \text{(by interpolation)}$$

and

$$Q_g = \frac{x_o(E_p I_p)^{3/5}(n_h)^{2/5}}{0.15L}$$

$$= \frac{(0.008)[(207 \times 10^6)(123 \times 10^{-6})]^{3/5}(12,000)^{2/5}}{(0.15)(25)} = 40.2 \text{ kN}$$

Hence, $Q_g = $ **40.2 kN (<152.4 kN).** ∎

11.19 *Pile-Driving Formulas*

To develop the desired load-carrying capacity, a point bearing pile must penetrate the dense soil layer sufficiently or have sufficient contact with a layer of rock. This requirement cannot always be satisfied by driving a pile to a predetermined depth, because soil profiles vary. For that reason, several equations have been developed to calculate the ultimate capacity of a pile during driving. These dynamic equations are widely used in the field to determine whether a pile has reached a satisfactory bearing value at the predetermined depth. One of the earliest such equations—commonly referred to as the *Engineering News (EN) Record formula*—is derived from the work—energy theory. That is,

Energy imparted by the hammer per blow =
(pile resistance)(penetration per hammer blow)

According to the EN formula, the pile resistance is the ultimate load Q_u, expressed as

$$Q_u = \frac{W_R h}{S + C} \qquad (11.107)$$

where

W_R = weight of the ram
h = height of fall of the ram
S = penetration of pile per hammer blow
C = a constant

The pile penetration, S, is usually based on the average value obtained from the last few driving blows. In the equation's original form, the following values of C were recommended:
For drop hammers,

$$C = \begin{cases} 25.4 \text{ mm if } S \text{ and } h \text{ are in mm} \\ 1 \text{ in. if } S \text{ and } h \text{ are in inches} \end{cases}$$

For steam hammers,

$$C = \begin{cases} 2.54 \text{ mm if } S \text{ and } h \text{ are in mm} \\ 0.1 \text{ in. if } S \text{ and } h \text{ are in inches} \end{cases}$$

Also, a factor of safety FS = 6 was recommended for estimating the allowable pile capacity. Note that, for single- and double-acting hammers, the term $W_R h$ can be replaced by $E H_E$, where E is the efficiency of the hammer and H_E is the rated energy of the hammer. Thus,

$$Q_u = \frac{E H_E}{S + C} \qquad (11.108)$$

The EN formula has been revised several times over the years, and other pile-driving formulas also have been suggested. Some of them are tabulated in Table 11.15.

Table 11.15 Pile-Driving Formulas

Name	Formula
Modified EN formula	$Q_u = \dfrac{EW_R h}{S + C} \dfrac{W_R + n^2 W_p}{W_R + W_p}$

where
E = efficiency of hammer
C = 2.54 mm if the units of S and h are in mm
C = 0.1 in. if the units of S and h are in in.
W_p = weight of the pile
n = coefficient of restitution between the ram and the pile cap

Typical values for E

Single- and double-acting hammers	0.7–0.85
Diesel hammers	0.8–0.9
Drop hammers	0.7–0.9

Typical values for n

Cast-iron hammer and concrete piles (without cap)	0.4–0.5
Wood cushion on steel piles	0.3–0.4
Wooden piles	0.25–0.3

Name	Formula
Michigan State Highway Commission formula (1965)	$Q_u = \dfrac{1.25 E H_E}{S + C} \dfrac{W_R + n^2 W_p}{W_R + W_p}$

where
H_E = manufacturer's maximum rated hammer energy (lb-in.)
E = efficiency of hammer
C = 0.1 in.
A factor of safety of 6 is recommended.

Name	Formula
Danish formula (Olson and Flaate, 1967)	$Q_u = \dfrac{E H_E}{S + \sqrt{\dfrac{E H_E L}{2 A_p E_p}}}$

where
E = efficiency of hammer
H_E = rated hammer energy
E_p = modulud of elasticity of the pile material
L = length of the pile
A_p = cross-sectional area of the pile

Name	Formula
Pacific Coast Uniform Building Code formula (International Conference of Building Officials, 1982)	$Q_u = \dfrac{(E H_E)\left(\dfrac{W_R + n W_p}{W_R + W_p}\right)}{S + \dfrac{Q_u L}{A E_p}}$

The value of n should be 0.25 for steel piles and 0.1 for all other piles. A factor of safety of 4 is generally recommended.

Table 11.15 (Continued)

Name	Formula
Janbu's formula (Janbu, 1953)	$Q_u = \dfrac{EH_E}{K'_u S}$

$$\text{where} \quad K'_u = C_d \left(1 + \sqrt{1 + \dfrac{\lambda'}{C_d}} \right)$$

$$C_d = 0.75 + 0.14 \left(\dfrac{W_p}{W_R} \right)$$

$$\lambda' = \left(\dfrac{EH_E L}{A_p E_p S^2} \right)$$

Gates's formula (Gates, 1957)

$Q_u = a\sqrt{EH_E}(b - \log S)$
If Q_u is in kips, then S is in in., $a = 27$, $b = 1$, and H_E is in kip-ft.
If Q_u is in kN, then S is in mm, $a = 104.5$, $b = 2.4$, and H_E is in kN-m.
$E = 0.75$ for drop hammer; $E = 0.85$ for all other hammers
Use a factor of safety of 3.

Navy-McKay formula

$$Q_u = \dfrac{EH_E}{S \left(1 + 0.3 \dfrac{W_p}{W_R} \right)}$$

Use a factor of safety of 6.

The maximum stress developed on a pile during the driving operation can be estimated from the pile-driving formulas presented in Table 11.15. To illustrate, we use the modified EN formula:

$$Q_u = \frac{EW_R h}{S + C} \frac{W_R + n^2 W_p}{W_R + W_p}$$

In this equation, S is the average penetration per hammer blow, which can also be expressed as

$$S = \frac{1}{N} \tag{11.109}$$

where

S is in inches
N = number of hammer blows per inch of penetration

Thus,

$$Q_u = \frac{EW_R h}{(1/N) + 0.1} \frac{W_R + n^2 W_p}{W_R + W_p} \tag{11.110}$$

Different values of N may be assumed for a given hammer and pile, and Q_u may be calculated. The driving stress Q_u/A_p can then be calculated for each value of N.

This procedure can be demonstrated with a set of numerical values. Suppose that a prestressed concrete pile 80 ft in length has to be driven by a hammer. The pile sides measure 10 in. From Table 11.3b, for this pile,

$$A_p = 100 \text{ in}^2$$

The weight of the pile is

$$A_p L \gamma_c = \left(\frac{100 \text{ in}^2}{144} \right) (80 \text{ ft}) (150 \text{ lb/ft}^3) = 8.33 \text{ kip}$$

If the weight of the cap is 0.67 kip, then

$$W_p = 8.33 + 0.67 = 9 \text{ kip}$$

For the hammer, let

$$\text{Rated energy} = 19.2 \text{ kip-ft} = H_E = W_R h$$

$$\text{Weight of ram} = 5 \text{ kip}$$

Assume that the hammer efficiency is 0.85 and that $n = 0.35$. Substituting these values into Eq. (11.110) yields

$$Q_u = \left[\frac{(0.85)(19.2 \times 12)}{\dfrac{1}{N} + 0.1} \right] \left[\frac{5 + (0.35)^2(9)}{5 + 9} \right] = \frac{85.37}{\dfrac{1}{N} + 0.1} \text{ kip}$$

Now the following table can be prepared:

N	Q_u (kip)	A_p (in²)	Q_u/A_p (kip/in²)
0	0	100	0
2	142.3	100	1.42
4	243.9	100	2.44
6	320.1	100	3.20
8	379.4	100	3.79
10	426.9	100	4.27
12	465.7	100	4.66
20	569.1	100	5.69

Both the number of hammer blows per inch and the stress can be plotted in a graph, as shown in Figure 11.38. If such a curve is prepared, the number of blows per inch of pile penetration corresponding to the allowable pile-driving stress can easily be determined.

Actual driving stresses in wooden piles are limited to about $0.7 f_u$. Similarly, for concrete and steel piles, driving stresses are limited to about $0.6 f'_c$ and $0.85 f_y$, respectively.

In most cases, wooden piles are driven with a hammer energy of less than 60 kN-m (≈ 45 kip-ft). Driving resistances are limited mostly to 4 to 5 blows per inch of pile penetration. For concrete and steel piles, the usual values of N are 6 to 8 and 12 to 14, respectively.

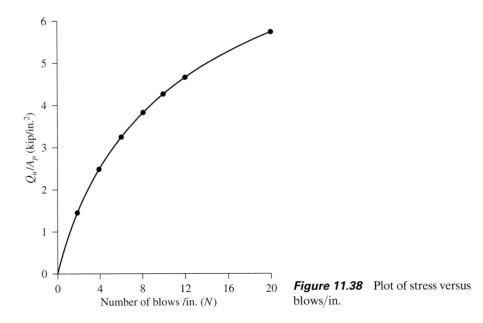

Figure 11.38 Plot of stress versus blows/in.

Example 11.9

A precast concrete pile 12 in. × 12 in. in cross section is driven by a hammer. Given

Maximum rated hammer energy = 30 kip-ft
Hammer efficiency = 0.8
Weight of ram = 7.5 kip
Pile length = 80 ft
Coefficient of restitution = 0.4
Weight of pile cap = 550 lb
$E_p = 3 \times 10^6$ kip/in^2
Number of blows for last 1 in. of penetration = 8

Estimate the allowable pile capacity by the

a. Modified EN formula (use FS = 6)
b. Danish formula (use FS = 4)
c. Gates formula (use FS = 3)

Solution
Part a

$$Q_u = \frac{EW_R h}{S + C} \frac{W_R + n^2 W_p}{W_R + W_p}$$

$$\text{Weight of pile + cap} = \left(\frac{12}{12} \times \frac{12}{12} \times 80\right)(150 \text{ lb/ft}^3) + 550$$

$$= 12{,}550 \text{ lb} = 12.55 \text{ kip}$$

Given: $W_R h = 30$ kip-ft.

$$Q_u = \frac{(0.8)(30 \times 12 \text{ kip-in.})}{\frac{1}{8} + 0.1} \times \frac{7.5 + (0.4)^2(12.55)}{7.5 + 12.55} = 607 \text{ kip}$$

$$Q_{\text{all}} = \frac{Q_u}{\text{FS}} = \frac{607}{6} \approx \mathbf{101 \text{ kip}}$$

Part b

$$Q_u = \frac{EH_E}{S + \sqrt{\dfrac{EH_E L}{2A_p E_p}}}$$

Use $E_p = 3 \times 10^6$ lb/in^2.

$$\sqrt{\frac{EH_E L}{2A_p E_p}} = \sqrt{\frac{(0.8)(30 \times 12)(80 \times 12)}{2(12 \times 12)\left(\dfrac{3 \times 10^6}{1000} \text{ kip/in}^2\right)}} = 0.566 \text{ in.}$$

$$Q_u = \frac{(0.8)(30 \times 12)}{\frac{1}{8} + 0.566} \approx 417 \text{ kip}$$

$$Q_{\text{all}} = \frac{417}{4} \approx \mathbf{104 \text{ kip}}$$

Part c

$$Q_u = a\sqrt{EH_E}(b - \log S) = 27\sqrt{(0.8)(30)}[1 - \log\left(\tfrac{1}{8}\right)] \approx 252 \text{ kip}$$

$$Q_{\text{all}} = \frac{252}{3} = \mathbf{84 \text{ kip}}$$ ∎

11.20   *Pile Capacity For Vibration-Driven Piles*

The principles of vibratory pile drivers (Figure 11.7e) were discussed briefly in Section 11.4. As mentioned there, the driver essentially consists of two counterrotating weights. The amplitude of the centrifugal driving force generated by a vibratory hammer can be given as

$$F_c = me\omega^2 \tag{11.111}$$

where

m = total eccentric rotating mass
e = distance between the center of each rotating mass and the center of rotation
ω = operating circular frequency

Vibratory hammers typically include an isolated *bias weight* that can range from 4 to 40 kN. The bias weight is isolated from oscillation by springs, so it acts as a net downward load helping the driving efficiency by increasing the penetration rate of the pile.

The use of vibratory pile drivers began in the early 1930s. Installing piles with vibratory drivers produces less noise and damage to the pile, compared with impact driving. However, because of a limited understanding of the relationships between the load, the rate of penetration, and the bearing capacity of piles, this method has not gained popularity in the United States.

Vibratory pile drivers are patented. Some examples are the Bodine Resonant Driver (BRD), the Vibro Driver of the McKiernan-Terry Corporation, and the Vibro Driver of the L. B. Foster Company. Davisson (1970) provided a relationship for estimating the ultimate pile capacity in granular soil:

In SI units,

$$Q_u(\text{kN}) = \frac{0.746(H_p) + 98(v_p\,\text{m/s})}{(v_p\,\text{m/s}) + (S_L\,\text{m/cycle})(f\,\text{Hz})} \tag{11.112}$$

In English units,

$$Q_u(\text{lb}) = \frac{550(H_p) + 22{,}000(v_p\,\text{ft/s})}{(v_p\,\text{ft/s}) + (S_L\,\text{ft/cycle})(f\,\text{Hz})} \tag{11.113}$$

where

H_p = horsepower delivered to the pile
v_p = final rate of pile penetration
S_L = loss factor
f = frequency, in Hz

The loss factor S_L for various types of granular soils is as follows (Bowles, 1996):

Closed-End Pipe Piles

- Loose sand: 0.244×10^{-3} m/cycle (0.0008 ft/cycle)
- Medium dense sand: 0.762×10^{-3} m/cycle (0.0025 ft/cycle)
- Dense sand: 2.438×10^{-3} m/cycle (0.008 ft/cycle)

H-Piles

- Loose sand: -0.213×10^{-3} m/cycle (-0.0007 ft/cycle)
- Medium dense sand: 0.762×10^{-3} m/cycle (0.0025 ft/cycle)
- Dense sand: 2.134×10^{-3} m/cycle (0.007 ft/cycle)

In 2000, Feng and Deschamps provided the following relationship for the ultimate capacity of vibrodriven piles in granular soil:

$$Q_u = \frac{3.6(F_c + 11W_B)}{1 + 1.8 \times 10^{10}\dfrac{v_p}{c}\sqrt{\text{OCR}}}\frac{L_E}{L} \tag{11.114}$$

Here,

F_c = centrifugal force
W_B = bias weight
v_p = final rate of pile penetration
c = speed of light $[1.8 \times 10^{10} \text{ m/min } (5.91 \times 10^{10} \text{ ft/min})]$
OCR = overconsolidation ratio
L_E = embedded length of pile
L = pile length

Example 11.10

Consider a 20-m-long steel pile driven by a Bodine Resonant Driver (Section HP 310×125) in a medium dense sand. If $H_p = 350$ horsepower, $v_p = 0.0016$ m/s, and $f = 115$ Hz, calculate the ultimate pile capacity, Q_u.

Solution
From Eq. (11.112),

$$Q_u = \frac{0.746 H_p + 98 v_p}{v_p + S_L f}$$

For an HP pile in medium dense sand, $S_L \approx 0.762 \times 10^{-3}$ m/cycle. So

$$Q_u = \frac{(0.746)(350) + (98)(0.0016)}{0.0016 + (0.762 \times 10^{-3})(115)} = \textbf{2928 kN}$$ ∎

11.21 *Negative Skin Friction*

Negative skin friction is a downward drag force exerted on a pile by the soil surrounding it. Such a force can exist under the following conditions, among others:

1. If a fill of clay soil is placed over a granular soil layer into which a pile is driven, the fill will gradually consolidate. The consolidation process will exert a downward drag force on the pile (see Figure 11.39a) during the period of consolidation.
2. If a fill of granular soil is placed over a layer of soft clay, as shown in Figure 11.39b, it will induce the process of consolidation in the clay layer and thus exert a downward drag on the pile.
3. Lowering of the water table will increase the vertical effective stress on the soil at any depth, which will induce consolidation settlement in clay. If a pile is located in the clay layer, it will be subjected to a downward drag force.

In some cases, the downward drag force may be excessive and cause foundation failure. This section outlines two tentative methods for the calculation of negative skin friction.

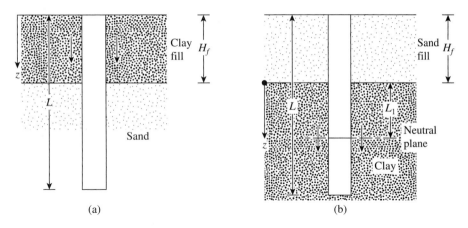

Figure 11.39 Negative skin friction

Clay Fill over Granular Soil (Figure 11.39a)

Similar to the β method presented in Section 11.13, the negative (downward) skin stress on the pile is

$$f_n = K'\sigma'_o \tan \delta' \tag{11.115}$$

where

K' = earth pressure coefficient = $K_o = 1 - \sin \phi'$
σ'_o = vertical effective stress at any depth $z = \gamma'_f z$
γ'_f = effective unit weight of fill
δ' = soil–pile friction angle ≈ 0.5–$0.7\phi'$

Hence, the total downward drag force on a pile is

$$Q_n = \int_0^{H_f} (pK'\gamma'_f \tan \delta')z\, dz = \frac{pK'\gamma'_f H_f^2 \tan \delta'}{2} \tag{11.116}$$

where H_f = height of the fill. If the fill is above the water table, the effective unit weight, γ'_f, should be replaced by the moist unit weight.

Granular Soil Fill over Clay (Figure 11.39b)

In this case, the evidence indicates that the negative skin stress on the pile may exist from $z = 0$ to $z = L_1$, which is referred to as the *neutral depth.* (See Vesic, 1977, pp. 25–26.) The neutral depth may be given as (Bowles, 1982)

$$L_1 = \frac{(L - H_f)}{L_1}\left[\frac{L - H_f}{2} + \frac{\gamma'_f H_f}{\gamma'}\right] - \frac{2\gamma'_f H_f}{\gamma'} \tag{11.117}$$

where γ'_f and γ' = effective unit weights of the fill and the underlying clay layer, respectively.

For end-bearing piles, the neutral depth may be assumed to be located at the pile tip (i.e., $L_1 = L - H_f$).

Once the value of L_1 is determined, the downward drag force is obtained in the following manner: The unit negative skin friction at any depth from $z = 0$ to $z = L_1$ is

$$f_n = K'\sigma'_o \tan \delta' \tag{11.118}$$

where

$K' = K_o = 1 - \sin \phi'$
$\sigma'_o = \gamma'_f H_f + \gamma' z$
$\delta' = 0.5\text{--}0.7\phi'$

$$Q_n = \int_0^{L_1} pf_n \, dz = \int_0^{L_1} pK'(\gamma'_f H_f + \gamma' z) \tan \delta' \, dz$$

$$= (pK'\gamma'_f H_f \tan \delta')L_1 + \frac{L_1^2 pK'\gamma' \tan \delta'}{2} \tag{11.119}$$

If the soil and the fill are above the water table, the effective unit weights should be replaced by moist unit weights. In some cases, the piles can be coated with bitumen in the downdrag zone to avoid this problem.

A limited number of case studies of negative skin friction is available in the literature. Bjerrum et al. (1969) reported monitoring the downdrag force on a test pile at Sorenga in the harbor of Oslo, Norway (noted as pile G in the original paper). The study of Bjerrum et al. (1969) was also discussed by Wong and Teh (1995) in terms of the pile being driven to bedrock at 40 m. Figure 11.40a shows the soil profile and the pile. Wong and Teh estimated the following quantities:

- *Fill:* Moist unit weight, $\gamma_f = 16 \text{ kN/m}^3$
  Saturated unit weight, $\gamma_{\text{sat}(f)} = 18.5 \text{ kN/m}^3$
  So

$$\gamma'_f = 18.5 - 9.81 = 8.69 \text{ kN/m}^3$$

  and

$$H_f = 13 \text{ m}$$

- *Clay:* $K' \tan \delta' \approx 0.22$
  Saturated effective unit weight, $\gamma' = 19 - 9.81 = 9.19 \text{ kN/m}^3$
- *Pile:* $L = 40 \text{ m}$
  Diameter, $D = 500 \text{ m}$

Thus, the maximum downdrag force on the pile can be estimated from Eq. (11.119). Since in this case the pile is a point bearing pile, the magnitude of $L_1 = 27 \text{ m}$, and

$$Q_n = (p)(K' \tan \delta')[\gamma_f \times 2 + (13 - 2)\gamma'_f](L_1) + \frac{L_1^2 p \gamma'(K' \tan \delta')}{2}$$

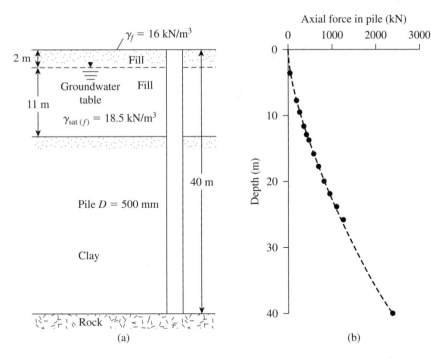

Figure 11.40   Negative skin friction on a pile in the harbor of Oslo, Norway [Based on Bjerrum et al. (1969) and Wong and Teh (1995)]

or

$$Q_n = (\pi \times 0.5)(0.22)[(16 \times 2) + (8.69 \times 11)](27) + \frac{(27)^2(\pi \times 0.5)(9.19)(0.22)}{2}$$

$$= 2348 \text{ kN}$$

The measured value of the maximum Q_n was about 2500 kN (Figure 11.40b), which is in good agreement with the calculated value.

Group Piles

11.22   Group Efficiency

In most cases, piles are used in groups, as shown in Figure 11.41, to transmit the structural load to the soil. A *pile cap* is constructed over *group piles*. The cap can be in contact with the ground, as in most cases (see Figure 11.41a), or well above the ground, as in the case of offshore platforms (see Figure 11.41b).

Determining the load-bearing capacity of group piles is extremely complicated and has not yet been fully resolved. When the piles are placed close to each other, a

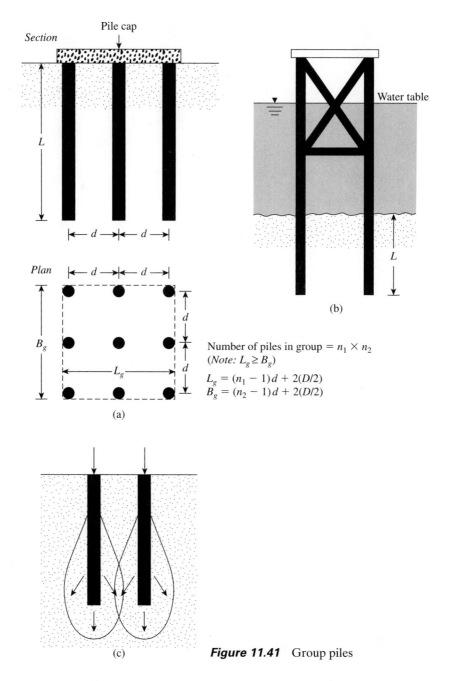

Number of piles in group $= n_1 \times n_2$
(*Note:* $L_g \geq B_g$)

$$L_g = (n_1 - 1)d + 2(D/2)$$
$$B_g = (n_2 - 1)d + 2(D/2)$$

Figure 11.41 Group piles

reasonable assumption is that the stresses transmitted by the piles to the soil will overlap (see Figure 11.41c), reducing the load-bearing capacity of the piles. Ideally, the piles in a group should be spaced so that the load-bearing capacity of the group is not less than the sum of the bearing capacity of the individual piles. In practice, the minimum center-to-center pile spacing, d, is $2.5D$ and, in ordinary situations, is actually about 3 to $3.5D$.

The efficiency of the load-bearing capacity of a group pile may be defined as

$$\eta = \frac{Q_{g(u)}}{\Sigma Q_u} \tag{11.120}$$

where

η = group efficiency
$Q_{g(u)}$ = ultimate load-bearing capacity of the group pile
Q_u = ultimate load-bearing capacity of each pile without the group effect

Many structural engineers use a simplified analysis to obtain the group efficiency for *friction piles*, particularly in sand. This type of analysis can be explained with the aid of Figure 11.41a. Depending on their spacing within the group, the piles may act in one of two ways: (1) as a *block*, with dimensions $L_g \times B_g \times L$, or (2) as *individual piles*. If the piles act as a block, the frictional capacity is $f_{av} p_g L \approx Q_{g(u)}$. [*Note:* p_g = perimeter of the cross section of block = $2(n_1 + n_2 - 2)d + 4D$, and f_{av} = average unit frictional resistance.] Similarly, for each pile acting individually, $Q_u \approx p L f_{av}$. (*Note:* p = perimeter of the cross section of each pile.) Thus,

$$\eta = \frac{Q_{g(u)}}{\Sigma Q_u} = \frac{f_{av}[2(n_1 + n_2 - 2)d + 4D]L}{n_1 n_2 p L f_{av}}$$
$$= \frac{2(n_1 + n_2 - 2)d + 4D}{p n_1 n_2} \tag{11.121}$$

Hence,

$$Q_{g(u)} = \left[\frac{2(n_1 + n_2 - 2)d + 4D}{p n_1 n_2}\right] \Sigma Q_u \tag{11.122}$$

From Eq. (11.122), if the center-to-center spacing d is large enough, $\eta > 1$. In that case, the piles will behave as individual piles. Thus, in practice, if $\eta < 1$, then

$$Q_{g(u)} = \eta \Sigma Q_u$$

and if $\eta \geq 1$, then

$$Q_{g(u)} = \Sigma Q_u$$

There are several other equations like Eq. (11.122) for calculating the group efficiency of friction piles. Some of these are given in Table 11.16.

Figure 11.42 shows the variation of the group efficiency η for a 3×3 group pile in sand (Kishida and Meyerhof, 1965). It can be seen that, for loose and medium sands, the magnitude of the group efficiency can be larger than unity. This is due primarily to the densification of sand surrounding the pile.

Table 11.16   Equations for Group Efficiency of Friction Piles

Name	Equation
Converse–Labarre equation	$\eta = 1 - \left[\dfrac{(n_1 - 1)n_2 + (n_2 - 1)n_1}{90 n_1 n_2} \right]\theta$ where $\theta(\deg) = \tan^{-1}(D/d)$
Los Angeles Group Action equation	$\eta = 1 - \dfrac{D}{\pi d n_1 n_2} [n_1(n_2 - 1)$ $+ \ n_2(n_1 - 1) + \sqrt{2}(n_1 - 1)(n_2 - 1)]$
Seiler–Keeney equation (Seiler and Keeney, 1944)	$\eta = \left\{ 1 - \left[\dfrac{11d}{7(d^2 - 1)} \right]\left[\dfrac{n_1 + n_2 - 2}{n_1 + n_2 - 1} \right] \right\} + \dfrac{0.3}{n_1 + n_2}$ where d is in ft

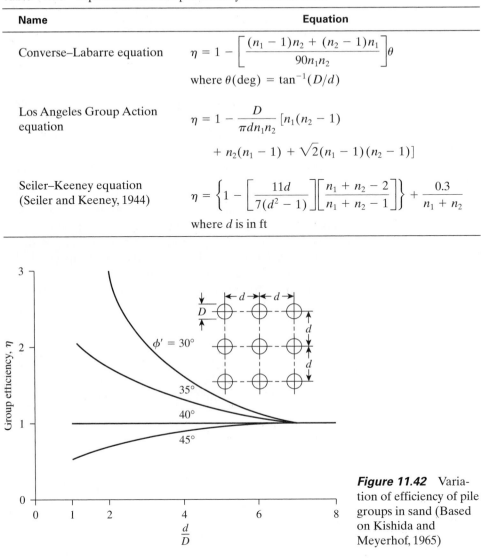

Figure 11.42 Variation of efficiency of pile groups in sand (Based on Kishida and Meyerhof, 1965)

11.23   *Ultimate Capacity of Group Piles in Saturated Clay*

Figure 11.43 shows a group pile in saturated clay. Using the figure, one can estimate the ultimate load-bearing capacity of group piles in the following manner:

Step 1.   Determine $\Sigma Q_u = n_1 n_2 (Q_p + Q_s)$. From Eq. (11.19),

$$Q_p = A_p [9 c_{u(p)}]$$

where $c_{u(p)}$ = undrained cohesion of the clay at the pile tip.

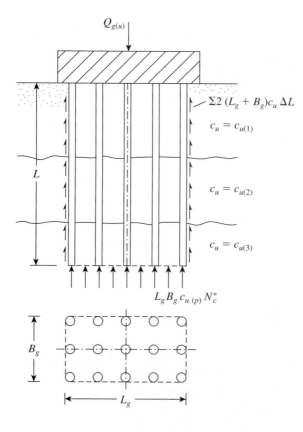

Figure 11.43 Ultimate capacity of group piles in clay

Also, from Eq. (11.50),

$$Q_s = \Sigma \, \alpha p c_u \Delta L$$

So,

$$\Sigma \, Q_u = n_1 n_2 [9 A_p c_{u(p)} + \Sigma \, \alpha p c_u \Delta L] \qquad (11.123)$$

Step 2. Determine the ultimate capacity by assuming that the piles in the group act as a block with dimensions $L_g \times B_g \times L$. The skin resistance of the block is

$$\Sigma \, p_g c_u \Delta L = \Sigma \, 2(L_g + B_g) c_u \Delta L$$

Calculate the point bearing capacity:

$$A_p q_p = A_p c_{u(p)} N_c^* = (L_g B_g) c_{u(p)} N_c^*$$

Obtain the value of the bearing capacity factor N_c^* from Figure 11.44. Thus, the ultimate load is

$$\Sigma \, Q_u = L_g B_g c_{u(p)} N_c^* + \Sigma \, 2(L_g + B_g) c_u \Delta L \qquad (11.124)$$

Step 3. Compare the values obtained from Eqs. (11.123) and (11.124). The *lower* of the two values is $Q_{g(u)}$.

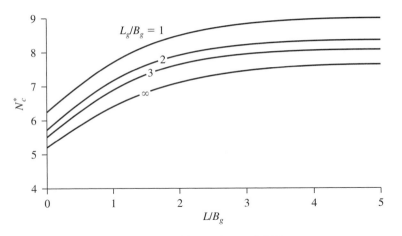

Figure 11.44   Variation of N_c^* with L_g/B_g and L/B_g

Example 11.11

The section of a 3×4 group pile in a layered saturated clay is shown in Figure 11.45. The piles are square in cross section (14 in. $\times$ 14 in.). The center-to-center spacing, d, of the piles is 35 in. Determine the allowable load-bearing capacity of the pile group. Use FS = 4. Note that the groundwater table coincides with the ground surface.

Solution
From Eq. (11.123),

$$\Sigma Q_u = n_1 n_2 \left[9A_p c_{u(p)} + \alpha_1 p c_{u(1)} L_1 + \alpha_2 p c_{u(2)} L_2 \right]$$

From Figure 11.45, $c_{u(1)} = 1050 \text{ lb/ft}^2$ and $c_{u(2)} = 1775 \text{ lb/ft}^2$.

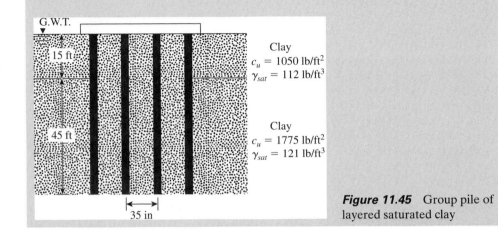

Figure 11.45   Group pile of layered saturated clay

In the top layer, the average value of effective stress is

$$\bar{\sigma}_o' = \left(\frac{15}{2}\right)(112 - 62.4) = 372 \text{ lb/ft}^2$$

Similarly, for the bottom layer,

$$\bar{\sigma}_o' = (15)(112 - 62.4) + \frac{45}{2}(121 - 62.4) = 2062.5 \text{ lb/ft}^2$$

Hence, for the top layer,

$$\frac{c_u}{\bar{\sigma}_o'} = \frac{1050}{372} = 2.82$$

From Figure 11.23, $\alpha_1 \approx 0.4$. Similarly, for the bottom layer,

$$\frac{c_u}{\bar{\sigma}_o'} = \frac{1775}{2062.5} = 0.86$$

From Figure 11.23, $\alpha_2 \approx 0.54$.

$$\Sigma Q_u = \frac{(3)(4)}{1000}\left[\begin{array}{c}(9)\left(\frac{14}{12}\right)^2(1775) + (0.4)\left(4 \times \frac{14}{12}\right)(1050)(15) \\ + (0.54)\left(4 \times \frac{14}{12}\right)(1775)(45)\end{array}\right] = 3029 \text{ kip}$$

For piles acting as a group.

$$L_g = (3)(35) + 14 = 119 \text{ in.} = 9.92 \text{ ft}$$
$$B_g = (2)(35) + 14 = 84 \text{ in.} = 7 \text{ ft}$$
$$\frac{L_g}{B_g} = \frac{9.92}{7} = 1.42$$
$$\frac{L}{B_g} = \frac{60}{7} = 8.57$$

From Figure 11.44, $N_c^* = 8.75$. From Eq. (11.124),

$$\Sigma Q_u = L_g B_g c_{u(p)} N_c^* + \Sigma 2(L_g + B_g) c_u \, \Delta L$$
$$= (9.92)(7)(1775)(8.75) + (2)(9.92 + 7)[(1050)(15) + (1775)(45)]$$
$$= 4313,000 \text{ lb} = 4313 \text{ kip}$$

Hence, $\Sigma Q_u = 3029$ kip.

$$\Sigma Q_{\text{all}} = \frac{3703}{\text{FS}} = \frac{3029}{4} \approx \textbf{757 kip} \qquad \blacksquare$$

11.24 Elastic Settlement of Group Piles

In general, the settlement of a group pile under a similar working load per pile increases with the width of the group (B_g) and the center-to-center spacing of the piles (d). Several investigations relating to the settlement of group piles have been reported in the literature, with widely varying results. The simplest relation for the settlement of group piles was given by Vesic (1969), namely,

$$S_{g(e)} = \sqrt{\frac{B_g}{D}} s_e \qquad (11.125)$$

where

$S_{g(e)}$ = elastic settlement of group piles
B_g = width of group pile section
D = width or diameter of each pile in the group
s_e = elastic settlement of each pile at comparable working load (see Section 11.17)

For group piles in sand and gravel, for elastic settlement, Meyerhof (1976) suggested the empirical relation

$$S_{g(e)}(\text{in.}) = \frac{2q\sqrt{B_g}I}{N_{60}} \qquad (11.126)$$

where

$$q = Q_g/(L_g B_g)\,(\text{in U.S. ton/ft}^2) \qquad (11.127)$$

and

L_g and B_g = length and width of the group pile section, respectively (ft)
N_{60} = average standard penetration number within seat of settlement ($\approx B_g$ deep below the tip of the piles)
I = influence factor = $1 - L/8B_g \geqslant 0.5$ $\qquad (11.128)$
L = length of embedment of piles

In SI units,

$$S_{g(e)}(\text{mm}) = \frac{0.96q\sqrt{B_g}I}{N_{60}} \qquad (11.129)$$

where q is in kN/m^2 and B_g and L_g are in m, and

$$I = 1 - \frac{L(m)}{8B_g(m)}$$

Similarly, the group pile settlement is related to the cone penetration resistance by the formula

$$S_{g(e)} = \frac{qB_gI}{2q_c} \tag{11.130}$$

where q_c = average cone penetration resistance within the seat of settlement. (Note that, in Eq. (11.130), all quantities are expressed in consistent units.)

11.25   *Consolidation Settlement of Group Piles*

The consolidation settlement of a group pile in clay can be estimated by using the 2:1 stress distribution method. The calculation involves the following steps (see Figure 11.46):

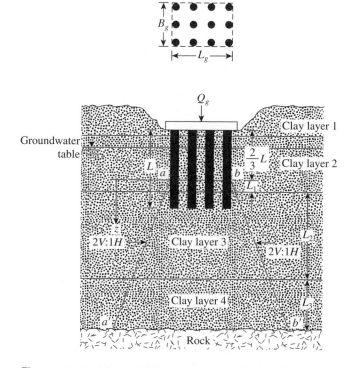

Figure 11.46   Consolidation settlement of group piles

Step 1. Let the depth of embedment of the piles be L. The group is subjected to a total load of Q_g. If the pile cap is below the original ground surface, Q_g equals the total load of the superstructure on the piles, minus the effective weight of soil above the group piles removed by excavation.

Step 2. Assume that the load Q_g is transmitted to the soil beginning at a depth of $2L/3$ from the top of the pile, as shown in the figure. The load Q_g spreads out along two vertical lines to one horizontal line from this depth. Lines aa' and bb' are the two 2:1 lines.

Step 3. Calculate the increase in effective stress caused at the middle of each soil layer by the load Q_g. The formula is

$$\Delta\sigma_i' = \frac{Q_g}{(B_g + z_i)(L_g + z_i)} \tag{11.131}$$

where

$$\Delta\sigma_i' = \text{increase in effective stress at the middle of layer } i$$
$$L_g, B_g = \text{length and width, respectively of the planned group piles}$$
$$z_i = \text{distance from } z = 0 \text{ to the middle of the clay layer } i$$

For example, in Figure 11.46, for layer 2, $z_i = L_1/2$; for layer 3, $z_i = L_1 + L_2/2$; and for layer 4, $z_i = L_1 + L_2 + L_3/2$. Note, however, that there will be no increase in stress in clay layer 1, because it is above the horizontal plane ($z = 0$) from which the stress distribution to the soil starts.

Step 4. Calculate the consolidation settlement of each layer caused by the increased stress. The formula is

$$\Delta s_{c(i)} = \left[\frac{\Delta e_{(i)}}{1 + e_{o(i)}} \right] H_i \tag{11.132}$$

where

$$\Delta s_{c(i)} = \text{consolidation settlement of layer } i$$
$$\Delta e_{(i)} = \text{change of void ratio caused by the increase in stress in layer } i$$
$$e_{o(i)} = \text{initial void ratio of layer } i \text{ (before construction)}$$
$$H_i = \text{thickness of layer } i \text{ (}Note: \text{ In Figure 11.46, for layer 2, } H_i = L_1;$$
$$\text{for layer 3, } H_i = L_2; \text{ and for layer 4, } H_i = L_3.)$$

Relationships involving $\Delta e_{(i)}$ are given in Chapter 1.

Step 5. The total consolidation settlement of the group piles is then

$$\Delta s_{c(g)} = \Sigma \Delta s_{c(i)} \tag{11.133}$$

Note that the consolidation settlement of piles may be initiated by fills placed nearby, adjacent floor loads, or the lowering of water tables.

Example 11.12

A group pile in clay is shown in Figure 11.47. Determine the consolidation settlement of the pile groups. All clays are normally consolidated.

Solution

The stress distribution pattern is shown in Figure 11.47. Hence

$$\Delta\sigma'_{(1)} = \frac{Q_g}{(L_g + z_1)(B_g + z_1)} = \frac{(500)(1000)}{\left(9 + \dfrac{21}{2}\right)\left(6 + \dfrac{21}{2}\right)} = 1554 \text{ lb/ft}^2$$

$$\Delta\sigma'_{(2)} = \frac{(500)(1000)}{(9 + 27)(6 + 27)} = 421 \text{ lb/ft}^2$$

$$\Delta\sigma'_{(3)} = \frac{(500)(1000)}{(9 + 36)(6 + 36)} = 265 \text{ lb/ft}^2$$

$$\Delta s_{c(1)} = \frac{C_{c(1)}H_1}{1 + e_{o(1)}} \log\left[\frac{\sigma'_{o(1)} + \Delta\sigma'_{(1)}}{\sigma'_{o(1)}}\right]$$

$$\sigma'_{o(1)} = (6)(105) + \left(27 + \frac{21}{2}\right)(115 - 62.4) = 2603 \text{ lb/ft}^2$$

$$\Delta s_{c(1)} = \frac{(0.3)(21)}{1 + 0.82} \log\left(\frac{2603 + 1554}{2603}\right) = 0.703 \text{ ft} = 8.45 \text{ in.}$$

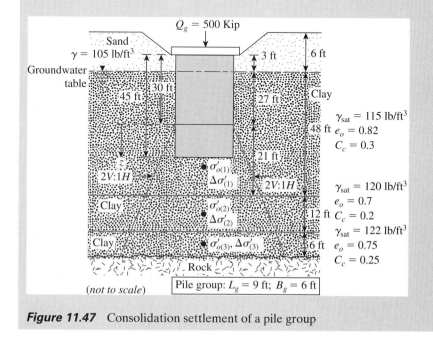

Figure 11.47 Consolidation settlement of a pile group

$$\Delta s_{c(2)} = \frac{C_{c(2)} H_2}{1 + e_{o(2)}} \log\left[\frac{\sigma'_{o(2)} + \Delta\sigma'_{(2)}}{\sigma'_{o(2)}}\right]$$

$$\sigma'_{o(2)} = (6)(105) + (27 + 21)(115 - 62.4) + (6)(120 - 62.4) = 3500 \text{ lb/ft}^2$$

$$\Delta s_{c(2)} = \frac{(0.2)(12)}{1 + 0.7} \log\left(\frac{3500 + 421}{3500}\right) = 0.07 \text{ ft} = 0.84 \text{ in.}$$

$$\sigma'_{o(3)} = (6)(105) + (48)(115 - 62.4) + (12)(120 - 62.4)$$
$$+ (3)(122 - 62.4) = 4025 \text{ lb/ft}^2$$

$$\Delta s_{c(3)} = \frac{(0.25)(6)}{1 + 0.75} \log\left(\frac{4025 + 265}{4025}\right) = 0.024 \text{ ft} \approx 0.29 \text{ in.}$$

Total settlement, $\Delta s_{c(g)} = 8.45 + 0.84 + 0.29 = \textbf{9.58 in.}$ ∎

11.26   Piles in Rock

For point bearing piles resting on rock, most building codes specify that $Q_{g(u)} = \Sigma Q_u$, provided that the minimum center-to-center spacing of the piles is $D + 300$ mm. For H-piles and piles with square cross sections, the magnitude of D is equal to the diagonal dimension of the cross section of the pile.

Problems

11.1 A prestressed concrete pile is 20 m long. The cross section of the pile is 460 mm × 460 mm. the pile is fully embedded in sand. Given for the sand: $\gamma = 18.6$ kN/m^3 and $\phi' = 30°$. Estimate the ultimate point load, Q_p, using
  a. Meyerhof's method
  b. Vesic's method (use $I_r = I_{rr} = 75$)
  c. Janbu's method (use $\eta' = 90°$)

11.2 Refer to Problem 11.1. Calculate the total frictional resistance for the pile. Use Eqs. (11.14), (11.37), (11.38), and (11.39). Given: $K = 1.5$ and $\delta' = 0.6\phi'$.

11.3 Refer to Problem 11.1. Calculate the allowable load-carrying capacity (that is, Q_p and Q_s) using Coyle and Castillo's method. Use FS = 4.

11.4 Redo Problem 11.1 with the following: length of pile = 80 ft, pile cross section = 12 in. × 12 in., $\gamma = 115$ lb/ft^3, and $\phi' = 35°$.

11.5 Refer to Problem 11.4. Calculate the friction resistance for the pile using Eqs. (11.14), (11.37), (11.38), and (11.39). Given: $K = 1.35$ and $\delta' = 0.75\ \phi'$.

11.6 Refer to Problem 11.4. Calculate the allowable load-carrying capacity of the pile tip (that is, Q_p) using Coyle and Castillo's method. Use FS = 4.

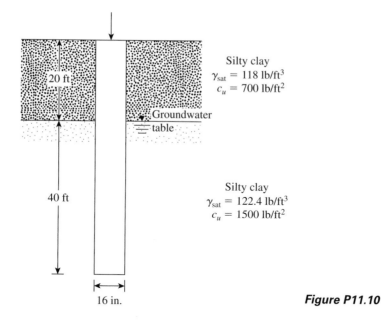

Silty clay
$\gamma_{sat} = 118$ lb/ft³
$c_u = 700$ lb/ft²

Groundwater table

Silty clay
$\gamma_{sat} = 122.4$ lb/ft³
$c_u = 1500$ lb/ft²

20 ft

40 ft

16 in.

Figure P11.10

11.7 A concrete pile 20 m long having a cross section of 381 mm × 381 mm is fully embedded in a saturated clay layer for which $\gamma_{sat} = 18.5$ kN/m³, $\phi = 0$, and $c_u = 70$ kN/m². Assume that the water table lies below the tip of the pile. Determine the allowable load that the pile can carry. (Let FS = 3.) Use the α method to estimate the skin friction.

11.8 Redo Problem 11.7 using the λ method for estimating the skin friction.

11.9 A concrete pile 60 ft long having a cross section of 15 in. × 15 in. is fully embedded in a saturated clay layer. For the clay, given: $\gamma_{sat} = 122$ lb/ft³, $\phi = 0$, and $c_u = 1450$ lb/ft². Assume that the groundwater table is located below the tip of the pile. Determine the allowable load that the pile can carry (FS = 3). Use the λ method to estimate the skin resistance.

11.10 A concrete pile 16 in. × 16 in. in cross section is shown in Figure P11.10. Calculate the ultimate skin friction resistance by using the
  a. α method
  b. λ method
  c. β method
  Use $\phi'_R = 20°$ for all clays, which are normally consolidated.

11.11 A steel pile (H-section; HP14 × 102; see Table 11.1b) is driven into a layer of sandstone. The length of the pile is 62 ft. Following are the properties of the sandstone: unconfined compression strength = $q_{u(lab)} = 11,400$ lb/in.² and angle of friction = 36°. Using a factor of safety of 3, estimate the allowable point load that can be carried by the pile. Use $[q_{u(design)} = q_{u(lab)}/5]$.

11.12 A concrete pile is 50 ft long and has a cross section of 16 in. × 16 in. The pile is embedded in a sand having $\gamma = 117$ lb/ft³ and $\phi' = 37°$. The allowable working load is 180 kip. If 110 kip are contributed by the frictional resistance and 70 kip are from the point load, determine the elastic settlement of the pile. Given: $E_p = 3 \times 10^6$ lb/in.^{2}, $E_s = 5 \times 10^3$ lb/in.^{2}, $\mu_s = 0.38$, and $\xi = 0.57$ [Eq. (11.63)].

11.13 Solve Problem 11.12 with the following: length of pile = 12 m, pile cross section = 0.305 m $\times$ 0.305 m, allowable working load = 338 kN, contribution of frictional resistance to working load = 240 kN, $E_p = 21 \times 10^6$ kN/m^2, $E_s = 30,000$ kN/m^2, $\mu_s = 0.3$, and $\xi = 0.6$ [Eq. (11.63)].

11.14 A 30-m long concrete pile is 305 mm $\times$ 305 mm in cross section and is fully embedded in a sand deposit. If $n_h = 9200$ kN/m^2, the moment at ground level, $M_g = 0$, the allowable displacement of pile head = 12 mm; $E_p = 21 \times 10^6$ kN/m^2; and $F_{Y\,(\text{pile})} = 21,000$ kN/m^2, calculate the allowable lateral load, Q_g, at the ground level. Use the elastic solution method.

11.15 Solve Problem 11.14 by Brom's method. Assume that the pile is flexible and free headed. Let the soil unit weight, $\gamma = 16$ kN/m^3; the soil friction angle, $\phi' = 30°$; and the yield stress of the pile material, $F_Y = 21$ MN/m^2.

11.16 A steel H-pile (section HP13 $\times$ 100) is driven by a hammer. The maximum rated hammer energy is 40 kip-ft, the weight of the ram is 12 kip, and the length of the pile is 90 ft. Also, we have coefficient of restitution = 0.35, weight of the pile cap = 2.4 kip, hammer efficiency = 0.85, number of blows for the last inch of penetration = 10, and $E_p = 30 \times 10^6$ lb/in.^{2}. Estimate the pile capacity using Eq. (11.108). Take FS = 6.

11.17 Solve Problem 11.16 using the modified EN formula. (See Table 11.15). Use FS = 4.

11.18 Solve Problem 11.16 using the Danish formula (See Table 11.15). Use FS = 3.

11.19 Figure 11.39a shows a pile. Let $L = 20$ m, D (pile diameter) = 450 mm, $H_f = 4$ m, $\gamma_{\text{fill}} = 17.5$ kN/m^3, and $\phi'_{\text{fill}} = 25°$. Determine the total downward drag force on the pile. Assume that the fill is located above the water table and that $\delta' = 0.5\phi'_{\text{fill}}$.

11.20 Redo Problem 11.19 assuming that the water table coincides with the top of the fill and that $\gamma_{\text{sat(fill)}} = 19.8$ kN/m^3. If the other quantities remain the same, what would be the downward drag force on the pile? Assume $\delta' = 0.5\phi'_{\text{fill}}$.

11.21 Refer to Figure 11.39b. Let $L = 60$ ft, $\gamma_{\text{fill}} = 105$ lb/ft^3, $\gamma_{\text{sat(clay)}} = 121$ lb/ft^3, $\phi'_{\text{clay}} = 28°$, $H_f = 12$ ft, and D (pile diameter) = 18 in. The water table coincides with the top of the clay layer. Determine the total downward drag force on the pile. Assume $\delta' = 0.5\phi'_{\text{clay}}$.

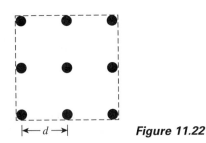

Figure 11.22

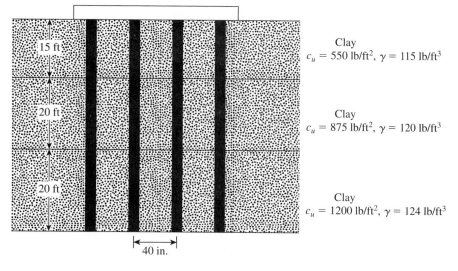

Clay
$c_u = 550$ lb/ft^2, $\gamma = 115$ lb/ft^3

Clay
$c_u = 875$ lb/ft^2, $\gamma = 120$ lb/ft^3

Clay
$c_u = 1200$ lb/ft^2, $\gamma = 124$ lb/ft^3

40 in.

Figure P11.26

11.22 The plan of a group pile (friction pile) in sand is shown in Figure P11.22. The piles are circular in cross section and have an outside diameter of 460 mm. The center-to-center spacings of the piles (d) are 920 mm. Determine the efficiency of pile group by
a. using Eq. (11.121)
b. using the Los Angles Group Action equation (Table 11.16).

11.23 Refer to Problem 11.22. If the center-to-center pile spacings are increased to 1200 mm, what will be the group efficiencies? (Solve parts a and b.)

11.24 The plan of a group pile is shown in Figure P11.22. Assume that the piles are embedded in a saturated homogeneous clay having a $c_u = 95.8$ kN/m^2. Given: diameter of piles (D) = 406 mm, center-to-center spacing of piles = 700 mm, and length of piles = 18.5 m. Find the allowable load-carrying capacity of the pile group. Use FS = 3. (*Note:* $\gamma_{sat} = 18$ kN/m^3 and the groundwater table is located at a depth of 23 m below the ground surface.)

11.25 Redo Problem 11.24 with the following: center-to-center spacing of piles = 30 in., length of piles = 45 ft, D = 12 in., c_u = 860 lb/ft^2, γ_{sat} = 122.4 lb/ft^3, and FS = 3. (*Note:* The ground water table coincides with the ground surface.)

11.26 The section of a 3 × 4 group pile in a layered saturated clay is shown in Figure P11.26. The piles are square in cross section (14 in. × 14 in.). The center-to-center spacing (d) of the piles is 40 in. Determine the allowable load-bearing capacity of the pile group. Use FS = 4.

11.27 Figure P11.27 shows a group pile in clay. Determine the consolidation settlement of the group. Use the 2:1 method of estimate the average effective stress in the clay layers.

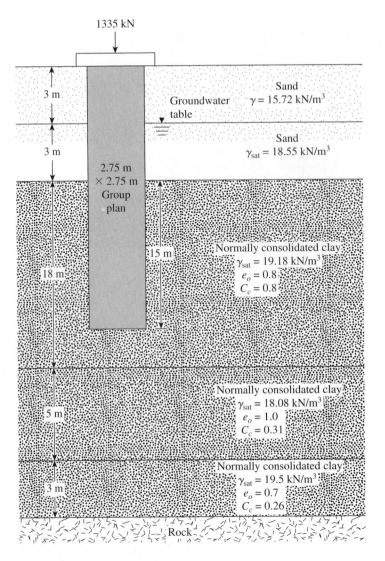

1335 kN

3 m

Sand
$\gamma = 15.72$ kN/m^3
Groundwater
table

3 m

Sand
$\gamma_{sat} = 18.55$ kN/m^3

2.75 m
× 2.75 m
Group
plan

15 m

18 m

Normally consolidated clay
$\gamma_{sat} = 19.18$ kN/m^3
$e_o = 0.8$
$C_c = 0.8$

5 m

Normally consolidated clay
$\gamma_{sat} = 18.08$ kN/m^3
$e_o = 1.0$
$C_c = 0.31$

3 m

Normally consolidated clay
$\gamma_{sat} = 19.5$ kN/m^3
$e_o = 0.7$
$C_c = 0.26$

Rock

Figure P11.27

References

American Petroleum Institute (API). *Recommended Practice of Planning, Designing, and Construction of Fixed Offshore Platforms,* Report No. API-RF-2A, Dallas, 115 pp.

American Society of Civil Engineers (1959). "Timber Piles and Construction Timbers," *Manual of Practice,* No. 17, American Society of Civil Engineers, New York.

American Society of Civil Engineers (1993). *Design of Pile Foundations (Technical Engineering and Design Guides as Adapted from the U.S. Army Corps of Engineers,* No. 1), American Society of Civil Engineers, New York.

Bjerrum, L., Johannessen, I. J., and Eide, O. (1969). "Reduction of Skin Friction on Steel Piles to Rock," *Proceedings, Seventh International Conference on Soil Mechanics and Foundation Engineering,* Mexico City, Vol. 2, pp. 27–34.

Bowles, J. E. (1982). *Foundation Analysis and Design,* McGraw-Hill, New York.

Bowles, J. E. (1996). *Foundation Analysis and Design,* McGraw-Hill, New York.

Briaud, J. L., Moore, B. H., and Mitchell, G. B. (1989). "Analysis of Pile Load Test at Lock and Dam 26," *Proceedings, Foundation Engineering: Current Principles and Practices, American Society of Civil Engineers,* Vol. 2, pp. 925–942.

Briaud, J. L., and Tucker, L. M. (1984). "Coefficient of Variation of *In-Situ* Tests in Sand," *Proceedings, Symposium on Probabilistic Characterization of Soil Properties,* Atlanta, pp. 119–139.

Briaud, J. L., Tucker, L., Lytton, R. L., and Coyle, H. M. (1985). *Behavior of Piles and Pile Groups,* Report No. FHWA/RD-83/038, Federal Highway Administration, Washington, DC.

Broms, B. B. (1965). "Design of Laterally Loaded Piles," *Journal of the Soil Mechanics and Foundations Division,* American Society of Civil Engineers, Vol. 91, No. SM3, pp. 79–99.

Bustamante, M., and Gianeselli, L. (1982). "Pile Bearing Capacity Prediction by Means of Static Penetrometer CPT," *Second European Symposium on Penetration Testing,* Vol. 2, pp. 493–500, Balkema, Amsterdam.

Coyle, H. M., and Castello, R. R. (1981). "New Design Correlations for Piles in Sand," *Journal of the Geotechnical Engineering Division,* American Society of Civil Engineers, Vol. 107, No. GT7, pp. 965–986.

Davisson, M. T. (1970). "BRD Vibratory Driving Formula," *Foundation Facts,* Vol. VI, No. 1, pp. 9–11.

Davisson, M. T., and Gill, H. L. (1963). "Laterally Loaded Piles in a Layered Soil System," *Journal of the Soil Mechanics and Foundations Division,* American Society of Civil Engineers, Vol. 89, No. SM3, pp. 63–94.

DeRuiter, J., and Beringen, F. L. (1979). "Pile Foundations for Large North Sea Structures," *Marine Geotechnology,* Vol. 3, No. 3, pp. 267–314.

Feng, Z., and Deschamps, R. J. (2000). "A Study of the Factors Influencing the Penetration and Capacity of Vibratory Driven Piles," *Soils and Foundations,* Vol. 40, No. 3, pp. 43–54.

Gates, M. (1957). "Empirical Formula for Predicting Pile Bearing Capacity." *Civil Engineering,* American Society of Civil Engineers, Vol. 27, No. 3, pp. 65–66.

Goodman, R. E. (1980). *Introduction to Rock Mechanics,* Wiley, New York.

Guang-Yu, Z. (1988). "Wave Equation Applications for Piles in Soft Ground," *Proceedings, Third International Conference on the Application of Stress-Wave Theory to Piles* (B. H. Fellenius, ed.), Ottawa, Ontario, Canada, pp. 831–836.

International Conference of Building Officials (1982). "Uniform Building Code," Whittier, CA.

Janbu, N. (1953). *An Energy Analysis of Pile Driving with the Use of Dimensionless Parameters,* Norwegian Geotechnical Institute, Oslo, Publication No. 3.

Janbu, N. (1976). "Static Bearing Capacity of Friction Piles," *Proceedings, Sixth European Conference on Soil Mechanics and Foundation Engineering,* Vol. 1.2, pp. 479–482.

Kishida, H., and Meyerhof, G. G. (1965). "Bearing Capacity of Pile Groups under Eccentric Loads in Sand," *Proceedings, Sixth International Conference on Soil Mechanics and Foundation Engineering,* Montreal, Vol. 2, pp. 270–274.

Matlock, H., and Reese, L. C. (1960). "Generalized Solution for Laterally Loaded Piles," *Journal of the Soil Mechanics and Foundations Division,* American Society of Civil Engineers, Vol. 86, No. SM5, Part I, pp. 63–91.

McClelland, B. (1974). "Design of Deep Penetration Piles for Ocean Structures," *Journal of the Geotechnical Engineering Division,* American Society of Civil Engineers, Vol. 100, No. GT7, pp. 709–747.

Meyerhof, G. G. (1976). "Bearing Capacity and Settlement of Pile Foundations," *Journal of the Geotechnical Engineering Division,* American Society of Civil Engineers, Vol. 102, No. GT3, pp. 197–228.

Meyerhof, G. G. (1995). "Behavior of Pile Foundations under Special Loading Conditions: 1994 R. M. Hardy Keynote Address," *Canadian Geotechnical Journal*, Vol. 32, No. 2, pp. 204–222.

Michigan State Highway Commission (1965). *A Performance Investigation of Pile Driving Hammers and Piles*, Lansing, MI, 338 pp.

Nottingham, L. C., and Schmertmann, J. H. (1975). *An Investigation of Pile Capacity Design Procedures*, Research Report No. D629, Department of Civil Engineering, University of Florida, Gainesville, FL.

Olson, R. E., and Flaate, K. S. (1967). "Pile Driving Formulas for Friction Piles in Sand," *Journal of the Soil Mechanics and Foundations Division*, American Society of Civil Engineers, Vol. 93, No. SM6, pp. 279–296.

Randolph, M. F., and Murphy, B. S. (1985). "Shaft Capacity of Driven Piles in Clay," *Proceedings, Offshore Technology Conference*, Vol. 1, Houston, pp. 371–378.

Schmertmann, J. H. (1978). *Guidelines for Cone Penetration Test: Performance and Design*, Report FHWA-TS-78-209, Federal Highway Administration, Washington, DC.

Seiler, J. F., and Keeney, W. D. (1944). "The Efficiency of Piles in Groups," *Wood Preserving News*, Vol. 22, No. 11 (November).

Skov, R., and Denver, H. (1988). "Time Dependence of Bearing Capacity of Piles," *Proceedings, Third International Conference on Application of Stress Wave Theory to Piles*, Ottawa, Canada, pp. 879–889.

Svinkin, M. (1996). Discussion on "Setup and Relaxation in Glacial Sand," *Journal of Geotechnical Engineering*, ASCE, Vol. 22, pp. 319–321.

Vesic, A. S. (1961). "Bending of Beams Resting on Isotropic Elastic Solids," *Journal of the Engineering Mechanics Division*, American Society of Civil Engineers, Vol. 87, No. EM2, pp. 35–53.

Vesic, A. S. (1970). "Tests on Instrumental Piles—Ogeechee River Site," *Journal of the Soil Mechanics and Foundations Division*, American Society of Civil Engineers, Vol. 96, No. SM2, pp. 561–584.

Vesic, A. S. (1977). *Design of Pile Foundations*, National Cooperative Highway Research Program Synthesis of Practice No. 42, Transportation Research Board, Washington, DC.

Vijayvergiya, V. N., and Focht, J. A., Jr. (1972). *A New Way to Predict Capacity of Piles in Clay*, Offshore Technology Conference Paper 1718, Fourth Offshore Technology Conference, Houston.

Wong, K. S., and Teh, C. I. (1995). "Negative Skin Friction on Piles in Layered Soil Deposit," *Journal of Geotechnical and Geoenvironmental Engineering*, American Society of Civil Engineers, Vol. 121, No. 6, pp. 457–465.

Woodward, R. J., Gardner, W. S., and Greer, D. M. (1972). *Drilled Pier Foundations*, McGraw-Hill, New York.

12

Drilled-Shaft Foundations

Introduction

The terms *caisson, pier, drilled shaft,* and *drilled pier* are often used interchangeably in foundation engineering; all refer to a *cast-in-place pile generally having a diameter of about 750 mm* ($\approx$2.5 ft) or more, with or without steel reinforcement and with or without an enlarged bottom. Sometimes the diameter can be as small as 305 mm ($\approx$1 ft).

To avoid confusion, we use the term *drilled shaft* for a hole drilled or excavated to the bottom of a structure's foundation and then filled with concrete. Depending on the soil conditions, casings may be used to prevent the soil around the hole from caving in during construction. The diameter of the shaft is usually large enough for a person to enter for inspection.

The use of drilled-shaft foundations has several advantages:

1. A single drilled shaft may be used instead of a group of piles and the pile cap.
2. Constructing drilled shafts in deposits of dense sand and gravel is easier than driving piles.
3. Drilled shafts may be constructed before grading operations are completed.
4. When piles are driven by a hammer, the ground vibration may cause damage to nearby structures. The use of drilled shafts avoids this problem.
5. Piles driven into clay soils may produce ground heaving and cause previously driven piles to move laterally. This does not occur during the construction of drilled shafts.
6. There is no hammer noise during the construction of drilled shafts; there is during pile driving.
7. Because the base of a drilled shaft can be enlarged, it provides great resistance to the uplifting load.
8. The surface over which the base of the drilled shaft is constructed can be visually inspected.
9. The construction of drilled shafts generally utilizes mobile equipment, which, under proper soil conditions, may prove to be more economical than methods of constructing pile foundations.
10. Drilled shafts have high resistance to lateral loads.

There are also a couple of drawbacks to the use of drilled-shaft construction. For one thing, the concreting operation may be delayed by bad weather and always needs close supervision. For another, as in the case of braced cuts, deep excavations for drilled shafts may induce substantial ground loss and damage to nearby structures.

12.2   *Types of Drilled Shafts*

Drilled shafts are classified according to the ways in which they are designed to transfer the structural load to the substratum. Figure 12.1a shows a drilled *straight shaft.* It extends through the upper layer(s) of poor soil, and its tip rests on a strong load-bearing soil layer or rock. The shaft can be cased with steel shell or pipe when required (as it is with cased, cast-in-place concrete piles; see Figure 11.4). For such shafts, the resistance to the applied load may develop from end bearing and also from side friction at the shaft perimeter and soil interface.

A *belled shaft* (see Figures 12.1b and c) consists of a straight shaft with a bell at the bottom, which rests on good bearing soil. The bell can be constructed in the shape of a dome (see Figure 12.1b), or it can be angled (see Figure 12.1c). For angled bells, the underreaming tools that are commercially available can make 30 to 45° angles with the vertical. For the majority of drilled shafts constructed in the United States, the entire load-carrying capacity is assigned to the end bearing only. However, under certain circumstances, the end-bearing capacity and the side friction are taken into account. In Europe, both the side frictional resistance and the end-bearing capacity are always taken into account.

Straight shafts can also be extended into an underlying rock layer. (See Figure 12.1d.) In the calculation of the load-bearing capacity of such shafts, the end bearing and the shear stress developed along the shaft perimeter and rock interface can be taken into account.

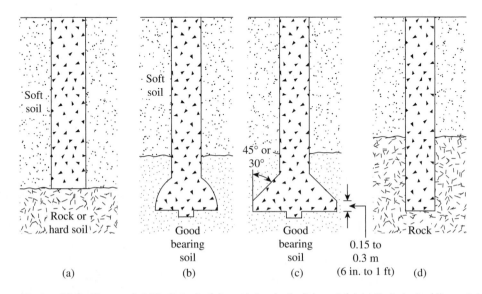

Figure 12.1   Types of drilled shaft: (a) straight shaft; (b) and (c) belled shaft; (d) straight shaft socketed into rock

Construction Procedures

The most common construction procedure used in the United States involves rotary drilling. There are three major types of construction methods: the dry method, the casing method, and the wet method.

Dry Method of Construction

This method is employed in soils and rocks that are above the water table and that will not cave in when the hole is drilled to its full depth. The sequence of construction, shown in Figure 12.2, is as follows:

Step 1. The excavation is completed (and belled if desired), using proper drilling tools, and the spoils from the hole are deposited nearby. (See Figure 12.2a.)

Step 2. Concrete is then poured into the cylindrical hole. (See Figure 12.2b.)

Step 3. If desired, a rebar cage is placed in the upper portion of the shaft. (See Figure 12.2c.)

Step 4. Concreting is then completed, and the drilled shaft will be as shown in Figure 12.2d.

Casing Method of Construction

This method is used in soils or rocks in which caving or excessive deformation is likely to occur when the borehole is excavated. The sequence of construction is shown in Figure 12.3 and may be explained as follows:

Step 1. The excavation procedure is initiated as in the case of the dry method of construction. (See Figure 12.3a.)

Step 2. When the caving soil is encountered, bentonite slurry is introduced into the borehole. (See Figure 12.3b.) Drilling is continued until the excavation goes past the caving soil and a layer of impermeable soil or rock is encountered.

Step 3. A casing is then introduced into the hole. (See Figure 12.3c.)

Step 4. The slurry is bailed out of the casing with a submersible pump. (See Figure 12.3d.)

Step 5. A smaller drill that can pass through the casing is introduced into the hole, and excavation continues. (See Figure 12.3e.)

Step 6. If needed, the base of the excavated hole can then be enlarged, using an underreamer. (See Figure 12.3f.)

Step 7. If reinforcing steel is needed, the rebar cage needs to extend the full length of the excavation. Concrete is then poured into the excavation and the casing is gradually pulled out. (See Figure 12.3g.)

Step 8. Figure 12.3h shows the completed drilled shaft.

Wet Method of Construction

This method is sometimes referred to as the *slurry displacement method*. Slurry is used to keep the borehole open during the entire depth of excavation. (See Figure 12.4.) Following are the steps involved in the wet method of construction:

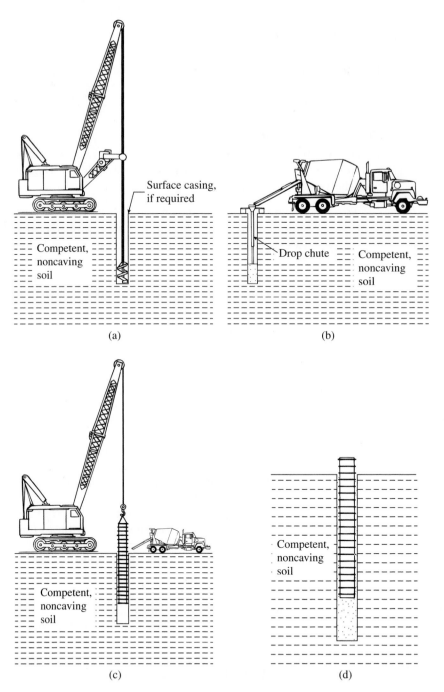

Figure 12.2 Dry method of construction: (a) initiating drilling; (b) starting concrete pour; (c) placing rebar cage; (d) completed shaft (After O'Neill and Reese, 1999)

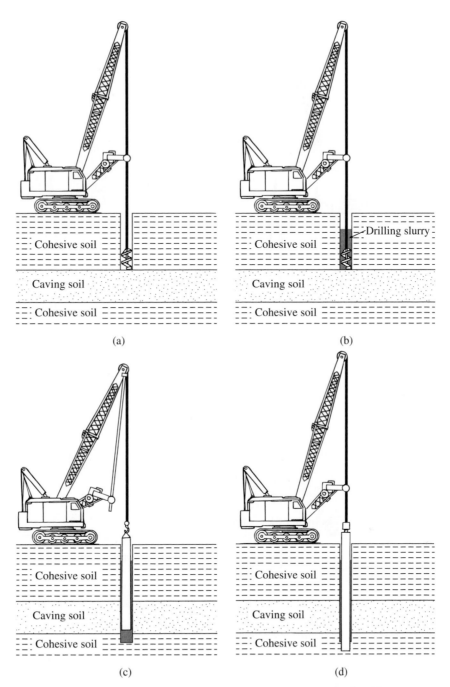

Figure 12.3 Casing method of construction: (a) initiating drilling; (b) drilling with slurry; (c) introducing casing; (d) casing is sealed and slurry is being removed from interior of casing; (e) drilling below casing; (f) underreaming; (g) removing casing; (h) completed shaft (After O'Neill and Reese, 1999)

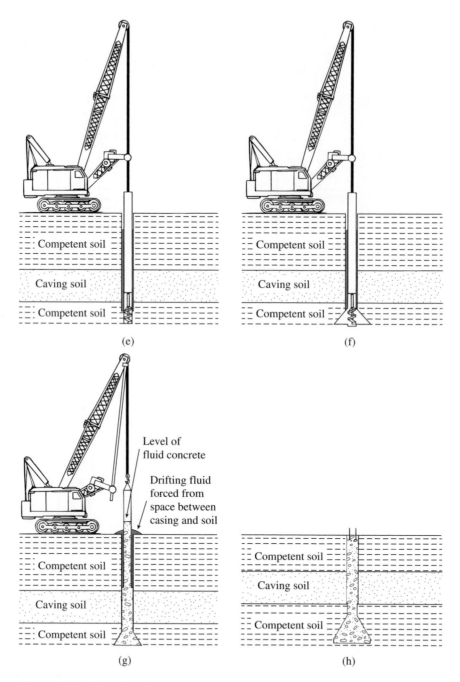

(e)

(f)

Level of
fluid concrete

Drifting fluid
forced from
space between
casing and soil

Competent soil

Caving soil

Competent soil

Competent soil

Caving soil

Competent soil

(g)

(h)

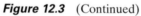

Figure 12.3   (Continued)

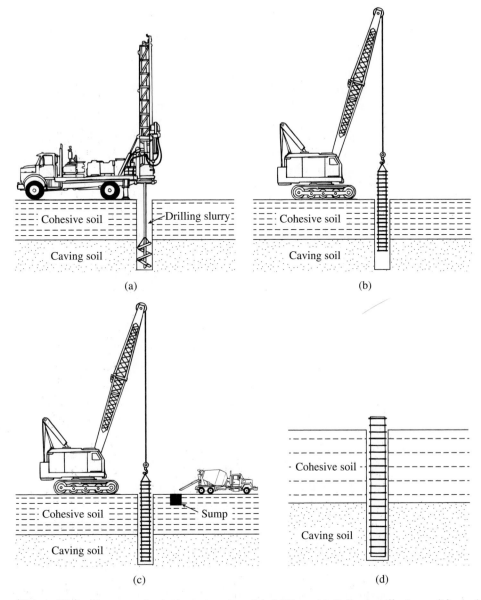

Figure 12.4 Slurry method of construction: (a) drilling to full depth with slurry; (b) placing rebar cage; (c) placing concrete; (d) completed shaft (After O'Neill and Reese, 1999)

Step 1. Excavation continues to full depth with slurry. (See Figure 12.4a.)

Step 2. If reinforcement is required, the rebar cage is placed in the slurry. (See Figure 12.4b.)

Step 3. Concrete that will displace the volume of slurry is then placed in the drill hole. (See Figure 12.4c.)

Step 4. Figure 12.4d shows the completed drilled shaft.

12.4   *Other Design Considerations*

For the design of ordinary drilled shafts without casings, a minimum amount of vertical steel reinforcement is always desirable. Minimum reinforcement is 1% of the gross cross-sectional area of the shaft. In California, a reinforcing cage having a length of about 3.65 m (12 ft) is used in the top part of the shaft, and no reinforcement is provided at the bottom. This procedure helps in the construction process, because the cage is placed after the concreting is almost complete.

For drilled shafts with nominal reinforcement, most building codes suggest using a design concrete strength, f_c, on the order of $f'_c/4$. Thus, the minimum shaft diameter becomes

$$f_c = 0.25f'_c = \frac{Q_w}{A_{gs}} = \frac{Q_w}{\frac{\pi}{4}D_s^2}$$

or

$$D_s = \sqrt{\frac{Q_w}{\left(\frac{\pi}{4}\right)(0.25)f'_c}} = 2.257\sqrt{\frac{Q_u}{f'_c}} \qquad (12.1)$$

where

D_s = diameter of the shaft
f'_c = 28-day concrete strength
Q_w = working load of the drilled shaft
A_{gs} = gross cross-sectional area of the shaft

If drilled shafts are likely to be subjected to tensile loads, reinforcement should be continued for the entire length of the shaft.

Concrete Mix Design

The concrete mix design for drilled shafts is not much different from that for any other concrete structure. When a reinforcing cage is used, consideration should be given to the ability of the concrete to flow through the reinforcement. In most cases, a concrete slump of about 15.0 mm (6 in.) is considered satisfactory. Also, the maximum size of the aggregate should be limited to about 20 mm (0.75 in.).

12.5   *Load Transfer Mechanism*

The load transfer mechanism from drilled shafts to soil is similar to that of piles, as described in Section 11.5. Figure 12.5 shows the load test results of a drilled shaft, that is, the load carried at the base, side, and the total load. It is important to note

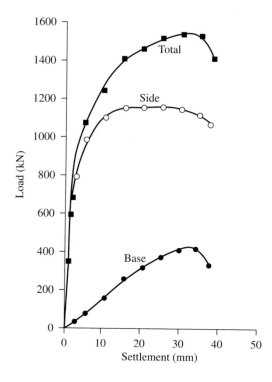

Figure 12.5 Load test on a drilled shaft

that the maximum load carried by the drilled shaft is about 1550 kN. The maximum load carried by side resistance is about 1125 kN, and this load is mobilized at a downward movement of about 15 mm (0.59 in.). However, a downward movement of about 35 mm ($\approx$1.38 in.) is required for full mobilization of the base resistance. This situation is similar to that observed in the case of piles.

12.6 *Estimation of Load-Bearing Capacity*

The ultimate load-bearing capacity of a drilled shaft (see Figure 12.6) is

$$Q_u = Q_p + Q_s \tag{12.2}$$

where

Q_u = ultimate load
Q_p = ultimate load-carrying capacity at the base
Q_s = frictional (skin) resistance

The ultimate base load Q_p can be expressed in a manner similar to the way it is expressed in the case of shallow foundations [Eq. (3.23)], or

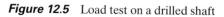

$$Q_p = A_p\left(c'N_cF_{cs}F_{cd}F_{cc} + q'N_qF_{qs}F_{qd}F_{qc} + \frac{1}{2}\gamma'N_\gamma F_{\gamma s}F_{\gamma d}F_{\gamma c}\right) \tag{12.3}$$

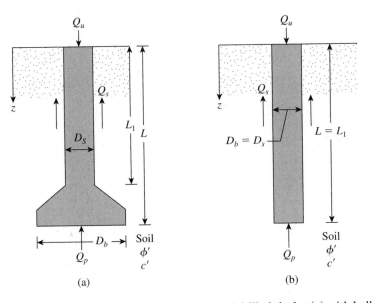

Figure 12.6   Ultimate bearing capacity of drilled shafts: (a) with bell and (b) straight shaft

where

$$c' = \text{cohesion}$$
$$N_c, N_q, N_\gamma = \text{bearing capacity factors (Table 3.3)}$$
$$F_{cs}, F_{qs}, F_{\gamma s} = \text{shape factors}$$
$$F_{cd}, F_{qd}, F_{\gamma d} = \text{depth factors}$$
$$F_{cc}, F_{qc}, F_{\gamma c} = \text{compressibility factors}$$
$$\gamma' = \text{effective unit weight of soil at the base of the shaft}$$
$$q' = \text{effective vertical stress at the base of the shaft}$$
$$A_p = \text{area of the base} = \frac{\pi}{4}D_b^2$$

In most instances, the last term (the one containing N_γ) is neglected, except in the case of a relatively short drilled shaft. With this assumption, the net load-carrying capacity at the base (i.e., the gross load minus the weight of the drilled shaft) may be approximated as

$$Q_{p(\text{net})} = A_p[c'N_cF_{cs}F_{cd}F_{cc} + q'(N_q - 1)F_{qs}F_{qd}F_{qc}] \tag{12.4}$$

Using Eqs. (3.27), (3.28), and (3.34), we obtain

$$F_{cs} = 1 + \frac{N_q}{N_c} \tag{12.5}$$

$$F_{qs} = 1 + \tan \phi' \tag{12.6}$$

$$F_{cd} = F_{qd} - \frac{1 - F_{qd}}{N_c \tan \phi'} \tag{12.7}$$

and

$$F_{qd} = 1 + 2 \tan \phi'(1 - \sin \phi')^2 \underbrace{\tan^{-1}\left(\frac{L}{D_b}\right)}_{\text{radians}} \tag{12.8}$$

According to Chen and Kulhawy (1994), F_{cc} and F_{qc} can be calculated in the following manner:

Step 1. Calculate the critical rigidity index as

$$I_{rc} = 0.5\exp\left[2.85\cot\left(45 - \frac{\phi'}{2}\right)\right] \tag{12.9}$$

Step 2. Calculate the reduced rigidity index as

$$I_{rr} = \frac{I_r}{1 + I_r\Delta} \tag{12.10}$$

where

$$I_r = \text{soil rigidity index}$$

$$= \frac{E_s}{2(1 + \mu_s)q' \tan \phi'} \tag{12.11}$$

in which

E_s = drained modulus of elasticity of soil
μ_s = drained Poisson's ratio of soil
Δ = volumetric strain within the plastic zone during loading

Step 3. If $I_{rr} \geq I_{rc}$ then

$$F_{cc} = F_{qc} = 1 \tag{12.12}$$

However, if $I_{rr} < I_{rc}$, then

$$F_{cc} = F_{qc} - \frac{1 - F_{qc}}{N_c \tan \phi'} \tag{12.13}$$

and

$$F_{qc} = \exp\left\{(-3.8 \tan \phi') + \left[\frac{(3.07 \sin \phi')(\log_{10} 2I_{rr})}{1 + \sin \phi'}\right]\right\} \tag{12.14}$$

The expression for the frictional, or skin, resistance Q_s is similar to that for piles; that is,

$$Q_s = \int_0^{L_1} pf \, dz \tag{12.15}$$

where

p = shaft perimeter = πD_s
f = unit frictional (or skin) resistance

The next two sections describe the procedures for obtaining the *ultimate* and *allowable* load-bearing capacities of drilled shafts in sand and saturated clay ($\phi = 0$).

12.7 Drilled Shafts in Granular Soil: Load-Bearing Capacity

For drilled shafts in sand, $c' = 0$; hence, Eq. (12.4) simplifies to

$$Q_{p(net)} = A_p[q'(N_q - 1)F_{qs}F_{qd}F_{qc}] \tag{12.16}$$

The value of N_q for a given soil friction angle ϕ' can be determined from Table 3.3. The shape factor F_{qs} and depth factor F_{qd} can be evaluated from Eqs. (12.6) and (12.8), respectively. To calculate the compressibility factor F_{qc}, Eqs. (12.9), (12.10), (12.11), (12.12), (12.13) and (12.14) will have to be used. The terms E_s, μ_s, and Δ can be estimated by the relationship (Chen and Kulhawy, 1994)

$$\frac{E_s}{p_a} = m \tag{12.17}$$

where

p_a = atmospheric pressure (≈ 100 kN/m^2 or 2000 lb/ft^2)
$m = \begin{cases} 100 \text{ to } 200 \text{ (loose soil)} \\ 200 \text{ to } 500 \text{ (medium dense soil)} \\ 500 \text{ to } 1000 \text{ (dense soil)} \end{cases}$
$\mu_s = 0.1 + 0.3\left(\dfrac{\phi' - 25}{20}\right)$ (for $25° \leqslant \phi' \leqslant 45°$) $\tag{12.18}$

and the formula

$$\Delta = 0.005\left(1 - \frac{\phi' - 25}{20}\right)\left(\frac{q'}{p_a}\right) \tag{12.19}$$

The magnitude of $Q_{p(net)}$ can also be reasonably estimated from a relationship based on the analysis of Berezantzev et al. (1961) that can be expressed as

$$Q_{p(net)} = A_p q'(\omega N_q^* - 1) \tag{12.20}$$

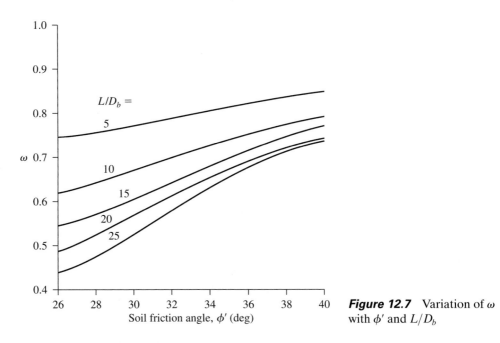

Figure 12.7 Variation of ω with ϕ' and L/D_b

where

$$N_q^* = \text{bearing capacity factor} = 0.21e^{0.17\phi'} \tag{12.21}$$
$$\omega = \text{correction factor} = f(L/D_b)$$

In Eq. (12.21), ϕ' is in degrees. The variation of ω with L/D_b is given in Figure 12.7.

The frictional resistance at ultimate load, Q_s, developed in a drilled shaft may be calculated from the relation given in Eq. (12.15), in which

$$p = \text{shaft perimeter} = \pi D_s$$

$$f = \text{unit frictional (or skin) resistance} = K\sigma_o' \tan \delta' \tag{12.22}$$

where

$K = \text{earth pressure coefficient} \approx K_o = 1 - \sin \phi'$
$\sigma_o' = \text{effective vertical stress at any depth } z$

Thus,

$$Q_s = \int_0^{L_1} pf\,dz = \pi D_s(1 - \sin \phi') \int_0^{L_1} \sigma_o' \tan \delta'\,dz \tag{12.23}$$

The value of σ_o' will increase to a depth of about $15D_s$ and will remain constant thereafter, as shown in Figure 11.18.

For cast-in-pile concrete and good construction techniques, a rough interface develops and, hence, δ'/ϕ' may be taken to be one. With poor slurry construction, $\delta'/\phi' \approx 0.7$ to 0.8.

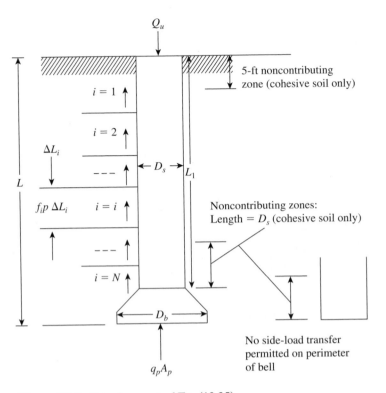

Figure 12.8 Development of Eq. (12.25)

An appropriate factor of safety should be applied to the ultimate load to obtain the net allowable load, or

$$Q_{\text{all(net)}} = \frac{Q_{p(\text{net})} + Q_s}{\text{FS}}$$ (12.24)

Load-Bearing Capacity Based on Settlement

On the basis of a database of 41 loading tests, Reese and O'Neill (1989) proposed a method for calculating the load-bearing capacity of drilled shafts that is based on settlement. The method is applicable to the following ranges:

1. Shaft diameter: D_s = 0.52 to 1.2 m (1.7 to 3.93 ft)
2. Bell depth: L = 4.7 to 30.5 m (15.4 to 100 ft)
3. Field standard penetration resistance: N_{60} = 5 to 60
4. Concrete slump = 100 to 225 mm (4 to 9 in.)

Reese and O'Neill's procedure (see Figure 12.8) gives

$$Q_{u(\text{net})} = \sum_{i=1}^{N} f_i p \Delta L_i + q_p A_p$$ (12.25)

where

f_i = ultimate unit shearing resistance in layer i
p = perimeter of the shaft = πD_s
q_p = unit point resistance
A_p = area of the base = $(\pi/4)D_b^2$

Following are the relationships for determining $Q_{u(net)}$ from Eq. (12.25) for sandy soils. We have

$$f_i = \beta\sigma'_{ozi} \leqslant 4 \text{ kip/ft}^2 \tag{12.26}$$

where

σ'_{ozi} = vertical effective stress at the middle of layer i
$\beta = 1.5 - 0.135z_i^{0.5}$ $\quad(0.25 \leqslant \beta \leqslant 1.2)$ $\tag{12.27}$
z_i = depth to the middle of layer i (ft)

The point bearing capacity is

$$q_p \text{ (kip/ft}^2) = 1.2N_{60} \leqslant 90 \text{ kip/ft}^2 \quad \text{(for } D_b < 50 \text{ in.)} \tag{12.28}$$

where N_{60} = mean *uncorrected* standard penetration number within a distance of $2D_b$ below the base of the drilled shaft. If D_b is equal to or greater than 50 in., excessive settlement may occur. In that case, q_p may be replaced by

$$q_{pr} = \frac{50}{D_b(\text{in.})}q_p \quad \text{(for } D_b \geqslant 50 \text{ in.)} \tag{12.29}$$

In SI units, Eqs. (12.26) through (12.29) will be of the form

$$f_i = \beta\sigma'_{ozi} \leqslant \quad 192 \text{ kN/m}^2 \tag{12.30}$$
$$\beta = 1.5 - 0.244z_i^{0.5} \quad (0.25 \leqslant \beta \leqslant 1.2) \tag{12.31}$$

(where z_i is in m)

$$q_p \text{ (kN/m}^2) = 57.5N_{60} \leqslant 4310 \text{ kN/m}^2 \quad \text{(for } D_b < 1.27 \text{ m)} \tag{12.32}$$

and

$$q_{pr} = \frac{1.27}{D_b(\text{m})}q_p \tag{12.33}$$

Based on the desired level of settlement, Figures 12.9 and 12.10 may now be used to calculate the allowable load, $Q_{all(net)}$. Note that the trend lines given in these figures is the average of all test results.

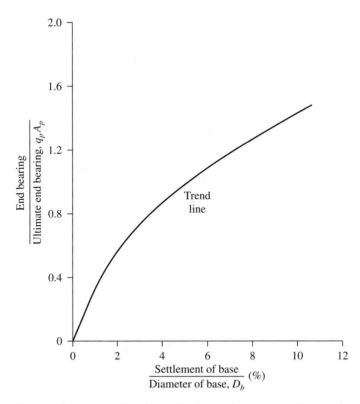

Figure 12.9   Normalized base-load transfer versus settlement in sand

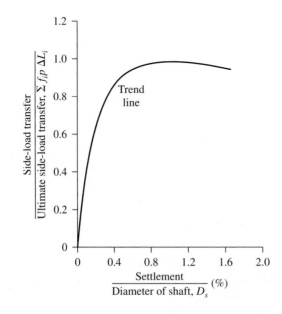

Figure 12.10   Normalized side-load transfer versus settlement in sand

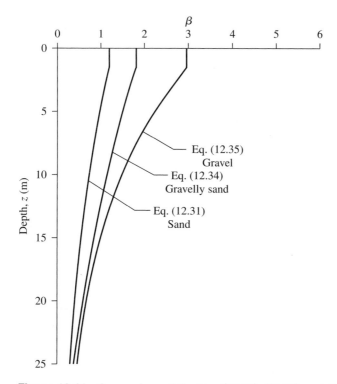

Figure 12.11 Comparison of β—Eqs. (12.31), (12.34), and (12.35)

More recently, Rollins et al. (2005) have modified Eq. (12.31) for gravelly sands as follows:

For sand with 25 to 50% gravel:

$$\beta = 2.0 - 0.15z_i^{0.75} \quad (0.25 \leq \beta \leq 1.8) \tag{12.34}$$

For sand with more than 50% gravel:

$$\beta = 3.4e^{-0.085z_i} \quad (0.25 \leq \beta \leq 3.0) \tag{12.35}$$

In Eqs. (12.34) and (12.35), z is in meters (m). In English units, Eqs. (12.34) and (12.35) can be expressed as

For sand with 25 to 50% gravel:

$$\beta = 2.0 - 0.062z_i^{0.75} \quad (0.25 \leq \beta \leq 1.8) \tag{12.36}$$

For sand with more than 50% gravel:

$$\beta = 3.4e^{-0.026z_i} \quad (0.25 \leq \beta \leq 3.0) \tag{12.37}$$

In Eqs. (12.36) and (12.37), z is in feet (ft).

Figure 12.11 shows a comparison of β expressed by Eqs. (12.31), (12.34), and (12.35). Also, Figure 12.12 provides the normalized side-load transfer trend based on the desired level of settlement for gravelly sand and gravel.

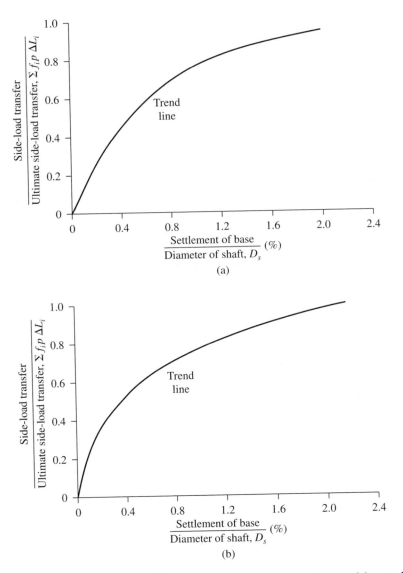

Figure 12.12 Normalized side-load transfer versus settlement: (a) gravelly sand (gravel 25–50%) and (b) gravel (more than 50%)

Example 12.1

A soil profile is shown in Figure 12.13. A point bearing drilled shaft with a bell is placed in a layer of dense sand and gravel. Determine the allowable load the drilled shaft could carry. Use Eq. (12.16) and a factor of safety of 4. Take $D_s = 1$ m and $D_b = 1.75$ m. For the dense sand layer, $\phi' = 36°$; $E_s = 500p_a$. Ignore the frictional resistance of the shaft.

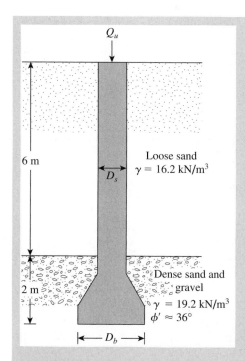

6 m

Loose sand
$\gamma = 16.2$ kN/m³

D_s

2 m

Dense sand and
gravel
$\gamma = 19.2$ kN/m³
$\phi' \approx 36°$

D_b

Q_u

Figure 12.13 Allowable load of drilled
shaft

Solution

We have

$$Q_{p(net)} = A_p[q'(N_q - 1)F_{qs}F_{qd}F_{qc}]$$

and

$$q' = (6)(16.2) + (2)(19.2) = 135.6 \text{ kN/m}^2$$

For $\phi' = 36°$, from Table 3.3, $N_q = 37.75$. Also,

$$F_{qs} = 1 + \tan \phi' = 1 + \tan 36 = 1.727$$

and

$$F_{qd} = 1 + 2 \tan \phi'(1 - \sin \phi')^2 \tan^{-1}\left(\frac{L}{D_b}\right)$$

$$= 1 + 2 \tan 36(1 - \sin 36)^2 \tan^{-1}\left(\frac{8}{1.75}\right) = 1.335$$

From Eq. (12.9),

$$I_{rc} = 0.5 \exp\left[2.85 \cot\left(45 - \frac{\phi'}{2}\right)\right] = 134.3$$

From Eq. (12.17), $E_s = mp_a$. With $m = 500$, we have

$$E_s = (500)(100) = 50,000 \text{ kN/m}^2$$

From Eq. (12.18),

$$\mu_s = 0.1 + 0.3\left(\frac{\phi' - 25}{20}\right) = 0.1 + 0.3\left(\frac{36 - 25}{20}\right) = 0.265$$

So

$$I_r = \frac{E_s}{2(1 + \mu_s)(q')(\tan \phi')} = \frac{50,000}{2(1 + 0.265)(135.6)(\tan 36)} = 200.6$$

From Eq. (12.10),

$$I_{rr} = \frac{I_r}{1 + I_r \Delta}$$

with

$$\Delta = 0.005\left(1 - \frac{\phi' - 25}{20}\right)\frac{q'}{p_a} = 0.005\left(1 - \frac{36 - 25}{20}\right)\left(\frac{135.6}{100}\right) = 0.0031$$

it follows that

$$I_{rr} = \frac{200.6}{1 + (200.6)(0.0031)} = 123.7$$

I_{rr} is less than I_{rc}. So, from Eq. (12.14),

$$F_{qc} = \exp\left\{(-3.8 \tan \phi') + \left[\frac{(3.07 \sin \phi')(\log_{10} 2I_{rr})}{1 + \sin \phi'}\right]\right\}$$

$$= \exp\left\{(-3.8 \tan 36) + \left[\frac{(3.07 \sin 36) \log(2 \times 123.7)}{1 + \sin 36}\right]\right\} = 0.958$$

Hence,

$$Q_{p(net)} = \left[\left(\frac{\pi}{4}\right)(1.75)^2\right](135.6)(37.75 - 1)(1.727)(1.335)(0.958) = 26,474 \text{ kN}$$

and

$$Q_{p(all)} = \frac{Q_{p(net)}}{FS} = \frac{26,474}{4} \approx \textbf{6619 kN} \qquad \blacksquare$$

Example 12.2

Solve Example 12.1 using Eq. (12.20).

Solution
Equation (12.20) asserts that

$$Q_{p(net)} = A_p q'(\omega N_q^* - 1)$$

We have

$$N_q^* = 0.21e^{0.17\phi'} = 0.21e^{(0.17)(36)} = 95.52$$

and

$$\frac{L}{D_b} = \frac{8}{1.75} = 4.57$$

From Figure 12.7, for $\phi' = 36°$ and $L/D_b = 4.57$, the value of ω is about 0.83. So

$$Q_{p(\text{net})} = \left[\left(\frac{\pi}{4}\right)(1.75)^2\right](135.6)[(0.83)(95.52) - 1] = 25,532 \text{ kN}$$

and

$$Q_{p(\text{all})} = \frac{25,532}{4} = \textbf{6383 kN}$$

■

Example 12.3

A drilled shaft is shown in Figure 12.14. The uncorrected average standard penetration number (N_{60}) within a distance of $2D_b$ below the base of the shaft is about 30. Determine

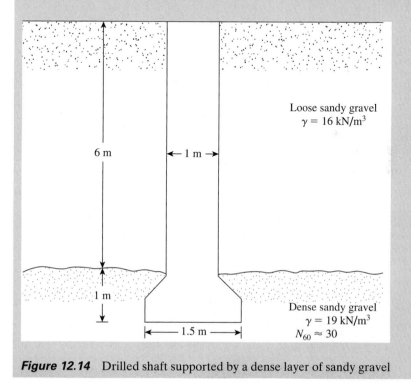

Figure 12.14 Drilled shaft supported by a dense layer of sandy gravel

a. The ultimate load-carrying capacity
b. The load-carrying capacity for a settlement of 12mm. Use Eq. (12.34).

Solution
Part a
From Eqs. (12.26) and (12.34),

$$f_i = \beta\sigma'_{ozi}$$

and

$$\beta = 2.0 - 0.15z^{0.75}$$

For this problem, $z_i = 6/2 = 3$ m, so

$$\beta = 2 - (0.15)(3)^{0.75} = 1.658$$

and

$$\sigma'_{ozi} = \gamma z_i = (16)(3) = 48 \text{ kN/m}^2$$

Thus,

$$f_i = (48)(1.658) = 79.58 \text{ kN/m}^2$$

and

$$\Sigma f_i p\, \Delta L_i = (79.58)(\pi \times 1)(6) = 1500 \text{ kN}$$

From Eq. (12.29)

$$q_p = 57.5N_{60} = (57.5)(30) = 1725 \text{ kN/m}^2$$

Note that D_b is greater than 1.27. So we will use Eq. (12.33).

$$q_{pr} = \left(\frac{1.27}{D_b}\right)q_p = \left(\frac{1.27}{1.5}\right)(1725) \approx 1461 \text{ kN/m}^2$$

Now

$$q_{pr}A_p = (14.61)\left(\frac{\pi}{4} \times 1.5^2\right) \approx 2582 \text{ kN}$$

Hence,

$$Q_{u(\text{net})} = q_{pr}A_p + \Sigma f_i p\, \Delta L_i = 2582 + 1500 = \textbf{4082 kN}$$

Part b
We have

$$\frac{\text{Allowable settlement}}{D_s} = \frac{12}{(1.0)(1000)} = 0.12 = 1.2\%$$

The trend line in Figure 12.12a shows that, for a normalized settlement of 1.2%, the normalized load is about 0.8 . Thus, the side load transfer is $(0.8)(1500) \approx 1200$ kN. Similarly,

$$\frac{\text{Allowable settlement}}{D_b} = \frac{12}{(1.5)(1000)} = 0.008 = 0.8\%$$

The trend line shown in Figure 12.9 indicates that, for a normalized settlement of 1.4%, the normalized base load is 0.317. So the base load is $(0.317)(2582) = 818.5$ kN. Hence, the total load is

$$Q = 1200 + 818.5 \approx \textbf{2018.5 kN}$$

∎

12.8   Drilled Shafts in Clay: Load-Bearing Capacity

For saturated clays with $\phi = 0$, the bearing capacity factor N_q in Eq. (12.4) is equal to unity. Thus, for this case, Eq. (12.4) will be of the form

$$Q_{p(net)} = A_p c_u N_c F_{cs} F_{cd} F_{cc} \tag{12.38}$$

where c_u = undrained cohesion.

Assuming that $L \geqslant 3D_b$, we can rewrite Eq. (12.38) as

$$Q_{p(net)} = A_p c_u N_c^* \tag{12.39}$$

where $N_c^* = N_c F_{cs} F_{cd} F_{cc} = 1.33[(\ln I_r) + 1]$ in which I_r = soil rigidity index. (12.40)

The soil rigidity index was defined in Eq. (12.11). For $\phi = 0$,

$$I_r = \frac{E_s}{3c_u} \tag{12.41}$$

O'Neill and Reese (1999) provided an approximate relationship between c_u and $E_s/3c_u$. This relationship is shown in Figure 12.15. For all practical purposes, if c_u/p_a is equal to or greater than unity (p_a = atmospheric pressure ≈ 100 kN/m^2 or 2000 lb/ft^2), then the magnitude of N_c^* can be taken to be 9.

Experiments by Whitaker and Cooke (1966) showed that, for belled shafts, the full value of $N_c^* = 9$ is realized with a base movement of about 10 to 15% of D_b. Similarly, for straight shafts ($D_b = D_s$), the full value of $N_c^* = 9$ is obtained with a base movement of about 20% of D_b.

The expression for the skin resistance of drilled shafts in clay is similar to Eq. (11.50), or

$$Q_s = \sum_{L=0}^{L=L_1} \alpha^* c_u p \, \Delta L \tag{12.42}$$

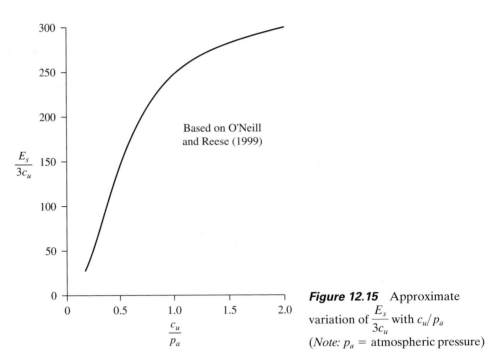

Figure 12.15 Approximate variation of $\dfrac{E_s}{3c_u}$ with c_u/p_a (*Note:* p_a = atmospheric pressure)

Kulhawy and Jackson (1989) reported the field-test result of 106 straight drilled shafts—65 in uplift and 41 in compression. The best correlation obtained from the results is

$$\alpha^* = 0.21 + 0.25\left(\frac{p_a}{c_u}\right) \leq 1 \tag{12.43}$$

where p_a = atmospheric pressure $\approx 1 \text{ ton/ft}^2$ ($\approx 100 \text{ kN/m}^2$).
So, conservatively, we may assume that

$$\alpha^* = 0.4 \tag{12.44}$$

Load-Bearing Capacity Based on Settlement

Reese and O'Neill (1989) suggested a procedure for estimating the ultimate and allowable (based on settlement) bearing capacities for drilled shafts in clay. According to this procedure, we can use Eq. (12.25) for the net ultimate load, or

$$Q_{u(\text{net})} = \sum_{i=1}^{n} f_i p\,\Delta L_i + q_p A_p$$

The unit skin friction resistance can be given as

$$f_i = \alpha_i^* c_{u(i)} \tag{12.45}$$

The following values are recommended for α_i^*:

$\alpha_i^* = 0$ for the top 1.5 m (5 ft) and bottom 1 diameter, D_s, of the drilled shaft. (*Note:* If $D_b > D_s$, then $\alpha^* = 0$ for 1 diameter above the top of the bell and for the peripheral area of the bell itself.)

$\alpha_i^* = 0.55$ elsewhere.

$$q_p = 6c_{ub}\left(1 + 0.2\frac{L}{D_b}\right) \leqslant 9c_{ub} \leqslant 80 \text{ kip/ft}^2 \qquad (12.46)$$

where c_{ub} = average undrained cohesion within $2D_b$ below the base.

If D_b is large, excessive settlement will occur at the ultimate load per unit area, q_p, as given by Eq. (12.46). Thus, for $D_b > 1.91$ m (75 in.), q_p may be replaced by

$$q_{pr} = F_r q_p \qquad (12.47)$$

where

$$F_r = \frac{2.5}{\psi_1 D_b \text{ (in.)} + \psi_2} \leqslant 1 \qquad (12.48)$$

in which

$$\psi_1 = 0.0071 + 0.0021\left(\frac{L}{D_b}\right) \leqslant 0.015 \qquad (12.49)$$

and

$$\psi_2 = 0.45(c_{ub})^{0.5} \quad (0.5 \leqslant \psi_2 \leqslant 1.5)$$
$$\uparrow \qquad\qquad\qquad (12.50)$$
$$\text{kip/ft}^2$$

In SI units, Eqs. (12.46), (12.48), (12.49), and (12.50) can be expressed as

$$q_p = 6c_{ub}\left(1 + 0.2\frac{L}{D_b}\right) \leqslant 9c_{ub} \leqslant 3830 \text{ kN/m}^2 \qquad (12.51)$$

$$F_r = \frac{2.5}{\psi_1 D_b \text{ (mm)} + \psi_2} \leqslant 1 \qquad (12.52)$$

$$\psi_1 = 2.78 \times 10^{-4} + 8.26 \times 10^{-5}\left(\frac{L}{D_b}\right) \leqslant 5.9 \times 10^{-4} \qquad (12.53)$$

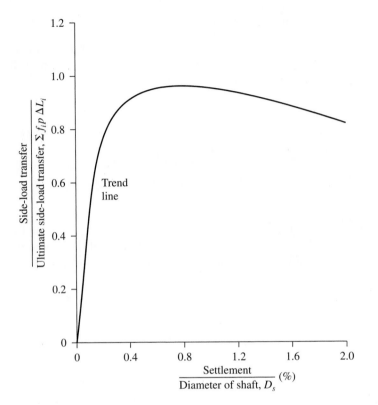

Figure 12.16 Normalized side-load transfer versus settlement in cohesive soil

and

$$\psi_2 = 0.065[c_{ub} \ (kN/m^2)]^{0.5} \tag{12.54}$$

Figures 12.16 and 12.17 may now be used to evaluate the allowable load-bearing capacity, based on settlement. (Note that the ultimate bearing capacity in Figure 12.16 is q_p, not q_{pr}.) To do so,

Step 1. Select a value of settlement, s.

Step 2. Calculate $\displaystyle\sum_{i=1}^{N} f_ip \ \Delta L_i$ and q_pA_p.

Step 3. Using Figures 12.16 and 12.17 and the calculated values in Step 2, determine the *side load* and the *end bearing load*.

Step 4. The sum of the side load and the end bearing load gives the total allowable load.

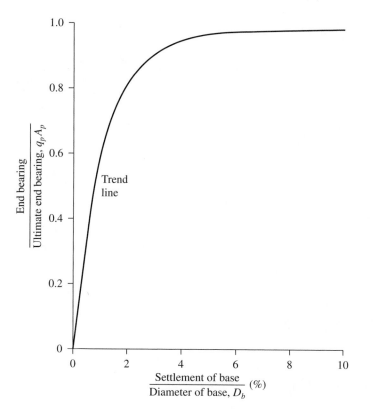

Figure 12.17 Normalized base-load transfer versus settlement in cohesive soil

---

Example 12.4

Figure 12.18 shows a drilled shaft without a bell. Here, $L_1 = 8$ m, $L_2 = 3$ m, $D_s = 1.5$ m, $c_{u(1)} = 50$ kN/m², and $c_{u(2)} = 105$ kN/m². Determine

a. The net ultimate point bearing capacity
b. The ultimate skin resistance
c. The working load, Q_w (FS = 3)

Use Eqs. (12.39), (12.42), and (12.44).

Solution
Part a
From Eq. (12.39),

$$Q_{p(\text{net})} = A_p c_u N_c^* = A_p c_{u(2)} N_c^* = \left[\left(\frac{\pi}{4} \right)(1.5)^2 \right](105)(9) \approx \mathbf{1670 \ kN}$$

(*Note:* Since $c_u/p_a > 1$, $N_c^* \approx 9$.)

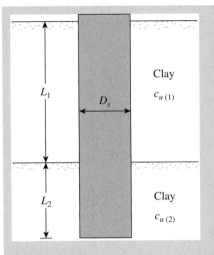

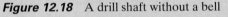

Figure 12.18 A drill shaft without a bell

Part b

From Eq. (12.42),

$$Q_s = \Sigma \alpha^* c_u p \Delta L$$

From Eq. (12.44),

$$\alpha^* = 0.4$$

$$p = \pi D_s = (3.14)(1.5) = 4.71 \text{ m}$$

and

$$Q_s = (0.4)(4.71)[(50 \times 8) + (105 \times 3)] \approx \mathbf{1347 \text{ kN}}$$

Part c

$$Q_w = \frac{Q_{p(\text{net})} + Q_s}{\text{FS}} = \frac{1670 + 1347}{3} = \mathbf{1005.7 \text{ kN}} \qquad \blacksquare$$

Example 12.5

A drilled shaft in a cohesive soil is shown in Figure 12.19. Use Reese and O'Neill's method to determine

a. The ultimate load-carrying capacity (Eqs. 12.45 through 12.50)
b. The load-carrying capacity for an allowable settlement of 0.5 in.

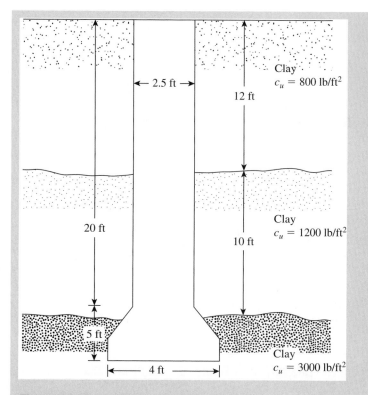

Figure 12.19 A drilled shaft in layered clay

Solution

Part a

From Eq. (12.45),

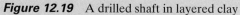
$$f_i = \alpha_i^* c_{u(i)}$$

From Figure 12.19,

$$\Delta L_1 = 12 - 5 = 7 \text{ ft}$$
$$\Delta L_2 = (20 - 12) - D_s = (20 - 12) - 2.5 = 5.5 \text{ ft}$$
$$c_{u(1)} = 800 \text{ lb/ft}^2$$

and

$$c_{u(2)} = 1200 \text{ lb/ft}^2$$

Hence,

$$\Sigma f_i p \Delta L_i = \Sigma \alpha_i^* c_{u(i)} p \Delta L_i$$
$$= (0.55)(800)(\pi \times 2.5)(7) + (0.55)(1200)(\pi \times 2.5)(5.5)$$
$$= 52{,}700 \text{ lb} = 52.7 \text{ kip}$$

Again, from Eq. (12.46),

$$q_p = 6c_{ub}\left(1 + 0.2\frac{L}{D_b}\right) = (6)(3000)\left[1 + 0.2\left(\frac{20 + 5}{4}\right)\right] = 40,500 \text{ lb/ft}^2$$

$$= 40.5 \text{ kip/ft}^2$$

A check reveals that

$$q_p = 9c_{ub} = (9)(3000) = 27,000 \text{ lb/ft}^2 = 27 \text{ kip/ft}^2 < 40.5 \text{ kip/ft}^2$$

So we use $q_p = 27 \text{ kip/ft}^2$:

$$q_p A_p = q_p\left(\frac{\pi}{4}D_b^2\right) = (27)\left[\left(\frac{\pi}{4}\right)(4)^2\right] \approx 339.3 \text{ kip}$$

Hence,

$$Q_u = \Sigma \alpha_i^* c_{u(i)} p \Delta L_i + q_p A_p = 52.7 + 339.3 = \mathbf{392 \ kip}$$

Part b
We have

$$\frac{\text{Allowable settlement}}{D_s} = \frac{0.5}{(2.5)(12)} = 0.0167 = 1.67\%$$

The trend line shown in Figure 12.16 indicates that, for a normalized settlement of 1.67%, the normalized side load is about 0.89. Thus, the side load is

$$(0.89)(\Sigma f_i p \Delta L_i) = (0.89)(52.7) = 46.9 \text{ kip}$$

Again,

$$\frac{\text{Allowable settlement}}{D_b} = \frac{0.5}{(4)(12)} = 0.0104 = 1.04\%$$

The trend line shown in Figure 12.17 indicates that, for a normalized settlement of 1.04%, the normalized end bearing is about 0.57, so

$$\text{Base load} = (0.57)(q_p A_p) = (0.57)(339.3) = 193.4 \text{ kip}$$

Thus, the total load is

$$Q = 46.9 + 193.4 = \mathbf{240.3 \ kip} \qquad \blacksquare$$

12.9   *Settlement of Drilled Shafts at Working Load*

The settlement of drilled shafts at working load is calculated in a manner similar to that outlined in Section 11.17. In many cases, the load carried by shaft resistance is small compared with the load carried at the base. In such cases, the contribution of s_3 may be ignored. (Note that in Eqs. (11.64) and (11.65) the term D should be replaced by D_b for drilled shafts.)

Example 12.6

For the drilled shaft of Example 12.4, estimate the elastic settlement at working loads (i.e., $Q_w = 1005$ kN). Use Eqs. (11.63), (11.65), and (11.66). Take $\xi = 0.65$, $E_p = 21 \times 10^6$ kN/m², $E_s = 14,000$ kN/m², $\mu_s = 0.3$, and $Q_{wp} = 250$ kN.

Solution
From Eq. (11.63),

$$S_{e(1)} = \frac{(Q_{wp} + \xi Q_{ws})L}{A_p E_p}$$

Now,

$$Q_{ws} = 1005 - 250 = 755 \text{ kN}$$

so

$$S_{e(1)} = \frac{[250 + (0.65)(755)](11)}{\left(\dfrac{\pi}{4} \times 1.5^2\right)(21 \times 10^6)} = 0.00022 \text{ m} = 0.22 \text{ mm}$$

From Eq. (11.65),

$$S_{e(2)} = \frac{Q_{wp} C_p}{D_b q_p}$$

From Table 11.12, for stiff clay, $C_p \approx 0.04$; also,

$$q_p = c_{u(b)} N_c^* = (105)(9) = 945 \text{ kN/m}^2$$

Hence,

$$S_{e(2)} = \frac{(250)(0.04)}{(1.5)(945)} = 0.0071 \text{ m} = 7.1 \text{ mm}$$

Again, from Eqs. (11.66) and (11.67),

$$S_{e(3)} = \left(\frac{Q_{ws}}{pL}\right)\left(\frac{D_s}{E_s}\right)(1 - \mu_s^2)I_{ws}$$

where

$$I_{ws} = 2 + 0.35\sqrt{\frac{L}{D_s}} = 2 + 0.35\sqrt{\frac{11}{1.5}} = 2.95$$

$$S_{e(3)} = \left[\frac{755}{(\pi \times 1.5)(11)}\right]\left(\frac{1.5}{14,000}\right)(1 - 0.3^2)(2.95) = 0.0042 \text{ m} = 4.2 \text{ mm}$$

The total settlement is

$$S_e = S_{e(1)} + S_{e(2)} + S_{e(3)} = 0.22 + 7.1 + 4.2 \approx \mathbf{11.52 \ mm} \qquad \blacksquare$$

12.10   *Lateral Load-Carrying Capacity—Characteristic Load and Moment Method*

Several methods for analyzing the lateral load-carrying capacity of piles, as well as the load-carrying capacity of drilled shafts, were presented in Section 11.18; therefore, they will not be repeated here. In 1994, Duncan et al. developed a *characteristic load method* for estimating the lateral load capacity for drilled shafts that is fairly simple to use. We describe this method next.

According to the characteristic load method, the *characteristic load* Q_c and *moment* M_c form the basis for the dimensionless relationship that can be given by the following correlations:

Characteristic Load

$$Q_c = 7.34D_s^2 \ (E_p R_I)\left(\frac{c_u}{E_p R_I}\right)^{0.68} \qquad \text{(for clay)} \qquad (12.55)$$

$$Q_c = 1.57D_s^2 \ (E_p R_I)\left(\frac{\gamma' D_s \phi' K_p}{E_p R_I}\right)^{0.57} \qquad \text{(for sand)} \qquad (12.56)$$

Characteristic Moment

$$M_c = 3.86D_s^3 \ (E_p R_I)\left(\frac{c_u}{E_p R_I}\right)^{0.46} \qquad \text{(for clay)} \qquad (12.57)$$

$$M_c = 1.33D_s^3(E_p R_I)\left(\frac{\gamma' D_s \phi' K_p}{E_p R_I}\right)^{0.40} \qquad \text{(for sand)} \qquad (12.58)$$

In these equations,

D_s = diameter of drilled shafts
E_p = modulus of elasticity of drilled shafts
R_I = ratio of moment of inertia of drilled shaft section to moment of inertia of a solid section (*Note:* $R_I = 1$ for uncracked shaft without central void)
γ' = effective unit weight of sand
ϕ' = effective soil friction angle (degrees)
K_p = Rankine passive pressure coefficient = $\tan^2(45 + \phi'/2)$

Deflection Due to Load Q_g Applied at the Ground Line

Figures 12.20 and 12.21 give the plot of Q_g/Q_c versus x_o/D_s for drilled shafts in sand and clay due to the load Q_g applied at the ground surface. Note that x_o is the ground

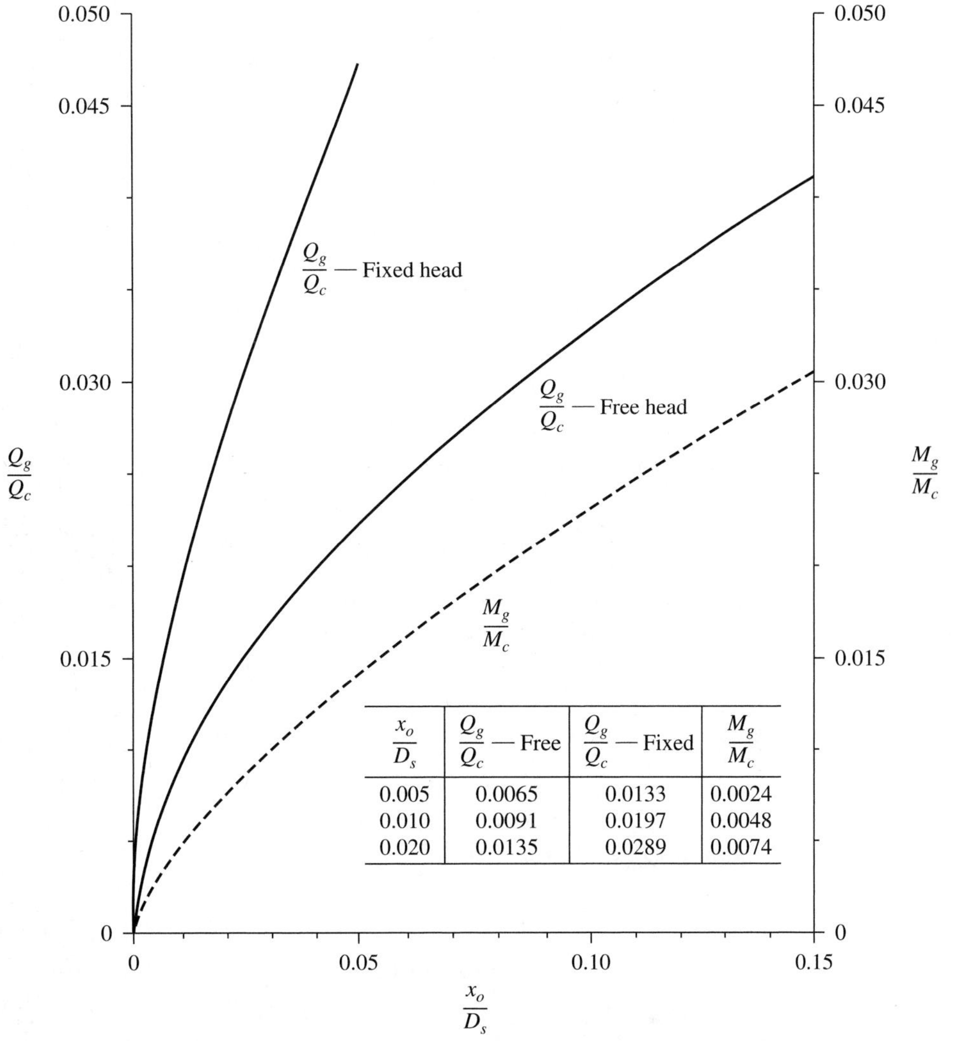

$\frac{x_o}{D_s}$	$\frac{Q_g}{Q_c}$ — Free	$\frac{Q_g}{Q_c}$ — Fixed	$\frac{M_g}{M_c}$
0.005	0.0065	0.0133	0.0024
0.010	0.0091	0.0197	0.0048
0.020	0.0135	0.0289	0.0074

Figure 12.20 Plot of $\frac{Q_g}{Q_c}$ and $\frac{M_g}{M_c}$ versus $\frac{x_o}{D_s}$ in clay

line deflection. If the magnitudes of Q_g and Q_c are known, the ratio Q_g/Q_c can be calculated. The figure can then be used to estimate the corresponding value of x_o/D_s and, hence, x_o.

Deflection Due to Moment Applied at the Ground Line

Figures 12.20 and 12.21 give the variation plot of M_g/M_c with x_o/D_s for drilled shafts in sand and clay due to an applied moment M_g at the ground line. Again, x_o is the ground line deflection. If the magnitudes of M_g, M_c, and D_s are known, the value of x_o can be calculated with the use of the figure.

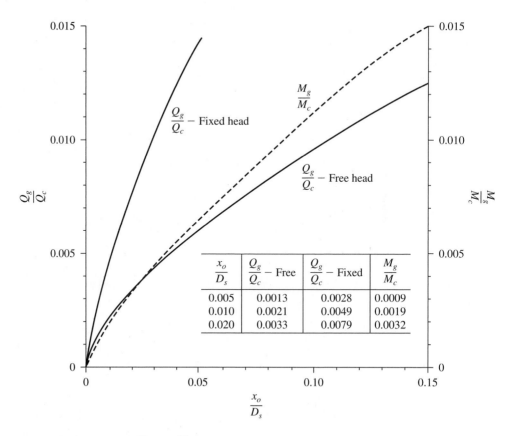

The table in the figure:

$\dfrac{x_o}{D_s}$	$\dfrac{Q_g}{Q_c}$ — Free	$\dfrac{Q_g}{Q_c}$ — Fixed	$\dfrac{M_g}{M_c}$
0.005	0.0013	0.0028	0.0009
0.010	0.0021	0.0049	0.0019
0.020	0.0033	0.0079	0.0032

Figure 12.21 Plot of $\dfrac{Q_g}{Q_c}$ and $\dfrac{M_g}{M_c}$ versus $\dfrac{x_o}{D_s}$ in sand

Deflection Due to Load Applied Above the Ground Line

When a load Q is applied above the ground line, it induces both a load $Q_g = Q$ and a moment $M_g = Qe$ at the ground line, as shown in Figure 12.22a. A superposition solution can now be used to obtain the ground line deflection. The step-by-step procedure is as follows (refer to Figure 12.22b):

Step 1. Calculate Q_g and M_g.
Step 2. Calculate the deflection x_{oQ} that would be caused by the load Q_g acting alone.
Step 3. Calculate the deflection x_{oM} that would be caused by the moment acting alone.
Step 4. Determine the value of a load Q_{gM} that would cause the same deflection as the moment (i.e., x_{oM}).
Step 5. Determine the value of a moment M_{gQ} that would cause the same deflection as the load (i.e., x_{oQ}).
Step 6. Calculate $(Q_g + Q_{gM})/Q_c$ and determine x_{oQM}/D_s.
Step 7. Calculate $(M_g + M_{gQ})/M_c$ and determine x_{oMQ}/D_s.

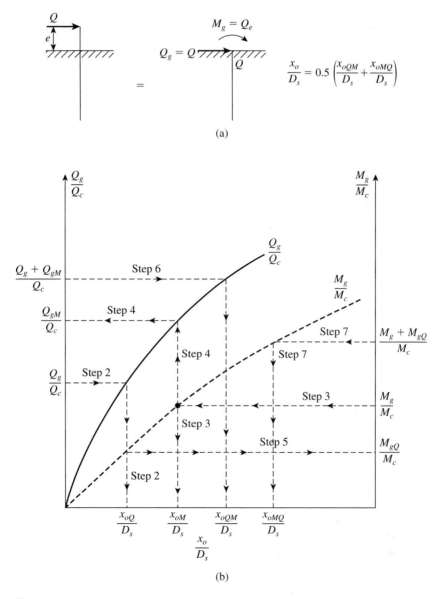

Figure 12.22 Superposition of deflection due to load and moment

Step 8.   Calculate the combined deflection:

$$x_{o(\text{combined})} = 0.5(x_{oQM} + x_{oMQ}) \qquad (12.59)$$

Maximum Moment in Drilled Shaft Due to Ground Line Load Only

Figure 12.23 shows the plot of Q_g/Q_c with $M_{\max}/M_c$ for fixed- and free-headed drilled shafts due only to the application of a ground line load Q_g. For fixed-headed

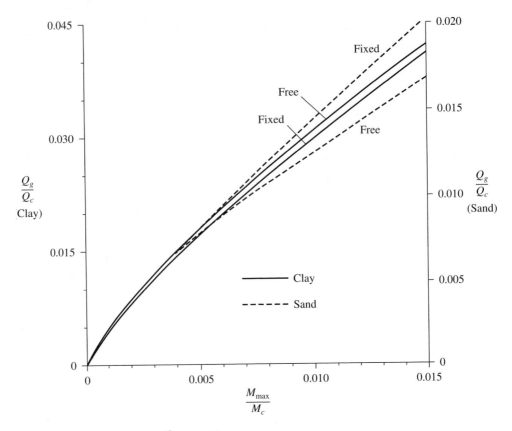

Figure 12.23   Variation of $\dfrac{Q_g}{Q_c}$ with $\dfrac{M_{\max}}{M_c}$

shafts, the maximum moment in the shaft, $M_{\max}$, occurs at the ground line. For this condition, if Q_c, M_c, and Q_g are known, the magnitude of $M_{\max}$ can be easily calculated.

Maximum Moment Due to Load and Moment at Ground Line

If a load Q_g and a moment M_g are applied at the ground line, the maximum moment in the drilled shaft can be determined in the following manner:

Step 1. Using the procedure described before, calculate $x_{o(\text{combined})}$ from Eq. (12.59).

Step 2. To solve for the characteristic length T, use the following equation:

$$x_{o(\text{combined})} = \frac{2.43 Q_g}{E_p I_p} T^3 + \frac{1.62 M_g}{E_p I_p} T^2 \tag{12.60}$$

Step 3. The moment in the shaft at a depth z below the ground surface can be calculated as

$$M_z = A_m Q_g T + B_m M_g \tag{12.61}$$

where A_m, B_m = dimensionless moment coefficients (Matlock and Reese, 1961); see Figure 12.24.

The value of the maximum moment M_{max} can be obtained by calculating M_z at various depths in the upper part of the drilled shaft.

The characteristic load method just described is valid only if L/D_s has a certain minimum value. If the actual L/D_s is less than $(L/D_s)_{min}$, then the ground line deflections will be underestimated and the moments will be overestimated. The values of $(L/D_s)_{min}$ for drilled shafts in sand and clay are given in the following table:

Clay		Sand	
$\dfrac{E_p R_i}{c_u}$	$(L/D_s)_{min}$	$\dfrac{E_p R_i}{\gamma' D_s \phi' K_p}$	$(L/D_s)_{min}$
1×10^5	6	1×10^4	8
3×10^5	10	4×10^4	11
1×10^6	14	2×10^5	14
3×10^6	18		

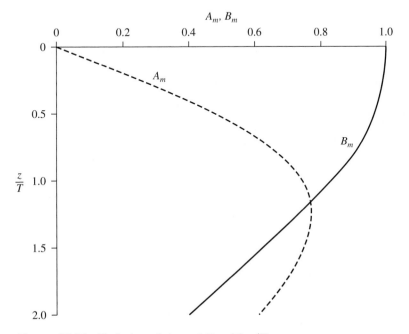

Figure 12.24 Variation of A_m and B_m with z/T

Example 12.7

A free-headed drilled shaft in clay is shown in Figure 12.25. Let $E_p = 22 \times 10^6$ kN/m². Determine

    a. The ground line deflection, $x_{o(\text{combined})}$
    b. The maximum bending moment in the drilled shaft
    c. The maximum tensile stress in the shaft
    d. The minimum penetration of the shaft needed for this analysis

Solution
We are given

$D_s = 1$ m
$c_u = 100$ kN/m²
$R_I = 1$
$E_p = 22 \times 10^6$ kN/m²

and

$$I_p = \frac{\pi D_s^4}{64} = \frac{(\pi)(1)^4}{64} = 0.049 \text{ m}^4$$

Part a
From Eq. (12.55),

$$Q_c = 7.34 D_s^2 \, (E_p R_I) \left(\frac{c_u}{E_p R_I} \right)^{0.68}$$

$$= (7.34)(1)^2 [(22 \times 10^6)(1)] \left[\frac{100}{(22 \times 10^6)(1)} \right]^{0.68}$$

$$= 37,607 \text{ kN}$$

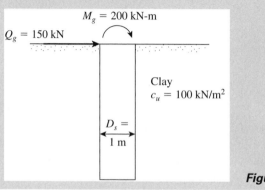

$M_g = 200$ kN-m
$Q_g = 150$ kN
Clay
$c_u = 100$ kN/m²
$D_s = 1$ m

Figure 12.25   Free-headed drilled shaft

From Eq. (12.57),

$$M_c = 3.86 D_s^3 (E_p R_I) \left(\frac{c_u}{E_p R_I} \right)^{0.46}$$

$$= (3.86)(1)^3 [(22 \times 10^6)(1)] \left[\frac{100}{(22 \times 10^6)(1)} \right]^{0.46}$$

$$= 296,139 \text{ kN-m}$$

Thus,

$$\frac{Q_g}{Q_c} = \frac{150}{37,607} = 0.004$$

From Figure 12.20, $x_{oQ} \approx (0.0025) D_s = 0.0025 \text{ m} = 2.5 \text{ mm}$. Also,

$$\frac{M_g}{M_c} = \frac{200}{296,139} = 0.000675$$

From Figure 12.20, $x_{oM} \approx (0.0014) D_s = 0.0014 \text{ m} = 1.4 \text{ mm}$, so

$$\frac{x_{oM}}{D_s} = \frac{0.0014}{1} = 0.0014$$

From Figure 12.20, for $x_{oM}/D_s = 0.0014$, the value of $Q_{gM}/Q_c \approx 0.002$. Hence,

$$\frac{x_{oQ}}{D_s} = \frac{0.0025}{1} = 0.0025$$

From Figure 12.20, for $x_{oQ}/D_s = 0.0025$, the value of $M_{gQ}/M_c \approx 0.0013$, so

$$\frac{Q_g}{Q_c} + \frac{Q_{gM}}{Q_c} = 0.004 + 0.002 = 0.006$$

From Figure 12.20, for $(Q_g + Q_{gM})/Q_c = 0.006$, the value of $x_{oQM}/D_s \approx 0.0046$. Hence,

$$x_{oQM} = (0.0046)(1) = 0.0046 \text{ m} = 4.6 \text{ mm}$$

Thus, we have

$$\frac{M_g}{M_c} + \frac{M_{gQ}}{M_c} = 0.000675 + 0.0013 \approx 0.00198$$

From Figure 12.20, for $(M_g + M_{gQ})/M_c = 0.00198$, the value of $x_{oMQ}/D_s \approx 0.0041$. Hence,

$$x_{oMQ} = (0.0041)(1) = 0.0041 \text{ m} = 4.1 \text{ mm}$$

Consequently,

$$x_{o \text{ (combined)}} = 0.5(x_{oQM} + x_{oMQ}) = (0.5)(4.6 + 4.1) = \mathbf{4.35 \ mm}$$

Part b
From Eq. (12.60),

$$x_{o \text{ (combined)}} = \frac{2.43 Q_g}{E_p I_p} T^3 + \frac{1.62 M_g}{E_p I_p} T^2$$

so

$$0.00435 \ \text{m} = \frac{(2.43)(150)}{(22 \times 10^6)(0.049)} T^3 + \frac{(1.62)(200)}{(22 \times 10^6)(0.049)} T^2$$

or

$$0.00435 \ \text{m} = 338 \times 10^{-6} \, T^3 + 300.6 \times 10^{-6} \, T^2$$

and it follows that

$$T \approx 2.05 \ \text{m}$$

From Eq. (12.61),

$$M_z = A_m Q_g T + B_m M_g = A_m(150)(2.05) + B_m(200) = 307.5 A_m + 200 B_m$$

Now the following table can be prepared:

$\dfrac{z}{T}$	A_m (Figure 12.24)	B_m (Figure 12.24)	M_z (kN-m)
0	0	1.0	200
0.4	0.36	0.98	306.7
0.6	0.52	0.95	349.9
0.8	0.63	0.9	373.7
1.0	0.75	0.845	399.6
1.1	0.765	0.8	395.2
1.25	0.75	0.73	376.6

So the maximum moment is 399.4 kN-m $\approx$ 400 kN-m and occurs at $z/T \approx 1$. Hence,

$$z = (1)(T) = (1)(2.05 \ \text{m}) = \mathbf{2.05 \ m}$$

Part c
The maximum tensile stress is

$$\sigma_{\text{tensile}} = \frac{M_{\text{max}}\left(\dfrac{D_s}{2}\right)}{I_p} = \frac{(400)\left(\dfrac{1}{2}\right)}{0.049} = \mathbf{4081.6 \ kN/m^2}$$

Part d
We have

$$\frac{E_p R_I}{c_u} = \frac{(22 \times 10^6)(1)}{100} = 2.2 \times 10^5$$

By interpolation, for $(E_p R_I)/c_u = 2.2 \times 10^5$, the value of $(L/D_s)_{\min} \approx 8.5$. So

$$L \approx (8.5)(1) = \mathbf{8.5\ m} \qquad\blacksquare$$

12.11 *Drilled Shafts Extending into Rock*

In Section 12.1, we noted that drilled shafts can be extended into rock. In the current section, we describe the principles of analysis of the load-bearing capacity of such drilled shafts, based on the procedure developed by Reese and O'Neill (1988, 1989). Figure 12.26 shows a drilled shaft whose depth of embedment in rock is equal to L. In the design process to be recommended, it is assumed that *there is either side resistance between the shaft and the rock or point resistance at the bottom, but not both.* Following is a step-by-step procedure for estimating the ultimate bearing capacity:

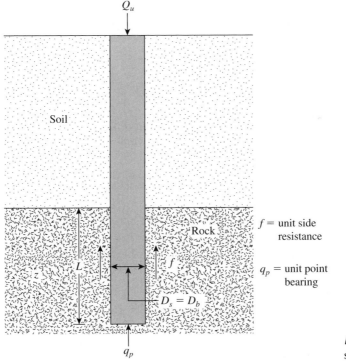

Figure 12.26 Drilled shaft socketed into rock

Step 1.   Calculate the ultimate unit side resistance as

$$f \text{ (lb/in}^2) = 2.5q_u^{0.5} \leqslant 0.15q_u \tag{12.62}$$

where q_u = unconfined compression strength of a rock core of NW size or larger, or of the drilled shaft concrete, whichever is smaller (in lb/in²). In SI units, Eq (12.62) can be expressed as

$$f \text{ (kN/m}^2) = 6.564q_u^{0.5} \text{ (kN/m}^2) \leqslant 0.15q_u \text{ (kN/m}^2) \tag{12.63}$$

Step 2.   Calculate the ultimate capacity based on side resistance only, or

$$Q_u = \pi D_s L f \tag{12.64}$$

Step 3.   Calculate the settlement s_e of the shaft at the top of the rock socket, or

$$s_e = s_{e(s)} + s_{e(b)} \tag{12.65}$$

where

$s_{e(s)}$ = elastic compression of the drilled shaft within the socket, assuming no side resistance
$s_{e(b)}$ = settlement of the base

However,

$$s_{e(s)} = \frac{Q_u L}{A_c E_c} \tag{12.66}$$

and

$$s_{e(b)} = \frac{Q_u I_f}{D_s E_{\text{mass}}} \tag{12.67}$$

where

Q_u = ultimate load obtained from Eq. (12.62) or Eq. (12.63) (this assumes that the contribution of the overburden to the side shear is negligible)
A_c = cross-sectional area of the drilled shaft in the socket $\qquad$ (12.68)

$$= \frac{\pi}{4}D_s^2$$

E_c = Young's modulus of the concrete and reinforcing steel in the shaft
E_{mass} = Young's modulus of the rock mass into which the socket is drilled
I_f = elastic influence coefficient (see Table 12.1)

The magnitude of E_{mass} can be taken as

$$\frac{E_{\text{mass}}}{E_{\text{core}}} \approx 0.0266(\text{RDQ}) - 1.66 \tag{12.69}$$

Table 12.1   Settlement Influence Factor, I_f

L/D_s	I_f			
	E_c/E_{mas}			
	10	50	100	5000
0	1.10	1.10	1.10	1.10
1	0.51	0.47	0.47	0.47
2	0.47	0.37	0.35	0.32
4	0.43	0.31	0.28	0.23
6	0.41	0.27	0.24	0.18
8	0.40	0.25	0.21	0.14
10	0.39	0.24	0.18	0.12
12	0.38	0.22	0.16	0.10
20	0.37	0.20	0.15	0.08

where

RDQ = rock quality designation, in %

E_{core} = Young's modulus of intact specimens of rock cores of NW size or larger

However, unless the socket is very long

$$s_e \approx s_{e(b)} = \frac{Q_u I_f}{D_s E_{mass}} \tag{12.70}$$

Step 4.   If s_e is less than 10 mm ($\approx$0.4 in.), then the ultimate load-carrying capacity is that calculated by Eq. (12.64). If $s_e \geq 10$ mm. (0.4 in.), then go to Step 5.

Step 5.   If $s_e \geq 10$ mm (0.4 in.), there may be rapid, progressive side shear failure in the rock socket, resulting in a complete loss of side resistance. In that case, the ultimate capacity is equal to the point resistance, or

$$Q_u = 3A_p \left[\frac{3 + \dfrac{c_s}{D_s}}{10\left(1 + 300\dfrac{\delta}{c_s}\right)^{0.5}} \right] q_u \tag{12.71}$$

where

c_s = spacing of discontinuities (same unit as D_s)

δ = thickness of individual discontinuity (same unit as D_s)

q_u = unconfined compression strength of the rock beneath the base of the socket, or the drilled shaft concrete, whichever is smaller

Note that Eq. (12.71) applies for horizontally stratified discontinuities with $c_s > 305$ mm (12 in.) and $\delta < 5$ mm (0.2 in.).

Example 12.8

Consider the case of a drilled shaft extending into rock, as shown in Figure 12.27. Let $L = 15$ ft, $D_s = 3$ ft, q_u (rock) $= 10,500$ lb/in^2, q_u (concrete) $= 3000$ lb/in^2, $E_c = 3 \times 10^6$ lb/in^2, RQD (rock) $= 80\%$, E_{core} (rock) $= 0.36 \times 10^6$ lb/in^2, $c_s = 18$ in., and $\delta = 0.15$ in. Estimate the allowable load-bearing capacity of the drilled shaft. Use a factor of safety (FS) $= 3$.

Solution

Step 1. From Eq. (12.62),

$$f \,(\text{lb/in}^2) = 2.5\, q_u^{0.5} \leqslant 0.15 q_u$$

Since q_u (concrete) $< q_u$ (rock), use q_u (concrete) in Eq. (12.62). Hence,

$$f = 2.5(3000)^{0.5} = 136.9 \text{ lb/in}^2$$

As a check, we have

$$f = 0.15 q_u = (0.15)(3000) = 450 \text{ lb/in}^2 > 136.9 \text{ lb/in}^2$$

So use $f = 136.9$ lb/in^2.

Step 2. From Eq. (12.64),

$$Q_u = \pi D_s L f = [(\pi)(3 \times 12)(15 \times 12)(136.9)]\frac{1}{1000} = 2787 \text{ kip}$$

Step 3. From Eqs. (12.65), (12.66), and (12.67),

$$s_e = \frac{Q_u L}{A_c E_c} + \frac{Q_u I_f}{D_s E_{mass}}$$

From Eq. (12.69), for RDQ $= 80\%$

$$\frac{E_{mass}}{E_{core}} = (0.0266)(80) - 1.66 = 0.468$$

$$E_{mass} = 0.468 E_{core} = (0.468)(0.36 \times 10^6) = 0.168 \times 10^6 \text{ lb/in}^2$$

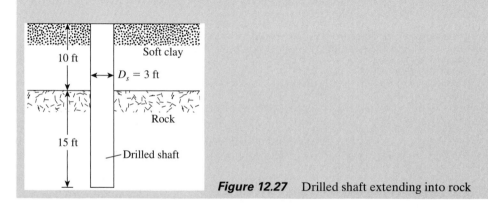

Figure 12.27   Drilled shaft extending into rock

so

$$\frac{E_c}{E_{\text{mass}}} = \frac{3 \times 10^6}{0.168 \times 10^6} \approx 17.9$$

and

$$\frac{L}{D_s} = \frac{15}{3} = 5$$

From Table 12.1, for $E_c/E_{\text{mass}} = 17.9$ and $L/D_s = 5$, the magnitude of I_f is about 0.35. Hence,

$$s_e = \frac{(2787 \times 10^3 \text{ lb})(15 \times 12 \text{ in.})}{\frac{\pi}{4}(3 \times 12 \text{ in.})^2(3 \times 10^6 \text{ lb/in}^2)} + \frac{(2787 \times 10^3 \text{ lb})(0.35)}{(3 \times 12 \text{ in.})(0.168 \times 10^6 \text{ lb/in}^2)}$$

$$= 0.325 \text{ in.} < 0.4 \text{ in.}$$

Therefore,

$$Q_u = 2787 \text{ kip}$$

and

$$Q_{\text{all}} = \frac{Q_u}{\text{FS}} = \frac{2787}{3} = \textbf{929 kip} \qquad \blacksquare$$

Problems

12.1 A drilled shaft is shown in Figure P12.1. Determine the net allowable point bearing capacity. Given

$D_b = 6$ ft	$\gamma_c = 100$ lb/ft^3
$D_s = 3.5$ ft	$\gamma_s = 112$ lb/ft^3
$L_1 = 18$ ft	$\phi' = 38°$
$L_2 = 10$ ft	$c_u = 720$ lb/ft^2

Factor of safety = 3
Use Eq. (12.20).

12.2 Redo Problem 12.1, this time using Eq. (12.16). Let $E_s = 400p_a$.

12.3 For the drilled shaft described in Problem 12.1, what skin resistance would develop in the top 18 ft, which are in clay? Use Eqs. (12.42) and (12.44).

12.4 Redo Problem 12.1 with the following:

$D_b = 1.75$ m	$\gamma_c = 17.8$ kN/m^3
$D_s = 1$ m	$\gamma_s = 18.2$ kN/m^3
$L_1 = 4$ m	$\phi' = 32°$
$L_2 = 2.5$ m	$c_u = 32$ kN/m^2

Factor of safety = 4

12.5 Redo Problem 12.4 using Eq. (12.16). Let $E_s = 600p_a$.

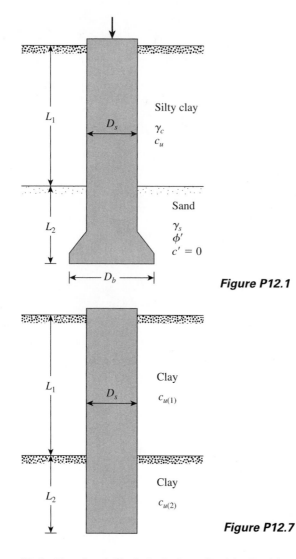

Silty clay
γ_c
c_u

Sand
γ_s
ϕ'
$c' = 0$

Figure P12.1

Clay
$c_{u(1)}$

Clay
$c_{u(2)}$

Figure P12.7

12.6 For the drilled shaft described in Problem 12,4, what skin friction would develop in the top 4 m?
a. Use Eqs. (12.42) and (12.44).
b. Use Eq. (12.45).

12.7 Figure P12.7 shows a drilled shaft without a bell. Assume the following values:

$L_1 = 20$ ft $c_{u(1)} = 1000$ lb/ft^2
$L_2 = 15$ ft $c_{u(2)} = 1800$ lb/ft^2
$D_s = 5$ ft

Determine:
a. The net ultimate point bearing capacity [use Eqs. (12.39) and (12.40)]
b. The ultimate skin friction [use Eqs. (12.42) and (12.44)]
c. The working load Q_w (factor of safety = 4)

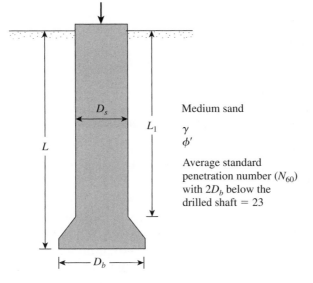

Figure P12.9

12.8 Repeat Problem 12.7 with the following data:

$L_1 = 25$ ft $c_{u(1)} = 1200$ lb/ft^2
$L_2 = 10$ ft $c_{u(2)} = 2000$ lb/ft^2
$D_s = 3.5$ ft

Use Eqs. (12.45) and (12.46).

12.9 A drilled shaft in a medium sand is shown in Figure P12.9. Using the method proposed by Reese and O'Neill, determine the following:

a. The net allowable point resistance for a base movement of 25 mm
b. The shaft frictional resistance for a base movement of 25 mm
c. The total load that can be carried by the drilled shaft for a total base movement of 25 mm

Assume the following values:

$L = 14$ m $\gamma = 19$ kN/m^3
$L_1 = 12.5$ m $\phi' = 36°$
$D_s = 1$ m $D_r = 65\%$ (medium sand)
$D_b = 2$ m

12.10 In Figure P12.9, let $L = 25$ ft, $L_1 = 18$ ft, $D_s = 3$ ft, $D_b = 4.5$ ft, $\gamma = 118$ lb/ft^3, and $\phi' = 35°$. The average uncorrected standard penetration number (N_{60}) within $2D_b$ below the drilled shaft is 29. Determine

a. The ultimate load-carrying capacity
b. The load-carrying capacity for a settlement of 1 in.
The sand has 35% gravel. Use Eq. (12.36) and Figures 12.9 and 12.12.

12.11 For the drilled shaft described in Problem 12.7, determine

a. The ultimate load-carrying capacity
b. The load carrying capacity for a settlement of 1 in.
Use the procedure outlined by Reese and O'Neill. (See Figures 12.16 and 12.17.)

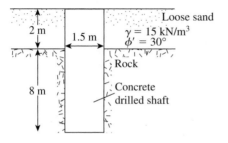

Figure P12.14

12.12 For the drilled shaft described in Problem 12.7, estimate the total elastic settlement at working load. Use Eqs. (11.63), (11.65), and (11.66). Assume that
$E_p = 3.2 \times 10^6 \text{ lb/in}^2$, $C_p = 0.03$, $\xi = 0.65$, $\mu_s = 0.3$, $E_s = 1800 \text{ lb/in}^2$, and
$Q_{ws} = 0.8 Q_w$. Use the value of Q_w from Part (c) of Problem 12.7.

12.13 For the drilled shaft described in Problem 12.8, estimate the total elastic settlement at working load. Use Eqs. (11.63), (11.65), and (11.66). Assume that
$E_p = 3 \times 10^6 \text{ lb/in}^2$, $C_p = 0.03$, $\xi = 0.65$, $\mu_s = 0.3$, $E_s = 2000 \text{ lb/in}^2$, and
$Q_{ws} = 0.83 Q_w$. Use the value of Q_w from Part (c) of Problem 12.8.

12.14 Figure P12.14 shows a drilled shaft extending to rock. Assume the following values:

$$q_{u\ (\text{concrete})} = 28{,}000 \text{ kN/m}^2 \qquad E_{(\text{concrete})} = 22 \text{ GN/m}^2$$

$$q_{u\ (\text{rock})} = 46{,}000 \text{ kN/m}^2 \qquad E_{\text{core(rock)}} = 12.1 \text{ GN/m}^2$$
$$\text{RDQ}_{(\text{rock})} = 75\%$$
Spacing of discontinuity in rock $= 550 \text{ mm}$
Thickness of individual discontinuity in rock $= 3 \text{ mm}$

Estimate the allowable load-bearing capacity of the drilled shaft. Use FS $= 3$.

12.15 A free-headed drilled shaft is shown in Figure P12.15. Let $Q_g = 260 \text{ kN}$,
$M_g = 0$, $\gamma = 17.5 \text{ kN/m}^3$, $\phi' = 35°$, $c' = 0$, and $E_p = 22 \times 10^6 \text{ kN/m}^2$.
Determine
  a. The ground line deflection, x_o
  b. The maximum bending moment in the drilled shaft
  c. The maximum tensile stress in the shaft
  d. The minimum penetration of the shaft needed for this analysis

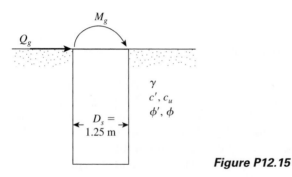

Figure P12.15

References

Berezantzev, V. G., Khristoforov, V. S., and Golubkov, V. N. (1961). "Load Bearing Capacity and Deformation of Piled Foundations," *Proceedings, Fifth International Conference on Soil Mechanics and Foundation Engineering,* Paris, Vol. 2, pp. 11–15.

Chen, Y.-J., and Kulhawy, F. H. (1994). "Case History Evaluation of the Behavior of Drilled Shafts under Axial and Lateral Loading," *Final Report, Project 1493-04, EPRI TR-104601,* Geotechnical Group, Cornell University, Ithaca, NY, December.

Duncan, J. M., Evans, L. T., Jr., and Ooi, P. S. K. (1994). "Lateral Load Analysis of Single Piles and Drilled Shafts," *Journal of Geotechnical Engineering,* ASCE, Vol. 120, No. 6, pp. 1018–1033.

Kulhawy, F. H., and Jackson, C. S. (1989). "Some Observations on Undrained Side Resistance of Drilled Shafts," *Proceedings, Foundation Engineering: Current Principles and Practices,* American Society of Civil Engineers, Vol. 2, pp. 1011–1025.

Matlock, H., and Reese, L.C. (1961). "Foundation Analysis of Offshore Pile-Supported Structures," in *Proceedings, Fifth International Conference on Soil Mechanics and Foundation Engineering,* Vol. 2, Paris, pp. 91–97.

O'Neill, M. W. (1997). Personal communication.

O'Neill, M.W., and Reese, L.C. (1999). *Drilled Shafts: Construction Procedure and Design Methods,* FHWA Report No. IF-99-025.

Reese, L. C., and O'Neill, M. W. (1988). *Drilled Shafts: Construction and Design,* FHWA, Publication No. HI-88-042.

Reese, L. C., and O'Neill, M. W. (1989). "New Design Method for Drilled Shafts from Common Soil and Rock Tests," *Proceedings, Foundation Engineering: Current Principles and Practices,* American Society of Civil Engineers, Vol. 2, pp. 1026–1039.

Rollins, K. M., Clayton, R. J., Mikesell, R. C., and Blaise, B. C. (2005). "Drilled Shaft Side Friction in Gravelly Soils," *Journal of Geotechnical and Geoenvironmental Engineering,* American Society of Civil Engineers, Vol. 131, No. 8, pp. 987–1003.

Whitaker, T., and Cooke, R. W. (1966). "An Investigation of the Shaft and Base Resistance of Large Bored Piles in London Clay," *Proceedings, Conference on Large Bored Piles,* Institute of Civil Engineers, London, pp. 7–49.

13

Foundations on Difficult Soils

13.1   Introduction

In many areas of the United States and other parts of the world, certain soils make the construction of foundations extremely difficult. For example, expansive or collapsible soils may cause high differential movements in structures through excessive heave or settlement. Similar problems can also arise when foundations are constructed over sanitary landfills. Foundation engineers must be able to identify difficult soils when they are encountered in the field. Although not all the problems caused by all soils can be solved, preventive measures can be taken to reduce the possibility of damage to structures built on them. This chapter outlines the fundamental properties of three major soil conditions—collapsible soils, expansive soils, and sanitary landfills—and methods of careful construction of foundations.

Collapsible Soil

13.2   Definition and Types of Collapsible Soil

Collapsible soils, which are sometimes referred to as *metastable soils,* are unsaturated soils that undergo a large change in volume upon saturation. The change may or may not be the result of the application of additional load. The behavior of collapsing soils under load is best explained by the typical void ratio effective pressure plot (e against $\log \sigma'$) for a collapsing soil, as shown in Figure 13.1. Branch ab is determined from the consolidation test on a specimen at its natural moisture content. At an effective pressure level of σ'_w, the equilibrium void ratio is e_1. However, if water is introduced into the specimen for saturation, the soil structure will collapse. After saturation, the equilibrium void ratio at the same effective pressure level σ'_w is e_2; cd is the branch of the e–$\log \sigma'$ curve under additional load after saturation. Foundations that are constructed on such soils may undergo large and sudden settlement if the soil under them becomes saturated with an unanticipated supply of moisture. The moisture may come from any of several sources, such as (a) broken water pipelines, (b) leaky sewers, (c) drainage from reservoirs and swimming pools,

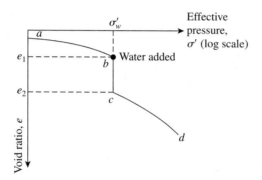

Figure 13.1   Nature of variation of void ratio with pressure for a collapsing soil

(d) a slow increase in groundwater, and so on. This type of settlement generally causes considerable structural damage. Hence, identifying collapsing soils during field exploration is crucial.

The majority of naturally occurring collapsing soils are *aeolian*—that is, wind-deposited sands or silts, such as loess, aeolic beaches, and volcanic dust deposits. The deposits have high void ratios and low unit weights and are cohesionless or only slightly cohesive. *Loess* deposits have silt-sized particles. The cohesion in loess may be the result of clay coatings surrounding the silt-size particles. The coatings hold the particles in a rather stable condition in an unsaturated state. The cohesion also may be caused by the presence of chemical precipitates leached by rainwater. When the soil becomes saturated, the clay binders lose their strength and undergo a structural collapse. In the United States, large parts of the Midwest and the arid West have such types of deposit. Loess deposits are also found over 15 to 20% of Europe and over large parts of China. The thickness of loess deposits can range up to about 10 m (33 ft) in the central United States. In parts of China it can be up to 100 m (330 ft). Figure 13.2 shows the extent of the loess deposits in the Mississippi River basin.

Many collapsing soils may be residual soils that are products of the weathering of parent rocks. Weathering produces soils with a large range of particle-size distribution. Soluble and colloidal materials are leached out by weathering, resulting in large void ratios and thus unstable structures. Many parts of South Africa and Zimbabwe have residual soils that are decomposed granites. Sometimes collapsing soil deposits may be left by flash floods and mudflows. These deposits dry out and are poorly consolidated. An excellent review of collapsing soils is that of Clemence and Finbarr (1981).

<table>
<tr><td>**13.3**</td><td>### *Physical Parameters for Identification*</td></tr>
</table>

Several investigators have proposed various methods for evaluating the physical parameters of collapsing soils for identification. Some of these methods are discussed briefly in Table 13.1.

Jennings and Knight (1975) suggested a procedure for describing the *collapse potential* of a soil: An undisturbed soil specimen is taken at its natural moisture content in a consolidation ring. Step loads are applied to the specimen up to a pressure level σ'_w of 200 kN/m^2 ($\approx$29 lb/ft^2). (In Figure 13.1, this is σ'_w.) At that pressure, the specimen is flooded for saturation and left for 24 hours. This test provides the void

Figure 13.2 Loess deposit in Mississippi River basin

ratios e_1 and e_2 before and after flooding, respectively. The collapse potential may now be calculated as

$$C_p = \Delta\varepsilon = \frac{e_1 - e_2}{1 + e_o} \qquad (13.1)$$

where

e_o = natural void ratio of the soil
$\Delta\varepsilon$ = vertical strain

The severity of foundation problems associated with a collapsible soil have been correlated with the collapse potential C_p by Jennings and Knight (1975). They were summarized by Clemence and Finbarr (1981) and are given in Table 13.2.

Holtz and Hilf (1961) suggested that a loessial soil that has a void ratio large enough to allow its moisture content to exceed its liquid limit upon saturation is susceptible to collapse. So, for collapse,

$$w_{(\text{saturated})} \geq \text{LL} \qquad (13.2)$$

where LL = liquid limit.

However, for saturated soils,

$$e_o = wG_s \tag{13.3}$$

where G_s = specific gravity of soil solids.

Table 13.1 Reported Criteria for Identification of Collapsing Soil[a]

Investigator	Year	Criteria
Denisov	1951	Coefficient of subsidence: $$K = \frac{\text{void ratio at liquid limit}}{\text{natural void ratio}}$$ $K = 0.5{-}0.75$: highly collapsible $K = 1.0$: noncollapsible loam $K = 1.5{-}2.0$: noncollapsible soils
Clevenger	1958	If dry unit weight is less than $12.6\ \text{kN/m}^3(\approx 80\ \text{lb/ft}^3)$, settlement will be large; if dry unit weight is greater than $14\ \text{kN/m}^3(\approx 90\ \text{lb/ft}^3)$ settlement will be small.
Priklonski	1952	$$K_D = \frac{\text{natural moisture content} - \text{plastic limit}}{\text{plasticity index}}$$ $K_D < 0$: highly collapsible soils $K_D > 0.5$: noncollapsible soils $K_D > 1.0$: swelling soils
Gibbs	1961	Collapse ratio, $R = \dfrac{\text{saturation moisture content}}{\text{liquid limit}}$ This was put into graph form.
Soviet Building Code	1962	$$L = \frac{e_o - e_L}{1 + e_o}$$ where e_o = natural void ratio and e_L = void ratio at liquid limit. For natural degree of saturation less than 60%, if $L > -0.1$, the soil is a collapsing soil.
Feda	1964	$$K_L = \frac{w_o}{S_r} - \frac{\text{PL}}{\text{PI}}$$ where w_o = natural water content, S_r = natural degree of saturation, PL = plastic limit, and PI = plasticity index. For $S_r < 100\%$, if $K_L > 0.85$, the soil is a subsident soil.
Benites	1968	A dispersion test in which 2 g of soil are dropped into 12 ml of distilled water and specimen is timed until dispersed; dispersion times of 20 to 30 s were obtained for collapsing Arizona soils.
Handy	1973	Iowa loess with clay (<0.002 mm) contents: $<16\%$: high probability of collapse $16{-}24\%$: probability of collapse $24{-}32\%$: less than 50% probability of collapse $>32\%$: usually safe from collapse

[a]Modified after Lutenegger and Saber (1988)

Table 13.2 Relation of Collapse Potential to the
Severity of Foundation Problems[a]

$C_p(\%)$	Severity of problem
0–1	No problem
1–5	Moderate trouble
5–10	Trouble
10–20	Severe trouble
20	Very severe trouble

[a]After Clemence and Finbarr (1981)

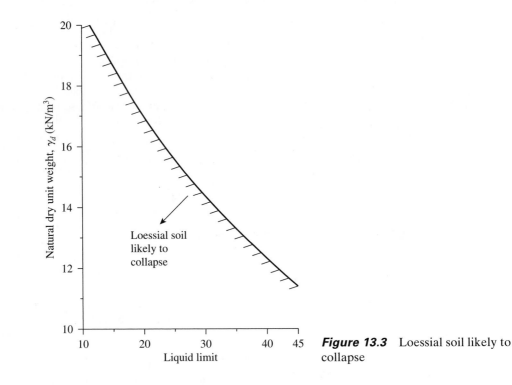

Loessial soil
likely to
collapse

Figure 13.3 Loessial soil likely to collapse

Combining Eqs. (13.2) and (13.3) for collapsing soils yields

$$e_o \geq (\text{LL})(G_s) \tag{13.4}$$

The natural dry unit weight of the soil required for its collapse is

$$\gamma_d \leq \frac{G_s \gamma_w}{1 + e_o} = \frac{G_s \gamma_w}{1 + (\text{LL})(G_s)} \tag{13.5}$$

For an average value of $G_s = 2.65$, the limiting values of γ_d for various liquid limits may now be calculated from Eq. (13.5).

Figure 13.3 shows a plot of the preceding limiting values of dry unit weights against the corresponding liquid limits. For any soil, if the natural dry unit weight falls below the limiting line, the soil is likely to collapse.

Care should be taken to obtain undisturbed samples for determining the collapse potentials and dry unit weights—preferably block samples cut by hand. The reason is that samples obtained by thin-walled tubes may undergo some compression during the sampling process. However, if cut block samples are used, the boreholes should be made *without water.*

13.4 *Procedure for Calculating Collapse Settlement*

Jennings and Knight (1975) proposed the following laboratory procedure for determining the collapse settlement of structures upon saturation of soil:

Step 1. Obtain *two* undisturbed soil specimens for tests in a standard consolidation test apparatus (oedometer).

Step 2. Place the two specimens under $1 \text{ kN/m}^2 (0.15 \text{ lb/in}^2)$ pressure for 24 hours.

Step 3. After 24 hours, saturate one specimen by flooding. Keep the other specimen at its natural moisture content.

Step 4. After 24 hours of flooding, resume the consolidation test on both specimens by doubling the load (the same as in the standard consolidation test) to the desired pressure level.

Step 5. Plot the e–log σ' graphs for both specimens (Figures 13.4a and b).

Step 6. Calculate the *in situ* effective pressure, σ'_o. Draw a vertical line corresponding to the pressure σ'_o.

Step 7. From the e–log σ'_o curve of the soaked specimen, determine the preconsolidation pressure, σ'_c. If $\sigma'_c/\sigma'_o = 0.8$–$1.5$, the soil is normally consolidated; however, if $\sigma'_c/\sigma'_o > 1.5$, the soil is preconsolidated.

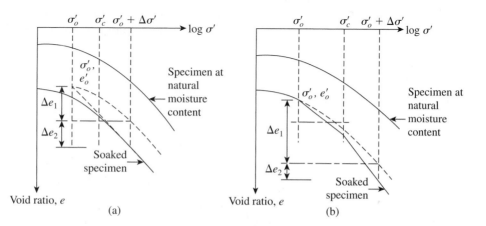

Figure 13.4 Settlement calculation from double oedometer test: (a) normally consolidated soil; (b) overconsolidated soil

Step 8. Determine e_o', corresponding to σ_o' from the e–log σ_o' curve of the soaked specimen. (This procedure for normally consolidated and over-consolidated soils is shown in Figures 13.4a and b, respectively.)

Step 9. Through point (σ_o', e_o') draw a curve that is similar to the e–log σ_o' curve obtained from the specimen tested at its natural moisture content.

Step 10. Determine the incremental pressure, $\Delta\sigma'$, on the soil caused by the construction of the foundation. Draw a vertical line corresponding to the pressure of $\sigma_o' + \Delta\sigma'$ in the e–log σ_o' curve.

Step 11. Now, determine Δe_1 and Δe_2. The settlement of soil without change in the natural moisture content is

$$S_{c(1)} = \frac{\Delta e_1}{1 + e_o'}(H) \tag{13.6}$$

where H = thickness of soil susceptible to collapse.
Also, the settlement caused by collapse in the soil structure is

$$S_{c(2)} = \frac{\Delta e_2}{1 + e_o'}(H) \tag{13.7}$$

13.5 Foundation Design in Soils Not Susceptible to Wetting

For actual foundation design purposes, some standard field load tests may also be conducted. Figure 13.5 shows the relation of the nature of load per unit area versus settlement in a field load test in a loessial deposit. Note that the load–settlement relationship is essentially linear up to a certain critical pressure, σ_{cr}', at which there is a breakdown of the soil structure and hence a large settlement. Sudden breakdown of soil structure is more common with soils having a high natural moisture content than with normally dry soils.

If enough precautions are taken in the field to prevent moisture from increasing under structures, spread foundations and mat foundations may be built on

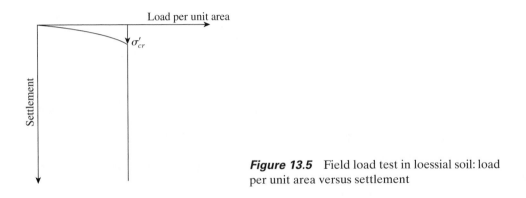

Figure 13.5 Field load test in loessial soil: load per unit area versus settlement

collapsible soils. However, the foundations must be proportioned so that the critical stresses (see Figure 13.5) in the field are never exceeded. A factor of safety of about 2.5 to 3 should be used to calculate the allowable soil pressure, or

$$\sigma'_{all} = \frac{\sigma'_{cr}}{FS} \tag{13.8}$$

where

σ'_{all} = allowable soil pressure
FS = factor of safety

The differential and total settlements of these foundations should be similar to those of foundations designed for sandy soils.

Continuous foundations may be safer than isolated foundations over collapsible soils in that they can effectively minimize differential settlement. Figure 13.6 shows a typical procedure for constructing continuous foundations. This procedure uses footing beams and longitudinal load-bearing beams.

In the construction of heavy structures, such as grain elevators, over collapsible soils, settlements up to about 0.3 m ($\approx$1 ft) are sometimes allowed (Peck, Hanson, and Thornburn, 1974). In this case, tilting of the foundation is not likely to occur, because there is no eccentric loading. The total expected settlement for such structures can be estimated from standard consolidation tests on specimens of field moisture content. Without eccentric loading, the foundations will exhibit uniform settlement over loessial deposits; however, if the soil is of residual or colluvial nature, settlement may not be uniform. The reason is the nonuniformity generally encountered in residual soils.

Extreme caution must be used in building heavy structures over collapsible soils. If large settlements are expected, drilled-shaft and pile foundations should be considered. These types of foundation can transfer the load to a stronger load-bearing stratum.

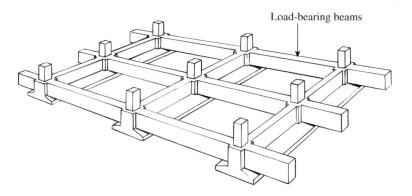

Figure 13.6 Continuous foundation with load-bearing beams (After Clemence and Finbarr, 1981)

13.6   *Foundation Design in Soils Susceptible to Wetting*

If the upper layer of soil is likely to get wet and collapse sometime after construction of the foundation, several design techniques to avoid failure of the foundation may be considered:

1. If the expected depth of wetting is about 1.5 to 2 m ($\approx$5 to 6.5 ft) from the ground surface, the soil may be moistened and recompacted by heavy rollers. Spread footings and mats may be constructed over the compacted soil. An alternative to recompaction by heavy rollers is *heavy tamping,* which is sometimes referred to as *dynamic compaction.* (See Chapter 14.) Heavy tamping consists primarily of dropping a heavy weight repeatedly on the ground. The height of the drop can vary from 8 to 30 m ($\approx$25 to 100 ft). The stress waves generated by the dropping weight help in the densification of the soil.

   Lutenegger (1986) reported the case history of stabilizing a thick layer of friable loess before construction by dynamic compaction in Russe, Bulgaria.

2. If conditions are favorable, foundation trenches can be flooded with solutions of sodium silicate and calcium chloride to stabilize the soil chemically. The soil will then behave like a soft sandstone and resist collapse upon saturation. This method is successful only if the solutions can penetrate to the desired depth; thus, it is most applicable to fine sand deposits. Silicates are rather costly and are not generally used. However, in some parts of Denver, silicates have been used successfully.

   The injection of a sodium silicate solution to stabilize collapsible soil deposits has been used extensively in the former Soviet Union and Bulgaria (Houston and Houston, 1989). This process, which is used for dry collapsible soils and for wet collapsible soils that are likely to compress under the added weight of the structure to be built, consists of three steps:

   Step 1. Injection of carbon dioxide to remove any water that is present and for preliminary activation of the soil
   Step 2. Injection of sodium silicate grout
   Step 3. Injection of carbon dioxide to neutralize alkalies.

3. When the soil layer is susceptible to wetting to a depth of about 10 m ($\approx$30 ft), several techniques may be used to cause collapse of the soil *before* construction of the foundation is begun. Two of these techniques are *vibroflotation* and *ponding* (also called *flooding*). Vibroflotation is used successfully in free-draining soil. (See Chapter 14.) Ponding—by constructing low dikes—is utilized at sites that have no impervious layers. However, even after saturation and collapse of the soil by ponding, some additional settlement of the soil may occur after construction of the foundation is begun. Additional settlement may also be caused by incomplete saturation of the soil at the time of construction. Ponding may be used successfully in the construction of earth dams.

4. If precollapsing the soil is not practical, foundations may be extended beyond the zone of possible wetting, although the technique may require drilled shafts and piles. The design of drilled shafts and piles must take into consideration the effect of negative skin friction resulting from the collapse of the soil structure and the associated settlement of the zone of subsequent wetting.

In some cases, a *rock-column type of foundation* (*vibroreplacement*) may be considered. Rock columns are built with large boulders that penetrate the collapsible soil layer. They act as piles in transferring the load to a more stable soil layer.

Expansive Soils

13.7 *General Nature of Expansive Soils*

Many plastic clays swell considerably when water is added to them and then shrink with the loss of water. Foundations constructed on such clays are subjected to large uplifting forces caused by the swelling. These forces induce heaving, cracking, and the breakup of both building foundations and slab-on-grade members. Expansive clays cover large parts of the United States, South America, Africa, Australia, and India. In the United States, these clays are predominant in Texas, Oklahoma, and the upper Missouri Valley. In general, expansive clays have liquid limits and plasticity indices greater than about 40 and 15, respectively.

As noted, an increase in moisture content causes clay to swell. The depth in a soil to which periodic changes of moisture occur is usually referred to as the *active zone* (see Figure 13.7). The depth of the active zone varies, depending on the location of the site. Some typical active-zone depths in American cities are given in Table 13.3. In some clays and clay shales in the western United States, the depth of the active zone can be as much as 15 m ($\approx$50 ft). The active-zone depth can easily be determined by plotting the liquidity index against the depth of the soil profile over several seasons. Figure 13.8 shows such a plot for the Beaumont formation in the Houston area.

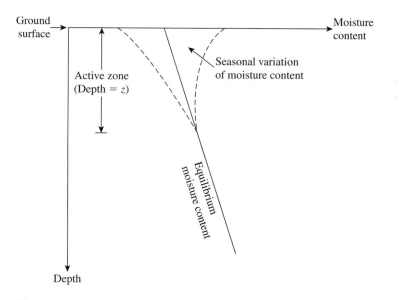

Figure 13.7 Definition of active zone

Table 13.3 Typical Active-Zone Depths in Some U.S. Cities[a]

City	Depth of active zone	
	(m)	(ft)
Houston	1.5 to 3	5 to 10
Dallas	2.1 to 4.6	7 to 15
San Antonio	3 to 9	10 to 30
Denver	3 to 4.6	10 to 15

[a] After O'Neill and Poormoayed (1980)

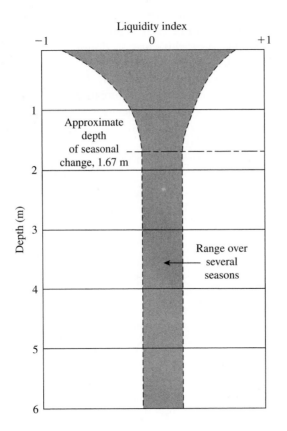

Figure 13.8 Active zone in Houston area, Beaumont formation (After O'Neill and Poormoayed, 1980)

13.8 *Laboratory Measurement of Swell*

To study the magnitude of possible swell in a clay, simple laboratory oedometer tests can be conducted on undisturbed specimens. Two common tests are the unrestrained swell test and the swelling pressure test.

Unrestrained Swell Test

In the *unrestrained swell test,* the specimen is placed in an oedometer under a small surcharge of about 6.9 kN/m^2 (1 lb/in^2). Water is then added to the specimen, and the expansion of the volume of the specimen (i.e., its height; the area of cross section is constant) is measured until equilibrium is reached. The percentage of free swell may than be expressed as the ratio

$$s_{w(\text{free})}(\%) = \frac{\Delta H}{H}(100) \tag{13.9}$$

where

$s_{w(\text{free})}$ = free swell, as a percentage
ΔH = height of swell due to saturation
H = original height of the specimen

Vijayvergiya and Ghazzaly (1973) analyzed various soil test results obtained in this manner and prepared a correlation chart of the free swell, liquid limit, and natural moisture content, as shown in Figure 13.9. O'Neill and Poormoayed (1980) developed a relationship for calculating the free surface swell from this chart:

$$\Delta S_F = 0.0033 Z s_{w(\text{free})} \tag{13.10}$$

where

ΔS_F = free surface swell
Z = depth of active zone
$s_{w(\text{free})}$ = free swell, as a percentage (see Figure 13.9)

Swelling Pressure Test

The *swelling pressure test* can be conducted by taking a specimen in a consolidation ring and applying a pressure equal to the effective overburden pressure, σ'_o, plus the approximate anticipated surcharge caused by the foundation, σ'_s. Water is then added to the specimen. As the specimen starts to swell, pressure is applied in small increments to prevent swelling. Pressure is maintained until full swelling pressure is developed on the specimen, at which time the total pressure is

$$\sigma'_T = \sigma'_o + \sigma'_s + \sigma'_1 \tag{13.11}$$

where

σ'_T = total pressure applied to prevent swelling, or zero swell pressure
σ'_1 = further pressure applied to prevent swelling after addition of water

Figure 13.10 shows the variation of the percentage of swell with effective pressure during a swelling pressure test. (For more information on this type of test, see Sridharan et al., 1986.)

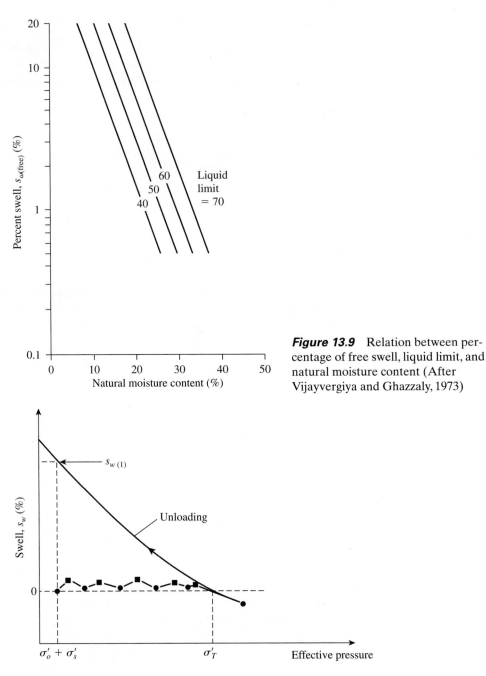

Figure 13.9 Relation between percentage of free swell, liquid limit, and natural moisture content (After Vijayvergiya and Ghazzaly, 1973)

Figure 13.10 Swelling pressure test

A σ'_T of about 20 to 30 kN/m^2 (400 to 650 lb/ft^2) is considered to be low, and a σ'_T of 1500 to 2000 kN/m^2 (30,000 to 40,000 lb/ft^2) is considered to be high. After zero swell pressure is attained, the soil specimen can be unloaded in steps to the level of the effective overburden pressure, σ'_o. The unloading process will cause the specimen to swell. The equilibrium swell for each pressure level is also recorded. The variation of the swell, s_w in percent, and the applied pressure on the specimen will be like that shown in Figure 13.10.

The swelling pressure test can be used to determine the surface heave, ΔS, for a foundation (O'Neill and Poormoayed, 1980) as given by the formula

$$\Delta S = \sum_{i=1}^{n} [s_{w(1)}\,(\%)](H_i)(0.01) \tag{13.12}$$

where

$s_{w(1)}(\%)$ = swell, in percent, for layer i under a pressure of $\sigma_o' + \sigma_s'$ (see Figure 13.10)

ΔH_i = thickness of layer i

Example 13.1

A soil profile has an active zone of expansive soil of 2 m. The liquid limit and the average natural moisture content during the construction season are 60% and 30%, respectively. Determine the free surface swell.

Solution

From Figure 13.9 for LL = 60% and w = 30%, $s_{w(\text{free})}$ = 1%. From Eq. (13.10),

$$\Delta S_F = 0.0033 Z s_{w(\text{free})}$$

Hence,

$$\Delta S_F = 0.0033(2)(1)(1000) = \textbf{6.6 mm} \qquad \blacksquare$$

Example 13.2

An expansive soil profile has an active-zone thickness of 5.2 m. A shallow foundation is to be constructed 1.2 m below the ground surface. A swelling pressure test provided the following data:

Depth below ground surface (m)	Swell under overburden and estimated foundation surcharge pressure, $s_{w(1)}$ (%)
1.2	3.0
2.2	2.0
3.2	1.2
4.2	0.55
5.2	0.0

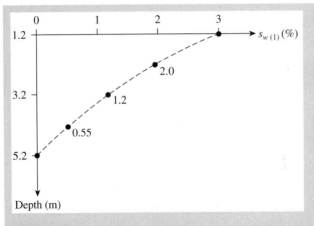

Figure 13.11

a. Estimate the total possible swell under the foundation.
b. If the allowable total swell is 15 mm, what would be the necessary undercut?

Solution
Part a
Figure 13.11 shows the plot of depth versus $s_{w(1)}$ (%). The area of this diagram will be the total swell. Thus

$$\Delta S = \frac{1}{100}\left[\left(\frac{1}{2}\right)(0.55 + 0)(1) + \left(\frac{1}{2}\right)(0.55 + 1.2)(1)\right.$$

$$\left. + \left(\frac{1}{2}\right)(1.2 + 2)(1) + \left(\frac{1}{2}\right)(2 + 3)(1)\right]$$

$$= 0.0525 \text{ m} = \textbf{52.5 m}$$

Part b
Total swell at various depths can be calculated as follows:

Depth (m)	Total swell, ΔS (m)
5.2	0
4.2	$0 + \frac{1}{2}(0.55 + 0)(1)(1/100) = 0.00275$
3.2	$0.00275 + \frac{1}{2}(1.2 + 0.55)(1)(1/100) = 0.0115$
2.2	$0.0115 + \frac{1}{2}(2 + 1.2)(1)(1/100) = 0.0275$
1.2	$0.0275 + \frac{1}{2}(2 + 3)(1)(1/100) = 0.0525$

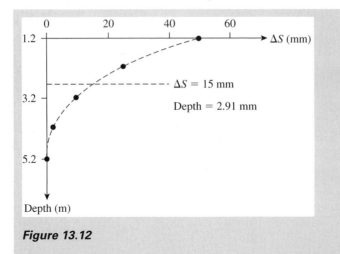

Figure 13.12

The plot of ΔS versus depth is shown in Figure 13.12. From this figure, the depth of undercut is $2.91 - 1.2 = \mathbf{1.71}$ **m below the bottom of the foundation.**

13.9 Classification of Expansive Soil on the Basis of Index Tests

Classification systems for expansive soils are based on the problems they create in the construction of foundations (potential swell). Most of the classifications contained in the literature are summarized in Figure 13.13 and Table 13.4. However, the classification system developed by the U.S. Army Waterways Experiment Station (Snethen et al., 1977) is the one most widely used in the United States. It has also been summarized by O'Neill and Poormoayed (1980); see Table 13.5. More recently, Sridharan (2005) proposed an index called the *free swell ratio* to predict the clay type, potential swell classification, and dominant clay minerals present in a given soil. The free swell ratio can be determined by finding the equilibrium sediment volumes of 10 grams of an oven-dried specimen passing No. 40 U.S. sieve (0.425 mm opening) in distilled water (V_d) and in CCl_4 or kerosene (V_K). The free swell ratio (FSR) is defined as

$$\text{FSR} = \frac{V_d}{V_K} \tag{13.13}$$

Table 13.6 gives the expansive soil classification based on free swell ratio. Figure 13.14 shows the classification of soil based on the free swell ratio.

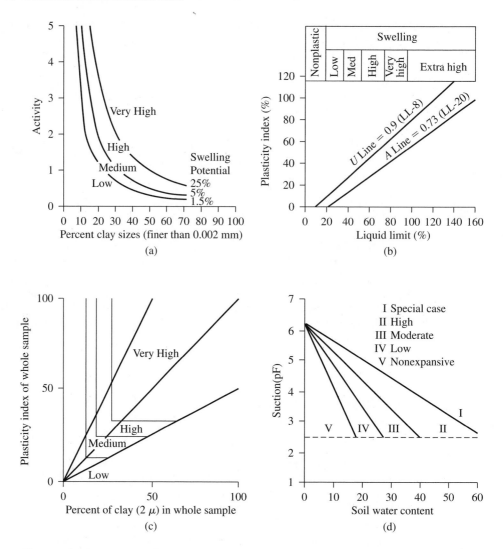

Figure 13.13 Commonly used criteria for determining swell potential (After Abduljauwad and Al-Sulaimani, 1993)

Foundation Considerations for Expansive Soils

If a soil has a low swell potential, standard construction practices may be followed. However, if the soil possesses a marginal or high swell potential, precautions need to be taken, which may entail

1. Replacing the expansive soil under the foundation
2. Changing the nature of the expansive soil by compaction control, prewetting, installation of moisture barriers, or chemical stabilization

Table 13.4 Summary of Some Criteria for Identifying Swell Potential (After Abduljauwad and Al-Sulaimani, 1993)

Reference	Criteria	Remarks
Holtz (1959)	CC > 28, PI > 35, and SL < 11 (very high) 20 ≤ CC ≤ 31, 25 ≤ PI ≤ 41, and 7 ≤ SL ≤ 12 (high) 13 ≤ CC ≤ 23, 15 ≤ PI ≤ 28, and 10 ≤ SL ≤ 16 (medium) CC ≤ 15, PI ≤ 18, and SL ≥ 15 (low)	Based on CC, PI, and SL
Seed et al. (1962)	See Figure 13.13a	Based on oedometer test using compacted specimen, percentage of clay <2 μm, and activity
Altmeyer (1955)	LS < 5, SI > 12, and PS < 0.5 (noncritical) 5 ≤ LS ≤ 8, 10 ≤ SL ≤ 12, and 0.5 ≤ PS ≤ 1.5 (marginal) LS > 8, SL < 10, and PS > 1.5 (critical)	Based on LS, SL, and PS Remolded sample ($\rho_{d(\max)}$ and w_{opt}) soaked under 6.9 kPa surcharge
Dakshanamanthy and Raman (1973)	See Figure 13.13b	Based on plasticity chart
Raman (1967)	PI > 32 and SI > 40 (very high) 23 ≤ PI ≤ 32 and 30 ≤ SI ≤ 40 (high) 12 ≤ PI ≤ 23 and 15 ≤ SI ≤ 30 (medium) PI < 12 and SI < 15 (low)	Based on PI and SI
Sowers and Sowers (1970)	SL < 10 and PI > 30 (high) 10 ≤ SL ≤ 12 and 15 ≤ PI ≤ 30 (moderate) SL > 12 and PI < 15 (low)	Little swell will occur when w_o results in LI of 0.25
Van Der Merwe (1964)	See Figure 13.13c	Based on PI, percentage of clay <2 μm, and activity
Uniform Building Code, 1968	EI > 130 (very high) and 91 ≤ EI ≤ 130 (high) 51 ≤ EI ≤ 90 (medium) and 21 ≤ EI ≤ 50 (low) 0 ≤ EI ≤ 20 (very low)	Based on oedometer test on compacted specimen with degree of saturation close to 50% and surcharge of 6.9 kPa
Snethen (1984)	LL > 60, PI > 35, τ_{nat} > 4, and SP > 1.5 (high) 30 ≤ LL ≤ 60, 25 ≤ PI ≤ 35, 1.5 ≤ τ_{nat} ≤ 4, and 0.5 ≤ SP ≤ 1.5 (medium) LL < 30, PI < 25, τ_{nat} < 1.5, and SP < 0.5 (low)	PS is representative for field condition and can be used without τ_{nat}, but accuracy will be reduced
Chen (1988)	PI ≥ 35 (very high) and 20 ≤ PI ≤ 55 (high) 10 ≤ PI ≤ 35 (medium) and PI ≤ 15 (low)	Based on PI
McKeen (1992)	Figure 13.13d	Based on measurements of soil water content, suction, and change in volume on drying
Vijayvergiya and Ghazzaly (1973)	log SP = (1/12)(0.44 LL − w_o + 5.5)	Empirical equations
Nayak and Christensen (1974)	SP = (0.00229 PI)(1.45C)/w_o + 6.38	Empirical equations
Weston (1980)	SP = $0.00411(\text{LL}_w)^{4.17} q^{-3.86} w_o^{-2.33}$	Empirical equations

(Continued)

Table 13.4 (Continued)

Note: C = clay, % CC = colloidal content, % EI = Expansion index = 100 $\times$ percent swell $\times$ fraction passing No. 4 sieve LI = liquidity index, % LL = liquid limit, % LL_w = weighted liquid limit, % LS = linear shrinkage, % PI = plasticity index, % PS = probable swell, %	q = surcharge SI = shrinkage index = LL $-$ SL, % SL = shrinkage limit, % SP = swell potential, % w_o = natural soil moisture w_{opt} = optimum moisture content, % τ_{nat} = natural soil suction in tsf $\rho_{d(max)}$ = max dry density

Table 13.5 Expansive Soil Classification System[a]

Liquid limit	Plasticity index	Potential swell (%)	Potential swell classification
<50	<25	<0.5	Low
50–60	25–35	0.5–1.5	Marginal
>60	>35	>1.5	High
Potential swell = vertical swell under a pressure equal to overburden pressure			

[a] Compiled from O'Neill and Poormoayed (1980)

Table 13.6 Expansive Soil Classification Based on Free Swell Ratio

Free swell ratio	Clay type	Potential swell classification	Dominant clay mineral
≤ 1.0	Non-swelling	Negligible	Kaolinite
1.0–1.5	Mixture of swelling and non-swelling	Low	Kaolinite and montmorillonite
1.5–2.0	Swelling	Moderate	Montmorillonite
2.0–4.0	Swelling	High	Montmorillonite
> 4.0	Swelling	Very High	Montmorillonite

3. Strengthening the structures to withstand heave, constructing structures that are flexible enough to withstand the differential soil heave without failure, or constructing isolated deep foundations below the depth of the active zone

One particular method may not be sufficient in all situations. Combining several techniques may be necessary, and local construction experience should always be considered. Following are some details regarding the commonly used techniques for dealing with expansive soils.

Replacement of Expansive Soil

When shallow, moderately expansive soils are present at the surface, they can be removed and replaced by less expansive soils and then compacted properly.

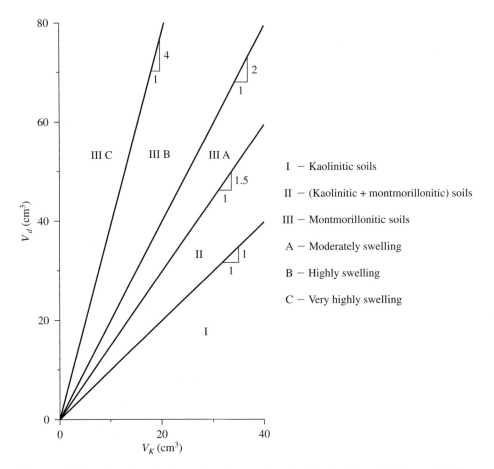

Figure 13.14 Classification based on free swell ratio (adapted from Sridharan, 2005)

Changing the Nature of Expansive Soil

1. *Compaction:* The heave of expansive soils decreases substantially when the soil is compacted to a lower unit weight on the high side of the optimum moisture content (possibly 3 to 4% above the optimum moisture content). Even under such conditions, a slab-on-ground type of construction should not be considered when the total probable heave is expected to be about 38 mm (1.5 in.) or more.

2. *Prewetting:* One technique for increasing the moisture content of the soil is ponding and hence achieving most of the heave before construction. However, this technique may be time consuming because the seepage of water through highly plastic clays is slow. After ponding, 4 to 5% of hydrated lime may be added to the top layer of the soil to make it less plastic and more workable (Gromko, 1974).

3. *Installation of moisture barriers:* The long-term effect of the differential heave can be reduced by controlling the moisture variation in the soil. This is achieved by providing vertical moisture barriers about 1.5 m ($\approx$5 ft) deep around the perimeter of slabs for the slab-on-grade type of construction. These moisture barriers may be constructed in trenches filled with gravel, lean concrete, or impervious membranes.

4. *Stabilization of soil:* Chemical stabilization with the aid of lime and cement has often proved useful. A mix containing about 5% lime is sufficient in most cases. Lime or cement and water are mixed with the top layer of soil and compacted. The addition of lime or cement will decrease the liquid limit, the plasticity index, and the swell characteristics of the soil. This type of stabilization work can be done to a depth of 1 to 1.5 m (≈3 to 5 ft). Hydrated high-calcium lime and dolomite lime are generally used for lime stabilization.

Another method of stabilization of expansive soil is the *pressure injection* of lime slurry or lime–fly-ash slurry into the soil, usually to a depth of 4 to 5 m or (12 to 16 ft) and occasionally deeper to cover the active zone. Further details of the pressure injection technique are presented in Chapter 14. Depending on the soil conditions at a site, single or multiple injections can be planned, as shown in Figure 13.15. Figure 13.16

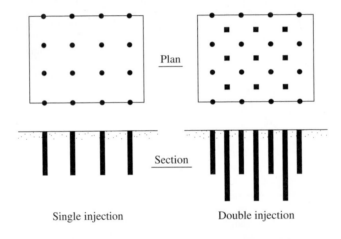

Figure 13.15 Multiple lime slurry injection planning for a building pad

Figure 13.16 Pressure injection of lime slurry for a building pad (Courtesy of GKN Hayward Baker, Inc., Woodbine Division, Ft. Worth, Texas)

Figure 13.17 Slope stabilization of a canal bank by pressure injection of lime–fly-ash slurry (Courtesy of GKN Hayward Baker, Inc., Woodbine Division, Ft. Worth, Texas)

shows the slurry pressure injection work for a building pad. The stakes that are marked are the planned injection points. Figure 13.17 shows lime–fly-ash stabilization by pressure injection of the bank of a canal that had experienced sloughs and slides.

13.11 *Construction on Expansive Soils*

Care must be exercised in choosing the type of foundation to be used on expansive soils. Table 13.7 shows some recommended construction procedures based on the total predicted heave, ΔS, and the length-to-height ratio of the wall panels. For example, the table proposes the use of waffle slabs as an alternative in designing rigid buildings that are capable of tolerating movement. Figure 13.18 shows a schematic diagram of a waffle slab. In this type of construction, the ribs hold the structural load. The waffle voids allow the expansion of soil.

Table 13.7 also suggests the use of a drilled shaft foundation with a suspended floor slab when structures are constructed independently of movement of the soil. Figure 13.19a shows a schematic diagram of such an arrangement. The bottom of the shafts should be placed below the active zone of the expansive soil. For the design of the shafts, the uplifting force, U, may be estimated (see Figure 13.19b) from the equation

$$U = \pi D_s Z \sigma'_T \tan \phi'_{ps} \tag{13.14}$$

where

D_s = diameter of the shaft
Z = depth of the active zone
ϕ'_{ps} = effective angle of plinth–soil friction
σ'_T = pressure for zero swell (see Figure 13.10;
$\qquad \sigma'_T = \sigma'_o + \sigma'_s + \sigma'_1$)

Table 13.7 Construction Procedures for Expansive Clay Soils[a]

Total predicted heave (mm)		Recommended construction	Method	Remarks
L/H = 1.25	*L/H* = 2.5			
0 to 6.35	12.7	No precaution		
6.35 to 12.7	12.7 to 50.8	Rigid building tolerating movement (steel reinforcement as necessary)	*Foundations:* Pads Strip footings mat (waffle)	Footings should be small and deep, consistent with the soil-bearing capacity. Mats should resist bending.
			Floor slabs: Waffle Tile	Slabs should be designed to resist bending and should be independent of grade beams.
			Walls:	Walls on a mat should be as flexible as the mat. There should be no vertical rigid connections. Brickwork should be strengthened with tie bars or bands.
12.7 to 50.8	50.8 to 101.6	Building damping movement	*Joints:* Clear Flexible	Contacts between structural units should be avoided, or flexible, waterproof material may be inserted in the joints.
			Walls: Flexible Unit construction Steel frame	Walls or rectangular building units should heave as a unit.
			Foundations: Three point Cellular Jacks	Cellular foundations allow slight soil expansion to reduce swelling pressure. Adjustable jacks can be inconvenient to owners. Three-point loading allows motion without duress.
>50.8	>101.6	Building independent of movement	*Foundation drilled shaft:* Straight shaft Bell bottom	Smallest-diameter and widely spaced shafts compatible with the load should be placed. Clearance should be allowed under grade beams.
			Suspended floor:	Floor should be suspended on grade beams 305 to 460 mm above the soil.

[a] After Gromko, 1974

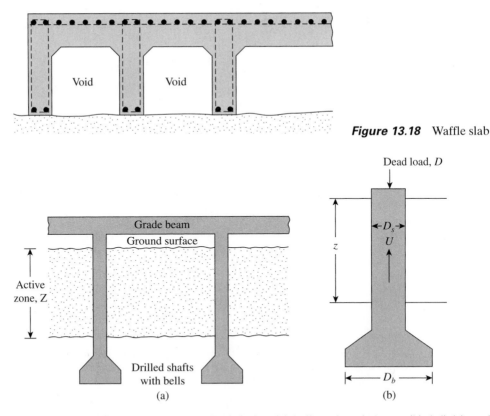

Figure 13.18 Waffle slab

Figure 13.19 (a) Construction of drilled shafts with bells and grade beam; (b) definition of parameters in Eq. (13.14)

In most cases, the value of ϕ'_{ps} varies between 10 and 20°. An average value of the zero horizontal swell pressure must be determined in the laboratory. In the absence of laboratory results, $\sigma'_T \tan \phi'_{ps}$ may be considered equal to the undrained shear strength of clay, c_u, in the active zone.

The belled portion of the drilled shaft will act as an anchor to resist the uplifting force. Ignoring the weight of the drilled shaft, we have

$$Q_{net} = U - D \tag{13.15}$$

where

Q_{net} = net uplift load
D = dead load

Now,

$$Q_{net} \approx \frac{c_u N_c}{FS}\left(\frac{\pi}{4}\right)(D_b^2 - D_s^2) \tag{13.16}$$

where

c_u = undrained cohesion of the clay in which the bell is located
N_c = bearing capacity factor

FS = factor of safety

D_b = diameter of the bell of the drilled shaft

Combining Eqs. (13.15) and (13.16) gives

$$U - D = \frac{c_u N_c}{FS}\left(\frac{\pi}{4}\right)(D_b^2 - D_s^2) \tag{13.17}$$

Conservatively, from Table 3.3 and Eq. (3.27),

$$N_c \approx N_{c(\text{strip})}F_{cs} = N_{c(\text{strip})}\left(1 + \frac{N_q B}{N_c L}\right) \approx 5.14\left(1 + \frac{1}{5.14}\right) = 6.14$$

A drilled-shaft design is examined in Example 13.3.

Example 13.3

Figure 13.20 shows a drilled shaft with a bell. The depth of the active zone is 5 m. The zero swell pressure of the swelling clay (σ'_T) is 450 kN/m². For the drilled shaft, the dead load (D) is 600 kN and the live load is 300 kN. Assume $\phi'_{ps} = 12°$.

a. Determine the diameter of the bell, D_b.
b. Check the bearing capacity of the drilled shaft assuming zero uplift force.

Solution
Part a: Determining the Bell Diameter, D_b
The uplift force, Eq. (13.14), is

$$U = \pi D_s Z \sigma'_T \tan \phi'_{ps}$$

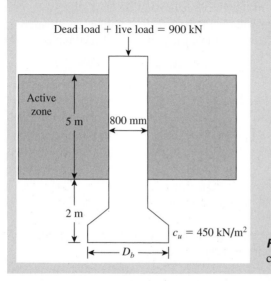

Dead load + live load = 900 kN

Active zone

5 m   800 mm

2 m

c_u = 450 kN/m²

D_b

Figure 13.20   Drill shaft in a swelling clay

Given: $Z = 5$ m and $\sigma'_T = 450$ kN/m^2. Then

$$U = \pi(0.8)(5)(450)\tan 12° = 1202 \text{ kN}$$

Assume the dead load and live load to be zero, and FS in Eq. (13.17) to be 1.25. So, from Eq. (13.17),

$$U = \frac{c_u N_c}{\text{FS}}\left(\frac{\pi}{4}\right)(D_b^2 - D_s^2)$$

$$1202 = \frac{(450)(6.14)}{1.25}\left(\frac{\pi}{4}\right)(D_b^2 - 0.8^2); \quad D_b = \textbf{1.15 m}$$

The factor of safety against uplift with the dead load also should be checked. A factor of safety of at least 2 is desirable. So, from Eq. (13.17)

$$\text{FS} = \frac{c_u N_c\left(\dfrac{\pi}{4}\right)(D_b^2 - D_s^2)}{U - D}$$

$$= \frac{(450)(6.14)\left(\dfrac{\pi}{4}\right)(1.15^2 - 0.8^2)}{1202 - 600} = \textbf{2.46} > \textbf{2—OK}$$

Part b: Check for Bearing Capacity
Assume that $U = 0$. Then

$$\text{Dead load + live load} = 600 + 300 = 900 \text{ kN}$$

$$\text{Downward load per unit area} = \frac{900}{\left(\dfrac{\pi}{4}\right)(D_b^2)} = \frac{900}{\left(\dfrac{\pi}{4}\right)(1.15^2)} = 866.5 \text{ kN/m}^2$$

Net bearing capacity of the soil under the bell $= q_{u(\text{net})} = c_u N_c$

$$= (450)(6.14) = 2763 \text{ kN/m}^2$$

Hence, the factor of safety against bearing capacity failure is

$$\text{FS} = \frac{2763}{866.5} = \textbf{3.19} > \textbf{3—OK} \qquad \blacksquare$$

Sanitary Landfills

13.12 General Nature of Sanitary Landfills

Sanitary landfills provide a way to dispose of refuse on land without endangering public health. Sanitary landfills are used in almost all countries, to varying degrees of success. The refuse disposed of in sanitary landfills may contain organic, wood,

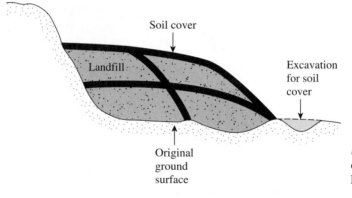

Soil cover

Landfill

Excavation
for soil
cover

Original
ground
surface

Figure 13.21   Schematic
diagram of a sanitary
landfill in progress

paper, and fibrous wastes, or demolition wastes such as bricks and stones. The refuse is dumped and compacted at frequent intervals and is then covered with a layer of soil, as shown in Figure 13.21. In the compacted state, the average unit weight of the refuse may vary between 5 and 10 kN/m^3 (32 to 64 lb/ft^3). A typical city in the United States, with a population of 1 million, generates about $3.8 \times 10^6 \text{ m}^3$ ($\approx 135 \times 10^6 \text{ ft}^3$) of compacted landfill material per year.

As property values continue to increase in densely populated areas, constructing structures over sanitary landfills becomes more and more tempting. In some instances, a visual site inspection may not be enough to detect an old sanitary landfill. However, construction of foundations over sanitary landfills is generally problematic because of poisonous gases (e.g., methane), excessive settlement, and low inherent bearing capacity.

13.13   *Settlement of Sanitary Landfills*

Sanitary landfills undergo large continuous settlements over a long time. Yen and Scanlon (1975) documented the settlement of several landfill sites in California. After completion of the landfill, the settlement rate (Figure 13.22) may be expressed as

$$m = \frac{\Delta H_f}{\Delta t} \tag{13.18}$$

where

m = settlement rate
H_f = maximum height of the sanitary landfill

On the basis of several field observations, Yen and Scanlon determined the following empirical correlations for the settlement rate:

$m = 0.0268 - 0.0116 \log t_1$ (for fill heights ranging from 12 to 24 m)   (13.19)

$m = 0.038 - 0.0155 \log t_1$ (for fill heights ranging from 24 to 30 m)   (13.20)

$m = 0.0433 - 0.0183 \log t_1$ (for fill heights larger than 30 m)   (13.21)

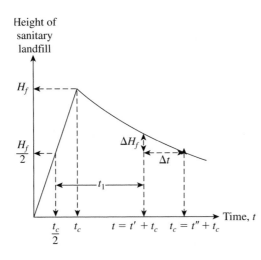

Figure 13.22 Settlement of sanitary landfills

In these equations,

m is in m/mo.
t_1 is the median fill age, in months

The median fill age may be defined from Figure 13.22 as

$$t_1 = t - \frac{t_c}{2} \qquad (13.22)$$

where

t = time from the beginning of the landfill
t_c = time for completion of the landfill

Equations (13.19), (13.20), and (13.21) are based on field data from landfills for which t_c varied from 70 to 82 months. To get an idea of the approximate length of time required for a sanitary landfill to undergo complete settlement, consider Eq. (13.19). For a fill 12 m high and for t_c = 72 months,

$$m = 0.0268 - 0.0116 \log t_1$$

so

$$\log t_1 = \frac{0.0268 - m}{0.0116}$$

If $m = 0$ (zero settlement rate), $\log t_1 = 2.31$, or $t_1 \approx 200$ months. Thus, settlement will continue for $t_1 - t_c/2 = 200 - 36 = 164$ months (≈ 14 years) after completion of the fill—a fairly long time. This calculation emphasizes the need to pay close attention to the settlement of foundations constructed on sanitary landfills.

A comparison of Eqs. (13.19) through (13.21) for rates of settlement shows that the value of m increases with the height of the fill. However, for fill heights greater than about 30 m, the rate of settlement should not be much different from that obtained from Eq. (13.21). The reason is that the decomposition of organic matter close to the surface is mainly the result of an anaerobic environment. For deeper fills, the decomposition is slower. Hence, for fill heights greater than about 30 m, the rate of settlement does not exceed that for fills that are about 30 m in height.

In English units, Eqs. (13.19)–(13.21) can be expressed in the following form:

$$m = 0.088 - 0.038 \log t_1 \qquad \text{(for fill heights 40 to 80 ft)} \qquad (13.23)$$

$$m = 0.125 - 0.051 \log t_1 \qquad \text{(for fill heights 80 to 100 ft)} \qquad (13.24)$$

$$m = 0.142 - 0.06 \log t_1 \qquad \text{(for fill heights greater than 100 ft)} \quad (13.25)$$

Sowers (1973) also proposed a formula for calculating the settlement of a sanitary landfill, namely,

$$\Delta H_f = \frac{\alpha H_f}{1 + e} \log \left(\frac{t''}{t'} \right) \qquad (13.26)$$

where

H_f = height of the fill
e = void ratio
α = a coefficient for settlement
t'', t' = times (see Figure 13.22)
ΔH_f = settlement between times t' and t''

The coefficients α fall between

$$\alpha = 0.09e \quad \text{(for conditions favorable to decomposition)} \qquad (13.27)$$

and

$$\alpha = 0.03e \quad \text{(for conditions unfavorable to decomposition)} \qquad (13.28)$$

Equation (13.26) is similar to the equation for secondary consolidation settlement [Equation (5.78)].

Problems

13.1   For leossial soil, given $G_s = 2.74$. Plot a graph of γ_d (kN/m³) versus liquid limit to identify the zone in which the soil is likely to collapse on saturation. If a soil has a liquid limit of 27, $G_s = 2.74$, and $\gamma_d = 14.5$ kN/m³, is collapse likely to occur?

13.2   A collapsible soil layer in the field has a thickness of 3 m. The average effective overburden pressure on the soil layer is 62 kN/m². An undisturbed specimen of this soil was subjected to a double oedometer test. The preconsolidation pressure of the specimen as determined from the soaked specimen was 84 kN/m². Is the soil in the field normally consolidated or preconsolidated?

13.3 An expansive soil has an active-zone thickness of 10 ft. The natural moisture content of the soil is 20% and its liquid limit is 50. Calculate the free surface swell of the expansive soil upon saturation.

13.4 An expansive soil profile has an active-zone thickness of 12 ft. A shallow foundation is to be constructed at a depth of 4 ft below the ground surface. Based on the swelling pressure test, the following are given.

Depth from ground surface (ft)	Swell under overburden and estimated foundation surcharge pressure, $s_{w(1)}$ (%)
4	4.75
6	2.75
8	1.5
10	0.6
12	0.0

Estimate the total possible swell under the foundation.

13.5 Refer to Problem 13.4. If the allowable total swell is 1 in., what would be the necessary undercut?

13.6 Repeat Problem 13.4 with the following: active zone thickness = 6 m; depth of shallow foundation = 1.5 m.

Depth from ground surface (m)	Swell under overburden and estimated foundation surcharge pressure, $s_{w(1)}$ (%)
1.5	5.5
2.0	3.1
3.0	1.5
4.0	0.75
5.0	0.4
6.0	0.0

13.7 Refer to Problem 13.6. If the allowable total swell is 30 mm, what would be the necessary undercut?

13.8 Refer to Figure 13.19b. For the drilled shaft with bell, given:

Thickness of active zone, $Z = 9$ m
Dead load = 1500 kN
Live load = 300 kN
Diameter of the shaft, $D_s = 1$ m
Zero swell pressure for the clay in the active zone = 600 kN/m^2
Average angle of plinth-soil friction, $\phi'_{ps} = 20°$
Average undrained cohesion of the clay around the bell = 150 kN/m^2

Determine the diameter of the bell, D_b. A factor of safety of 3 against uplift is required with the assumption that dead load plus live load is equal to zero.

13.9 Refer to Problem 13.8. If an additional requirement is that the factor of safety against uplift is at least 4 with the dead load on (live load = 0), what should be the diameter of the bell?

References

Abduljauwad, S. N., and Al-Sulaimani, G. J. (1993). "Determination of Swell Potential of Al-Qatif Clay," *Geotechnical Testing Journal,* American Society for Testing and Materials, Vol. 16, No. 4, pp. 469–484.

Altmeyer, W. T. (1955). "Discussion of Engineering Properties of Expansive Clays," *Journal of the Soil Mechanics and Foundations Division,* American Society of Civil Engineers, Vol. 81, No. SM2, pp. 17–19.

Benites, L. A. (1968). "Geotechnical Properties of the Soils Affected by Piping near the Benson Area, Cochise County, Arizona," M. S. Thesis, University of Arizona, Tucson.

Chen, F. H. (1988). *Foundations on Expansive Soils,* Elsevier, Amsterdam.

Clemence, S. P., and Finbarr, A. O. (1981). "Design Considerations for Collapsible Soils," *Journal of the Geotechnical Engineering Division,* American Society of Civil Engineers, Vol. 107, No. GT3, pp. 305–317.

Clevenger, W. (1958). "Experience with Loess as Foundation Material," *Transactions,* American Society of Civil Engineers, Vol. 123, pp. 151–170.

Dakshanamanthy, V., and Raman, V. (1973). "A Simple Method of Identifying an Expansive Soil," *Soils and Foundations,* Vol. 13, No. 1, pp. 97–104.

Denisov, N. Y. (1951). *The Engineering Properties of Loess and Loess Loams,* Gosstroiizdat, Moscow.

Feda, J. (1964). "Colloidal Activity, Shrinking and Swelling of Some Clays," *Proceedings, Soil Mechanics Seminar,* Loda, Illinois, pp. 531–546.

Gibbs, H. J. (1961). *Properties Which Divide Loose and Dense Uncemented Soils,* Earth Laboratory Report EM-658, Bureau of Reclamation, U.S. Department of the Interior, Washington, DC.

Gromko, G. J. (1974). "Review of Expansive Soils," *Journal of the Geotechnical Engineering Division,* American Society of Civil Engineers, Vol. 100, No. GT6, pp. 667–687.

Handy, R. L. (1973). "Collapsible Loess in Iowa," *Proceedings, Soil Science Society of America,* Vol. 37, pp. 281–284.

Holtz, W. G. (1959). "Expansive Clays—Properties and Problems," *Journal of the Colorado School of Mines,* Vol. 54, No. 4, pp. 89–125.

Holtz, W. G., and Hilf, J. W. (1961). "Settlement of Soil Foundations Due to Saturation," *Proceedings, Fifth International Conference on Soil Mechanics and Foundation Engineering,* Paris, Vol. 1, 1961, pp. 673–679.

Houston, W. N., and Houston, S. L. (1989). "State-of-the-Practice Mitigation Measures for Collapsible Soil Sites," *Proceedings, Foundation Engineering: Current Principles and Practices,* American Society of Civil Engineers, Vol. 1, pp. 161–175.

Jennings, J. E., and Knight, K. (1975). "A Guide to Construction on or with Materials Exhibiting Additional Settlements Due to 'Collapse' of Grain Structure," *Proceedings, Sixth Regional Conference for Africa on Soil Mechanics and Foundation Engineering,* Johannesburg, pp. 99–105.

Lutenegger, A. J. (1986). "Dynamic Compaction in Friable Loess," *Journal of Geotechnical Engineering,* American Society of Civil Engineers, Vol. 112, No. GT6, pp. 663–667.

Lutenegger, A. J., and Saber, R. T. (1988). "Determination of Collapse Potential of Soils," *Geotechnical Testing Journal,* American Society for Testing and Materials, Vol. 11. No. 3, pp. 173–178.

McKeen, R. G. (1992). "A Model for Predicting Expansive Soil Behavior," *Proceedings, Seventh International Conference on Expansive Soils,* Dallas, Vol. 1, pp. 1–6.

Nayak, N. V., and Christensen, R. W. (1974). "Swell Characteristics of Compacted Expansive Soils," *Clay and Clay Minerals,* Vol. 19, pp. 251–261.

O'Neill, M. W., and Poormoayed, N. (1980). "Methodology for Foundations on Expansive Clays," *Journal of the Geotechnical Engineering Division,* American Society of Civil Engineers, Vol. 106, No. GT12, pp. 1345–1367.

Peck, R. B., Hanson, W. E., and Thornburn, T. B. (1974). *Foundation Engineering,* Wiley, New York.

Priklonskii, V. A. (1952). *Gruntovedenic, Vtoraya Chast',* Gosgeolzdat, Moscow.

Raman, V. (1967). "Identification of Expansive Soils from the Plasticity Index and the Shrinkage Index Data," *The Indian Engineer,* Vol. 11, No. 1, pp. 17–22.

Seed, H. B., Woodward, R. J., Jr., and Lundgren, R. (1962). "Prediction of Swelling Potential for Compacted Clays," *Journal of the Soil Mechanics and Foundations Division,* American Society of Civil Engineers, Vol. 88, No. SM3, pp. 53–87.

Semkin, V. V., Ermoshin, V. M., and Okishev, N. D. (1986). "Chemical Stabilization of Loess Soils in Uzbekistan," *Soil Mechanics and Foundation Engineering* (trans. from Russian), Vol. 23, No. 5, pp. 196–199.

Snethen, D. R. (1984). "Evaluation of Expedient Methods for Identification and Classification of Potentially Expansive Soils," *Proceedings, Fifth International Conference on Expansive Soils,* Adelaide, pp. 22–26.

Snethen, D. R., Johnson, L. D., and Patrick, D. M. (1977). *An Evaluation of Expedient Methodology for Identification of Potentially Expansive Soils,* Report No. FHWA-RD-77-94, U.S. Army Engineers Waterways Experiment Station, Vicksburg, MS.

Sowers, G. F. (1973). "Settlement of Waste Disposal Fills," *Proceedings, Eighth International Conference on Soil Mechanics and Foundation Engineering,* Moscow, pp. 207–210.

Sowers, G. B., and Sowers, G. F. (1970). *Introductory Soil Mechanics and Foundations,* 3d ed. Macmillan, New York.

Sridharan, A. (2005). "On Swelling Behaviour of Clays," *Proceedings, International Conference on Problematic Soils,* North Cyprus, Vol. 2, pp. 499–516.

Sridharan, A., Rao, A. S., and Sivapullaiah, P. V. (1986), "Swelling Pressure of Clays," *Geotechnical Testing Journal,* American Society for Testing and Materials, Vol. 9, No. 1, pp. 24–33.

Uniform Building Code (1968). *UBC Standard No. 29-2.*

Van Der Merwe, D. H. (1964), "The Prediction of Heave from the Plasticity Index and Percentage Clay Fraction of Soils," *Civil Engineer in South Africa,* Vol. 6, No. 6, pp. 103–106.

Vijayvergiya, V. N., and Ghazzaly, O. I. (1973). "Prediction of Swelling Potential of Natural Clays," *Proceedings, Third International Research and Engineering Conference on Expansive Clays,* pp. 227–234.

Weston, D. J. (1980). "Expansive Roadbed Treatment for Southern Africa," *Proceedings, Fourth International Conference on Expansive Soils,* Vol. 1, pp. 339–360.

Yen, B. C., and Scanlon, B. (1975). "Sanitary Landfill Settlement Rates," *Journal of the Geotechnical Engineering Division,* American Society of Civil Engineers, Vol. 101, No. GT5, pp. 475–487.

14

Soil Improvement and Ground Modification

14.1 Introduction

The soil at a construction site may not always be totally suitable for supporting structures such as buildings, bridges, highways, and dams. For example, in granular soil deposits, the *in situ* soil may be very loose and indicate a large elastic settlement. In such a case, the soil needs to be densified to increase its unit weight and thus its shear strength.

Sometimes the top layers of soil are undesirable and must be removed and replaced with better soil on which the structural foundation can be built. The soil used as fill should be well compacted to sustain the desired structural load. Compacted fills may also be required in low-lying areas to raise the ground elevation for construction of the foundation.

Soft saturated clay layers are often encountered at shallow depths below foundations. Depending on the structural load and the depth of the layers, unusually large consolidation settlement may occur. Special soil-improvement techniques are required to minimize settlement.

In Chapter 13, we mentioned that the properties of expansive soils could be altered substantially by adding stabilizing agents such as lime. Improving *in situ* soils by using additives is usually referred to as *stabilization.*

Various techniques are used to

1. Reduce the settlement of structures
2. Improve the shear strength of soil and thus increase the bearing capacity of shallow foundations
3. Increase the factor of safety against possible slope failure of embankments and earth dams
4. Reduce the shrinkage and swelling of soils

This chapter discusses some of the general principles of soil improvement, such as compaction, vibroflotation, precompression, sand drains, wick drains, stabilization by admixtures, jet grouting, and deep mixing, as well as the use of stone columns and sand compaction piles in weak clay to construct foundations.

| **14.2** | *General Principles of Compaction* |

If a small amount of water is added to a soil that is then compacted, the soil will have a certain unit weight. If the moisture content of the same soil is gradually increased and the energy of compaction is the same, the dry unit weight of the soil will gradually increase. The reason is that water acts as a lubricant between the soil particles, and under compaction it helps rearrange the solid particles into a denser state. The increase in dry unit weight with increase of moisture content for a soil will reach a limiting value beyond which the further addition of water to the soil will result in a *reduction* in dry unit weight. The moisture content at which the *maximum dry unit weight* is obtained is referred to as the *optimum moisture content.*

The standard laboratory tests used to evaluate maximum dry unit weights and optimum moisture contents for various soils are

- The Standard Proctor test (ASTM designation D-698)
- The Modified Proctor test (ASTM designation D-1557)

The soil is compacted in a mold in several layers by a hammer. The moisture content of the soil, w, is changed, and the dry unit weight, γ_d, of compaction for each test is determined. The maximum dry unit weight of compaction and the corresponding optimum moisture content are determined by plotting a graph of γ_d against w (%). The standard specifications for the two types of Proctor test are given in Tables 14.1 and 14.2.

Table 14.1 Specifications for Standard Proctor Test (Based on ASTM Designation 698)

Item	Method A	Method B	Method C
Diameter of mold	101.6 mm (4 in.)	101.6 mm (4 in.)	152.4 mm (6 in.)
Volume of mold	944 cm^3 ($\frac{1}{30}$ ft^3)	944 cm^3 ($\frac{1}{30}$ ft^3)	2124 cm^3 (0.075 ft^3)
Weight of hammer	2.5 kg (5.5 lb)	2.5 kg (5.5 lb)	2.5 kg (5.5 lb)
Height of hammer drop	304.8 mm (12 in.)	304.8 mm (12 in.)	304.8 mm (12 in.)
Number of hammer blows per layer of soil	25	25	56
Number of layers of compaction	3	3	3
Energy of compaction	600 kN·m/m^3 (12,400 ft·lb/ft^3)	600 kN·m/m^3 (12,400 ft·lb/ft^3)	600 kN·m/m^3 (12,400 ft·lb/ft^3)
Soil to be used	Portion passing No. 4 (4.57-mm) sieve. May be used if 20% *or less* by weight of material is retained on No. 4 sieve.	Portion passing $\frac{3}{8}$-in. (9.5-mm) sieve. May be used if soil retained on No. 4 sieve *is more* than 20% and 20% *or less* by weight is retained on 9.5-mm ($\frac{3}{8}$-in.) sieve.	Portion passing $\frac{3}{4}$-in. (19.0-mm) sieve. May be used if *more than* 20% by weight of material is retained on 9.5 mm ($\frac{3}{8}$ in.) sieve and *less than* 30% by weight is retained on 19.00-mm ($\frac{3}{4}$-in.) sieve.

Table 14.2 Specifications for Modified Proctor Test (Based on ASTM Designation 1557)

Item	Method A	Method B	Method C
Diameter of mold	101.6 mm (4 in.)	101.6 mm (4 in.)	152.4 mm (6 in.)
Volume of mold	944 cm³ ($\frac{1}{30}$ ft³)	944 cm³ ($\frac{1}{30}$ ft³)	2124 cm³ (0.075 ft³)
Weight of hammer	4.54 kg (10 lb)	4.54 kg (10 lb)	4.54 kg (10 lb)
Height of hammer drop	457.2 mm (18 in.)	457.2 mm (18 in.)	457.2 mm (18 in.)
Number of hammer blows per layer of soil	25	25	56
Number of layers of compaction	5	5	5
Energy of compaction	2700 kN·m/m³ (56,000 ft·lb/ft³)	2700 kN·m/m³ (56,000 ft·lb/ft³)	2700 kN·m/m³ (56,000 ft·lb/ft³)
Soil to be used	Portion passing No. 4 (4.57-mm) sieve. May be used if 20% *or less* by weight of material is retained on No. 4 sieve.	Portion passing 9.5-mm ($\frac{3}{8}$-in.) sieve. May be used if soil retained on No. 4 sieve *is more* than 20% and 20% *or less* by weight is retained on 9.5-mm ($\frac{3}{8}$-in.) sieve.	Portion passing 19.0-mm ($\frac{3}{4}$-in.) sieve. May be used if *more than* 20% by weight of material is retained on 9.5-mm ($\frac{3}{8}$-in.) sieve and *less than* 30% by weight is retained on 19-mm ($\frac{3}{4}$-in.) sieve.

Figure 14.1 shows a plot of γ_d against w (%) for a clayey silt obtained from standard and modified Proctor tests (method A). The following conclusions may be drawn:

1. The maximum dry unit weight and the optimum moisture content depend on the degree of compaction.
2. The higher the energy of compaction, the higher is the maximum dry unit weight.
3. The higher the energy of compaction, the lower is the optimum moisture content.
4. No portion of the compaction curve can lie to the right of the zero-air-void line. The zero-air-void dry unit weight, γ_{zav}, at a given moisture content is the theoretical maximum value of γ_d, which means that all the void spaces of the compacted soil are filled with water, or

$$\gamma_{zav} = \frac{\gamma_w}{\dfrac{1}{G_s} + w} \qquad (14.1)$$

where

γ_w = unit weight of water
G_s = specific gravity of the soil solids
w = moisture content of the soil

5. The maximum dry unit weight of compaction and the corresponding optimum moisture content will vary from soil to soil.

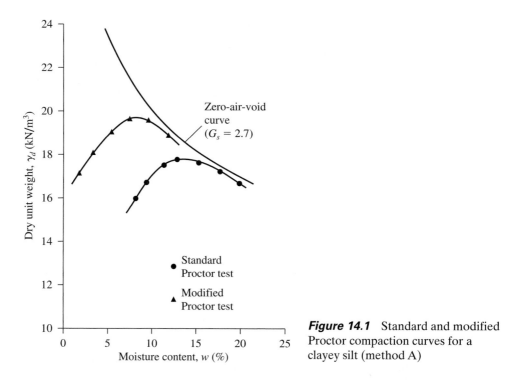

Figure 14.1 Standard and modified Proctor compaction curves for a clayey silt (method A)

Using the results of laboratory compaction (γ_d versus w), specifications may be written for the compaction of a given soil in the field. In most cases, the contractor is required to achieve a relative compaction of 90% or more on the basis of a specific laboratory test (either the standard or the modified Proctor compaction test). The relative compaction is defined as

$$\text{RC} = \frac{\gamma_{d(\text{field})}}{\gamma_{d(\text{max})}} \qquad (14.2)$$

Chapter 1 introduced the concept of relative density (for the compaction of granular soils), defined as

$$D_r = \left[\frac{\gamma_d - \gamma_{d(\text{min})}}{\gamma_{d(\text{max})} - \gamma_{d(\text{min})}} \right] \frac{\gamma_{d(\text{max})}}{\gamma_d}$$

where

γ_d = dry unit weight of compaction in the field
$\gamma_{d(\text{max})}$ = maximum dry unit weight of compaction as determined in the laboratory
$\gamma_{d(\text{min})}$ = minimum dry unit weight of compaction as determined in the laboratory

For granular soils in the field, the degree of compaction obtained is often measured in terms of relative density. Comparing the expressions for relative density and relative compaction reveals that

$$RC = \frac{A}{1 - D_r(1 - A)} \tag{14.3}$$

where $A = \dfrac{\gamma_{d(min)}}{\gamma_{d(max)}}$.

Lee and Singh (1971) reviewed 47 different soils and, on the basis of their review, presented the following correlation:

$$D_r(\%) = \frac{(RC - 80)}{0.2} \tag{14.4}$$

Gurtug and Sridharan (2004) analyzed a number of laboratory-compaction test results on fine-grained (cohesive) soils. Based on this study, the following correlations were developed:

$$w_{opt} = [1.95 - 0.38 \log(CE)](PL) \tag{14.5}$$

$$\gamma_{d(max)} = [0.5 + 0.173 \log(CE)]\gamma_{d(PL)} \tag{14.6}$$

$$\gamma_{d(max)} = 22.26 - 0.28w_{opt} \tag{14.7}$$

$$\gamma_{d(max)} = 22.68e^{-0.0183w_{opt}} \tag{14.8}$$

where

w_{opt} = optimum moisture content (%)
PL = plastic limit (%)
CE = compaction energy (kN·m/m³)
$\gamma_{d(max)}$ = maximum dry unit weight (kN/m³)
$\gamma_{d(PL)}$ = dry unit weight at moisture content equal to plastic limit (kN/m³)

Figure 14.2 shows the correlation from which Eq. (14.8) was developed.

14.3 Correction for Compaction of Soils with Oversized Particles

As shown in Tables 14.1 and 14.2, depending on the method used, certain oversized particles (such as material retained on a no. 4 sieve or a $\frac{3}{4}$-in. sieve) may need to be removed from the soil to conduct laboratory compaction tests. ASTM test designation

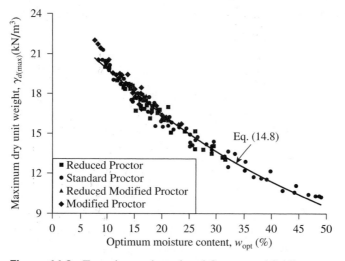

Figure 14.2 Experimental results of Gurtug and Sridharan on which Eq. (14.8) is derived

D-4718 provides a method for making corrections for maximum dry unit weight and optimum moisture content in the presence of oversized particles. This is helpful in writing specifications for field compaction. According to the method, the corrected maximum dry unit weight may be calculated as

$$\gamma_{d(\max)\text{-}C} = \frac{100\gamma_w}{\dfrac{P_C}{G_m} + \dfrac{\gamma_w(100 - P_C)}{\gamma_{d(\max)\text{-}F}}} \tag{14.9}$$

where

γ_w = unit weight of water
P_C = percentage of oversized particles by weight
G_m = bulk specific gravity of the oversized particles
$\gamma_{d(\max)\text{-}F}$ = maximum dry unit weight of the finer fraction used in the laboratory compaction

The corrected optimum moisture content can be expressed as

$$w_{\text{optimum-}C}(\%) = w_{\text{optimum-}F}(100 - P_C) + w_C P_C \tag{14.10}$$

where

$w_{\text{optimum-}F}$ = optimum moisture content of the finer fraction determined in the laboratory
w_C = saturated surface dry moisture content of the oversized particles (fraction)

The preceding equations for corrected maximum dry unit weight and optimum moisture content are valid when the oversized particles are about 30% or less (by weight) of the total soil sample.

14.4   *Field Compaction*

Ordinary compaction in the field is done by rollers. Of the several types of roller used, the most common are

1. Smooth-wheel rollers (or smooth drum rollers)
2. Pneumatic rubber-tired rollers
3. Sheepsfoot rollers
4. Vibratory rollers

Figure 14.3 shows a *smooth-wheel roller* that can also create vertical vibration during compaction. Smooth-wheel rollers are suitable for proof-rolling subgrades and for finishing the construction of fills with sandy or clayey soils. They provide 100% coverage under the wheels, and the contact pressure can be as high as 300 to 400 kN/m^2 ($\approx$45 to 60 lb/in^2). However, they do not produce a uniform unit weight of compaction when used on thick layers.

Pneumatic rubber-tired rollers (Figure 14.4) are better in many respects than smooth-wheel rollers. Pneumatic rollers, which may weigh as much as 2000 kN (450 kip), consist of a heavily loaded wagon with several rows of tires. The tires are closely spaced—four to six in a row. The contact pressure under the tires may range up to 600 to 700 kN/m^2 ($\approx$85 to 100 lb/in^2), and they give about 70 to 80% coverage. Pneumatic rollers, which can be used for sandy and clayey soil compaction, produce a combination of pressure and kneading action.

Sheepsfoot rollers (Figure 14.5) consist basically of drums with large numbers of projections. The area of each of the projections may be 25 to 90 cm^2 (4 to 14 in^2).

Figure 14.3   Vibratory smooth-wheel rollers (Courtesy of Tampo Manufacturing Co., Inc., San Antonio, Texas)

Figure 14.4   Pneumatic rubber-tired roller (Courtesy of Tampo Manufacturing Co., Inc., San Antonio, Texas)

These rollers are *most effective in compacting cohesive soils.* The contact pressure under the projections may range from 1500 to 7500 kN/m^2($\approx$215 to 1100 lb/in^2). During compaction in the field, the initial passes compact the lower portion of a lift. Later, the middle and top of the lift are compacted.

Vibratory rollers are efficient in compacting granular soils. Vibrators can be attached to smooth-wheel, pneumatic rubber-tired or sheepsfoot rollers to send vibrations into the soil being compacted. Figures 14.3 and 14.5 show vibratory smooth-wheel rollers and a vibratory sheepsfoot roller, respectively.

In general, compaction in the field depends on several factors, such as the type of compactor, type of soil, moisture content, lift thickness, towing speed of the compactor, and number of passes the roller makes.

Figure 14.6 shows the variation of the unit weight of compaction with depth for a poorly graded dune sand compacted by a vibratory drum roller. Vibration was produced by mounting an eccentric weight on a single rotating shaft within the drum cylinder. The weight of the roller used for this compaction was 55.7 kN (12.5 kip), and the drum diameter was 1.19 m (3.9 ft). The lifts were kept at 2.44 m (8 ft). Note that, at any depth, the dry unit weight of compaction increases with the number of passes the roller makes. However, the rate of increase in unit weight gradually decreases after about 15 passes. Note also the variation of dry unit weight with depth by the number of roller passes. The dry unit weight and hence the relative density, D_r, reach maximum values at a depth of about 0.5 m ($\approx$1.6 ft) and then gradually decrease as the depth increases. The reason is the lack of confining pressure toward the surface. Once the relation between depth and relative density (or dry unit weight) for a soil for a given number of passes is determined, for satisfactory compaction based on a given specification, the approximate thickness of each lift can be easily estimated.

Figure 14.5   Vibratory sheepsfoot roller (Courtesy of Tampo Manufacturing Co., Inc., San Antonio, Texas)

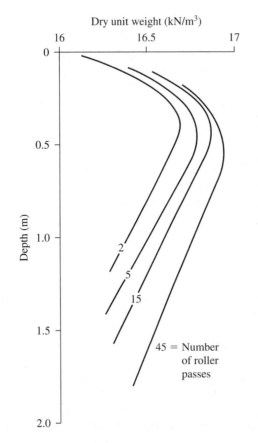

Figure 14.6   Vibratory compaction of a sand: Variation of dry unit weight with depth and number of roller passes; lift thickness = 2.44 m (After D'Appolonia et al., 1969)

Figure 14.7 Hand-held vibrating plate

Hand-held vibrating plates (Figure 14.7) can be used for effective compaction of granular soils over a limited area. Vibrating plates are also gang-mounted on machines. These can be used in less restricted areas.

14.5 *Compaction Control for Clay Hydraulic Barriers*

Compacted clays are commonly used as hydraulic barriers in cores of earth dams, liners and covers of landfills, and liners of surface impoundments. Since the primary purpose of a barrier is to minimize flow, the hydraulic conductivity, k, is the

Table 14.3   Characteristics of Soils Reported in Figures 14.8, 14.9, and 14.10

Soil	Classification	Liquid limit	Plasticity index	Percent finer than No. 200 sieve (0.075 mm)
Wisconsin A	CL	34	16	85
Wisconsin B	CL	42	19	99
Wisconsin C	CH	84	60	71

controlling factor. In many cases, it is desired that the hydraulic conductivity be less than 10^{-7}cm/s. This can be achieved by controlling the minimum degree of saturation during compaction, a relation that can be explained by referring to the compaction characteristics of three soils described in Table 14.3 (Othman and Luettich, 1994).

Figures 14.8, 14.9, and 14.10 show the standard and modified Proctor test results and the hydraulic conductivities of compacted specimens. Note that the solid symbols represent specimens with hydraulic conductivities of 10^{-7}cm/s or less. As can be seen from these figures, the data points plot generally parallel to the line of full saturation. Figure 14.11 shows the effect of the degree of saturation during compaction on the hydraulic conductivity of the three soils. It is evident from the figure that, if it is desired that the maximum hydraulic conductivity be 10^{-7}cm/s, then all soils should be compacted at a minimum degree of saturation of 88%.

In field compaction at a given site, soils of various composition may be encountered. Small changes in the content of fines will change the magnitude of hydraulic conductivity. Hence, considering the various soils likely to be encountered at a given site the procedure just described aids in developing a minimum-degree-of-saturation criterion for compaction to construct hydraulic barriers.

14.6   *Vibroflotation*

Vibroflotation is a technique developed in Germany in the 1930s for *in situ* densification of thick layers of loose granular soil deposits. Vibroflotation was first used in the United States about 10 years later. The process involves the use of a *vibroflot* (called the *vibrating unit*), as shown in Figure 14.12. The device is about 2 m (6 ft) in length. This vibrating unit has an eccentric weight inside it and can develop a centrifugal force. The weight enables the unit to vibrate horizontally. Openings at the bottom and top of the unit are for water jets. The vibrating unit is attached to a follow-up pipe. The figure shows the vibroflotation equipment necessary for compaction in the field.

The entire compaction process can be divided into four steps (see Figure 14.13):

Step 1.   The jet at the bottom of the vibroflot is turned on, and the vibroflot is lowered into the ground.

Step 2.   The water jet creates a quick condition in the soil, which allows the vibrating unit to sink.

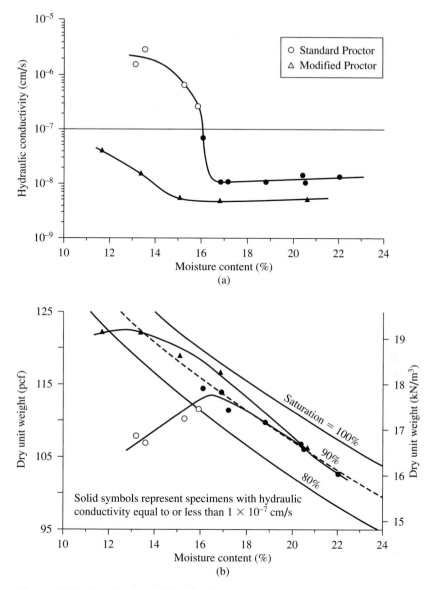

Figure 14.8 Standard and Modified Proctor test results and hydraulic conductivity of Wisconsin A soil (After Othman and Luettich, 1994)

Step 3. Granular material is poured into the top of the hole. The water from the lower jet is transferred to the jet at the top of the vibrating unit. This water carries the granular material down the hole.

Step 4. The vibrating unit is gradually raised in about 0.3-m (1-ft) lifts and is held vibrating for about 30 seconds at a time. This process compacts the soil to the desired unit weight.

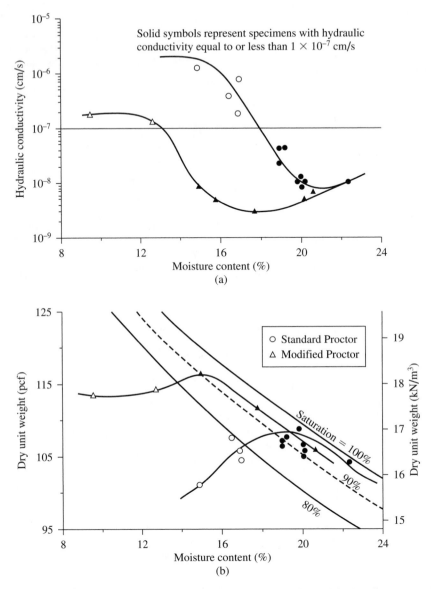

Figure 14.9 Standard and Modified Proctor test results and hydraulic conductivity of Wisconsin B soil (After Othman and Luettich, 1994)

Table 14.4 gives the details of various types of vibroflot unit used in the United States. The 30-HP electric units have been used since the latter part of the 1940s. The 100-HP units were introduced in the early 1970s. The zone of compaction around a single probe will vary according to the type of vibroflot used. The cylindrical zone of compaction will have a radius of about 2 m (6 ft) for a 30-HP unit and about 3 m (10 ft) for a 100-HP unit. Compaction by vibroflotation involves various probe spacings,

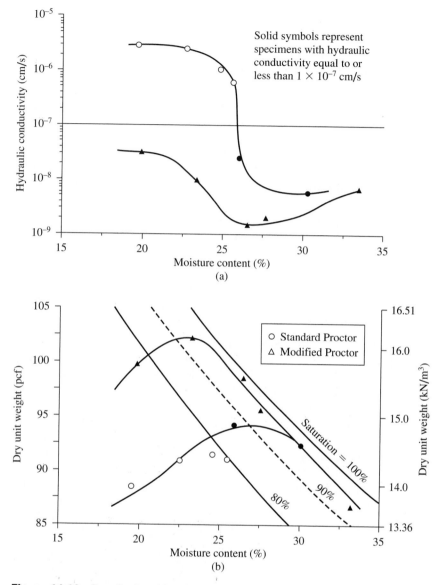

Figure 14.10 Standard and Modified Proctor test results and hydraulic conductivity of Wisconsin C soil (After Othman and Luettich, 1994)

depending on the zone of compaction. (See Figure 14.14.) Mitchell (1970) and Brown (1977) reported several successful cases of foundation design that used vibroflotation.

The success of densification of *in situ* soil depends on several factors, the most important of which are the grain-size distribution of the soil and the nature of the backfill used to fill the holes during the withdrawal period of the vibroflot. The range

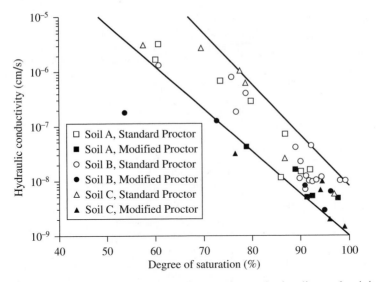

Figure 14.11 Effect of degree of saturation on hydraulic conductivity of Wisconsin A, B, and C soils (After Othman and Luettich, 1994)

of the grain-size distribution of *in situ* soil marked Zone 1 in Figure 14.15 is most suitable for compaction by vibroflotation. Soils that contain excessive amounts of fine sand and silt-size particles are difficult to compact; for such soils, considerable effort is needed to reach the proper relative density of compaction. Zone 2 in Figure 14.15 is the approximate lower limit of grain-size distribution for compaction by vibroflotation. Soil deposits whose grain-size distribution falls into Zone 3 contain appreciable amounts of gravel. For these soils, the rate of probe penetration may be rather slow, so compaction by vibroflotation might prove to be uneconomical in the long run.

The grain-size distribution of the backfill material is one of the factors that control the rate of densification. Brown (1977) defined a quantity called *suitability number* for rating a backfill material. The suitability number is given by the formula

$$S_N = 1.7\sqrt{\frac{3}{(D_{50})^2} + \frac{1}{(D_{20})^2} + \frac{1}{(D_{10})^2}} \tag{14.11}$$

where D_{50}, D_{20}, and D_{10} are the diameters (in mm) through which 50%, 20%, and 10%, respectively, of the material is passing. The smaller the value of S_N, the more desirable is the backfill material. Following is a backfill rating system proposed by Brown (1977):

Range of S_N	Rating as backfill
0–10	Excellent
10–20	Good
20–30	Fair
30–50	Poor
>50	Unsuitable

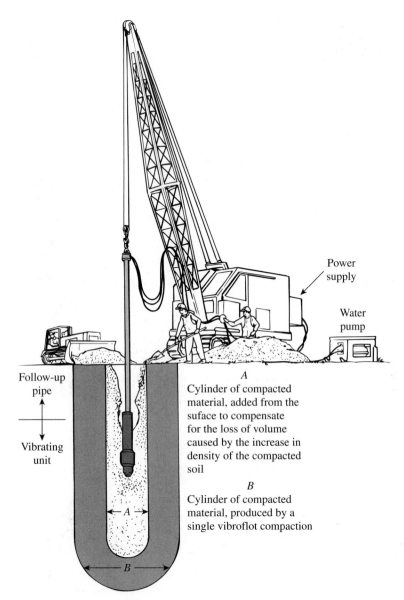

Power
supply

Water
pump

Follow-up
pipe

Vibrating
unit

A
Cylinder of compacted
material, added from the
suface to compensate
for the loss of volume
caused by the increase in
density of the compacted
soil

B
Cylinder of compacted
material, produced by a
single vibroflot compaction

A

B

Figure 14.12   Vibroflotation unit (After Brown, 1977)

An excellent case study that evaluated the benefits of vibroflotation was presented by Basore and Boitano (1969). Densification of granular subsoil was necessary for the construction of a three-story office building at the Treasure Island Naval
Station in San Francisco, California. The top 9 m ($\approx$30 ft) of soil at the site was loose
to medium-dense sand fill that had to be compacted. Figure 14.16a shows the nature
of the layout of the vibroflotation points. Sixteen compaction points were arranged in
groups of four, with 1.22 m (4 ft), 1.52 m (5 ft), 1.83 m (6 ft), and 2.44 m (8 ft) spacing.

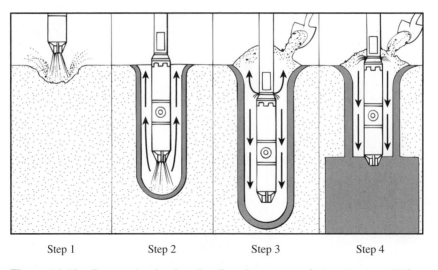

Step 1   Step 2   Step 3   Step 4

Figure 14.13 Compaction by the vibroflotation process (After Brown, 1977)

Prior to compaction, standard penetration tests were conducted at the centers of groups of three compaction points. After the completion of compaction by vibroflotation, the variation of the standard penetration resistance with depth was determined at the same points.

Figure 14.16b shows the variation of standard penetration resistance, N_{60}, with depth before and after compaction for vibroflotation point spacings $S' = 1.22$ m (4 ft) and 2.44 m (8 ft). From this figure, the following general conclusions can be drawn:

- For any given S', the magnitude of N_{60} after compaction decreases with an increase in depth.
- An increase in N_{60} indicates an increase in the relative density of sand.
- The degree of compaction decreases with the increase in S'. At $S' = 1.22$ m (4 ft), the degree of compaction at any depth is the largest. However, at $S' = 2.44$ m (8 ft), the vibroflotation had practically no effect in compacting soil.

During the past 30 to 35 years, the vibroflotation technique has been used successfully on large projects to compact granular subsoils, thereby controlling structural settlement.

14.7 Precompression

When highly compressible, normally consolidated clayey soil layers lie at a limited depth and large consolidation settlements are expected as the result of the construction of large buildings, highway embankments, or earth dams, precompression of soil may be used to minimize postconstruction settlement. The principles of precompression are best explained by reference to Figure 14.17. Here, the proposed structural

Table 14.4 Types of Vibrating Units[a]

	100-HP electric and hydraulic motors	30-HP electric motors
(a) Vibrating tip		
Length	2.1 m (7 ft)	1.86 m (6.11 ft)
Diameter	406 mm (16 in.)	381 mm (15 in.)
Weight	18 kN (4000 lb)	18 kN (4000 lb)
Maximum movement when free	12.5 mm (0.49 in.)	7.6 mm (0.3 in.)
Centrifugal force	160 kN (18 ton)	90 kN (10 ton)
(b) Eccentric		
Weight	1.16 kN (260 lb)	0.76 kN (170 lb)
Offset	38 mm (1.5 in.)	32 mm (1.25 in.)
Length	610 mm (24 in.)	387 mm (15.3 in.)
Speed	1800 rpm	1800 rpm
(c) Pump		
Operating flow rate	0–1.6 m³/min (0–400 gal/min)	0–6 m³/min (0–150 gal/min)
Pressure	690–1035 kN/m² (100–150 lb/in²)	690–1035 kN/m² (100–150 lb/in²)
(d) Lower follow-up pipe and extensions		
Diameter	305 mm (12 in.)	305 mm (12 in.)
Weight	3.65 kN/m (250 lb/ft)	3.65 kN/m (250 lb/ft)

[a] After Brown (1977)

load per unit area is $\Delta\sigma'_{(p)}$, and the thickness of the clay layer undergoing consolidation is H_c. The maximum primary consolidation settlement caused by the structural load is then

$$S_{c(p)} = \frac{C_c H_c}{1 + e_o} \log \frac{\sigma'_o + \Delta\sigma'_{(p)}}{\sigma'_o} \tag{14.12}$$

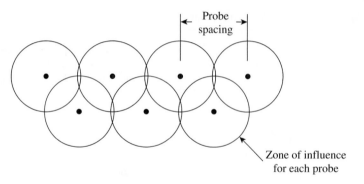

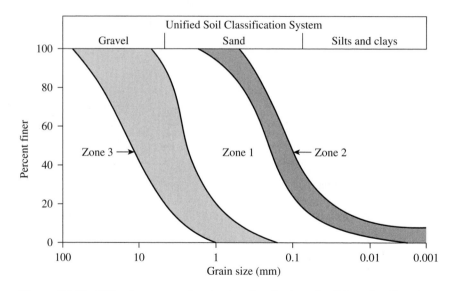

Figure 14.14 Nature of probe spacing for vibroflotation

Figure 14.15 Effective range of grain-size distribution of soil for vibroflotation

The settlement–time relationship under the structural load will be like that shown in Figure 14.17b. However, if a surcharge of $\Delta\sigma'_{(p)} + \Delta\sigma'_{(f)}$ is placed on the ground, the primary consolidation settlement will be

$$S_{c(p+f)} = \frac{C_c H_c}{1 + e_o} \log \frac{\sigma'_o + [\Delta\sigma'_{(p)} + \Delta\sigma'_{(f)}]}{\sigma'_o} \qquad (14.13)$$

The settlement–time relationship under a surcharge of $\Delta\sigma'_{(p)} + \Delta\sigma'_{(f)}$ is also shown in Figure 14.17b. Note that a total settlement of $S_{c(p)}$ would occur at time t_2, which

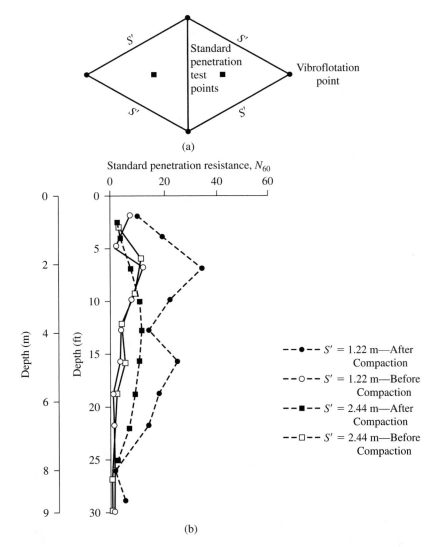

Figure 14.16 (a) Layout of vibroflotation compaction points; (b) variation of standard penetration resistance (N_{60}) before and after compaction (Adapted from Basore and Boitano, 1969)

is much shorter than t_1. So, if a temporary total surcharge of $\Delta\sigma'_{(p)} + \Delta\sigma'_{(f)}$ is applied on the ground surface for time t_2, the settlement will equal $S_{c(p)}$. At that time, if the surcharge is removed and a structure with a permanent load per unit area of $\Delta\sigma'_{(p)}$ is built, no appreciable settlement will occur. The procedure just described is called *precompression*. The total surcharge $\Delta\sigma'_{(p)} + \Delta\sigma'_{(f)}$ can be applied by means of temporary fills.

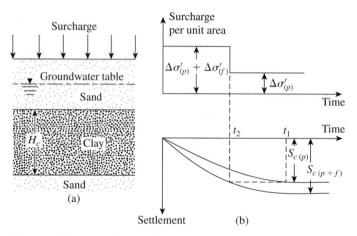

Figure 14.17 Principles of precompression

Derivation of Equations for Obtaining $\Delta\sigma'_{(f)}$ and t_2

Figure 14.17b shows that, under a surcharge of $\Delta\sigma'_{(p)} + \Delta\sigma'_{(f)}$, the degree of consolidation at time t_2 after the application of the load is

$$U = \frac{S_{c(p)}}{S_{c(p+f)}}\qquad(14.14)$$

Substitution of Eqs. (14.12) and (14.13) into Eq. (14.14) yields

$$U = \frac{\log\left[\dfrac{\sigma'_o + \Delta\sigma'_{(p)}}{\sigma'_o}\right]}{\log\left[\dfrac{\sigma'_o + \Delta\sigma'_{(p)} + \Delta\sigma'_{(f)}}{\sigma'_o}\right]} = \frac{\log\left[1 + \dfrac{\Delta\sigma'_{(p)}}{\sigma'_o}\right]}{\log\left\{1 + \dfrac{\Delta\sigma'_{(p)}}{\sigma'_o}\left[1 + \dfrac{\Delta\sigma'_{(f)}}{\Delta\sigma'_{(p)}}\right]\right\}}\qquad(14.15)$$

Figure 14.18 gives magnitudes of U for various combinations of $\Delta\sigma'_{(p)}/\sigma'_o$, and $\Delta\sigma'_{(f)}/\Delta\sigma'_{(p)}$. The degree of consolidation referred to in Eq. (14.15) is actually the average degree of consolidation at time t_2, as shown in Figure 14.17b. However, if the average degree of consolidation is used to determine t_2, some construction problems might occur. The reason is that, after the removal of the surcharge and placement of the structural load, the portion of clay close to the drainage surface will continue to swell, and the soil close to the midplane will continue to settle. (See Figure 14.19.) In some cases, net continuous settlement might result. A conservative approach may solve the problem; that is, assume that U in Eq. (14.15) is the midplane degree of consolidation (Johnson, 1970a). Now, from Eq. (1.59),

$$U = f(T_v)\qquad(1.59)$$

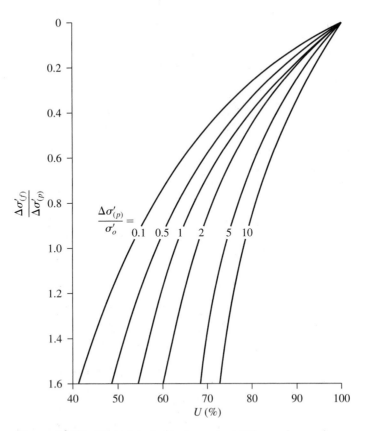

Figure 14.18 Plot of $\Delta\sigma'_{(f)}/\Delta\sigma'_{(p)}$ against U for various values of $\Delta\sigma'_{(p)}/\sigma'_o$—Eq. (14.15)

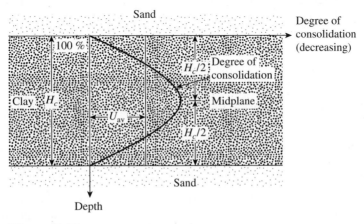

Figure 14.19

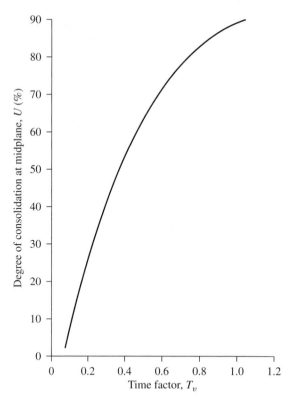

Figure 14.20 Plot of midplane degree of consolidation against T_v

where

T_v = time factor = $C_v t_2/H^2$
C_v = coefficient of consolidation
t_2 = time
H = maximum drainage path ($=H_c/2$ for two-way drainage and H_c for one-way drainage)

The variation of U (the midplane degree of consolidation) with T_v is given in Figure 14.20.

Procedure for Obtaining Precompression Parameters

Two problems may be encountered by engineers during precompression work in the field:

1. The value of $\Delta\sigma'_{(f)}$ is known, but t_2 must be obtained. In such a case, obtain σ'_o, $\Delta\sigma_{(p)}$, and solve for U, using Eq. (14.15) or Figure 14.18. For this value of U, obtain T_v from Figure 14.20. Then

$$t_2 = \frac{T_v H^2}{C_v} \tag{14.16}$$

2. For a specified value of t_2, $\Delta\sigma'_{(f)}$ must be obtained. In such a case, calculate T_v. Then use Figure 14.20 to obtain the midplane degree of consolidation, U. With the estimated value of U, go to Figure 14.18 to get the required value of $\Delta\sigma'_{(f)}/\Delta\sigma'_{(p)}$, and then calculate $\Delta\sigma'_{(f)}$.

Several case histories on the successful use of precompression techniques for improving foundation soil are available in the literature (for example, Johnson, 1970a).

Example 14.1

Examine Figure 14.17. During the construction of a highway bridge, the average permanent load on the clay layer is expected to increase by about 115 kN/m². The average effective overburden pressure at the middle of the clay layer is 210 kN/m². Here, $H_c = 6$ m, $C_c = 0.28$, $e_o = 0.9$, and $C_v = 0.36$ m²/mo. The clay is normally consolidated. Determine

a. The total primary consolidation settlement of the bridge without precompression
b. The surcharge, $\Delta\sigma'_{(f)}$, needed to eliminate the entire primary consolidation settlement in nine months by precompression.

Solution

Part a
The total primary consolidation settlement may be calculated from Eq. (14.12):

$$S_{c(p)} = \frac{C_c H_c}{1 + e_o} \log \left[\frac{\sigma'_o + \Delta\sigma'_{(p)}}{\sigma'_o} \right] = \frac{(0.28)(6)}{1 + 0.9} \log \left[\frac{210 + 115}{210} \right]$$

$$= 0.1677 \text{ m} = \textbf{167.7 mm}$$

Part b
We have

$$T_v = \frac{C_v t_2}{H^2}$$

$$C_v = 0.36 \text{ m}^2/\text{mo.}$$

$$H = 3 \text{ m (two-way drainage)}$$

$$t_2 = 9 \text{ mo.}$$

Hence,

$$T_v = \frac{(0.36)(9)}{3^2} = 0.36$$

According to Figure 14.20, for $T_v = 0.36$, the value of U is 47%. Now,

$$\Delta\sigma'_{(p)} = 115 \text{ kN/m}^2$$

and

$$\sigma'_o = 210 \text{ kN/m}^2$$

so

$$\frac{\Delta\sigma'_{(p)}}{\sigma'_o} = \frac{115}{210} = 0.548$$

According to Figure 14.18, for $U = 47\%$ and $\Delta\sigma'_{(p)}/\sigma'_o = 0.548$, $\Delta\sigma'_{(f)}/\Delta\sigma'_{(p)} \approx 1.8$; thus,

$$\Delta\sigma'_{(f)} = (1.8)(115) = \textbf{207 kN/m}^2 \qquad \blacksquare$$

14.8   *Sand Drains*

The use of sand drains is another way to accelerate the consolidation settlement of soft, normally consolidated clay layers and achieve precompression before the construction of a desired foundation. Sand drains are constructed by drilling holes through the clay layer(s) in the field at regular intervals. The holes are then back-filled with sand. This can be achieved by several means, such as (a) rotary drilling and then backfilling with sand; (b) drilling by continuous-flight auger with a hol-low stem and backfilling with sand (through the hollow steam); and (c) driving hollow steel piles. The soil inside the pile is then jetted out, after which backfilling with sand is done. Figure 14.21 shows a schematic diagram of sand drains. After backfilling the drill holes with sand, a surcharge is applied at the ground surface. The surcharge will increase the pore water pressure in the clay. The excess pore water pressure in the clay will be dissipated by drainage—both vertically and radi-ally to the sand drains—thereby accelerating settlement of the clay layer. In Fig-ure 14.21a, note that the radius of the sand drains is r_w. Figure 14.21b shows the

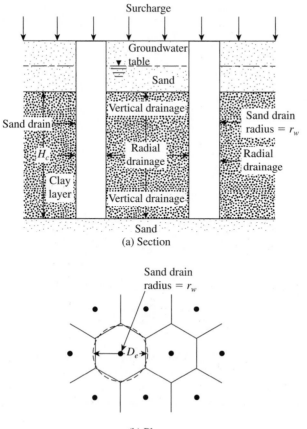

(a) Section

(b) Plan

Figure 14.21   Sand drains

plan of the layout of the sand drains. The effective zone from which the radial drainage will be directed toward a given sand drain is approximately cylindrical, with a diameter of d_e.

To determine the surcharge that needs to be applied at the ground surface and the length of time that it has to be maintained, see Figure 14.17 and use the corresponding equation, Eq. (14.15):

$$U_{v,r} = \frac{\log\left[1 + \dfrac{\Delta\sigma'_{(p)}}{\sigma'_o}\right]}{\log\left\{1 + \dfrac{\Delta\sigma'_{(p)}}{\sigma'_o}\left[1 + \dfrac{\Delta\sigma'_{(f)}}{\Delta\sigma'_{(p)}}\right]\right\}} \tag{14.17}$$

The notations $\Delta\sigma'_{(p)}$, σ'_o, and $\Delta\sigma'_{(f)}$ are the same as those in Eq. (14.15); however, the left-hand side of Eq. (14.17) is the *average degree* of consolidation instead of the degree of consolidation at midplane. Both *radial* and *vertical* drainage contribute to the average degree of consolidation. If $U_{v,r}$ can be determined for any time t_2 (see Figure 14.17b), the total surcharge $\Delta\sigma'_{(f)} + \Delta\sigma'_{(p)}$ may be obtained easily from Figure 14.18. The procedure for determining the average degree of consolidation ($U_{v,r}$) follows:

For a given surcharge and duration, t_2, the average degree of consolidation due to drainage in the vertical and radial directions is

$$U_{v,r} = 1 - (1 - U_r)(1 - U_v) \tag{14.18}$$

where

U_r = average degree of consolidation with radial drainage only
U_v = average degree of consolidation with vertical drainage only

The successful use of sand drains has been described in detail by Johnson (1970b). As with precompression, constant field settlement observations may be necessary during the period the surcharge is applied.

Average Degree of Consolidation Due to Radial Drainage Only

Figure 14.22 shows a schematic diagram of a sand drain. In the figure, r_w is the radius of the sand drain and $r_e = d_e/2$ is the radius of the effective zone of drainage. It is also important to realize that, during the installation of sand drains, a certain zone of clay surrounding them is smeared, thereby changing the hydraulic conductivity of the clay. In the figure, r_s is the radial distance from the center of the sand drain to the farthest point of the smeared zone. Now, for the average-degree-of-consolidation relationship, we will use the *theory of equal strain*. Two cases may arise that relate to the nature of the application of surcharge, and they are shown in Figure 14.23. (See the notations shown in Figure 14.17). Either (a) the entire surcharge is applied instantaneously

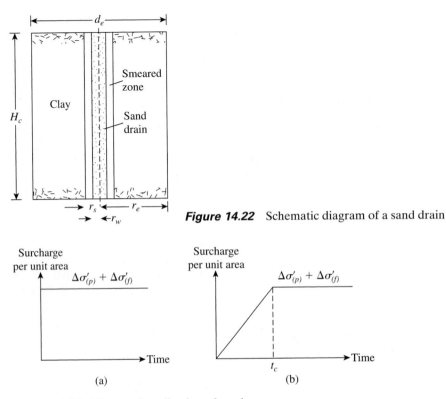

Figure 14.22 Schematic diagram of a sand drain

Figure 14.23 Nature of application of surcharge

(see Figure 14.23a), or (b) the surcharge is applied in the form of a ramp load (see Figure 14.23b). When the entire surcharge is applied instantaneously (Barron, 1948),

$$U_r = 1 - \exp\left(\frac{-8T_r}{m}\right) \tag{14.19}$$

where

$$m = \frac{n^2}{n^2 - S^2} \ln\left(\frac{n}{S}\right) - \frac{3}{4} + \frac{S^2}{4n^2} + \frac{k_h}{k_s}\left(\frac{n^2 - S^2}{n^2}\right) \ln S \tag{14.20}$$

in which

$$n = \frac{d_e}{2r_w} = \frac{r_e}{r_w} \tag{14.21}$$

$$S = \frac{r_s}{r_w} \tag{14.22}$$

and

k_h = hydraulic conductivity of clay in the horizontal direction in the unsmeared zone

k_s = horizontal hydraulic conductivity in the smeared zone

T_r = nondimensional time factor for radial drainage only = $\dfrac{C_{vr}t_2}{d_e^2}$ (14.23)

C_{vr} = coefficient of consolidation for radial drainage

$$= \frac{k_h}{\left[\dfrac{\Delta e}{\Delta\sigma'(1 + e_{av})}\right]\gamma_w} \qquad (14.24)$$

For a *no-smear case*, $r_s = r_w$ and $k_h = k_s$, so $S = 1$ and Eq. (14.20) becomes

$$m = \left(\frac{n^2}{n^2 - 1}\right)\ln(n) - \frac{3n^2 - 1}{4n^2} \qquad (14.25)$$

Table 14.5 gives the values of U_r for various values of T_r and n.

If the surcharge is applied in the form of a *ramp* and *there is no smear*, then (Olson, 1977)

$$U_r = \frac{T_r - \dfrac{1}{A}[1 - \exp(-AT_r)]}{T_{rc}} \qquad \text{(for } T_r \leqslant T_{rc}) \qquad (14.26)$$

and

$$U_r = 1 - \frac{1}{AT_{rc}}[\exp(AT_{rc}) - 1]\exp(-AT_{rc}) \qquad \text{(for } T_r \geqslant T_{rc}) \qquad (14.27)$$

where

$$T_{rc} = \frac{C_{vr}t_c}{d_e^2} \text{ (see Figure 14.23b for the definition of } t_c) \qquad (14.28)$$

and

$$A = \frac{2}{m} \qquad (14.29)$$

Average Degree of Consolidation Due to Vertical Drainage Only

Using Figure 14.23a, for instantaneous application of a surcharge, we may obtain the average degree of consolidation due to vertical drainage only from Eqs. (1.60) and (1.61). We have

$$T_v = \frac{\pi}{4}\left[\frac{U_v(\%)}{100}\right]^2 \qquad \text{(for } U_v = 0 \text{ to } 60\%) \qquad \text{[Eq. (1.60)]}$$

Table 14.5 Variation of U_r for Various Values of T_r and n, No-Smear Case [Eqs. (14.19) and (14.25)]

Degree of consolidation U_r (%)	Time factor T_r for value of $n (= r_e/r_w)$				
	5	10	15	20	25
0	0	0	0	0	0
1	0.0012	0.0020	0.0025	0.0028	0.0031
2	0.0024	0.0040	0.0050	0.0057	0.0063
3	0.0036	0.0060	0.0075	0.0086	0.0094
4	0.0048	0.0081	0.0101	0.0115	0.0126
5	0.0060	0.0101	0.0126	0.0145	0.0159
6	0.0072	0.0122	0.0153	0.0174	0.0191
7	0.0085	0.0143	0.0179	0.0205	0.0225
8	0.0098	0.0165	0.0206	0.0235	0.0258
9	0.0110	0.0186	0.0232	0.0266	0.0292
10	0.0123	0.0208	0.0260	0.0297	0.0326
11	0.0136	0.0230	0.0287	0.0328	0.0360
12	0.0150	0.0252	0.0315	0.0360	0.0395
13	0.0163	0.0275	0.0343	0.0392	0.0431
14	0.0177	0.0298	0.0372	0.0425	0.0467
15	0.0190	0.0321	0.0401	0.0458	0.0503
16	0.0204	0.0344	0.0430	0.0491	0.0539
17	0.0218	0.0368	0.0459	0.0525	0.0576
18	0.0232	0.0392	0.0489	0.0559	0.0614
19	0.0247	0.0416	0.0519	0.0594	0.0652
20	0.0261	0.0440	0.0550	0.0629	0.0690
21	0.0276	0.0465	0.0581	0.0664	0.0729
22	0.0291	0.0490	0.0612	0.0700	0.0769
23	0.0306	0.0516	0.0644	0.0736	0.0808
24	0.0321	0.0541	0.0676	0.0773	0.0849
25	0.0337	0.0568	0.0709	0.0811	0.0890
26	0.0353	0.0594	0.0742	0.0848	0.0931
27	0.0368	0.0621	0.0776	0.0887	0.0973
28	0.0385	0.0648	0.0810	0.0926	0.1016
29	0.0401	0.0676	0.0844	0.0965	0.1059
30	0.0418	0.0704	0.0879	0.1005	0.1103
31	0.0434	0.0732	0.0914	0.1045	0.1148
32	0.0452	0.0761	0.0950	0.1087	0.1193
33	0.0469	0.0790	0.0987	0.1128	0.1239
34	0.0486	0.0820	0.1024	0.1171	0.1285
35	0.0504	0.0850	0.1062	0.1214	0.1332
36	0.0522	0.0881	0.1100	0.1257	0.1380
37	0.0541	0.0912	0.1139	0.1302	0.1429
38	0.0560	0.0943	0.1178	0.1347	0.1479
39	0.0579	0.0975	0.1218	0.1393	0.1529
40	0.0598	0.1008	0.1259	0.1439	0.1580
41	0.0618	0.1041	0.1300	0.1487	0.1632
42	0.0638	0.1075	0.1342	0.1535	0.1685
43	0.0658	0.1109	0.1385	0.1584	0.1739
44	0.0679	0.1144	0.1429	0.1634	0.1793

(Continued)

Table 14.5 (Continued)

Degree of consolidation U_r (%)	Time factor T_r for value of n (= r_e/r_w)				
	5	10	15	20	25
45	0.0700	0.1180	0.1473	0.1684	0.1849
46	0.0721	0.1216	0.1518	0.1736	0.1906
47	0.0743	0.1253	0.1564	0.1789	0.1964
48	0.0766	0.1290	0.1611	0.1842	0.2023
49	0.0788	0.1329	0.1659	0.1897	0.2083
50	0.0811	0.1368	0.1708	0.1953	0.2144
51	0.0835	0.1407	0.1758	0.2020	0.2206
52	0.0859	0.1448	0.1809	0.2068	0.2270
53	0.0884	0.1490	0.1860	0.2127	0.2335
54	0.0909	0.1532	0.1913	0.2188	0.2402
55	0.0935	0.1575	0.1968	0.2250	0.2470
56	0.0961	0.1620	0.2023	0.2313	0.2539
57	0.0988	0.1665	0.2080	0.2378	0.2610
58	0.1016	0.1712	0.2138	0.2444	0.2683
59	0.1044	0.1759	0.2197	0.2512	0.2758
60	0.1073	0.1808	0.2258	0.2582	0.2834
61	0.1102	0.1858	0.2320	0.2653	0.2912
62	0.1133	0.1909	0.2384	0.2726	0.2993
63	0.1164	0.1962	0.2450	0.2801	0.3075
64	0.1196	0.2016	0.2517	0.2878	0.3160
65	0.1229	0.2071	0.2587	0.2958	0.3247
66	0.1263	0.2128	0.2658	0.3039	0.3337
67	0.1298	0.2187	0.2732	0.3124	0.3429
68	0.1334	0.2248	0.2808	0.3210	0.3524
69	0.1371	0.2311	0.2886	0.3300	0.3623
70	0.1409	0.2375	0.2967	0.3392	0.3724
71	0.1449	0.2442	0.3050	0.3488	0.3829
72	0.1490	0.2512	0.3134	0.3586	0.3937
73	0.1533	0.2583	0.3226	0.3689	0.4050
74	0.1577	0.2658	0.3319	0.3795	0.4167
75	0.1623	0.2735	0.3416	0.3906	0.4288
76	0.1671	0.2816	0.3517	0.4021	0.4414
77	0.1720	0.2900	0.3621	0.4141	0.4546
78	0.1773	0.2988	0.3731	0.4266	0.4683
79	0.1827	0.3079	0.3846	0.4397	0.4827
80	0.1884	0.3175	0.3966	0.4534	0.4978
81	0.1944	0.3277	0.4090	0.4679	0.5137
82	0.2007	0.3383	0.4225	0.4831	0.5304
83	0.2074	0.3496	0.4366	0.4992	0.5481
84	0.2146	0.3616	0.4516	0.5163	0.5668
85	0.2221	0.3743	0.4675	0.5345	0.5868
86	0.2302	0.3879	0.4845	0.5539	0.6081
87	0.2388	0.4025	0.5027	0.5748	0.6311
88	0.2482	0.4183	0.5225	0.5974	0.6558
89	0.2584	0.4355	0.5439	0.6219	0.6827
90	0.2696	0.4543	0.5674	0.6487	0.7122

(*Continued*)

Table 14.5 (Continued)

Degree of consolidation U_r (%)	Time factor T_r for value of n ($= r_e/r_w$)				
	5	10	15	20	25
91	0.2819	0.4751	0.5933	0.6784	0.7448
92	0.2957	0.4983	0.6224	0.7116	0.7812
93	0.3113	0.5247	0.6553	0.7492	0.8225
94	0.3293	0.5551	0.6932	0.7927	0.8702
95	0.3507	0.5910	0.7382	0.8440	0.9266
96	0.3768	0.6351	0.7932	0.9069	0.9956
97	0.4105	0.6918	0.8640	0.9879	1.0846
98	0.4580	0.7718	0.9640	1.1022	1.2100
99	0.5391	0.9086	1.1347	1.2974	1.4244

and

$$T_v = 1.781 - 0.933 \log (100 - U_v(\%)) \quad (\text{for } U_v > 60\%) \text{ [Eq. (1.61)]}$$

where U_v = average degree of consolidation due to vertical drainage only, and

$$T_v = \frac{C_v t_2}{H^2} \qquad \text{[Eq. (1.55)]}$$

where C_v = coefficient of consolidation for vertical drainage.

For the case of ramp loading, as shown in Figure 14.23b, the variation of U_v with T_v can be expressed as (Olson, 1977):

For $T_v \leq T_c$:

$$U_v = \frac{T_v}{T_c} \left\{ 1 - \frac{2}{T_v} \Sigma \frac{1}{M^4} [1 - \exp(-M^2 T_v)] \right\} \qquad (14.30)$$

For $T_v \geq T_c$:

$$U_v = 1 - \frac{2}{T_c} \Sigma \frac{1}{M^4} [\exp(-M^2 T_c) - 1] \exp(-M^2 T_v) \qquad (14.31)$$

where

$$M = \frac{\pi}{2}(2m' + 1)$$
$$m' = 0, 1, 2, \ldots$$

$$T_c = \frac{C_v t_c}{H^2} \qquad (14.32)$$

where H = length of maximum vertical drainage path. Figure 14.24 shows the variation of $U_v(\%)$ with T_c and T_v.

Aboshi and Monden (1963) provided details on the field performance of 2700 sand drains used to construct the Toya Quay Wall on reclaimed land in Japan. Results of this study were summarized by Johnson (1970b) and this analysis is recommended for study.

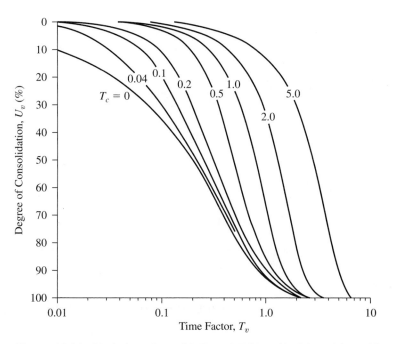

Figure 14.24 Variation of U_v with T_v and T_c [Eqs. (14.30) and (14.31)]

Example 14.2

Redo Example 14.1, with the addition of some sand drains. Assume that $r_w = 0.1$ m, $d_e = 3$ m, $C_v = C_{vr}$, and the surcharge is applied instantaneously. (See Figure 14.23a.) Also assume that this is a no-smear case.

Solution
Part a
The total primary consolidation settlement will be 167.7 mm, as before.
Part b
From Example 14.1, $T_v = 0.36$. Using Eq. (1.60), we obtain

$$T_v = \frac{\pi}{4}\left[\frac{U_v(\%)}{100}\right]^2$$

or

$$U_v = \sqrt{\frac{4T_v}{\pi}} \times 100 = \sqrt{\frac{(4)(0.36)}{\pi}} \times 100 = 67.7\%$$

Also,

$$n = \frac{d_e}{2r_w} = \frac{3}{2 \times 0.1} = 15$$

Again,

$$T_r = \frac{C_{vr}t_2}{d_e^2} = \frac{(0.36)(9)}{(3)^2} = 0.36$$

From Table 14.5 for $n = 15$ and $T_r = 0.36$, the value of U_r is about 77%, Hence,

$$U_{v,r} = 1 - (1 - U_v)(1 - U_r) = 1 - (1 - 0.67)(1 - 0.77)$$
$$= 0.924 = 92.4\%$$

Now, from Figure 14.18, for $\Delta\sigma_p'/\sigma_o' = 0.548$ and $U_{v,r} = 92.4\%$, the value of $\Delta\sigma_f'/\Delta\sigma_p' \approx 0.12$. Hence,

$$\Delta\sigma_{(f)}' = (115)(0.12) = \textbf{13.8 kN/m}^2 \qquad \blacksquare$$

Example 14.3

Suppose that, for the sand drain project of Figure 14.21, the clay is normally consolidated. We are given the following data:

Clay: $H_c = 15$ ft (two-way drainage)

$C_c = 0.31$

$e_o = 1.1$

Effective overburden pressure at the middle of the clay layer

$= 1000$ lb/ft^2

$C_v = 0.115$ ft^2/day

Sand drain: $r_w = 0.3$ ft

$d_e = 6$ ft

$C_v = C_{vr}$

A surcharge is applied as shown in Figure 14.25. Assume this to be a no-smear case. Calculate the degree of consolidation 30 days after the surcharge is first applied. Also, determine the consolidation settlement at that time due to the surcharge.

Solution
From Eq. (14.32),

$$T_c = \frac{C_v t_c}{H^2} = \frac{(0.115 \text{ ft}^2/\text{day})(60)}{\left(\dfrac{15}{2}\right)^2} = 0.123$$

and

$$T_v = \frac{C_v t_2}{H^2} = \frac{(0.115)(30)}{\left(\dfrac{15}{2}\right)^2} = 0.061$$

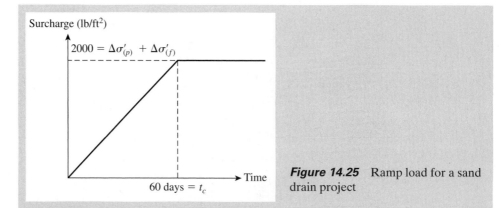

Figure 14.25   Ramp load for a sand drain project

Using Figure 14.24 for $T_c = 0.123$ and $T_v = 0.061$, we have $U_v \approx 9\%$. For the sand drain,

$$n = \frac{d_e}{2r_w} = \frac{6}{(2)(0.3)} = 10$$

From Eq. (14.28),

$$T_{rc} = \frac{C_{vr}t_c}{d_e^2} = \frac{(0.115)(60)}{(6)^2} = 0.192$$

and

$$T_r = \frac{C_{vr}t_2}{d_e^2} = \frac{(0.115)(30)}{(6)^2} = 0.096$$

Again, from Eq. (14.26),

$$U_r = \frac{T_r - \dfrac{1}{A}[1 - \exp(-AT_r)]}{T_{rc}}$$

Also, for the no-smear case,

$$m = \frac{n^2}{n^2 - 1}\ln(n) - \frac{3n^2 - 1}{4n^2} = \frac{10^2}{10^2 - 1}\ln(10) - \frac{3(10)^2 - 1}{4(10)^2} = 1.578$$

and

$$A = \frac{2}{m} = \frac{2}{1.578} = 1.267$$

so

$$U_r = \frac{0.096 - \dfrac{1}{1.267}[1 - \exp(-1.267 \times 0.096)]}{0.192} = 0.03 = 3\%$$

From Eq. (14.18),

$$U_{v,r} = 1 - (1 - U_r)(1 - U_v) = 1 - (1 - 0.03)(1 - 0.09) = 0.117 = \mathbf{11.7\%}$$

The total primary settlement is thus

$$S_{c(p)} = \frac{C_c H_c}{1 + e_o} \log \left[\frac{\sigma_o' + \Delta\sigma_{(p)}' + \Delta\sigma_f'}{\sigma_o'} \right]$$

$$= \frac{(0.31)(15)}{1 + 1.1} \log \left(\frac{1000 + 2000}{1000} \right) = 1.056 \text{ ft}$$

and the settlement after 30 days is

$$S_{c(p)} U_{v,r} = (1.056)(0.117)(12) = \mathbf{1.48 \text{ in.}} \qquad \blacksquare$$

14.9   Prefabricated Vertical Drains

Prefabricated vertical drains (PVDs), also referred to as *wick* or *strip drains,* were originally developed as a substitute for the commonly used sand drain. With the advent of materials science, these drains began to be manufactured from synthetic polymers such as polypropylene and high-density polyethylene. PVDs are normally manufactured with a corrugated or channeled synthetic core enclosed by a geotextile filter, as shown schematically in Figure 14.26. Installation rates reported in the literature are on the order of 0.1 to 0.3 m/s, excluding equipment mobilization and setup time. PVDs have been used extensively in the past for expedient consolidation of low-permeability soils under surface surcharge. The main advantage of PVDs over sand drains is that they do not require drilling; thus, installation is much faster. Figures 14.27a and b are photographs of the installation of PVDs in the field.

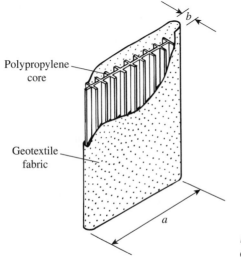

Polypropylene core

Geotextile fabric

b

a

Figure 14.26   Prefabricated vertical drain (PVD)

(a)

(b)

Figure 14.27 Installation of PVDs in the field (*Note:* (b) is a closeup view of (a)) (Courtesy of E. C. Shin, Incheon, South Korea)

Design of PVDs

The relationships for the average degree of consolidation due to radial drainage into sand drains are given in Eqs. (14.19) through (14.24) for equal-strain cases. Yeung (1997) used these relationships to develop design curves for PVDs. The theoretical developments used by Yeung are given next.

Figure 14.28 shows the layout of a square-grid pattern of prefabricated vertical drains. See also Figure 14.26 for the definition of a and b). The equivalent diameter of a PVD can be given as

$$d_w = \frac{2(a + b)}{\pi} \tag{14.33}$$

Now, Eq. (14.19) can be rewritten as

$$U_r = 1 - \exp\left(-\frac{8C_{vr}t}{d_w^2} \frac{d_w^2}{d_e^2 m}\right) = 1 - \exp\left(-\frac{8T_r'}{\alpha'}\right) \tag{14.34}$$

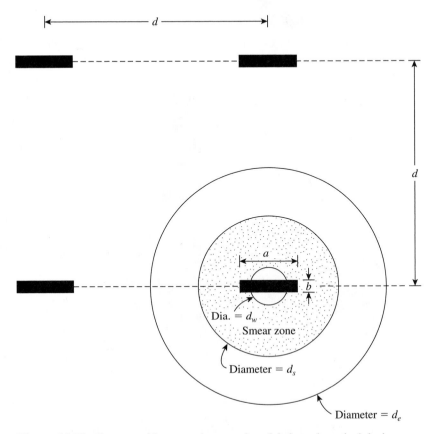

Figure 14.28 Square-grid pattern layout of prefabricated vertical drains

where d_e = diameter of the effective zone of drainage = $2r_e$. Also,

$$T_r' = \frac{C_{vr}t}{d_w^2} \tag{14.35}$$

$$\alpha' = n^2 m = \frac{n^4}{n^2 - S^2} \ln\left(\frac{n}{S}\right) - \left(\frac{3n^2 - S^2}{4}\right) + \frac{k_h}{k_s}(n^2 - S^2) \ln S \tag{14.36}$$

and

$$n = \frac{d_e}{d_w} \tag{14.37}$$

From Eq. (14.34),

$$T_r' = -\frac{\alpha'}{8} \ln (1 - U_r)$$

or

$$(T_r')_1 = \frac{T_r'}{\alpha'} = -\frac{\ln (1 - U_r)}{8} \tag{14.38}$$

Table 14.6 gives the variation of $(T_r')_1$ with U_r. Also, Figure 14.29 shows plots of α' versus n for k_h/k_s = 5 and 10 and S = 2 and 3.

Table 14.6 Variation of $(T_r')_1$ with U_r [Eq. (14.38)]

U_r (%)	$(T_r')_1$
0	0
5	0.006
10	0.013
15	0.020
20	0.028
25	0.036
30	0.045
35	0.054
40	0.064
45	0.075
50	0.087
55	0.100
60	0.115
65	0.131
70	0.150
75	0.173
80	0.201
85	0.237
90	0.288
95	0.374

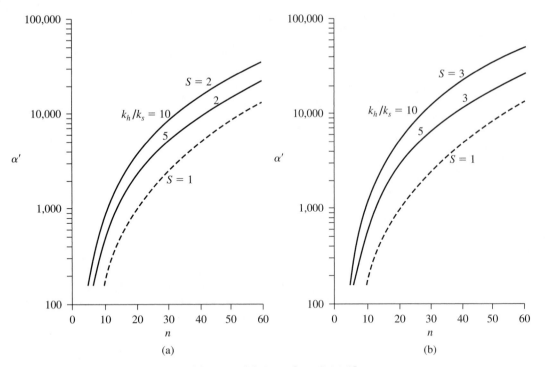

Figure 14.29 Plot of α' versus n: (a) $S = 2$: (b) $S = 3$ [Eq. (14.36)]

Following is a step-by-step procedure for the design of prefabricated vertical drains:

Step 1. Determine time t_2 available for the consolidation process and the $U_{v,r}$ required therefore [Eq. (14.17)]

Step 2. Determine U_v at time t_2 due to vertical drainage. From Eq. (14.18)

$$U_r = 1 - \frac{1 - U_{v,r}}{1 - U_v} \qquad (14.39)$$

Step 3. For the PVD that is to be used, calculate d_w from Eq. (14.33).

Step 4. Determine $(T'_r)_1$ from Eqs. (14.38) and (14.39).

Step 5. Determine T'_r from Eq. (14.35).

Step 6. Determine

$$\alpha' = \frac{T'_r}{(T'_r)_1}.$$

Step 7. Using Figure 14.29 and α' determined from Step 6, determine n.

Step 8. From Eq. (14.37),

$$d_e = n \qquad d_w$$
$$\uparrow \qquad \uparrow$$
$$\text{Step 7} \quad \text{Step 3}$$

Step 9. Choose the drain spacing:

$$d = \frac{d_e}{1.05} \quad \text{(for triangular pattern)}$$

$$d = \frac{d_e}{1.128} \quad \text{(for square pattern)}$$

A Case History

The installation of PVDs combined with preloading is an efficient way to gain strength in soft clays for construction of foundations. An example of a field study can be found in the works of Shibuya and Hanh (2001) which describes a full-scale test embankment 40 m × 40 m in plan constructed over a soft clay layer located at Nong Ngu Hao, Thailand. PVDs were installed in the soft clay layer in a triangular pattern (Figure 14.30a). Figure 14.30b shows the pattern of preloading at the site along with the settlement-time plot at the ground surface below the center of the test embankment. Maximum settlement was reached after about four months. The variation of the undrained shear strength (c_u) with depth in the soft clay layer before and after the soil improvement is shown in Figure 14.30c. The variation of c_u with depth is based on field vane shear tests. The undrained shear strength increases by about 50 to 100% at various depths.

14.10   Lime Stabilization

As mentioned in Section 14.1, admixtures are occasionally used to stabilize soils in the field—particularly fine-grained soils. The most common admixtures are lime, cement, and lime–fly ash. The main purposes of stabilizing the soil are to (a) modify the soil, (b) expedite construction, and (c) improve the strength and durability of the soil.

The types of *lime* commonly used to stabilize fine-grained soils are hydrated high-calcium lime $[Ca(OH)_2]$, calcitic quicklime (CaO), monohydrated dolomitic lime $[Ca(OH)_2 \cdot MgO]$, and dolomitic quicklime. The quantity of lime used to stabilize most soils usually is in the range from 5 to 10%. When lime is added to clayey soils, two *pozzolanic* chemical reactions occur: *cation exchange* and *flocculation–agglomeration*. In the cation exchange and flocculation–agglomeration reactions, the *monovalent* cations generally associated with clays are replaced by the *divalent* calcium ions. The cations can be arranged in a series based on their affinity for exchange:

$$Al^{3+} > Ca^{2+} > Mg^{2+} > NH_4^+ > K^+ > Na^+ > Li^+$$

Any cation can replace the ions to its right. For example, calcium ions can replace potassium and sodium ions from a clay. Flocculation–agglomeration produces a change in the texture of clay soils. The clay particles tend to clump together to form larger particles, thereby (a) decreasing the liquid limit, (b) increasing the plastic limit, (c) decreasing the plasticity index, (d) increasing the shrinkage limit, (e) increasing the workability, and (f) improving the strength and deformation properties of a soil.

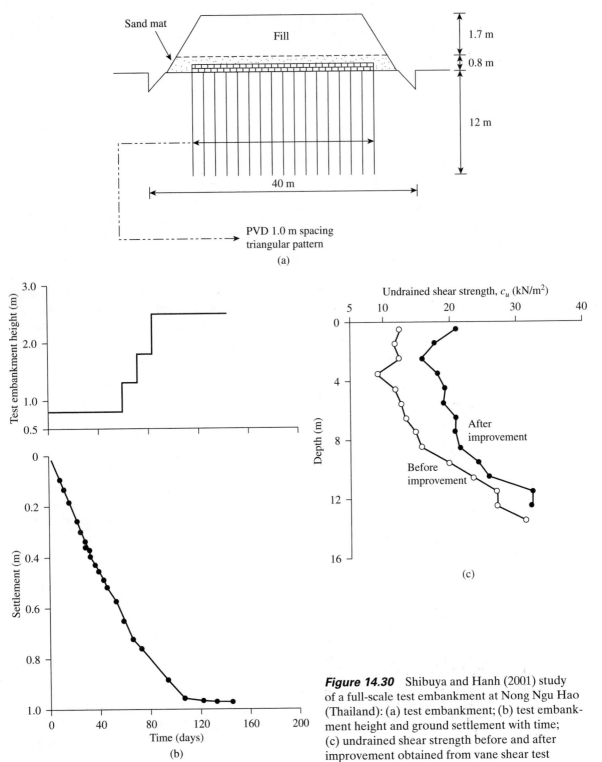

Figure 14.30 Shibuya and Hanh (2001) study of a full-scale test embankment at Nong Ngu Hao (Thailand): (a) test embankment; (b) test embankment height and ground settlement with time; (c) undrained shear strength before and after improvement obtained from vane shear test

Pozzolanic reaction between soil and lime involves a reaction between lime and the silica and alumina of the soil to form cementing material. One such reaction is

$$Ca(OH)_2 + SiO_2 \rightarrow CSH$$
$$\uparrow$$
Clay silica

where

$C = CaO$
$S = SiO_2$
$H = H_2O$

The pozzolanic reaction may continue for a long time.

Figure 14.31 shows the variation of the liquid limit, the plasticity index, and the shrinkage limit of a clay with the percentage of lime admixture. The first 2 to 3% lime (on the dry-weight basis) substantially influences the workability and the property (such as plasticity) of the soil. The addition of lime to clayey soils also affects their compaction characteristics. Lime stabilization in the field can be done in three ways:

1. The *in situ* material or the borrowed material can be mixed with the proper amount of lime at the site and then compacted after the addition of moisture.
2. The soil can be mixed with the proper amount of lime and water at a plant and then hauled back to the site for compaction.
3. Lime slurry can be pressure injected into the soil to a depth of 4 to 5 m (12 to 16 ft). Figure 14.32 shows a vehicle used for pressure injection of lime slurry.

The slurry-injection mechanical unit is mounted to the injection vehicle. A common injection unit is a hydraulic-lift mast with crossbeams that contain the injection rods. The rods are pushed into the ground by the action of the lift mast beams. The slurry is generally mixed in a batching tank about 3 m (10 ft) in diameter and 12 m (36 ft) long

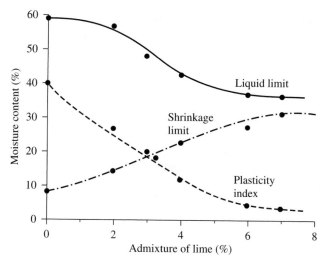

Figure 14.31 Variation of liquid limit, plasticity index, and shrinkage of a clay with lime additive

Figure 14.32 Equipment for pressure injection of lime slurry (Courtesy of GKN Hayward Baker, Inc., Woodbine Division, Ft. Worth, Texas)

and is pumped at high pressure to the injection rods. Figure 14.33 is a photograph of the lime slurry pressure-injection process. The ratio typically specified for the preparation of lime slurry is 1.13 kg (2.5 lb) of dry lime to a gallon of water.

Because the addition of hydrated lime to soft clayey soils immediately increases the plastic limit, thus changing the soil from plastic to solid and making it appear to "dry up," limited amounts of the lime can be thrown on muddy and troublesome construction sites. This action improves trafficability and may save money and time. Quicklimes have also been successfully used in drill holes having diameters of 100 to 150 mm (4 to 6 in.) for stabilization of subgrades and slopes. For this type of work, holes are drilled in a grid pattern and then filled with quicklime.

14.11 Cement Stabilization

Cement is being increasingly used as a stabilizing material for soil, particularly in the construction of highways and earth dams. The first controlled soil–cement construction in the United States was carried out near Johnsonville, South Carolina, in 1935. Cement can be used to stabilize sandy and clayey soils. As in the case of lime, cement helps decrease the liquid limit and increase the plasticity index and workability of clayey soils. Cement stabilization is effective for clayey soils when the liquid limit is less than 45 to 50 and the plasticity index is less than about 25. The optimum requirements of cement by volume for effective stabilization of various types of soil are given in Table 14.7.

Figure 14.33 Pressure injection of lime slurry (Courtesy of GKN Hayward Baker, Inc., Woodbine Division, Ft. Worth, Texas)

Table 14.7 Cement Requirement by Volume for Effective Stabilization of Various Soils[a]

Soil type		Percent cement by volume
AASHTO classification	Unified classification	
A-2 and A-3	GP, SP, and SW	6–10
A-4 and A-5	CL, ML, and MH	8–12
A-6 and A-7	CL, CH	10–14

[a] After Mitchell and Freitag (1959)

Table 14.8 Typical Compressive Strengths of Soils and Soil–Cement Mixtures[a]

Material	Unconfined compressive strength range	
	kN/m²	lb/in²
Untreated soil:		
Clay, peat	Less than 350	Less than 50
Well-compacted sandy clay	70–280	10–40
Well-compacted gravel, sand, and clay mixtures	280–700	40–100
Soil–cement (10% cement by weight):		
Clay, organic soils	Less than 350	Less than 50
Silts, silty clays, very poorly graded sands, slightly organic soils	350–1050	50–150
Silty clays, sandy clays, very poorly graded sands, and gravels	700–1730	100–250
Silty sands, sandy clays, sands, and gravels	1730–3460	250–500
Well-graded sand–clay or gravel–sand–clay mixtures and sands and gravels	3460–10,350	500–1500

[a] After Mitchell and Freitag (1959)

Like lime, cement helps increase the strength of soils, and strength increases with curing time. Table 14.8 presents some typical values of the unconfined compressive strength of various types of untreated soil and of soil–cement mixtures made with approximately 10% cement by weight.

Granular soils and clayey soils with low plasticity obviously are most suitable for cement stabilization. Calcium clays are more easily stabilized by the addition of cement, whereas sodium and hydrogen clays, which are expansive in nature, respond better to lime stabilization. For these reasons, proper care should be given in the selection of the stabilizing material.

For field compaction, the proper amount of cement can be mixed with soil either at the site or at a mixing plant. If the latter approach is adopted, the mixture can then be carried to the site. The soil is compacted to the required unit weight with a predetermined amount of water.

Similar to lime injection, cement slurry made of portland cement and water (in a water–cement ratio of 0.5:5) can be used for pressure grouting of poor soils under foundations of buildings and other structures. Grouting decreases the hydraulic conductivity of soils and increases their strength and load-bearing capacity. For the design of low-frequency machine foundations subjected to vibrating forces, stiffening the foundation soil by grouting and thereby increasing the resonant frequency is sometimes necessary.

14.12 Fly-Ash Stabilization

Fly ash is a by-product of the pulverized coal combustion process usually associated with electric power-generating plants. It is a fine-grained dust and is composed primarily of silica, alumina, and various oxides and alkalies. Fly ash is pozzolanic in nature and can react with hydrated lime to produce cementitious products. For that reason, lime–fly-ash mixtures can be used to stabilize highway bases and subbases. Effective mixes can be prepared with 10 to 35% fly ash and 2 to 10% lime. Soil–lime–fly-ash mixes are compacted under controlled conditions, with proper amounts of moisture to obtain stabilized soil layers.

A certain type of fly ash, referred to as "Type C" fly ash, is obtained from the burning of coal primarily from the western United States. This type of fly ash contains a fairly large proportion (up to about 25%) of free lime that, with the addition of water, will react with other fly-ash compounds to form cementitious products. Its use may eliminate the need to add manufactured lime.

14.13 *Stone Columns*

A method now being used to increase the load-bearing capacity of shallow foundations on soft clay layers is the construction of stone columns. This generally consists of water-jetting a vibroflot (see Section 14.6) into the soft clay layer to make a circular hole that extends through the clay to firmer soil. The hole is then filled with an imported gravel. The gravel in the hole is gradually compacted as the vibrator is withdrawn. The gravel used for the stone column has a size range of 6 to 40 mm (0.25 to 1.6 in.). Stone columns usually have diameters of 0.5 to 0.75 m (1.6 to 2.5 ft) and are spaced at about 1.5 to 3 m (5 to 10 ft) center to center. Figure 14.34 shows the construction of a stone column.

After stone columns are constructed, a fill material should always be placed over the ground surface and compacted before the foundation is constructed. The stone columns tend to reduce the settlement of foundations at allowable loads. Several case histories of construction projects using stone columns are presented in Hughes and Withers (1974), Hughes et al. (1975), Mitchell and Huber (1985), and other works.

Stone columns work more effectively when they are used to stabilize a large area where the undrained shear strength of the subsoil is in the range of 10 to 50 kN/m^2 (200 to 1000 lb/ft^2) than to improve the bearing capacity of structural foundations (Bachus and Barksdale, 1989). Subsoils weaker than that may not provide sufficient lateral support for the columns. For large-site improvement, stone columns are most effective to a depth of 6 to 10 m (20 to 30 ft). However, they have been constructed to a depth of 31 m (100 ft). Bachus and Barksdale provided the following general guidelines for the design of stone columns to stabilize large areas.

Figure 14.35a shows the plan view of several stone columns. The area replacement ratio for the stone columns may be expressed as

$$a_s = \frac{A_s}{A} \tag{14.40}$$

where

A_s = area of the stone column
A = total area within the unit cell

Figure 14.34   Construction of a stone column (Courtesy of The Reinforced Earth Company, Vienna, Virginia and Menard Soil Treatment Inc., Orange, California)

For an *equilateral triangular pattern* of stone columns,

$$a_s = 0.907\left(\frac{D}{s}\right)^2 \tag{14.41}$$

where

D = diameter of the stone column
s = spacing between the columns

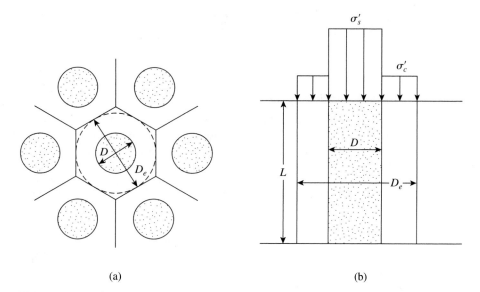

Figure 14.35   (a) Stone columns in a triangular pattern; (b) stress concentration due to change in stiffness

When a uniform stress by means of a fill operation is applied to an area with stone columns to induce consolidation, a stress concentration occurs due to the change in the stiffness between the stone columns and the surrounding soil. (See Figure 14.35b.) The stress concentration factor is defined as

$$n' = \frac{\sigma'_s}{\sigma'_c} \tag{14.42}$$

where

σ'_s = effective stress in the stone column
σ'_c = effective stress in the subgrade soil

The relationships for σ'_s and σ'_c are

$$\sigma'_s = \sigma' \left[\frac{n'}{1 + (n' - 1)a_s} \right] = \mu_s \sigma' \tag{14.43}$$

and

$$\sigma'_c = \sigma' \left[\frac{1}{1 + (n' - 1)a_s} \right] = \mu_c \sigma' \tag{14.44}$$

where

σ' = average effective vertical stress
μ_s, μ_c = stress concentration coefficients

The improvement in the soil owing to the stone columns may be expressed as

$$\frac{S_{e(t)}}{S_e} = \mu_c \tag{14.45}$$

where

$S_{e(t)}$ = settlement of the treated soil
S_e = total settlement of the untreated soil

Load-Bearing Capacity of Stone Columns

If a foundation is constructed over a stone column as shown in Figure 14.36, failure will occur by bulging of the column at ultimate load. The bulging will occur within a length of 2.5D to 3D, measured from the top of the stone column, where D is the diameter of the column.

Hughes et al. (1975) provided an approximate relationship for the ultimate bearing capacity of stone columns, which can be given as

$$q_u = \tan^2\left(45 + \frac{\phi'}{2}\right)(4c_u + \sigma'_r) \qquad (14.46)$$

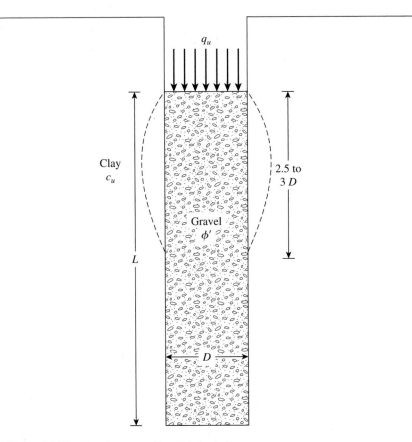

Figure 14.36   Bearing capacity of stone column

where

c_u = undrained shear strength of the clay
σ_r' = effective radial stress as measured by a pressuremeter ($\approx 2c_u$)
ϕ' = effective stress friction angle of the stone column material

Thus, assuming that the stone column carries the entire load of the foundation, the ultimate load can be given as

$$Q_u = \frac{\pi}{4} D^2 \tan^2\left(45 + \frac{\phi'}{2}\right)(4c_u + \sigma_r') \qquad (14.47)$$

On the basis of large-scale-model tests, Christoulas et al. (2000) suggested that

$$Q_u = \pi D L c_u \qquad (14.48)$$

In the opinion of the author, the lower of the two values of Q_u obtained from Eqs. (14.47) and (14.48) should be used for actual design purposes. The allowable load can then be given as

$$Q_{\text{all}} = \frac{Q_u}{\text{FS}} \qquad (14.49)$$

where FS = factor of safety (≈ 1.5 to 2).

Christoulas et al. (2000) also suggested a relationship between the load Q and the elastic settlement S_e for stone columns that can be expressed as

$$S_e = \left(\frac{Q}{LE_{\text{clay}}}\right)I_d \qquad (\text{for } Q \leq Q_1) \qquad (14.50)$$

and

$$S_e = \left(\frac{Q_1}{LE_{\text{clay}}}\right)I_d + \left(\frac{Q - Q_1}{4LE_{\text{clay}}}\right)I_d \qquad (\text{for } Q_1 \leq Q \leq Q_u) \qquad (14.51)$$

where

$$Q_1 = \frac{0.1DLE_{\text{clay}}}{I_d} \qquad (14.52)$$

in which

E_{clay} = modulus of elasticity of clay
I_d = influence factor (Mattes and Poulos, 1969)

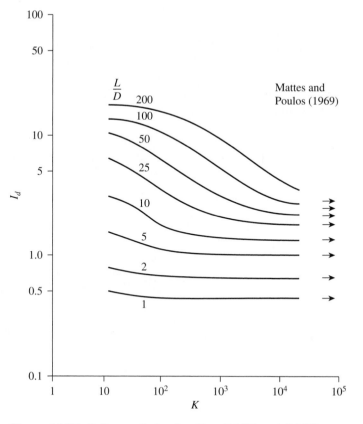

Figure 14.37   Influence factor I_d—Eqs. (14.50) and (14.51)

The influence factor proposed by Mattes and Poulos is a function of three quantities:

1. $K = \dfrac{E_{\text{col}}}{E_{\text{clay}}}$

where E_{col} = modulus of elasticity of the stone column material.

2. $\dfrac{L}{D}$

3. Poisson's ratio of clay, μ_{clay}. A value of $\mu_{\text{clay}} = 0.5$ will give a conservative result.

The variation of I_d (for $\mu_{\text{clay}} = 0.5$) with K is shown in Figure 14.37.

14.14   *Sand Compaction Piles*

Sand compaction piles are similar to stone columns, and they can be used in marginal sites to improve stability, control liquefaction, and reduce the settlement of various structures. Built in soft clay, these piles can significantly accelerate the pore water pressure-dissipation process and hence the time for consolidation.

Sand piles were first constructed in Japan between 1930 and 1950 (Ichimoto, 1981). Large-diameter compacted sand columns were constructed in 1955, using the Compozer technique (Aboshi et al., 1979). The Vibro-Compozer method of sand pile construction was developed by Murayama in Japan in 1958 (Murayama, 1962).

Sand compaction piles are constructed by driving a hollow mandrel with its bottom closed during driving. On partial withdrawal of the mandrel, the bottom doors open. Sand is poured from the top of the mandrel and is compacted in steps by applying air pressure as the mandrel is withdrawn. The piles are usually 0.46 to 0.76 m (1.5 to 2.5 ft) in diameter and are placed at about 1.5 to 3 m (5 to 10 ft) center to center. The pattern of layout of sand compaction piles is the same as for stone columns. Figure 14.38 shows the construction of sand compaction piles in the harbor of Yokohama, Japan.

Figure 14.38 Construction of sand compaction pile in Yokohama, Japan, harbor (Courtesy of E.C. Shin, Korea)

Basore and Boitano (1969) reported a case history on the densification of a granular subsoil having a thickness of about 9 m (30 ft) at the Treasure Island Naval Station in San Francisco, California, using sand compaction piles. The sand piles had diameters of 356 mm (14 in.). Figure 14.39a shows the layout of the sand piles. The

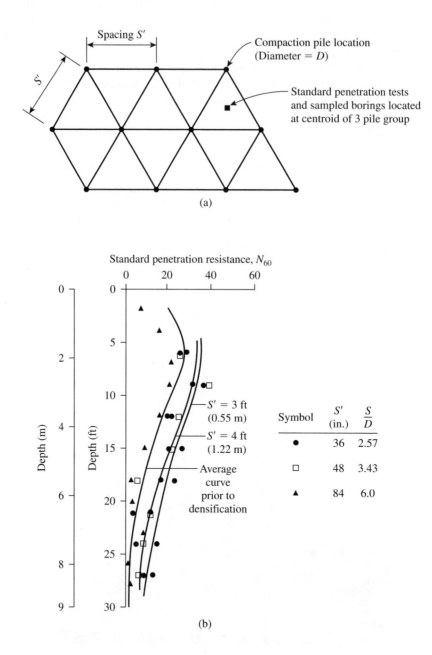

Figure 14.39   Sand compaction pile test of Basore and Boitano (1969): (a) Layout of the compaction piles; (b) Standard penetration resistance variation with depth and S'

spacing, S', between the piles was varied. The standard penetration resistance, N_{60}, before and after the construction of piles are shown in Figure 14.39b (see location of SPT test in Figure 14.39a). From this figure, it can be seen that the effect of densification at any given depth decreases with the increase in S' (or S'/D). These tests show that when S'/D exceeds about 4 to 5, the effect of densification is practically negligible.

14.15 *Dynamic Compaction*

Dynamic compaction is a technique that is beginning to gain popularity in the United States for densification of granular soil deposits. The process primarily involves dropping a heavy weight repeatedly on the ground at regular intervals. The weight of the hammer used varies from 8 to 35 metric tons, and the height of the hammer drop varies between 7.5 and 30.5 m (≈ 25 and 100 ft). The stress waves generated by the hammer drops help in the densification. The degree of compaction achieved depends on

- The weight of the hammer
- The height of the drop
- The spacing of the locations at which the hammer is dropped

Leonards et al. (1980) suggested that the significant depth of influence for compaction is approximately

$$DI \simeq \tfrac{1}{2}\sqrt{W_H h} \qquad (14.53)$$

where

DI = significant depth of densification (m)
W_H = dropping weight (metric ton)
h = height of drop (m)

In English units, Eq. (14.53) becomes

$$DI = 0.61\sqrt{W_H h} \qquad (14.54)$$

where DI and h are in ft and W_H is in kip.

Partos et al. (1989) provided several case histories of site improvement that used dynamic compaction. In 1992, Poran and Rodriguez suggested a rational method for conducting dynamic compaction for granular soils in the field. According to their method, for a hammer of width D having a weight W_H and a drop h, the approximate shape of the densified area will be of the type shown in Figure 14.40

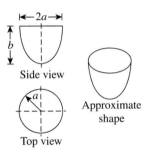

Figure 14.40 Approximate shape of the densified area due to dynamic compaction (After Poran and Rodriguez, 1992)

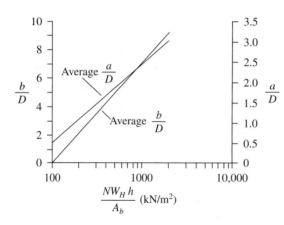

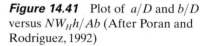

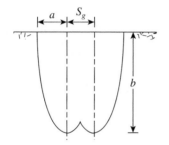

Figure 14.41 Plot of a/D and b/D versus $NW_H h/Ab$ (After Poran and Rodriguez, 1992)

Figure 14.42 Approximate grid spacing for dynamic compaction

(i.e., a semiprolate spheroid). Note that in this figure $b = DI$. Figure 14.41 gives the design chart for a/D and b/D versus $NW_H h/Ab$ (D = width of the hammer if not circular in cross section; A = area of cross section of the hammer; N = number of required hammer drops). The method uses the following steps:

Step 1. Determine the required significant depth of densification, $DI(=b)$.

Step 2. Determine the hammer weight (W_H), height of drop (h), dimensions of the cross section, and thus the area A and the width D.

Step 3. Determine $DI/D = b/D$.

Step 4. Use Figure 14.41 and determine the magnitude of $NW_H h/Ab$ for the value of b/D obtained in Step 3.

Step 5. Since the magnitudes of W_H, h, A, and b are known (or assumed) from Step 2, the number of hammer drops can be estimated from the value of $NW_H h/Ab$ obtained from Step 4.

Step 6. With known values of $NW_H h/Ab$, determine a/D and thus a from Figure 14.41.

Step 7. The grid spacing, S_g, for dynamic compaction may now be assumed to be equal to or somewhat less than a. (See Figure 14.42.)

14.16   *Jet Grouting*

Jet grouting is a soil stabilization process whereby cement slurry in injected into soil at a high velocity to form a soil–concrete matrix. Conceptually, the process of jet grouting was first developed in the 1960s. Most of the research work after that was conducted in Japan (Ohta and Shibazaki, 1982). The technique was introduced into Europe in the late 1970s, whereas the process was first used in the United States in the early 1980s (Welsh, Rubright, and Coomber, 1986).

Three basic systems of jet grouting have been developed—single, double, and triple rod systems. In all cases, hydraulic rotary drilling is used to reach the design depth at which the soil has to be stabilized. Figure 14.43a shows the *single rod system* in which a cement slurry is injected at a high velocity to form a soil–cement matrix. In the *double rod system* (Figure 14.43b), the cement slurry is injected at a high velocity sheathed in a cone of air at an equally high velocity to erode and mix the soil well. The *triple rod system* (Figure 14.43c) uses high-pressure water shielded in a cone of air to erode the soil. The void created in this process is then filled with a pre-engineering cement slurry.

The effectiveness of the jet grouting is very much influenced by the nature of erodibility of soil. Gravelly soil and clean sand are highly erodible, whereas highly plastic clays are difficult to erode. A summary of the range of parameters generally encountered for the three systems above follows (Welsh and Burke, 1991; Burke, 2004):

Single Rod System:

   A. Grout slurry

      Pressure . . . . 0.4–$0.7 \ \mathrm{MN/m^2}$
      Volume . . . . 100–$300 \ \mathrm{l/min}$
      Specific gravity 1.25–1.6
      Number of nozzles 1–6

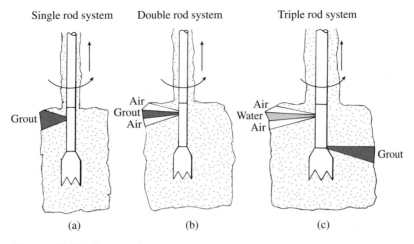

      (a)   (b)   (c)

Figure 14.43   Jet grouting

B. Lift

Step height . . . . 5–600 mm
Step time . . . . 4–30 sec

C. Rotation . . . . 7–20 rpm
D. Stabilized soil column diameter

Soft clay . . . . 0.4–0.9 m
Silt . . . . 0.6–1.1 m
Sand . . . . 0.8–1.2 m

Double Rod System:

A. Grout slurry

Pressure . . . . 0.3–0.7 MN/m^2
Volume . . . . 100 –600 l/min
Specific gravity 1.25–1.8
Number of nozzles 1–2

B. Air

Pressure . . . . 700–1500 kN/m^2
Volume . . . . 8–30 m^3/min

C. Lift

Step height . . . . 25–400 mm
Step time . . . . 4–30 sec

D. Rotation . . . . 7–15 rpm
E. Stabilized soil column diameter

Soft clay . . . . 0.9–1.8 m
Silt . . . . 0.9–1.8 m
Sand . . . . 1.2–2.1 m

Triple Rod System:

A. Grout slurry

Pressure . . . . 700 kN/m^2–0.1 MN/m^2
Volume . . . . 120–200 l/min
Specific gravity 1.5–2.0
Number of nozzles 1–3

B. Air

Pressure . . . . 700–1500 kN/m^2
Volume . . . . 4–15 m^3/min

C. Water

Pressure . . . . 0.3–0.4 MN/m^2
Volume . . . . 80–200 l/min

D. Lift

> Step height . . . . 20–50 mm
> Step time . . . . 4–20 sec

E. Rotation . . . . 7–15 rpm
F. Stabilized soil column diameter

> Soft clay . . . . 0.9–1.2 m
> Silt . . . . 0.9–1.4 m
> Sand . . . . 0.9–2.5 m

Problems

14.1 Make the necessary calculations and prepare the zero-air-void unit-weight curves (in kN/m^3) as related to a Proctor compaction test for $G_s = 2.6, 2.65, 2.7,$ and 2.75.

14.2 A sandy soil has a maximum dry unit weight of $112 \ lb/ft^3$ and a dry unit weight of compaction in the field of $100 \ lb/ft^3$. Estimate the following:
   a. The relative compaction in the field
   b. The relative density in the field
   c. The minimum dry unit weight of the soil

14.3 A silty clay soil has a plastic limit (PL) of 14. Estimate the optimum moisture content and the maximum dry unit weight of the soil when compacted using the procedure of:
   a. Standard Proctor test
   b. Modified Proctor test
   Use the Gurtug and Sridharan correlations.

14.4 The following are given for a natural soil deposit:
   Moist unit weight, $\gamma = 105 \ lb/ft^3$
   Moisture content, $w = 14\%$
   $G_s = 2.71$
   This soil is to be excavated and transported to a construction site for use in a compacted fill. If the specification calls for the soil to be compacted to a minimum dry unit weight of $108 \ lb/ft^3$ at the same moisture content of 14%, how many cubic yards of soil from the excavation site are needed to produce 20,000 cubic yards of compacted fill? How many twenty-ton truckloads are needed to transport the excavated soil?

14.5 A proposed embankment fill required $8000 \ m^3$ of compacted soil. The void ratio of the compacted fill is specified to be 0.6. Four available borrow pits are shown below along with the void ratios of the soil and the cost per cubic meter for moving the soil to the proposed construction site.

Borrow pit	Void ratio	Cost ($/m³)
A	0.82	9
B	0.91	7
C	0.95	8
D	0.75	11

Make the necessary calculations to select the pit from which the soil should be brought to minimize the cost. Assume G_s to be the same for all borrow-pit soil.

14.6 For a vibroflotation work, the backfill to be used has the following characteristics:

$D_{50} = 2$ mm
$D_{20} = 0.7$ mm
$D_{10} = 0.65$ mm

Determine the suitability number of the backfill. How would you rate the material?

14.7 Repeat Problem 14.6 with the following:

$D_{50} = 3.2$ mm
$D_{20} = 0.91$ mm
$D_{10} = 0.72$ mm

14.8 Refer to Figure 14.17. For a large fill operation, the average permanent load $[\Delta\sigma'_{(p)}]$ on the clay layer will increase by about 75 kN/m². The average effective overburden pressure on the clay layer before the fill operation is 110 kN/m². For the clay layer, which is normally consolidated and drained at top and bottom, given: $H_c = 8$ m, $C_c = 0.27$, $e_o = 1.02$, $C_v = 0.52$ m²/month. Determine the following:

a. The primary consolidation settlement of the clay layer caused by the addition of the permanent load $\Delta\sigma'_{(p)}$
b. The time required for 80% of primary consolidation settlement under the additional permanent load only
c. The temporary surcharge, $\Delta\sigma'_{(f)}$, that will be required to eliminate the entire primary consolidation settlement in 12 months by the precompression technique

14.9 Repeat Problem 14.8 with the following: $\Delta\sigma'_{(p)} = 1200$ lb/ft², average effective overburden pressure on the clay layer $= 1500$ lb/ft², $H_c = 15$ ft, $C_c = 0.3$, $e_o = 1.0$, and $C_v = 2.3 \times 10^{-2}$ in²/min.

14.10 The diagram of a sand drain is shown in Figures 14.21 and 14.22. Given: $r_w = 0.25$ m, $r_s = 0.35$ m, $d_e = 4.5$ m, $C_v = C_{vr} = 0.3$ m²/month, $k_h/k_s = 2$, and $H = 9$ m. Determine:

a. The degree of consolidation for the clay layer caused only by the sand drains after six months of surcharge application
b. The degree of consolidation for the clay layer that is caused by the combination of vertical drainage (drained on top and bottom) and radial drainage after six months of the application of surcharge. Assume that the surcharge is applied instantaneously.

14.11 A 10-ft thick clay layer is drained at the top and bottom. Its characteristics are $C_{vr} = C_v$ (for vertical drainage) $= 0.042$ ft²/day, $r_w = 8$ in., and $d_e = 6$ ft. Estimate the degree of consolidation of the clay layer caused by the combination of vertical and radial drainage at $t = 0.2, 0.4, 0.8$, and 1 year. Assume that the surcharge is applied instantaneously, and there is no smear.

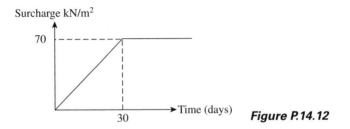

Surcharge kN/m²

70

30

Time (days)

Figure P.14.12

14.12 For a sand drain project (Figure 14.21), the following are given:

Clay: Normally consolidated
$H_c = 5.5$ m (one-way drainage)
$C_c = 0.3$
$e_o = 0.76$
$C_v = 0.015$ m²/day
Effective overburden pressure at the middle of clay
layer = 80 kN/m²

Sand drain: $r_w = 0.07$ m
$r_w = r_s$
$d_e = 2.5$ m
$C_v = C_{vr}$

A surcharge is applied as shown in Figure P14.12. Calculate the degree of consolidation and the consolidation settlement 50 days after the beginning of the surcharge application.

References

Aboshi, H., Ichimoto, E., and Harada, K. (1979). "The Compozer—a Method to Improve Characteristics of Soft Clay by Inclusion of Large Diameter Sand Column," *Proceedings, International Conference on Soil Reinforcement, Reinforced Earth and Other Techniques,* Vol. 1, Paris, pp. 211–216.

Aboshi, H., and Monden, H. (1963). "Determination of the Horizontal Coefficient of Consolidation of an Alluvial Clay," *Proceedings, Fourth Australia–New Zealand Conference on Soil Mechanics and Foundation Engineering,* pp. 159–164.

American Society for Testing and Materials (1997). *Annual Book of Standards,* Vol. 04.08, West Conshohocken, PA.

Bachus, R. C., and Barksdale, R. D. (1989). "Design Methodology for Foundations on Stone Columns," *Proceedings, Foundation Engineering: Current Principles and Practices* American Society of Civil Engineers, Vol. 1, pp. 244–257.

Barron, R. A. (1948). "Consolidation of Fine-Grained Soils by Drain Wells," *Transactions,* American Society of Civil Engineers, Vol. 113, pp. 718–754.

Basore, C. E., and Boitano, J. D. (1969). "Sand Densification by Piles and Vibroflotation," *Journal of the Soil Mechanics and Foundations Division,* American Society of Civil Engineers, Vol. 95, No. SM6, pp. 1303–1323.

Brown, R. E. (1977). "Vibroflotation Compaction of Cohesionless Soils," *Journal of the Geotechnical Engineering Division,* American Society of Civil Engineers, Vol. 103, No. GT12, pp. 1437–1451.

Burke, G. K. (2004). "Jet Grouting Systems: Advantages and Disadvantages," *Proceedings, GeoSupport 2004: Drilled Shafts, Micropiling, Deep Mixing, Remedial Methods, and Special Foundation Systems,* American Society of Civil Engineers, pp. 875–886.

Christoulas, S., Bouckovalas, G., and Giannaros, C. (2000). "An Experimental Study on Model Stone Columns," *Soils and Foundations,* Vol. 40, No. 6, pp. 11–22.

D'Appolonia, D. J., Whitman, R. V., and D'Appolonia, E. (1969). "Sand Compaction with Vibratory Rollers," *Journal of the Soil Mechanics and Foundations Division,* American Society of Civil Engineers, Vol. 95, No. SM1, pp. 263–284.

Gurtug, Y., and Sridharan, A. (2004). "Compaction Behaviour and Prediction of Its Characteristics of Fine Grained Soils with Particular Reference to Compaction Energy," *Soils and Foundations,* Vol. 44, No. 5, pp. 27–36.

Hughes, J. M. O., and Withers, N. J. (1974). "Reinforcing of Soft Cohesive Soil with Stone Columns," *Ground Engineering,* Vol. 7, pp. 42–49.

Hughes, J. M. O., Withers, N. J., and Greenwood, D. A. (1975). "A Field Trial of Reinforcing Effects of Stone Columns in Soil," *Geotechnique,* Vol. 25, No. 1, pp. 31–34.

Ichimoto, A. (1981). "Construction and Design of Sand Compaction Piles," *Soil Improvement, General Civil Engineering Laboratory* (in Japanese), Vol. 5. pp. 37–45.

Johnson, S. J. (1970a). "Precompression for Improving Foundation Soils," *Journal of the Soil Mechanics and Foundations Division,* American Society of Civil Engineers. Vol. 96, No. SM1, pp. 114–144.

Johnson, S. J. (1970b). "Foundation Precompression with Vertical Sand Drains," *Journal of the Soil Mechanics and Foundations Division,* American Society of Civil Engineers. Vol. 96, No. SM1, pp. 145–175.

Lee, K. L., and Singh, A. (1971). "Relative Density and Relative Compaction," *Journal of the Soil Mechanics and Foundations Division,* American Society of Civil Engineers, Vol. 97, No. SM7, pp. 1049–1052.

Leonards, G. A., Cutter, W. A., and Holtz, R. D. (1980). "Dynamic Compaction of Granular Soils," *Journal of Geotechnical Engineering Division,* ASCE, Vol. 96, No. GT1, pp. 73–110.

Mattes, N. S., and Poulos, H. G. (1969). "Settlement of Single Compressible Pile," *Journal of the Soil Mechanics and Foundations Division,* ASCE, Vol. 95, No. SM1, pp. 189–208.

Mitchell, J. K. (1970). "In-Place Treatment of Foundation Soils," *Journal of the Soil Mechanics and Foundations Division,* American Society of Civil Engineers, Vol. 96, No. SM1, pp. 73–110.

Mitchell, J. K., and Freitag, D. R. (1959). "A Review and Evaluation of Soil–Cement Pavements," *Journal of the Soil Mechanics and Foundations Division,* American Society of Civil Engineers, Vol. 85, No. SM6, pp. 49–73.

Mitchell, J. K., and Huber, T. R. (1985). "Performance of a Stone Column Foundation," *Journal of Geotechnical Engineering,* American Society of Civil Engineers, Vol. 111, No. GT2, pp. 205–223.

Murayama, S. (1962). "An Analysis of Vibro-Compozer Method on Cohesive Soils," *Construction in Mechanization* (in Japanese), No. 150, pp. 10–15.

Ohta, S., and Shibazaki, M. (1982). "A Unique Underpinning of Soil Specification Utilizing Super-High Pressure Liquid Jet," *Proceedings, Conference on Grouting in Geotechnical Engineering,* New Orleans, Louisiana.

Olson, R. E. (1977). "Consolidation under Time-Dependent Loading," *Journal of Geotechnical Engineering Division,* ASCE, Vol. 102, No. GT1, pp. 55–60.

Othman, M. A., and Luettich, S. M. (1994). "Compaction Control Criteria for Clay Hydraulic Barriers," *Transportation Research Record,* No. 1462, National Research Council, Washington, DC, pp. 28–35.

Poran, C. J., and Rodriguez, J. A. (1992). "Design of Dynamic Compaction," *Canadian Geotechnical Journal,* Vol. 2, No. 5, pp. 796–802.

Shibuya, S., and Hanh, L. T. (2001). "Estimating Undrained Shear Strength of Soft Clay Ground Improved by Preloading with PVD—Case History in Bangkok," *Soils and Foundations,* Vol. 41, No. 4, pp. 95–101.

Welsh, J. P., and Burke, G. K. (1991). "Jet Grouting–Uses for Soil Improvement," *Proceedings, Geotechnical Engineering Congress,* American Society of Civil Engineers, Vol. 1, pp. 334–345.

Welsh, J. P., Rubright, R. M., and Coomber, D. B. (1986). "Jet Grouting for support of Structures," presented at the Spring Convention of the American Society of Civil Engineers, Seattle, Washington.

Yeung, A. T. (1997). "Design Curves for Prefabricated Vertical Drains," *Journal of Geotechnical and Geoenvironmental Engineering,* Vol. 123, No. 8, pp. 755–759.

Appendix A

Field Instruments

This appendix provides brief descriptions of some common instruments used in the field in geotechnical engineering projects: the piezometer, earth pressure cell, load cell, cone pressuremeter, dilatometer, and inclinometer.

Piezometer

The piezometer measures water levels and pore water pressure. The simplest piezometer is the Casagrande piezometer (shown schematically in Figure 2.16) which is installed in a borehole of 75 to 150 mm diameter. Figure A.1 shows a photograph of the

Figure A.1 Components of Casagrande-type piezometer (Courtesy of N. Sivakugan, James Cook University, Australia)

735

components of the Casagrande piezometer. The device consists of a plastic riser pipe joined to a filter tip that is placed in sand. A bentonite seal is placed above the sand to isolate the pore water pressure at the filter tip. The annular space between the riser pipe and the borehole is backfilled with a bentonite–cement grout to prevent vertical migration of water. Although they are simple, reliable, and inexpensive, the standpipe piezometers are slow to reflect the change in pore water pressures and require someone at the site to monitor their operation. The other types are pneumatic piezometers, vibrating wire piezometers, and hydraulic piezometers. Piezometers are used to monitor pore water pressures to determine safe rates of fill or excavation; to monitor pore pressure dissipation in ground improvement techniques such as PVDs, sand drains, or dynamic compaction; to monitor pore pressures to check the performance of embankments, landfills, and tailings dams; and to monitor water drawdown during pumping tests. Its nonmetallic construction makes the piezometer suitable in any adverse environment for long-term operations in most soils.

Earth Pressure Cell

An earth pressure cell (Figure A.2) measures the total earth pressure, which includes the effective stress and the pore water pressure. It is made by welding two circular stainless-steel diaphragm plates around their periphery. The space between the plates is filled with a de-aired fluid. The cell is installed with its sensitive surface in direct contact with the soil. Any change in earth pressure is transmitted to the fluid inside the cell, which is measured by a pressure transducer that is similar to a diaphragm-type piezometer. Earth pressure cells can be used to measure stresses within embankments and subgrades, foundation bearing pressures, and contact pressures on retaining walls, tunnel linings, railroad bases, piers, and bridge abutments.

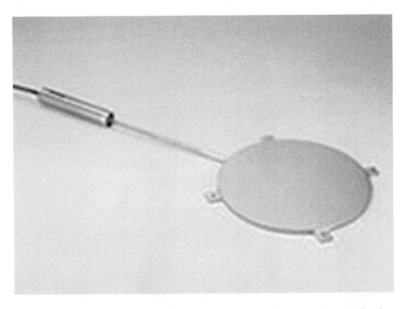

Figure A.2 Earth pressure cell (Courtesy of N. Sivakugan, James Cook University, Australia)

Figure A.3 Load cell (Courtesy of N. Sivakugan, James Cook University, Australia)

Load Cells

Load cells (Figure A.3) measure the loads acting on them. The load cell can be of the mechanical, hydraulic, vibrating wire, or electrical-resistant strain gauge type. Load cells measure strains or displacements under the applied loads that are translated into loads through calibration. Load cells are used in proof testing and performance monitoring of tiebacks, rock bolts, soil nails and other anchor systems, and pile load tests. They are also used to measure strut loads in braced excavations. In the laboratory, they are employed in triaxial devices and in load tests on model piles and footings.

Cone Pressuremeter

The cone gives a reliable classification of soils and good estimates of strength, but is poor in estimating soil stiffness. The pressuremeter, on the other hand, is well suited to measuring stiffness and strength parameters (Houlsby and Withers 1988). The cone pressuremeter (Figure A.4) exploits the features of a piezocone and a pressuremeter. It combines the continuous profiling capability of the cone with the ability of the pressuremeter to measure the load-deformation characteristics of soils—particularly the shear modulus. Approximately 2 m in length, the probe consists of a piezocone at the lower end of the penetrometer shaft and a pressuremeter of the

Figure A.5 Inclinometer system (Courtesy of N. Sivakugan, James Cook University, Australia)

Figure A.4 Cone pressuremeter (Courtesy of N. Sivakugan, James Cook University, Australia)

same diameter above, as shown in the figure. The top of the probe can be connected by the cone penetration test rods to the ground, and the probe is generally pushed into the ground by a cone truck or by jacking against a kentledge. The historical developments of the cone pressuremeter are given in Dalton (1997).

Inclinometer

Inclinometers are quite popular for measuring lateral displacement in slopes, landslides, retaining walls, piles, and bridge abutments. An inclinometer system (Figure A.5) is composed of a torpedo-shaped probe fitted with guide wheels and a tilt sensor

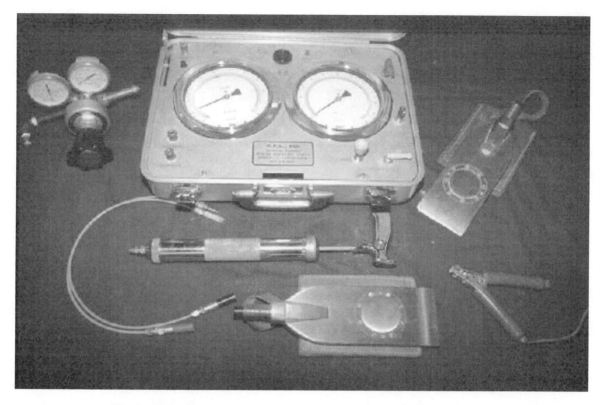

Figure A.6 Dilatometer and other equipment (Courtesy of N. Sivakugan, James Cook University, Australia)

and connected by a graduated cable to a readout unit. The probe is sent through the inclinometer casing installed vertically in a borehole, and the tilt sensor records the deviation between the probe axis and the vertical plane. By installing the inclinometer horizontally, the profiles of settlement or heave beneath embankments, storage tanks, and landfills can be obtained.

Dilatometer

The dilatometer test was discussed in detail in Section 2.18. Figure A.6 shows two flat dilatometers with other instruments for conducting the test.

References

Dalton, C. (1997). "Pressure for Change," *Ground Engineering,* Vol. 30, No. 9, pp. 22–23.
Houslby, G. T., and Withers, N. J. (1988). "Analysis of the Cone Penetrometer Tests in Clay," *Geotechnique,* Vol. 38, No. 4, pp. 575–587.

Answers to Selected Problems

Chapter 1

1.1 **a.** 0.76
  b. 0.43
  c. 14.93 kN/m^3
  d. 17.17 kN/m^3
  e. 53%

1.3 **a.** 0.45
  b. 69.5%
  c. 17.58 kN/m^3
  d. 14.53 kN/m^3

1.5 **a.** 129.2 lb/ft^3
  b. 7.2 lb/ft^3
  c. 124.6 lb/ft^3

1.7
Soil	Classification
A	A-7-6(9)
B	A-6(5)
C	A-3(0)
D	A-4(5)
E	A-2-6(1)
F	A-7-6(19)

1.9 0.08 cm/sec
1.11 1.73×10^{-6} cm/sec
1.13 -161.2 lb/ft^2
1.15 87 mm
1.17 **a.** 7.8 in.
  b. 509.5 days
1.19 6.23 days
1.21 After 30 days—7.5 mm
  After 120 days—40.5 mm

1.23 38°
1.25 387.8 kN/m^2
1.27 $c = 0; \varphi = 25°$
  $c' = 0; \varphi' = 34°$
1.29 14.7 kN/m^2

Chapter 2

2.1 **a.** 13.78%
  b. 1.907 in.

2.3
Depth (m)	$(N_1)_{60}$
1.5	10
3.0	10
4.6	10
6.0	8
7.5	11
9.0	10

2.5
Depth (m)	D_r (%)
1.5	61.2
3.0	50.0
4.5	43.3
6.0	35.4
7.5	42.1
9.0	41.2

2.7 40%
2.9 **a.** 738.1 lb/ft^2
  b. 639.3 lb/ft^2
2.11 40°

2.13   **a.** 726 lb/ft^2
      b. 3.12
2.15   **a.** 0.65
      b. 1.36
      c. 2131 kN/m^2
2.17   64%
2.19 $Z_1 = 2.6$ m; $v_1 = 492$ m/sec
      $Z_2 = 7.24$ m; $v_2 = 1390$ m/sec
      $v_3 = 3390$ m/sec

Chapter 3

3.1   **a.** 5195 lb/ft^2
     b. 372.8 kN/m^2
     c. 280 kN/m^2
3.3   **a.** 5879 lb/ft^2
     b. 373.7 kN/m^2
     c. 368.8 kN/m^2
3.5   3721 kN
3.7   **a.** 40°
     b. 6563.9 kip
3.9   707.3 kN
3.11   373 kip
3.13   77.86 kip

Chapter 4

4.1   9209 kN
4.3   568.2 kN/m^2
4.5   495.2 kN
4.7   677.25 kN
4.9   **a.** 3300 lb/ft^2
     b.

b (ft)	q_u (lb/ft^2)
0	7350
4	8550
8	9900
12	10,500
16	10,500
20	10,500

4.11   28.2 kip

Chapter 5

5.1   1191 lb/ft^2
5.3   619 lb/ft^2

5.5

z (ft)	$\Delta\sigma$ (lb/^{2})
0	2500
5	2125
10	1375
15	875
20	600

5.7   706.3 lb/ft^2
5.9   56 mm
5.11   0.419 in.
5.13   246 mm
5.15   19.3 mm
5.17   20.85 mm
5.19   5.18 kip/ft^2
5.21   635.6 kN/m^2
5.23   3.45 in.

Chapter 6

6.1   **a.** 771 kN/m^2
     b. 16,321 lb/ft^2
6.3   181.4 kN/m^2
6.5   3.39 m
6.7   0.193 m
6.9

Location	q (kN/m^2)
A	36.81
B	31.86
C	26.91
D	25.19
E	30.14
F	35.09

6.11   14.9 lb/in^3
6.13   18 kN/m^3

Chapter 7

7.1 $P_o = 5497.5$ lb/ft; $\bar{z} = 4$ ft
7.3 $P_o = 159.92$ kN/m; $\bar{z} = 1.77$ m
7.5   45.64 kN/m
7.7   12.6 kip/ft
7.9   5598 lb/ft
7.11 $P_{ae} = 103.3$ kN/m; $\bar{z} = 1.87$ m

7.13

z (m)	σ'_a (kN/m^2)
1.5	12.01
3.0	18.3
4.5	21.23
6.0	22.32

7.15 92.38 kip/ft
7.17 638.88 kN/m

Chapter 8

8.1 Overturning—3.47
Sliding—1.49
Bearing capacity—3.81
8.3 Overturning—2.47
Sliding—1.06
Bearing capacity—1.73
8.5 Overturning—6.2
Sliding—2.35
8.7

z (ft)	σ'_o (lb/ft^2)
5	1525
10	16.7
15	2115
20	2576.2

8.9 **a.** 0.201 in.
b. 43.52 ft
8.11 **a.** 0.161 in.
b. 40.33 ft
8.13 Overturning—3.43
Sliding—1.35
Bearing capacity—9.79

Chapter 9

9.1 **a.** 13.31 m
b. 29.3 m
c. 2762 kN-m/m
9.3 $D = 11.9$ ft; $M_{max} = 32{,}391$ lb-lb/ft
9.5 **a.** 7 m
b. 16.8 m
c. 367.04 kN-m/m
9.7 $D = 1.6$ m; $M_{max} = 51.32$ kN-m/m
9.9 **a.** 759 kN-m/m
b. PZ-35
9.11 PZ-27

9.13 $D = 4.6$ m
$F = 127.17$ kN/m
$M_{max} = 206$ kN-m/m
9.15 **a.** 1.15 ft
b. 3319 lb/ft
9.17

B (m)	P_u (kN)
0.3	22
0.6	36.3
0.9	48.6

Chapter 10

10.1 @ A—131.4 kN
@ B—69.3 kN
@ C—178.8 kN
10.3 @ A—148.5 kN
@ B—78.4 kN
@ C—202 kN
10.5 **a.** $\gamma_{av} = 17.94$ kN/m^3;
$c_{av} = 19.53$ kN/m^3
b. $\sigma_a = 65.4$ kN/m^2
10.7 @ A—78.19 kip
@ B—93.29 kip
@ C—43.88 kip
10.9 @ A—78.19 kip
@ B—89.56 kip
@ C—83.53 kip
10.11 0.88

Chapter 11

11.1 **a.** 332 kN
b. 2362 kN
c. 1448 kN
11.3 605 kN
11.5 333 kip
11.7 571 kN
11.9 128 kip
11.11 110.6 kip
11.13 10.42 mm
11.15 32.5 kN
11.17 292 kip
11.19 25.3 kN
11.21 56.4 kip
11.23 **a.** 89.96%
b. 76.06%

11.25 230 kip
11.27 217.7 mm

Chapter 12

12.1 3037 kip
12.3 57 kip
12.5 3442 kN
12.7 **a.** 304 kip
    b. 295.3 kip
    c. 149.9 kip
12.9 **a.** 1028.9 kN
    b. 3735.4 kN
    c. 4764.3 kN
12.11 **a.** 689.6 kip
    b. 568.1 kip
12.13 0.258 in.
12.15 **a.** 3.13 mm
    b. 594.3 kN-m
    c. 3104 kN/m^2
    d. 7.5 m

Chapter 13

13.1 Collapse is likely to occur.
13.3 1.19 in.
13.5 1.4 ft below the bottom of the foundation
13.7 1 m below the bottom of the foundation
13.9 5.2 m

Chapter 14

14.1

	γ_{zav} (kN/m^3)			
w (%)	$G_s = 2.60$	$G_s = 2.65$	$G_s = 2.70$	$G_s = 2.75$
5	22.57	22.95	23.34	23.72
10	20.24	20.55	20.86	21.16
15	18.35	18.60	18.85	19.1
20	16.78	16.99	17.20	17.40

14.3 **a.** $w_{opt} = 12.5\%$;
    $\gamma_{d(max)}$ 18.76 kN/m^3
    b. $w_{opt} = 9.05\%$;
    $\gamma_{d(max)}$ 19.73 kN/m^3
14.5 B
14.7 $S_N = 3.15 -$ Excellent
14.9 **a.** 6.9 in.
    b. 4.62 months
    c. 120 lb/ft^2

14.11

t (years)	$U_{r, v}$
0.2	0.72
0.4	0.91
0.8	0.99
1.0	0.997

Index

A

A parameter, Skempton:
  definition of, 48
  typical values, 48
AASHTO classification system, 14–15
Active earth pressure:
  Coulomb, 323–328
  earthquake condition, 328–331
  Rankine, 312–318
  rotation about top, 333–335
  translation, 334–336
Active zone, expansive soil, 649
Aeolian deposit, 61
Allowable bearing capacity, shallow
    foundation:
  based on settlement, 241–242
  correlation with cone penetration
    resistance, 241–242
  general, 128–129
Alluvial deposit, 61, 62–65
Anchor:
  factor of safety, 458
  holding capacity, clay, 460
  holding capacity, sand, 454–458
  placement of, 454
  plate, 452
  spacing, 458
Anchored sheet pile wall:
  computational pressure diagram
    method, 443–446
  design charts, free earth support
    method, 435–438
  fixed earth support method, 430
  general, 429–430
  moment reduction, sand, 440–442
  penetrating clay, 448–450
  penetrating sand, 430–433
  relative flexibility, 441
Angle of friction, 43

Apparent cohesion, 43
Approximate flexible method, mat,
    289–296
Area ratio, 76
At-rest earth pressure, 309–311
At–rest earth pressure coefficient, 310
Atterberg limits, 12–13
Average degree of consolidation, 34
Average vertical stress, rectangular load,
    213–215

B

B parameter, Skempton, 48
Backswamp deposit, 65
Bearing capacity:
  allowable, 128–129
  closely spaced, 185–188
  drilled shaft, settlement, 604–608,
    614–617
  drilled shaft, ultimate, 599–604,
    613–614
  eccentric inclined loading, 161–163
  eccentric loading, 148–154, 154–156
  effect of compressibility, 142–144
  effect of water table, 130–131
  factor, Terzaghi, 126
  factor of safety, 128–129
  failure, mode of, 122–124
  field load test, 260–262
  general equation, 132
  layered soil, 177–183
  modified factors, Terzaghi, 127
  on a slope, 191–193
  on top of a slope, 188–190
  seismic, 247–251
  theory, Terzaghi, 124–128
  ultimate, local shear failure, 122
Boring depth, 70–71
Boring log, 105, 106

Braced cut:
    bottom heave, 482–484
    design of, 472–475
    ground settlement, 487–488
    lateral yielding, 487–488
    pressure envelope, clay, 470, 471
    pressure envelope, layered soil, 471–472
    pressure envelope, sand, 470, 471
Braided-stream deposit, 63

C

Calcite, 61
Cantilever footing, 275
Cantilever retaining wall, general, 331
Cantilever sheet pile wall:
    penetrating clay, 423–426
    penetrating sand, 415–419
Cement stabilization, 714–718
Chalk, 61
Chemical bonding, geotextile, 381
Chemical weathering, 60
Circular load, stress, 205–206
Clay mineral, 5
Coefficient:
    consolidation, 34
    gradation, 3
    subgrade reaction, 291–293
    uniformity, 3
    volume compressibility, 34
Cohesion, 43
Collapse potential, 642
Collapsible soil:
    chemical stabilization of, 648
    criteria for identification, 643
    densification of, 648
    foundation design in, 648–649
    settlement, 645–646
Combined footing, 272–278
Compaction:
    control for hydraulic barriers, 681–682
    correction for oversized particles,
      676–677
    curves, 674, 675
    maximum dry unit weight, 673, 674, 676
    optimum moisture content, 673, 676
    Proctor test, 673–675
    relative, 675
    relative density of, 675
    specification for, 675
Compensated foundation, mat, 281, 283
Compressibility, effect on bearing capacity,
    142–144
Compression index:
    correlations for, 30–31
    definition of, 29
Concentrated load, stress, 204

Concrete mix, drilled shaft, 598
Cone penetration test, 90–96
Consolidation:
    average degree of, 34
    definition of, 27–28
    maximum drainage path, 34
    settlement, group pile, 581–582
    settlement calculation, 32–33, 252–256
    time rate of, 33–38
Construction joint, 374
Contact stress, dilatometer, 99
Continuous flight auger, 72
Contraction joint, 374
Conventional rigid method, mat, 285–289
Core barrel, 102
Coring, 101–105
Correction, vane shear strength, 88
Corrosion, reinforcement, 380
Coulomb's earth pressure:
    active, 323–324
    passive, 345–347
Counterfort retaining wall, 353
Critical hydraulic gradient, 26
Critical rigidity index, 142
Cross-hole seismic survey, 111–112
Curved failure surface, passive pressure,
    347–348

D

Darcy's law, 21
Darcy's velocity, 21
Deflocculating agent, 4
Degree of saturation, 7
Depth factor, bearing capacity, 134
Depth of tensile crack, 314
Dilatometer modulus, 100
Dilatometer test, 99–101
Direct shear test, 43–44
Displacement pile, 505
Dolomite, 61
Double-tube core barrel, 102
Drained friction angle:
    variation with plasticity index, 50
    variation with void ratio and pressure,
      49–50
Dredge line, 413
Drilled shaft:
    bearing capacity, settlement, 604–608,
      614–617
    bearing capacity, ultimate, 599–604,
      613–614
    concrete mix, 598
    construction procedure, 593–597
    lateral load, 622–627
    load transfer, 598–599
    rock, 631–633

settlement, working load, 620
    types of, 592
Drilling mud, 73
Drop, flow net, 25
Dry unit weight, 7
Dune sand, 67
Dynamic compaction:
    collapsible soil, 648
    design, 725–726
    general principles, 725
    significant depth of densification, 725–726

E

Earth pressure coefficient:
    at-rest, 309–310
    Coulomb, active, 324
    Coulomb, passive, 347
    Rankine active, horizontal backfill, 314
    Rankine active, inclined backfill, 315
    Rankine passive, horizontal backfill, 339
    Rankine passive, inclined backfill, 344
Eccentric load, bearing capacity, 148–154
Effective area, 148
Effective length, 148
Effective stress, 25–26
Effective width, 148
Elastic settlement:
    flexible foundation, 220–228
    general, 221
    rigid, 227
    strain influence factor method, 236–238
Elasticity modulus, typical values for, 240
Electric friction-cone penetrometer, 90
Embankment loading, stress, 216–217
Equipotential line, 24
Expansion stress, dilatometer, 99
Expansive soil:
    classification of, 655–658
    construction on, 656, 658–664
    criteria for identification, 657–658
    free swell ratio, 655
    general definition, 649
    swell, laboratory measurement, 651–653
    swell pressure test, 651–653

F

Factor of safety, shallow foundation, 128–129
Field load test, shallow foundation, 260–262
Field vane, dimensions of, 88
Filter, 376
Filter design criteria, 376
Flexible foundation, elastic settlement, 220–227

Flow channel, 25
Flow line, 24
Flow net, 24
Fly ash stabilization, 716–717
Foundation design, collapsible soil, 648–649
Free swell, expansive soil, 651
Friction angle, cone penetration test, 94
Friction pile, 503–504
Friction ratio, 90
Function, geotextile, 381

G

General bearing capacity, shallow foundation:
    bearing capacity factors, 132
    depth factor, 134
    equation, 132
    inclination factor, 134
    shape factor, 133
General shear failure, bearing capacity, 121
Geogrid:
    biaxial, 381
    function, 381
    general, 381–382
    properties, 382
    uniaxial, 381
Geotextile, general, 380
Glacial deposit, 65–66
Glacial till, 65
Glaciofluvial deposit, 66
Gradation coefficient, 3
Grain-size distribution, 2–5
Gravity retaining wall, 353
Ground moraine, 66
Group index, 15
Group name:
    coarse-grained soil, 18
    fine-grained soil, 19
    organic soil, 20
Group pile:
    efficiency, 573–576
    ultimate capacity, 576–578
Guard cell, pressuremeter test, 97

H

Hammer, pile driving, 504
Heave, 26
Helical auger, 71
Horizontal stress index, 100
Hydraulic conductivity:
    constant head test, 22
    definition of, 21
    falling head test, 22
    relationship with void ratio, 22–23
    typical values for, 22

Hydraulic gradient, 21
Hydrometer analysis, 4–5

I
Illite, 5
Inclination factor, bearing capacity, 134
Influence factor:
    embankment loading, 217, 218
    rectangular loading, 207

J
Jet grouting, 727–729
Joints, retaining wall, 374

K
Kaolinite, 5
Knitted geotextile, 380

L
Laplace's equation, 24
Lateral earth pressure, surcharge, 385,
    387–388
Lateral load:
    drilled shaft, 622–627
    elastic solution for pile, 546–554
    ultimate load analysis, pile, 554–560
Layered soil, bearing capacity, 177–183
Lime stabilization, 711, 713–714
Liquid limit, 13
Liquidity index, 53
Load transfer mechanism, pile, 508–509
Local shear failure, bearing capacity, 122
Loess, 67

M
Mat foundation:
    bearing capacity, 277–279
    compensated, 281, 283
    differential settlement of, 280–281
    gross ultimate bearing capacity,
        277–278
    net ultimate bearing capacity, 278
    rigidity factor, 280
    types, 275–276
Material index, 100
Meandering belt of stream, 63–64
Mechanical bonding, geotextile, 381
Mechanical friction cone penetrometer,
    90
Mechanical weathering, 60
Mesquite, 69
Modes of failure, 122–124
Modulus of elasticity, 240–241
Mohr-Coulomb failure criteria, 43
Moist unit weight, 7
Moisture content, 7

Montmorillonite, 5
Moraine, 65
Mudline, 413

N
Natural levee, 64
Needle-punched nonwoven geotextile, 381
Negative skin friction, pile, 570–573
Nondisplacement pile, 505
Nonwoven geotextile, 380
Normally consolidated soil, 29

O
Optimum moisture content, 673
Oversized particle correction, compaction,
    676–677
Overturning, retaining wall, 359–361
Organic soil, 61
Outwash plains, 64
Oxbow lake, 64

P
P-wave, 106
Passive pressure:
    Coulomb, 345–347
    curved failure surface, 347–348
    Rankine, horizontal backfill, 338–342
    Rankine, inclined backfill, 344
Percent finer, 2
Percussion drilling, 74
Pile capacity:
    Coyle and Castello's method, 520–521,
        525–526
    frictional resistance, 524–531
    Janbu's method, 516
    Meyerhof's method, 512–514
    rock, 531–532
    Vesic's method, 515–516, 517–518
Pile driving formula, 562–567
Pile installation, 504–507
Pile load test, 538–542
Pile type:
    composite, 502
    concrete, 496–499
    steel, 493–496
    timber, 500–502
Piston sampler, 85
Plastic limit, 13
Plasticity chart, 16
Plasticity index, 16
Pneumatic rubber-tired roller, 678
Point bar deposit, 63
Point bearing pile, 502–503
Point load, stress, 204
Poisson's ratio, typical values for, 240
Pore water pressure, 25

Pore water pressure parameter, 48
Porosity, 6
Post hole auger, 71
Pozzolanic reaction, 713
Precompression:
    general consideration, 688–691
    midplane degree of consolidation, 692
Preconsolidated soil, 29
Preconsolidation pressure, 29, 30
Prefabricated vertical drain, 706–711
Pressuremeter modulus, 98
Pressuremeter test, 97–99
Proportioning, retaining wall, 355
Punching shear coefficient, 179, 180
Punching shear failure, bearing capacity, 122

Q
Quick condition, 26

R
Radial shear zone, bearing capacity, 126
Rankine active earth pressure:
    horizontal backfill, 312–315
    inclined backfill, 315–318
Recompression curve, consolidation, 31
Reconnaissance, 69
Recovery ratio, 104
Rectangular combined footing, 272–273
Rectangular load, stress, 206–212
Refraction survey, 105–108
Reinforced earth, 379
Relative compaction, 675
Relative density, 9
Residual friction angle, 51, 52
Residual soil, 61
Residual strength envelope, 51
Resistivity, 112–113
Retaining wall:
    application of earth pressure theories, 356–357
    cantilever, 353
    counterfort, 353
    deep shear failure, 358, 359
    drainage, backfill, 374–376
    geogrid reinforcement 399–402
    geotextile reinforcement, 395–397
    gravity, 353
    joint, 374
    proportioning, 355
    stability check, 358–366
    strip reinforcement, 383–392
Rigidity index, 142
Rock quality designation, 104
Roller:
    pneumatic rubber-tired, 678
    sheepsfoot, 679

smooth wheel, 678
    vibratory, 679
Rotary drilling, 73

S
S-wave, 111
Sand compaction pile, 722–725
Sand drain:
    average degree of consolidation, radial drainage, 697–699
    general, 696–697
    radius of effective zone of drainage, 697
    smear zone, 697, 698
    theory of equal strain, 697–699, 700–702
Sanitary landfill:
    general, 665–666
    settlement of, 666–668
Saturated unit weight, 8
Saturation, degree of, 7
Seismic refraction survey, 105–108
Sensitivity, 53
Settlement, pile:
    elastic, 543–545
    group, 580–581
Settlement calculation, shallow foundation:
    consolidation, 252–256
    elastic, 220–228, 230–234
    tolerable, 241–244
Shape factor, bearing capacity, 133
Sheepsfoot roller, 679
Sheet pile:
    precast concrete, 410
    steel, 410–412
    wall construction method, 413–414
    wooden, 409–410
Shelby tube, 83
Shrinkage limit, 13
Sieve analysis, 2–4
Sieve size, 2
Single-tube core barrel, 102
Size limit, 5
Skempton-Bjerrum modification, consolidation settlement, 254–256
Skin, 383
Sliding, retaining wall, 361–363
Smear zone, sand drain, 697, 698
Smooth wheel roller, 678
Soil classification systems, 13–18
Soil compressibility factor, bearing capacity, 142–144
Spacing, boring, 71
Specific gravity, 9
Split-spoon sampler, 75–86
Spring core catcher, 77

Stability check, retaining wall:
  bearing capacity, 364–366
  overturning, 359–361
  sliding, 361–363
Stability number, 189
Stabilization:
  cement, 714–716
  fly ash, 716
  lime, 716–717
  pozzolanic reaction, 711, 713–714
Standard penetration number:
  correlation, consistency of clay, 78–79
  correlation, friction angle, 82
  correlation, overconsolidation ratio, 79
  correlation, relative density, 81
  correlation, sand, 80
Static penetration test, 90
Stone column:
  allowable bearing capacity, 720–721
  equivalent triangular pattern, 718
  general, 717
  stress concentration factor, 717, 719
Strain influence factor, 236
Stress:
  circular load, 205–206
  concentrated load, 204
  embankment load, 216–218
  rectangular load, 206–212
Structural design, mat:
  approximate flexible method, 289–295
  conventional rigid method, 286–289
Subgrade reaction coefficient, 291, 292–293
Suitability number, vibroflotation, 686
Swell pressure test, 651–653
Swell test, unrestrained, 651
Swelling index, 31

T

Tensile crack, 314
Terminal moraine, 65
Thermal bonding, geotextile, 381
Tie failure, retaining wall, 388–389
Tie force, retaining wall, 388
Time factor, 34
Time rate of consolidation, 33–38
Tolerable settlement, shallow foundation, 241–244

Trapezoidal footing, 273–275
Triaxial test:
  consolidated drained, 46
  consolidated undrained, 46–47
  unconsolidated undrained, 47

U

Ultimate bearing capacity, Terzaghi, 126–127
Unconfined compression strength, 48
Unconfined compression test, 48–49
Undrained cohesion, 47
Unified classification system, 15–20
Uniformity coefficient, 3
Unit weight:
  dry, 7, 8
  moist, 7, 8
  saturated, 7, 8
Unrestrained swell test, 651
Uplift capacity, shallow foundation, 193–198

V

Vane shear test, 86–90
Velocity, P-wave, 106
Vertical stress, average, 213–215
Vibratory roller, 679
Vibroflotation:
  backfill suitability number, 686
  construction method, 382, 383
  effective range, backfill, 688–686
  vibratory unit, 682
Virgin compression curve, 30
Void ratio, 5
Volume, coefficient of compressibility, 34

W

Waffle slab, 661, 663
Wash boring, 73
Water table, effect on bearing capacity, 130–131
Water table observation, 85–86
Weight-volume relationship, 5–9
Wenner method, resistivity survey, 112–113
Winker foundation, 289

Z

Zero-air-void unit weight, 674